THE EVOLUTION OF BEGGING

THE EVOLUTION OF BEGGING

COMPETITION, COOPERATION AND COMMUNICATION

EDITED BY

Jonathan Wright

School of Biological Sciences, University of Wales, Bangor, U.K.

AND

Marty L. Leonard

Department of Biology, Dalhousie University, Halifax, Canada

KLUWER ACADEMIC PUBLISHERS

DORDRECHT/BOSTON/LONDON

A C.I.P. Catalogue record for this book is available from the Library of Congress.

ISBN 1-4020-0571-7

Published by Kluwer Academic Publishers,
P.O. Box 17, 3300 AA Dordrecht, The Netherlands.

Sold and distributed in North, Central and South America
by Kluwer Academic Publishers,
101 Philip Drive, Norwell, MA 02061, U.S.A.

In all other countries, sold and distributed
by Kluwer Academic Publishers,
P.O. Box 322, 3300 AH Dordrecht, The Netherlands.

Printed on acid-free paper

Printed in the Netherlands

CONTENTS

PREFACE

Begging in nestling birds has become the model system for investigations into evolutionary conflicts of interest within families and the honest signalling of offspring need. This area of study was inspired by theoretical treatments of parent-offspring conflict and its resolution. However, it remained largely impenetrable to empirical study until handicap models for costly honest signals of need were introduced. Since then a wealth of scientific papers have been published on the nature of nestling begging signals and their potential honesty, which in turn have inspired a number of theoretical advances. The result has been an upsurge of interest in nestling begging systems, which has diversified into a range of exciting and challenging areas, from adaptive sex ratios and hormonal physiology to the evolution of mimicry in brood-parasitic nestlings.

Given the recent developments in this field, we decided that the time was right for a book focused on nestling begging. We asked scientists from almost every research group working in this area to contribute a chapter on a topic they found particularly interesting and exciting. We received an enthusiastic response, and the result is a volume that we feel encompasses the full range of ideas, natural systems and research projects to date on the evolution of begging. We hope that these contributions will inspire new researchers and stimulate further investigations in this dynamic field.

The book begins with a section on 'Theoretical Approaches' to the study of begging. Rufus Johnstone and Charles Godfray review the handicap models that stimulated the initial interest in begging signals. New theoretical approaches are also included in the form of the first genetic algorithm model of begging by Karen Price, Ron Ydenberg and Dave Daust, and a chapter by David Sloan Wilson and Anne Clarke on the application of multilevel selection theory to problems in begging. The final theoretical chapter is a detailed model by Kate Lessells on adaptive sex ratios and the allocation of care to male and female offspring.

The second section deals with 'Begging as a Signal and the Issue of Costs', which is a topic central to our understanding of begging behaviour. Becky Kilner begins this section with her chapter on the complex nature of begging signals and the relationship between different components of the display. Alex Roulin then describes his novel hypothesis concerning sibling negotiation in barn owl nestlings, whilst Andy Horn and Marty Leonard explore the efficacy of begging signals and how they are designed for effective signal transmission. This section concludes with comprehensive reviews of the theoretically all-important evidence for begging costs. Mark

Chappell and Gwen Bachman examine the evidence for an energetic cost of begging, whilst David Haskell tackles the issue of predation costs.

'Begging and Nestling Physiology' represents an exciting new area of growth in begging research. This section begins with Anne Clarke's chapter on the mechanisms of hunger and appetite and how they apply to begging. Bill Karasov and Jon Wright continue this theme, by examining nestling digestive capacity as the critical physiological link between 'need' and begging behaviour. A review of the hormonal control of begging is provided by Hubert Schwabl and Joe Lipar, and Nicola Saino and Anders Møller describe the relationship between immunocompetence and begging signals.

The section on 'Sibling Competition' explores the competitive nature of begging in a range of passerine and non-passerine systems. Barb Glassey and Scott Forbes begin it by examining how nestling begging strategies are affected by phenotypic handicaps, such as size asymmetries within broods. Bob Ricklefs asks how patterns of sibling competition affect nestling growth and food requirements and how, in turn, strategic variation in growth affects the fitness of both parents and young. Javier Bustamante, Dee Boersma and Lloyd Davis then describe feeding chases and sibling competition in Pygoscelid penguins, and Elsie Krebs discusses the complex systems of sibling competition, begging and food allocation in parrots. This section ends with Hugh Drummond asking why direct aggression amongst siblings is not more common in avian begging systems.

'Begging and Brood Parasitism' is a section which examines how begging systems can be exploited by brood parasites, thereby providing a valuable perspective concerning the evolution of honest begging signals. Don Dearborn and Gabriela Lichtenstein begin the section with a thorough review of begging behaviour in both generalist and specialist parasitic cowbirds. This is followed by two chapters examining the dishonest begging of *Clamator* cuckoos: one by Tomas Redondo and Jesus Zuñiga presenting experimental data on how parasitic cuckoos exploit the parental feeding rules of their magpie hosts, and another from Manuel Soler explaining how cuckoos respond to variation in corvid host breeding strategies. Bob and Laura Payne conclude the section with a discussion of mimetic begging vocalizations in host-specific viduline finches.

The final section of the book, 'Statistical Approaches', highlights the statistical challenges posed by begging data and provides advice on how to handle the often troublesome data sets associated with begging studies. Daniela Monk's chapter provides an introduction to logistic regression models and their application to begging data, and a customized multivariate model used to examine complex data on parental food allocation within broods. Scott Forbes ends the section, and the book, with a review of the statistical pitfalls involved in the study of nestling begging.

One of the rewards of organizing a volume of this type has been the opportunity to interact with so many exceptional scientists. We thank all our authors for their excellent contributions and their cooperation with us as editors. We would also like to thank the participants of the Gregynog 2000 Begging Workshop for their interesting presentations, stimulating ideas and good company. Workshop participants in addition to contributors to the book include: Trond Amundsen, Gabrielle Archard, Ruthi Brandt, Amber Budden, Camilla Hinde, Sarah Hunt, Anahita Kazem, Hilla Kedar, Andrew MacColl, Ben Mines, Samuel Neuenschwander, Kevin Pilz and Bonnie Ploger. For assistance in reviewing chapters, we are very grateful to: Dale Clayton, Pete Cotton, Nick Davies, Don Dearborn, Scott Forbes, Andy Horn, David Houston, Bill Karasov, Becky Kilner, Elsie Krebs, Owen Lyne, Rob Magrath, Peter McGregor, Sue McRae, Jonathan Newman, Dave Noble, Andrew Pacejka, Geoff Parker, Spencer Sealy, Ben Sheldon and Wes Weathers. We are particularly indebted to Anahita Kazem for proof-reading the entire volume, and for logistic support during the workshop. Thanks also to our editors at Kluwer, Noeline Gibson and Ursula Hertling, for advice. The wonderful cover was designed by Daniela Monk, using a photo-illustration by Ian C. Tait.

Jon Wright and Marty Leonard (October 2001)

THEORETICAL APPROACHES

1.

MODELS OF BEGGING AS A SIGNAL OF NEED

Rufus A. Johnstone[1] & H. Charles J. Godfray[2]
Department of Zoology, University of Cambridge, Cambridge CB2 3EJ, UK. (r.a.johnstone@zoo.cam.ac.uk)
NERC Centre for Population Biology, Imperial College at Silwood Park, Ascot, Berks SL5 7PY, UK (c.godfray@ic.ac.uk)

ABSTRACT

Parental feeding of young is often accompanied by striking begging displays on the part of offspring. These displays are now widely thought to function as signals of need; in other words, they serve to elicit additional resources from parents by advertising the benefits that offspring thereby stand to gain. Sib-sib and parent-offspring conflicts over resource allocation, however, favour misrepresentation of need. Consequently, signalling models have focused on signal cost as a means to maintain the reliability of offspring displays. Recently, however, the possibility of alternative, cheaper signalling equilibria has emerged. We review costly and cost-free signalling models, and suggest that both face difficulties in accounting for observed begging behaviour. We conclude with the suggestion that these difficulties may best be tackled by developing more realistic models that incorporate more of the complexities of parent-offspring interaction revealed by empirical studies of begging.

INTRODUCTION

Parental feeding and care of young is often preceded or accompanied by begging displays on the part of their offspring. Such displays occur in many different taxa, and take many different forms (e.g. Weary & Fraser 1995; Furlow 1997; Kilner & Johnstone 1997; Mock & Parker 1997; Weary et al. 1997; Rauter & Moore 1999; Manser & Avey 2000; Zhang & Jiang 2000). What is their function? At first glance, they seem extravagant, in terms of both energy expenditure and the risk of attracting unwanted attention.

J. Wright and M.L. Leonard (eds.), The Evolution of Begging, 1–20.

The most widespread explanation of extravagant begging behaviour invokes family conflicts over resource allocation. As Trivers (1974) first pointed out, offspring are selected to demand more resources than parents are selected to provide. This evolutionary conflict of interest will favour young that are able to manipulate their parents into bringing more food (Macnair & Parker 1978; Parker & Macnair 1978, 1979; Parker 1985; Godfray 1995a, 1999; Mock & Parker 1997; Lessells & Parker 1999). In addition, each individual offspring in a brood is selected to demand a greater share of the resources that are brought than its siblings are selected to yield (Macnair & Parker 1979; Mock & Parker 1997). In light of parent-offspring conflict and sibling competition over food allocation, extravagant begging displays seem less surprising – they are simply the means by which offspring attempt to claim more for themselves.

This explanation for begging is, however, incomplete. It leaves unanswered the question of how and why offspring are able to influence provisioning by means of begging. This is an issue that needs to be addressed, because offspring have no direct control over the rate of provisioning by parents. With regard to food allocation within a brood, offspring may in some cases be able to wrest a food item away from a parent (e.g. Mock & Parker 1997), or to monopolize access, thereby obtaining a disproportionate share of resources (e.g. McRae et al. 1993; Kacelnik et al. 1995). In other species, however, parents appear to have a high level of control over resource allocation (e.g. Krebs & Magrath 2000). Thus, the chief means by which offspring obtain extra resources will often be through manipulation of parental behaviour. If begging displays serve this purpose, we must consider why parents allow themselves to be manipulated. In other words, why does selection favour parents that respond to offspring begging?

BEGGING AS A SIGNAL

Perhaps the most popular hypothesis, which forms the subject of this chapter, is that begging serves as a signal of need (Godfray 1991). According to this theory, parents respond to solicitation displays because these traits and behaviours provide information about the state of their young. Vigorous begging indicates that an offspring stands to gain a substantial benefit from additional food. Consequently, it pays to provide more resources in response to more intense display, and to favour young that beg more strongly over their less demanding siblings (for reviews see Kilner & Johnstone 1997; Godfray & Johnstone 2000).

The idea of begging as a signal had been common prior to Trivers (1974). But with the growth of interest in parent-offspring conflict and sibling competition, the notion of signalling was largely abandoned. The classical models of Parker and Macnair (Macnair & Parker 1978; Parker & Macnair 1978, 1979), for instance, did not incorporate variability in offspring state, and hence ruled out the possibility that parents might acquire information by observing offspring begging behaviour.

Family conflicts over resource allocation appeared to cast doubt on the signalling hypothesis, because it seemed that selection would always favour offspring that exaggerated their need for food (i.e. that adopted begging behaviour typical of individuals in greater need than themselves). This 'dishonest' behaviour would be favoured because it would lead to the acquisition of additional resources. Reliable or honest signalling of need could not, therefore, be expected to persist (see Dawkins & Krebs 1978; Krebs & Dawkins 1984).

More recently, however, models of begging have resurrected the signalling hypothesis. The 'honesty' of begging in most of these analyses is maintained by the cost of the signal – an application of Zahavi's (1975, 1977a) handicap principle. Cost can maintain honesty, because it is only for young in poor condition that the benefits of additional food are likely to outweigh the expense of vigorous begging. For less needy offspring, there is less to gain from acquiring extra resources, and intense display is therefore unprofitable. This brief verbal argument has been fleshed out in a number of formal models of honest, costly signalling of need (Godfray 1991, 1995b; Maynard Smith 1991; Johnstone & Grafen 1992a), which demonstrate that reliable advertisement of offspring condition can indeed prove evolutionarily stable. The costly signalling hypothesis thus promises to explain simultaneously both parental responsiveness to offspring begging, and the extravagant nature of offspring displays. An illustrative (game theoretical) model of this type is presented below, closely based on Godfray (1991).

SIGNAL COST AND THE MAINTENANCE OF HONESTY: A SIMPLE MODEL

We focus on one episode of communication between a parent and its sole offspring. The 'condition' of the offspring, denoted c, is initially unknown to the parent. However, the offspring may invest any (non-negative) level of effort x in begging, where x denotes the additive fitness cost incurred. In response to this solicitation, the parent may then provide any (non-negative) amount of food y.

The direct fitness of the offspring increases with the amount of food provided, but the better its condition, the lower the marginal benefit of additional food. Formally, its direct fitness is equal to $f(c,y) - x$, where $f_y > 0$

and $f_{yc} < 0$. We also assume that the parent's future fitness decreases with the amount of food provided. Formally, it is equal to $g(y)$, where $g_y < 0$. Finally, we assume that the coefficient of relatedness of the current offspring to the parent's future young is r. Given these assumptions, the inclusive fitness pay-offs to parent and offspring, $W_p(c,x,y)$ and $W_o(c,x,y)$, are given by:

$$W_p(c,x,y) = [f(c,y) - x] + g(y)$$

$$W_o(c,x,y) = [f(c,y) - x] + rg(y)$$

(emphasising that the source of parent-offspring conflict is the relative weighting that each party gives to the present versus future reproductive attempts; while the parent weights both equally, the offspring places less weight on future reproduction, particularly when its relatedness to the parent's future young is low).

We are interested in the possibility of a signalling equilibrium, at which begging intensity is a smoothly decreasing function of c and the signal thus provides an unambiguous indication of offspring state. Let $x^*(c)$ denote the equilibrium begging intensity of an offspring in condition c, and $y^*(x)$ the equilibrium amount of food supplied by the parent to an offspring that begs at intensity x. Since the parent can correctly infer the state of its young from the intensity of the signal, the amount of food that it provides to an offspring in state c (which is given by $y^*(x^*(c))$) will strike the optimal balance (from the parent's point of view) between present and future reproductive success. Formally, this optimality condition implies that:

$$\frac{\partial W_p\left(c, x^*(c), y^*\left(x^*(c)\right)\right)}{\partial y} = 0$$

$$\Rightarrow f_y\left(c, y^*\left(x^*(c)\right)\right) = -g_y\left(y^*\left(x^*(c)\right)\right) \qquad (1)$$

(i.e. at equilibrium, the marginal benefit to an offspring of receiving more food must be exactly balanced by the marginal cost to the parent of providing that food).

What level of begging cost is required for the maintenance of the signalling equilibrium (or in other words, what form does $x^*(c)$ take)? The inclusive fitness pay-off to a mutant offspring of true condition c, that adopts the begging behaviour typical of an individual of condition c', can be written $W_o\left(c, x^*(c'), y^*\left(x^*(c')\right)\right)$. For stability of the signalling equilibrium, we require that an offspring should not gain by misrepresenting its true level of need, which implies that:

$$\frac{\partial W_o\left(c, x^*(c'), y^*\left(x^*(c')\right)\right)}{\partial c'} = 0 \quad \text{for } c' = c \tag{2}$$

In other words, for an offspring that signals honestly, misrepresentation of its true condition must yield no marginal gain. Writing out (2) fully yields:

$$\frac{\partial y^*\left(x^*(c)\right)}{\partial c}\left[f_y\left(c, y^*\left(x^*(c)\right)\right) + r g_y\left(y^*\left(x^*(c)\right)\right)\right] - \frac{\partial x^*(c)}{\partial c} = 0 \tag{3}$$

Combining (1) and (3) we obtain, after some rearrangement:

$$\frac{\partial x^*(c)}{\partial c} = -(1-r)\frac{\partial g\left(y^*\left(x^*(c)\right)\right)}{\partial c} \tag{4}$$

If we assume that an offspring in the best possible condition does not beg (i.e. that $x^*(c_{max}) = 0$, where c_{max} denotes the maximum attainable value of c), then equation (4) implies that:

$$x^*(c) = (1-r)\left[g\left(y^*\left(x^*(c_{\max})\right)\right) - g\left(y^*\left(x^*(c)\right)\right)\right] \tag{5}$$

Equation (5) reveals that at equilibrium, the begging cost incurred by an offspring of condition c is equal to $(1 - r)$ times the extra provisioning cost incurred by the parent, through feeding at the rate appropriate to an offspring of that condition compared to an offspring of condition c_{max} (see Nöldeke & Samuelson 1999). A sample solution, for illustrative functions *f()* and *g()*, is shown in Figure 1.

The Question of Cost

Theoretical demonstrations that signal cost might serve to maintain reliable signalling of need (such as the simple model presented above) have encouraged renewed empirical interest in begging. Since this chapter is concerned primarily with models of begging, we will not attempt a thorough review of this rapidly growing body of work. It suffices to say that an increasing number of studies (mostly of avian begging) have found evidence that begging does serve, in at least some cases, as a signal.

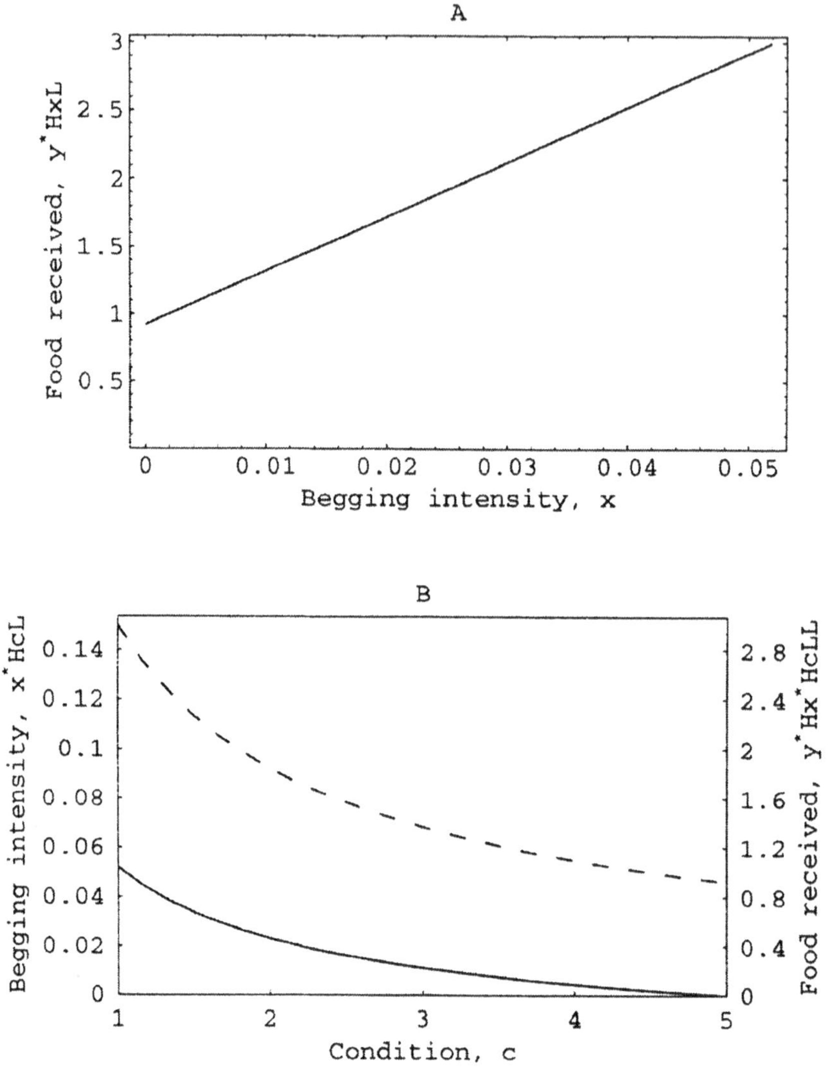

Figure 1. Graph A shows the equilibrium level of parental provisioning $y^*(x)$ in response to a begging intensity (or cost) of x; graph B the equilibrium begging intensity (or cost) of the offspring $x^*(c)$ (solid line) and the amount of food it obtains $y^*(x^*(c))$ (dashed line), as a function of its condition c (note different scales). These results were obtained for the case where $r = 0.5$, $f(c,y) = 1 - e^{-cy}$, $g(y) = 1 - 0.05\,y$ and c varies between 1 and 5.

Firstly, experimental manipulation of offspring state (e.g. by means of food deprivation or satiation) has been found to influence several aspects of the begging display, including calling (e.g. Redondo & Castro 1992; Cotton et al. 1996; Price et al. 1996; Lotem 1998), posture (e.g. Redondo & Castro 1992; Kacelnik et al. 1995; Kilner 1995; Leonard & Horn 1998) and mouth coloration (Kilner 1997; Kilner & Davies 1998). Secondly, experimental manipulation of the begging display, either indirectly through alteration of offspring state or directly (e.g. by playback of begging calls to supplement a brood's vocal display), has shown that begging can influence parental provisioning rate and, in some cases, food allocation at the nest (e.g. Bengtsson & Rydén 1981; Kilner 1995, 1997; Price 1996, 1998; Ottoson et al. 1997; Kilner et al. 1999; Leonard & Horn 2001).

Despite the above evidence, however, doubts have nevertheless been raised about the idea of offspring solicitation as an honest, costly signal of need. Most of these doubts focus on the cost of begging, which handicap models suggest is necessary for the maintenance of honesty. This unease stems from two sources: on the one hand, empirical studies have yielded varying conclusions regarding the magnitude of this cost, some suggesting that it may be rather small (e.g. McCarty 1996; M.A. Chappell & G.C. Bachman this volume); on the other hand, some analyses of handicap models have suggested that the expenditure required to maintain honesty may, in theory, prove excessively large (Rodríguez-Gironés et al. 1996; Bergstrom & Lachmann 1997).

Empirical efforts to measure the cost of begging have focused either on the energetic expenditure involved in begging, or on the risk of attracting unwanted attention from predators (M.A. Chappell & G.C. Bachman this volume; D. Haskell this volume). McCarty (1996) provided the first estimates of the former cost, and concluded that "compared to the energy requirements for other avian behaviours, the cost of begging is low", and that therefore "the assumption that begging is energetically costly needs to be re-examined". Leech and Leonard (1996) obtained similar results, but interpreted their findings rather differently, stating that although the energetic cost of begging "seems rather low, it may be important when the total energy budget of the nestling is considered" (and see Verhulst & Wiersma 1997; but also M.A. Chappell & G.C. Bachman this volume). In the absence of data on the fitness consequences of begging, it is difficult to decide what constitutes a significant level of energetic expenditure. Metabolic measurements thus do not settle the question of whether begging is as costly as handicap models predict. Recent studies have begun to explore the impact of begging effort on nestling growth, which provides a much clearer idea of fitness costs (e.g. Kilner 2001; Rodríguez-Gironés et al. 2001). However, there are too few data at present to draw general conclusions. Rodríguez-Gironés et al. (2001), for instance, found that begging had a significant impact on the growth of magpie (*Pica pica*) nestlings, but not on ring dove (*Streptopelia risoria*) squabs, while Kilner (2001) found a significant impact on the growth of canaries (*Serinus canaria*).

Turning to the risk of attracting unwanted attention, Haskell (1994) found that playback of recorded begging calls increased the rate of predation on artificial open-cup nests, but not on artificial cavity nests, while Leech and Leonard (1997) found an impact of playback on predation for both nest types (and see also Dearborn 1999). In addition to this direct evidence, comparative analyses suggest that species subject to greater nest predation have begging calls of lower volume and higher pitch, which makes it more difficult for predators to localize them (Briskie et al. 1999). These studies indicate that predation risk can play a role in shaping begging behaviour, but once again cannot confirm whether or not the predation cost of begging is as large as that predicted by handicap models.

In addition to equivocal empirical data, theoretical analysis of handicap models has also raised concerns about the level of cost needed to maintain honesty. Rodríguez-Gironés et al. (1996) showed that in Godfray's (1991) model of begging, both parent and offspring may experience lower fitness at a signalling equilibrium than they would do if no signal were given, and the parent were left uninformed about offspring need (and see Bergstrom & Lachmann 1997 for an equivalent result in an alternative signalling model). It is not clear how generally this result applies - while Rodríguez-Gironés et al. (1996, 1998) argued that lower fitness at a signalling equilibrium is the norm rather than the exception, Godfray and Johnstone (2000) found that a simple increase in the range of offspring condition was enough to render signalling profitable in Godfray's (1991) model. In at least some cases, however, it is clear that the cost required for the maintenance of honesty can exceed the benefit to be gained from a reliable signalling system.

The above finding does not call into question the stability of costly signalling against invasion by individuals that adopt alternative strategies. Given that the parent responds to the begging display, the offspring does best to signal (despite the costs this entails); and given that the offspring signals, the parent does best to respond. However, it does raise questions about how a costly signalling system can become established in the first place. Non-signalling is stable too: if the parent does not respond, there is nothing to be gained by begging, and if the offspring does not beg, there is nothing to be gained by responding. What selective pressures could drive the transition from a non-signalling equilibrium to a signalling equilibrium of lower fitness? Rodríguez-Gironés et al. (1998) have argued, on the basis of an evolutionary simulation, that this transition is likely to prove difficult, and that evolving populations are unlikely to converge on a signalling equilibrium that is costly and inefficient.

What are we to make of these results? Theoretical analyses suggest that a substantial (and sometimes excessive) cost of begging is required for the maintenance of honesty. At the same time, empirical efforts to demonstrate that begging does incur such a cost have yielded equivocal results. One possible resolution of this problem lies in the recent demonstration that alternative, cheaper signalling equilibria are possible.

ALTERNATIVE SIGNALLING EQUILIBRIA

All of the signalling models we have so far described focus on the possibility of a 'separating' equilibrium, at which begging intensity is a smoothly decreasing function of offspring condition. At a separating equilibrium, the parent obtains full information about offspring state, because young that differ in condition always differ in the intensity with which they beg. Bergstrom and Lachmann (1997; Lachmann & Bergstrom 1998) have shown, however, that this is not the only kind of signalling equilibrium possible. An alternative is a 'pooling' equilibrium, at which offspring condition is subdivided into a series of intervals or pools in which all individuals adopt the same begging behaviour. In fact, there are infinitely many possible pooling equilibria, differing in the number of pools they feature and the cost of begging incurred by offspring in each of them.

We can illustrate the possibility of pooling equilibria within the framework of the simple begging model presented above. As before, the condition (c) of the offspring is initially unknown to the parent, but may be revealed by its begging behaviour. Now, however, we are interested in a pooling, rather than a separating, equilibrium. Let us suppose that condition is randomly distributed between c_{min} = 1 to c_{max} = 5 with probability density $p(c)$, and that this range is subdivided into n pools, within each of which offspring adopt the same signal. For the boundaries of these pools we will write $t_0, t_1, \ldots, t_n$ (where $t_o = c_{min}$ and $t_n = c_{max}$), and for the (additive) costs associated with the n signals we will write $x_1, x_2, \ldots, x_n$. The parent cannot distinguish among offspring in any one pool, and so must perforce provide the same amount of food y_i to all offspring in pool i. At equilibrium, this amount will strike the optimal balance, on average, between present and future reproductive success. Formally, this implies that:

$$\int_{t_{i-1}}^{t_i} \frac{\partial W_p\left(c, x_i^*, y_i^*\right)}{\partial y} p(c)\,\mathrm{d}c = 0$$

$$\Rightarrow \frac{\int_{t_{i-1}}^{t_i} f_y\left(c, y_i^*\right) p(c)\,\mathrm{d}c}{\int_{t_{i-1}}^{t_i} p(c)\,\mathrm{d}c} = -g_y\left(y_i^*\right) \qquad (6)$$

(i.e. at equilibrium, the average marginal benefit to an offspring in pool i of receiving more food must be exactly balanced by the marginal cost to the parent of providing that food).

What levels of begging cost are required for the maintenance of this pooling equilibrium? At equilibrium, an offspring on the boundary between pools i and $i + 1$ (with condition t_i) must be indifferent between adopting signal i or signal $i + 1$. In other words, the offspring must obtain the same fitness pay-off regardless of which signal it adopts (were this not the case, selection would favour a shift in offspring strategy). Formally:

$$W_O\left(t_i, x_i^*, y_i^*\right) = W_O\left(t_i, x_{i+1}^*, y_{i+1}^*\right) \tag{7}$$

Together, equations (6) and (7) allow us to determine the levels of begging cost and food provision that characterize an equilibrium with any given number and positioning of contiguous pools. Some of these possible equilibria are illustrated in Figure 2, for the case in which c is evenly distributed between c_{min} and c_{max} (for the same functions *f()* and *g()* used to generate Figure 1).

Pooling equilibria share some qualitative features with separating equilibria, so that both can be considered instances of honest signalling. Lachmann and Bergstrom (1998), for example, show that the subdivision into pools must always be contiguous; in other words, there can never be a case in which signallers of low and high need give the same signal, while a signaller of intermediate need adopts a different display. However, pooling equilibria with a small number of pools typically entail lower signalling costs (averaging over all possible levels of signaller need) than do separating equilibria. Indeed, Bergstrom and Lachmann (1998) have shown that a stable pooling equilibrium is possible without any signal cost at all.

THE POSSIBILITY OF COST-FREE SIGNALLING

The first demonstration that cost-free signalling of need can be stable despite a conflict of interest between signaller and receiver is due to Maynard Smith (1994). His analysis, however, suggested that cost-free communication could only be stable given certain restrictions on the distribution of signaller need. Bergstrom and Lachmann (1998) subsequently showed that stability was possible in a much wider range of situations.

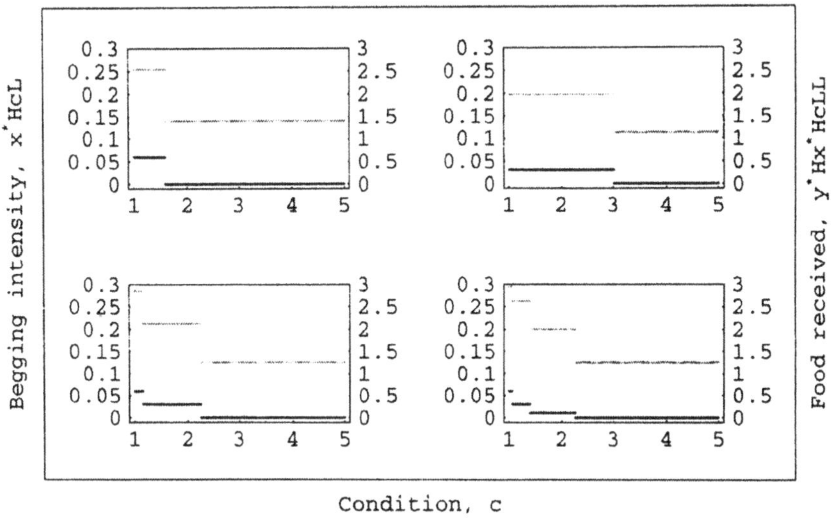

Figure 2. Each graph shows the equilibrium levels of begging $\dot{x}_1 \cdots \dot{x}_n$ (dark line) and the corresponding levels of parental provisioning $\dot{y}_1 \cdots \dot{y}_n$ (pale line) in relation to offspring condition (c), for one possible pooling equilibrium (note different scales for begging intensity and food received). These results were obtained for the case where $r = 0.5$, $f(c,y) = 1 - e^{-cy}$, $g(y) = 1 - 0.05\,y$ and c is evenly distributed between 1 and 5.

The possibility of cost-free signalling arises because offspring stand to lose by soliciting excessive resources from parents. Although a nestling gains by exaggerating its hunger to some degree, too great a misrepresentation of its state may lead the parent to expend so much effort on providing food that the indirect cost to the offspring (in terms of the loss to future siblings) outweighs the direct benefit that it gains. Selection will thus act against offspring that are overly deceptive. Figure 3 illustrates how this indirect cost of deception can stabilize a cost-free, pooling equilibrium.

Consider the simplest possible case, in which there are only two intervals of condition and two corresponding signalling options (to beg, or not to beg). The parent cannot distinguish among offspring in the first interval, and will perforce treat them all as though they are of condition c_1 (roughly speaking, the 'average' value for this interval). Similarly, it will treat all offspring in the second interval as though they are of condition c_2. An offspring of true condition c, as shown in Figure 3, stands to gain by exaggerating its hunger slightly. However, it cannot opt to do so – it faces the choice between slight underestimation of its need if it refrains from begging, or gross overestimation of its need if it chooses to beg. For offspring in good condition, the indirect costs of this overestimation of need (due to extracting excessive resources from the parent) may be great enough that it pays to refrain from begging, even if the signal entails no direct cost.

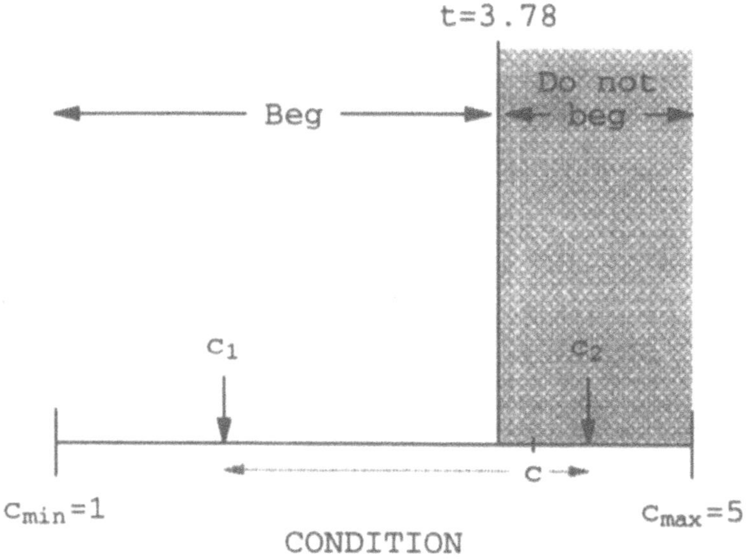

Figure 3. A cost-free pooling equilibrium can be stabilized by the indirect cost of extracting excessive resources from the parent. The figure illustrates a cost-free equilibrium with two pools, in which offspring whose condition lies between c_{min} (= 1) and t ($\approx$ 3.78) beg, while those whose condition lies between t and c_{max} (= 5) do not (this solution is based on the assumption that c is evenly distributed between 1 and 5, $r = 0.5$, $f(c,y) = 1 - e^{-cy}$ and $g(y) = 1 - 0.05\,y$). The parent treats offspring in the former pool as if they were of condition c_1 ($\approx$ 2.06), and offspring in the latter pool as if they were of condition c_2 ($\approx$ 4.35). An offspring of need c in the non-signalling pool stands to gain by slightly exaggerating its need for food, but cannot do so; as illustrated in the figure, it can only opt to beg, leading the parent to grossly underestimate its condition as c_1, or to refrain from begging, leading the parent to slightly overestimate its condition as c_2. The indirect costs of extracting excessive resources from the parent make the latter option better (for offspring in the non-signalling pool).

THE PROBLEM OF EQUILIBRIUM SELECTION

Given that an infinite number of signalling equilibria may be possible under the same conditions, ranging from a costly separating equilibrium to a cost-free pooling equilibrium to a non-signalling equilibrium, the question arises: which of these solutions will evolution converge upon? This question is hard to answer, because it deals with evolutionary dynamics rather than statics. Game theoretical models of signalling, such as those presented here, can offer little insight into the issue. Simulations of signal evolution (e.g. Rodríguez-Gironés et al. 1998) provide an alternative approach, but there remain many technical problems to address (Lachmann & Bergstrom 1998). In particular, the stability of any particular signalling equilibrium is likely to depend upon parental responses to signals that are not employed at that equilibrium. These 'out-of-equilibrium' responses are important because they determine the fate of novel signalling mutants. Models that incorporate 'noise' or uncertainty offer some hope of tackling this issue, because perceptual error can give rise to a greater range of perceived signals at equilibrium than the range of signals actually given, allowing one to determine evolutionarily stable responses to signals never employed at the ESS (Johnstone & Grafen 1992b). Alternatively, such responses may drift or fluctuate because they are rarely tested, allowing the possible exploitation of 'hidden preferences' (Grafen 1990; Arak & Enquist 1993, 1995).

The absence of a well-developed theory of signal evolution leaves us with no clear predictions as to what kind of signalling equilibria are most readily attainable. As stated above, the fact that separating equilibria can entail 'excessive' costs has led some to argue that cheaper, pooling equilibria are more likely end-points of evolution (e.g. Rodríguez-Gironés et al. 1996, 1998; Bergstrom & Lachmann 1997). But, as also stated, it is not clear how broad is the range of conditions over which separating equilibria do entail such high costs (Godfray & Johnstone 2000), nor that these costs will necessarily render the equilibria unattainable. Moreover, 'cheap-talk' models also face problems in accounting for observed begging behaviour.

At a pooling equilibrium, begging is inevitably less informative than at a separating equilibrium, since the parent is unable to discriminate among offspring within each pool or sub-range of state. At a cost-free pooling equilibrium, moreover, the information content of the signal is particularly low. As relatedness declines, the proportion of young in each successive pool, from most to the least needy, decreases. Fewer pools may thus be possible, as shown in Figure 4, and an ever larger proportion of young employ the signal indicative of highest need. The parent is therefore left with an increasingly limited ability to discriminate among offspring that differ in condition (Bergstrom & Lachmann 1998; B.O. Brilot & R.A. Johnstone unpublished data). This theoretical prediction does not appear to match the findings of empirical studies, which suggest that begging exhibits continuous variation in intensity (e.g. Iacovides & Evans 1998), and may provide considerable information about offspring state (e.g. Redondo & Castro 1992;

Kacelnik et al. 1995; Kilner 1995, 1997; Cotton et al. 1996; Price et al. 1996; Kilner & Davies 1998; Leonard & Horn 1998; Lotem 1998).

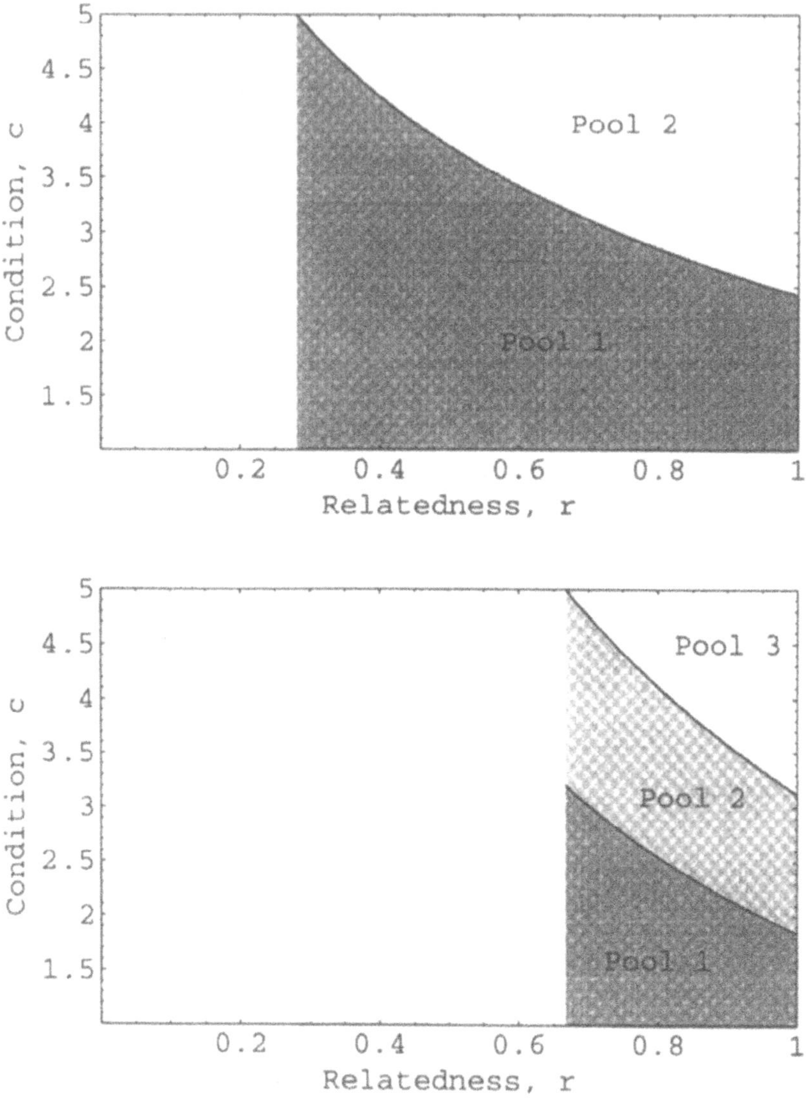

Figure 4. Sub-ranges of need within which offspring employ distinct signals, as a function of relatedness between current and future young (r), for a cost-free signalling equilibrium with 2 pools (upper graph) and with 3 pools (lower graph). The former type of equilibrium is possible only for $r > 0.281$, the latter only for $r > 0.668$. These results were obtained for cases where $r = 0.5$, $f(c,y) = 1 - e^{-cy}$, $g(y) = 1 - 0.05\,y$ and c is evenly distributed between 1 and 5.

We thus face something of a paradox: costly signalling models, which can account for the stability of informative and continuously variable begging behaviour, cannot easily explain the apparently low costs of begging; 'cheap-talk' models, which can account for low costs, cannot easily explain the high information content and continuous variability of begging. Perhaps it would be rash to rule out either possibility. While models of begging have assumed that offspring employ a single signal, in reality begging often combines several different signal components: posturing, calling and gaping may all, for instance, play a role (e.g. Kilner et al. 1999). R. Kilner (this volume) suggests that some aspects of the display, such as gaping, may represent a cost-free signal, while other components entail some expenditure. The combination of these different components in a single, composite display might provide parents with additional information about nestling state (i.e. they may function as 'back-up signals'; Johnstone 1996a). Alternatively, visual and vocal components of begging might reflect different aspects of offspring condition (i.e. they might serve as 'multiple messages'; Johnstone 1996a). Saino et al. (2000), for instance, suggest that the mouth colour of nestling barn swallows (*Hirundo rustica*) might provide information about their state of health (as opposed to hunger). These possibilities have yet to receive much theoretical attention.

FUTURE DIRECTIONS

The failure of signalling models to consider multiple signalling components illustrates some of the shortcomings of current theory. Debate over alternative signalling equilibria, and the theoretical complexities of modelling signal evolution, have distracted from the complexities revealed by empirical work on begging. Field and laboratory studies have increasingly shown that basic signalling models overlook many aspects of parent-offspring interaction. One obvious instance is sibling competition; while Godfray's (1991) model focuses on an exchange between a parent and a single nestling, begging typically involves interaction and competition among several young (e.g. Stamps et al. 1989; Price et al. 1996; Iacovides & Evans 1998; Lotem 1998; Price 1998).

More recent begging models have begun to address the issue of sibling competition. Godfray (1995b), for instance, extended his earlier analysis to consider a family with two young. Each varies in condition, and may advertise its state to the parent by means of a costly and continuously variable begging display, as in the basic model; but each may also respond to the begging of the other. This analysis suggests that a separating equilibrium is possible, which is qualitatively similar to the single offspring solution.

However, each young is expected to beg more intensely when its nestmate is more needy (i.e. in poorer condition), a prediction for which there is mixed support (see Smith & Montgomerie 1991; Cotton et al. 1996; Price et al. 1996; Leonard & Horn 1998; Leonard et al. 2000; Roulin et al. 2000). Extending this approach to larger broods is technically difficult. Johnstone (1999), however, presents a signalling model with multiple signallers in which all must signal simultaneously (thus ruling out response to each other's display). Once again, the incorporation of sibling competition does not alter the basic form of the signalling equilibrium, but may yield some quantitative effects (e.g. the cost required for the maintenance of honesty declined with brood size in this analysis).

The above models demonstrate that signalling theory can be extended to incorporate complexities such as sibling competition. However, interaction among broodmates also raises the possibility that begging may function as more than a simple signal. Where parents lack direct control over the division of food, offspring may settle this themselves through physical competition, perhaps by jockeying for position at the entrance to a cavity nest or seizing and monopolizing food items that a parent deposits (e.g. McRae et al. 1993; Kacelnik et al. 1995; Mock & Parker 1997). Begging may thus represent a form of direct competition between young, rather than a display aimed at parents. This situation matches the assumptions of classical models of sibling competition (Macnair & Parker 1979; Godfray & Parker 1992; Mock & Parker 1997) as compared to competitive signalling models.

Signalling and direct competition do not, however, represent mutually exclusive possibilities. Parental control over food allocation within a brood may range from complete (as signalling models assume) to totally absent (as classical sibling competition models assume), through all degrees between these extremes. Begging might thus combine competitive and signalling functions. It would be of great interest to determine how patterns of begging and resource division might be expected to change as the level of parental control alters, and to assess the selective pressures that favour parental versus offspring control over allocation. Rodríguez-Gironés (1999; see also Lotem et al. 1999) has also suggested that, even if parents lack control over food allocation, the level of direct sibling competition may nevertheless serve as a signal of offspring state in response to which a parent can adjust the total amount of food brought to the nest. This illustrates again the possibility that begging might combine competitive and signalling functions, and provides a simple explanation for how the begging display may originate (Rodríguez-Gironés et al. 1996; Rodríguez-Gironés 1999).

Sibling competition is not the only feature of parent-offspring interaction that suggests an alternative perspective on begging behaviour. Just as they overlook competition, the simple signalling models presented here effectively ignore the dynamic aspects of raising young. In reality, however, parent-offspring interaction can extend over a long period of time, and each begging exchange takes place in the context of this larger process (see Johnstone 1996b; Price et al. 1996; Iacovides & Evans 1998; Lotem 1998). Johnstone (1996b) demonstrates how dynamic considerations can modify the predictions of signalling models, with a simple extension of Godfray's (1991) nestling begging model in which the parent may transfer resources prior to as well as immediately after begging. In this analysis, the threat of costly begging induces the parent to over-allocate resources in order to reduce the subsequent level of solicitation by its offspring. This outcome is similar to the predictions of 'blackmail' models of begging (Zahavi 1977b; Macnair & Parker 1978, 1979; Parker & Macnair 1978, 1979; Parker 1985; Eshel & Feldman 1991), which suggest that begging serves as a means to induce parents to allocate more resources than would otherwise be optimal, rather than as a signal of need. In Johnstone's (1996b) analysis, however, begging still serves to convey information about offspring state. Once again, therefore, the model indicates that begging may combine a signalling function with some other purpose.

To sum up, we suggest that the most promising avenue for future studies of begging lies in the integration of signalling theory with alternative explanations for begging. 'Pure' signalling models, which focus on a simplistic one-shot encounter between a single parent and its offspring cannot hope to capture the full complexity of real parent-offspring interaction. The resolution of theoretical problems, such as equilibrium selection, may depend less on the development of new mathematical approaches than on the formulation of more realistic models that allow for sibling competition, blackmail and other such possibilities.

REFERENCES

Arak, A. & Enquist, M. 1993. Hidden preferences and the evolution of signals. *Philosophical Transactions of the Royal Society of London, Series B* **340**, 207-213.

Arak, A. & Enquist, M. 1995. Conflict, receiver bias and the evolution of signal form. *Philosophical Transactions of the Royal Society of London, Series B* **349**, 337-344.

Bengtsson, H. & Rydén, O. 1981. Development of parent-young interaction in asynchronously hatched broods of altricial birds. *Zeitschrift für Tierpsychologie* **56**, 255-272.

Bergstrom, C.T. & Lachmann, M. 1997. Signalling among relatives. I. Is costly signalling too costly? *Philosophical Transactions of the Royal Society of London, Series B* **352**, 609-617.

Bergstrom, C.T. & Lachmann, M. 1998. Signaling among relatives. III. Talk is cheap. *Proceedings of the National Academy of Sciences USA* **95**, 5100-5105.

Briskie, J.V., Martin, P.R. & Martin, T.E. 1999. Nest predation and the evolution of nestling begging calls. *Proceedings of the Royal Society of London, Series B* **266**, 2153-2159.

Cotton, P.A., Kacelnik, A. & Wright, J. 1996. Chick begging as a signal: are nestlings honest? *Behavioral Ecology* **7**, 178-182.

Dawkins, R. & Krebs, J.R. 1978. Animal signals: information or manipulation? In: *Behavioural Ecology: An Evolutionary Approach*, 1st Edition (Ed. by J.R. Krebs & N.B. Davies). Oxford: Blackwell Scientific Publications.

Dearborn, D.C. 1999. Brown headed cowbird nestling vocalizations and risk of nest predation. *The Auk* **116**, 448-457.

Eshel, I. & Feldman, M.W. 1991. The handicap principle in parent-offspring conflict: comparison of optimality and population genetic analyses. *American Naturalist* **137**, 167-185.

Furlow, F.B. 1997. Human neonatal cry quality as an honest signal of fitness. *Evolution and Human Behavior* **18**, 175-193.

Godfray, H.C.J. 1991. Signalling of need by offspring to their parents. *Nature* **352**, 328-330.

Godfray, H.C.J. 1995a. Evolutionary theory of parent-offspring conflict. *Nature* **376**, 133-138.

Godfray, H.C.J. 1995b. Signalling of need between parents and young: parent-offspring conflict and sibling rivalry. *American Naturalist* **146**, 1-24.

Godfray, H.C.J. 1999. Parent-offspring conflict. In: *Levels of Selection in Evolution* (Ed. by L. Keller). Princeton: Princeton University Press.

Godfray, H.C.J. & Johnstone, R.A. 2000. Begging and bleating: the evolution of parent-offspring signalling. *Philosophical Transactions of the Royal Society of London, Series B* **355**, 1581-1591.

Godfray, H.C.J. & Parker, G.A. 1992. Sibling competition, parent-offspring conflict and clutch size. *Animal Behaviour* **43**, 473-490.

Grafen, A. 1990. Biological signals as handicaps. *Journal of Theoretical Biology* **144**, 517-546.

Haskell, D. 1994. Experimental evidence that nestling begging behaviour incurs a cost due to nest predation. *Proceedings of the Royal Society of London, Series B* **257**, 161-164.

Iacovides, S. & Evans, R.M. 1998. Begging as graded signals of need for food in young ring-billed gulls. *Animal Behaviour* **56**, 79-85.

Johnstone, R.A. 1996a. Multiple displays in animal communication: 'backup signals' and 'multiple messages'. *Philosophical Transactions of the Royal Society of London, Series B* **351**, 329-338.

Johnstone, R.A. 1996b. Begging signals and parent-offspring conflict: do parents always win? *Proceedings of the Royal Society of London, Series B* **263**, 1677-1681.

Johnstone, R.A. 1999. Signaling of need, sibling competition, and the cost of honesty. *Proceedings of the National Academy of Sciences USA* **96**, 12644-12649.

Johnstone, R.A. & Grafen, A. 1992a. The continuous Sir Philip Sidney game: a simple model of biological signalling. *Journal of Theoretical Biology* **156**, 215-234.

Johnstone, R.A. & Grafen, A. 1992b. Error-prone signalling. *Proceedings of the Royal Society of London, Series B* **248**, 229-233.

Kacelnik, A., Cotton, P.A., Stirling, L. & Wright, J. 1995. Food allocation among nestling starlings – sibling competition and the scope of parental choice. *Proceedings of the Royal Society of London, Series B* **259**, 259-263.

Kilner, R. 1995. When do canary parents respond to nestling signals of need? *Proceedings of the Royal Society of London, Series B* **260**, 343-348.

Kilner, R. 1997. Mouth colour is a reliable signal of need in begging canary nestlings. *Proceedings of the Royal Society of London, Series B* **264**, 963-968.

Kilner, R.M. 2001. A growth cost of begging in captive canary chicks. *Proceedings of the National Academy of Sciences USA* **98**, 11394-11398.

Kilner, R. & Davies, N.B. 1998. Nestling mouth colour: ecological correlates of a begging signal. *Animal Behaviour* **56**, 705-712.

Kilner, R. & Johnstone, R.A. 1997. Begging the question: are offspring solicitation behaviours signals of need? *Trends in Ecology and Evolution* **12**, 11-15.

Kilner, R.M., Noble, D.G. & Davies, N.B. 1999. Signals of need in parent-offspring communication and their exploitation by the cuckoo. *Nature* **397**, 667-672.

Krebs, J.R. & Dawkins, R. 1984. Animal signals: mind-reading and manipulation. In: *Behavioural Ecology: an Evolutionary Approach*, 2nd Edition (Ed. by J.R. Krebs & N.B. Davies). Oxford: Blackwell Scientific Publications.

Krebs, E.A. & Magrath, R.D. 2000. Food allocation in crimson rosella broods: parents differ in their responses to chick hunger. *Animal Behaviour* **59**, 739-751.

Lachmann, M. & Bergstrom, C.T. 1998. Signalling among relatives. II. Beyond the tower of Babel. *Theoretical Population Biology* **54**, 146-160.

Leech, S.M. & Leonard, M.L. 1996. Is there an energetic cost to begging in nestling tree swallows (*Tachycineta bicolor*)? *Proceedings of the Royal Society of London, Series B* **263**, 983-987.

Leech, S.M. & Leonard, M.L. 1997. Begging and the risk of predation in nestling birds. *Behavioral Ecology* **8**, 644-646.

Leonard, M.L. & Horn, A.G. 1998. Need and nestmates affect begging in tree swallows. *Behavioral Ecology and Sociobiology* **42**, 431-436.

Leonard, M.L. & Horn, A.G. 2001. Begging calls and parental feeding decisions in tree swallows (*Tachycineta bicolor*). *Behavioral Ecology and Sociobiology* **49**, 170-175.

Leonard, M.L., Horn, A.G., Gozna, A. & Ramen, S. 2000. Brood size and begging intensity in nestling birds. *Behavioral Ecology* **11**, 196-201.

Lessells, C.M. & Parker, G.A. 1999. Parent-offspring conflict: the full-sib-half-sib fallacy. *Proceedings of the Royal Society of London, Series B* **266**, 1637-1643.

Lotem, A. 1998. Differences in begging behaviour between barn swallow, *Hirundo rustica*, nestlings. *Animal Behaviour* **55**, 809-818.

Lotem, A., Wagner, R.H. & Balshine-Earn, S. 1999. The overlooked signaling component of non-signaling behaviour. *Behavioral Ecology* **10**, 209-212.

Macnair, M.R. & Parker, G.A. 1978. Models of parent-offspring conflict. II. Promiscuity. *Animal Behaviour* **26**, 111-122.

Macnair, M.R. & Parker, G.A. 1979. Models of parent-offspring conflict. III. Intra-brood conflict. *Animal Behaviour* **27**, 1202-1209.

Manser, M.B. & Avey, G. 2000. The effect of pup vocalisations on food allocation in a cooperative mammal, the meerkat (*Suricata suricatta*). *Behavioral Ecology and Sociobiology* **48**, 429-437.

Maynard Smith, J. 1991. Honest signalling – the Philip Sidney Game. *Animal Behaviour* **42**, 1034-1035.

Maynard Smith, J. 1994. Must reliable signals always be costly? *Animal Behaviour* **42**, 1034-1035.

McCarty, J.P. 1996. The energetic cost of begging in nestling passerines. *The Auk* **113**, 178-188.

McRae, S.B., Weatherhead, P.J. & Montgomerie, R. 1993. American robin nestlings compete by jockeying for position. *Behavioral Ecology and Sociobiology* **33**, 102-106.

Mock, D.W. & Parker, G.A. 1997. *The Evolution of Sibling Rivalry*. Oxford: Oxford University Press.

Nöldeke, G. & Samuelson, L. 1999. How costly is the honest signaling of need? *Journal of Theoretical Biology* **197**, 527-539.

Ottoson, U., Backman, J. & Smith, H.G. 1997. Begging affects parental effort in the pied flycatcher, *Ficedula hypoleuca*. *Behavioral Ecology and Sociobiology* **41**, 381-384.

Parker, G.A. 1985. Models of parent-offspring conflict. V. Effects of the behaviour of the two parents. *Animal Behaviour* **33**, 519-533.

Parker, G.A. & Macnair, M.R. 1978. Models of parent-offspring conflict. I. Monogamy. *Animal Behaviour* **26**, 97-110.

Parker, G.A. & Macnair, M.R. 1979. Models of parent-offspring conflict. IV. Suppression: evolutionary retaliation by the parent. *Animal Behaviour* **27**, 1210-1235.

Price, K. 1996. Begging as competition for food in yellow-headed blackbirds. *The Auk* **113**, 963-968.

Price, K. 1998. Benefits of begging for yellow-headed blackbird nestlings. *Animal Behaviour* **56**, 571-577.

Price, K., Harvey, H. & Ydenberg, R.C. 1996. Begging tactics of nestling yellow-headed blackbirds, *Xanthocephalus xanthocephalus*, in relation to need. *Animal Behaviour* **51**, 421-435.

Rauter, C.M. & Moore, A.J. 1999. Do honest signalling models of offspring solicitation apply to insects? *Proceedings of the Royal Society of London, Series B* **266**, 1691-1696.

Redondo, T. & Castro, F. 1992. Signalling of nutritional need by magpie nestlings. *Ethology* **92**, 193-204.

Rodríguez-Gironés, M.A. 1999. Sibling competition stabilizes signalling resolution models of parent-offspring conflict. *Proceedings of the Royal Society of London, Series B* **266**, 2399-2402.

Rodríguez-Gironés, M.A., Cotton, P.A. & Kacelnik, A. 1996. The evolution of begging: signaling and sibling competition. *Proceedings of the National Academy of Sciences USA* **93**, 14637-14641.

Rodríguez-Gironés, M.A., Enquist, M. & Cotton, P.A. 1998. Instability of signaling resolution models of parent-offspring conflict. *Proceedings of the National Academy of Sciences USA* **95**, 4453-4457.

Rodríguez-Gironés, M.A., Zuñiga, J.M. & Redondo, T. 2001. Effects of begging on growth rates of nestling chicks. *Behavioral Ecology* **12**, 269-274.

Roulin, A., Ducrest, A.-L. & Dijkstra, C. 2000. Barn owl (*Tyto alba*) siblings vocally negotiate resources. *Proceedings of the Royal Society of London, Series B* **267**, 459-463.

Saino, N., Ninni, P., Calza, S., Martinelli, R., de Bernardi, F. & Møller, A.P. 2000. Better red than dead: carotenoid-based mouth colouration reveals infection in barn swallow nestlings. *Proceedings of the Royal Society of London, Series B* **267**, 57-61.

Smith, H.G. & Montgomerie, R. 1991. Nestling American robins compete with siblings by begging. *Behavioral Ecology and Sociobiology* **29**, 307-312.

Stamps, J., Clark, A., Arrowood, P. & Kus, B. 1989. Begging behaviour in budgerigars. *Ethology* **81**, 177-192.

Trivers, R.L. 1974. Parent-offspring conflict. *American Zoologist* **14**, 249-264.

Verhulst, S. & Wiersma, P. 1997. Is begging cheap? *The Auk* **114**, 134.

Weary, D.M. & Fraser, D. 1995. Calling by domestic piglets – reliable signals of need. *Animal Behaviour* **50**, 1047-1055.

Weary, D.M., Ross, S. & Fraser, D. 1997. Vocalisations by isolated piglets: a reliable indicator of piglet need directed towards the sow. *Applied Animal Behaviour Science* **53**, 249-257.

Zahavi, A. 1975. Mate selection – a selection for a handicap. *Journal of Theoretical Biology* **53**, 205-214.

Zahavi, A. 1977a. The cost of honesty (further remarks on the handicap principle). *Journal of Theoretical Biology* **67**, 603-605.

Zahavi, A. 1977b. Reliability in communication systems and the evolution of altruism. In: *Evolutionary Ecology* (Ed. by B. Stonehouse & C. Perrins). Baltimore: University Park Press.

Zhang, D.Y. & Jiang, X.H. 2000. Costly solicitation, timing of offspring conflict, and resource allocation in plants. *Annals of Botany* **86**, 123-131.

2.

STATE-DEPENDENT BEGGING WITH ASYMMETRIES AND COSTS: A GENETIC ALGORITHM APPROACH

Karen Price[1,2], Ron Ydenberg[1] & Dave Daust[2]
[1]*Behavioural Ecology Research Group, Department of Biological Sciences, Simon Fraser University, Burnaby, BC V5A 1S6, Canada (ydenberg@sfu.ca)*
[2]*RR #2 S23b C1, Burns Lake, BC V0J 1E0, Canada (kprice@futurenet.bc.ca)*

ABSTRACT

We develop a genetic algorithm model that evolves state-dependent begging strategies of nestlings and provisioning strategies of a parent. The model compares begging by nestlings with different abilities and needs, with and without predation costs. It shows that parents respond to cost-free begging as a signal of hunger in related nestlings. It suggests that competitively superior nestlings should be highly state-sensitive, begging less than smaller siblings when nearly satiated, but more when hungry. Such sensitivity evolves under two conditions: under low starvation risk, large siblings increase inclusive fitness by reducing the risk of sibling death; under high starvation and predation risk, large siblings decrease the probability of predation by reducing overall begging levels. Thus, loud begging by small nestlings may honestly signal need, or may manipulate their sibling's behaviour.

INTRODUCTION

Models of begging behaviour must simultaneously surmount several obstacles. First, begging strategies of nestlings and provisioning strategies of parents must be able to evolve freely (Godfray 1995a). Second, because the tactics adopted by one family member may affect the tactical choice of the others, begging needs to be modelled as a game. Third, because behaviour likely depends on nestling state (e.g. hunger, body condition, age; Kilner & Johnstone 1997), models must allow for state-dependence. Fourth, because

J. Wright and M.L. Leonard (eds.), The Evolution of Begging, 21–42.

state and food availability change over time, models must be dynamic (Godfray 1995a; Cotton et al. 1996; Godfray & Johnstone 2000). Fifth, inclusive fitness must be properly calculated. Sixth, because differences in competitive ability and need can interact in complex ways, models must include, minimally, one parent and two unequal nestlings. Genetic algorithms make possible a begging model that encompasses all of the crucial aspects listed above: individuals are related explicitly, strategies can be dynamic and state-dependent, interdependent parent-offspring strategies play the field in an implicit evolutionary game and multi-nestling families can be considered.

This chapter describes a genetic algorithm model in which both the begging strategies of nestlings and the food allocation strategy of parents evolve. Genetic algorithms were developed to optimize complex engineering design problems (Goldberg 1989) and have recently been adopted by behavioural ecologists (e.g. Sumida et al. 1990). Although numerical approaches are not easily related to analytical models of begging behaviour, the advantages seemed sufficiently compelling to develop the model we report here. We illustrate the model's utility by highlighting three issues discussed in recent begging literature: (1) Can cost-free begging evolve? (2) Does begging reflect nestling need when competitive abilities of the nestlings vary? (3) Is begging manipulative or honest?

The Issues

Begging has long been assumed to carry costs, and several authors have concluded that costs are necessary for evolutionary stability (Stamps et al. 1978; Macnair & Parker 1979; Parker et al. 1989; Godfray 1995b; but see R.A. Johnstone & H.C.J. Godfray this volume). The available empirical evidence for costs, however, is skimpy, suggesting that they are smaller and harder to document than initially believed (D.G. Haskell this volume; M.A. Chappell & G.C. Bachman this volume). Recent theoretical work suggests that cost-free signalling can be stable under some conditions (Maynard Smith 1991, 1994; Bergstrom & Lachmann 1998; Johnstone 1999). We model begging with and without predation costs to see if the dynamic nature of the model broadens the conditions for cost-free begging.

Most begging models consider a nestling's begging strategy in relation to hunger level. We refer to hunger as short-term need, because it may be reduced quickly by a few food deliveries. But long-term need - the total amount of provisioning a nestling requires to reach fledging - may also affect begging strategies, and does not change very quickly in response to feedings. Long-term need depends on condition (Richter 1984; Litovich &

Power 1992), gender (Fiala & Congdon 1983; Teather 1987; Teather & Weatherhead 1988) and size (Price et al. 1996; Lotem 1998; Cotton et al. 1999). All of these factors may affect a nestling's ability to beg, and if siblings compete for food by begging (Smith & Montgomerie 1991; Price 1996), begging may reflect nestling competitive ability (Zahavi 1977a; Grafen 1990) rather than conveying reliable information about hunger. From the parental perspective, information about nestling need and ability could both be valuable in deciding how to allocate food, but begging may not be able to signal ability and need simultaneously (Johnstone & Grafen 1993). Investigations into the evolution of begging as a signal of need to parents must consider situations in which parents are faced with offspring unequal in need and ability.

The literature on begging is rich in metaphor. There are battlegrounds with winners and losers, manipulative cheats and blackmailers fighting honest offspring. Though colourful, this language has sometimes obscured the issues and has led to confusion and unhelpful dichotomies (Mock & Forbes 1992). Discussions about the honesty of begging ignore whether begging honestly signals need or honestly signals ability. They also obscure the possibility that absolute begging levels may escalate due to competition, whereas relative begging levels within a brood may still contain information about nestling need or ability (Harper 1986). Focusing on metaphor can distract attention from model mechanics. For example, Zahavi (1977b) suggests that competitively subordinate nestlings manipulate their parents into bringing more food by attracting predators to the nest. Other models (Parker & Macnair 1979; Harper 1986) do not support the idea that parental response to begging level is affected by predation risk (Harper 1986), but they do not allow for the asymmetry in nestling ability that forms the basis for Zahavi's idea i.e. they pay attention to the language rather than to the model's underpinnings.

THE MODEL

The model is based on our work with yellow-headed blackbirds (*Xanthocephalus xanthocephalus*, Price 1994), which are marsh-nesting members of the Icteridae. Mothers bring most food. Nest predation is high. Nestlings beg very intensely, and show a strong size gradation with hatching order. Male nestlings are much larger than females.

Genetic algorithms work by creating variants in the design of a feature (here, behavioural strategies), testing their performance in a computer simulation, selecting the best performers, introducing new variation and repeating these steps many times to find the overall best performance. The strategies we examine specify how much each of a pair of siblings (alpha and beta nestlings) begs in relation to their hunger, and how the parent allocates food in relation to begging level.

Our model works as follows: (1) Initial (random) strategies for state-dependent begging and provisioning are encoded on a 'chromosome'. (2) The chromosomes are randomly paired and produce two offspring. Before reproduction they undergo procedures analogous to mutation and recombination to introduce new variation. (3) The chromosomes enter a simulation in which nestlings beg and parents provision according to the strategies they carry. Nestlings may starve, be depredated or survive. The probabilities of these events are influenced by nestling state and behaviour. (4) Surviving nestlings (i.e. the most successful strategies) enter the next generation as parents. Steps (2) - (4) are repeated 500 times ('generations').

Definitions

We represent short-term need as a state variable, equivalent to nestling hunger level. Long-term need is set as a parameter, representing a nestling's hatch position, gender or body condition (Price et al. 1996). Begging strategy, effort, ability and level are related terms: the encoded begging strategy specifies state-dependent begging effort, which is modified by begging ability to result in an expressed begging level. For a given effort, an able beggar can beg at a higher level (Parker et al. 1989). Level includes the intensity, loudness and duration of calls. We refer to a strategy as stable when it remains constant over 200 generations of the simulation (see below). Readers should note that this definition of stability is not, strictly speaking, equivalent to an ESS as defined in game theory.

Readers uninterested in details of the model can at this point skip to the Results section.

Encoding Strategies in the Model

The strategies are encoded as a sequence of five binary numbers (a chromosome). Each of the five binary numbers represents a 'gene' consisting of a five bit (0/1) string, so the entire strategy is encoded as 25 bits. Two of the genes specify the begging strategy employed when the bearer is an older, larger or competitively superior (alpha) nestling, two represent the strategy if the bearer is a younger, smaller or weaker (beta) nestling. The fifth gene governs the allocation of provisioned prey when the bearer is a parent.

The begging strategy specifies the relationship between begging effort and hunger. Begging effort ranges between 0 and 1. Within both the alpha and beta pairs of genes (alpha and beta roles are randomly assigned in each family), one gene determines begging effort when satiated, and the other determines begging effort when maximally hungry. There are 32 (i.e. 2^5) possible levels of hunger in between, and linear interpolation is used to find the begging effort at intermediate hunger states (Figure 1).

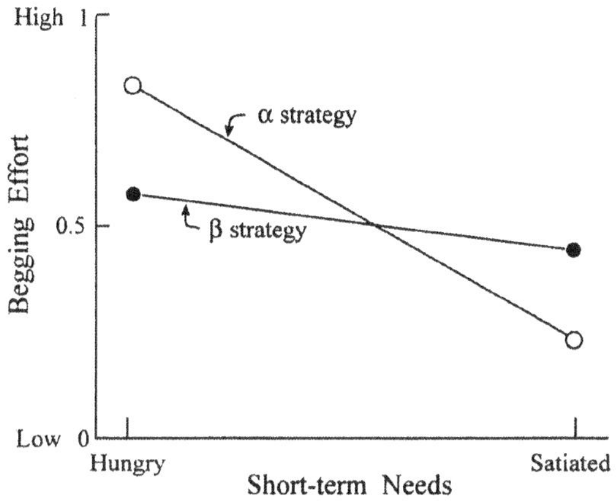

Figure 1. An example of begging strategies of alpha and beta nestlings. A pair of genes for each strategy encodes begging effort when hungry and when satiated. Begging effort is interpolated linearly at intermediate hunger levels.

The parental strategy is the probability that the noisier (higher begging level) of the two nestlings is fed. The smallest value of the gene causes the parent to always feed the quieter nestling, while the largest value causes the parent to always feed the noisier nestling. The midpoint causes random feeding in relation to begging level.

Recombination and Mutation

To initialize a run of the model, each bit is set randomly on each of an initial population of 80 chromosomes. Parent chromosomes are randomly assigned to pairs, aligned and crossed over at a single random location to recombine their genes. One of the new chromosomes is retained (selection random) and the other discarded. The parent chromosomes recombine once more to produce a second new chromosome, and these are assigned to the two offspring.

This procedure produces offspring chromosomes that are related, on average, by 0.5 ('full siblings'). Retaining both chromosomes resulting from one crossing-over produces offspring chromosomes related to each parent by an average of 0.5, but unrelated to each other ($r = 0$). Duplicating one chromosome from a single crossing-over creates identical ($r = 1$) nestlings. We ran the model with all three procedures to examine the effect of relatedness on begging behaviour.

In each generation, each bit 'mutates' (changes from 0 to 1, or 1 to 0), with probability $\mu = 0.01$ or 0.02. We found the appropriate mutation rate by numerical experimentation with the model. With mutation rate set too high, strategies do not converge; set too low, they easily become trapped on local optima.

Reproduction

Parental chromosomes are selected for mating with replacement. Using the above procedures, 40 nests each with two nestlings are created. One parent is randomly selected to provision the nestlings in its nest, and the families enter the nesting simulation stage of the model. During the simulation, each nestling begs as specified by its begging strategy, and is fed according to its parent's strategy. It grows if fed, and starves if not fed. In some model runs the nest can be depredated. As nestlings die, replacement nests are created. The simulation continues until there are 80 survivors to enter the next generation as parents. Nestlings gain no extra fitness benefits from fledging well fed, so success is determined solely by survival.

Nesting Simulation

The nesting simulation is composed of 100 periods ($t = 1, 2, 3, \ldots, 100$), with begging and provisioning occurring in each. We define the state of the *i*th nestling in each period t as the difference between its mass, $m_i(t)$, and the expected mass in period t, $\overline{m}(t)$ representing the linear, average growth trajectory of the population. State is calculated as:

$$X_i(t) = m_i(t) - \overline{m}(t) \tag{1}$$

Nestlings starve at $X_i(t) = X_{min}$ and are satiated at $X_i(t) = X_{max}$. At $t = 0$ (newly-hatched nestlings), $X_i(0) = 0$ (i.e. no deviation from the set growth curve). Each period begins with nestlings begging as specified by their strategy and hunger state. The parent then allocates a meal according to its provisioning strategy. We set meal size as a model parameter. In most cases, there is just sufficient food to raise two nestlings.

Nestlings pay a maintenance cost (c) each period. Unfed nestlings therefore lose mass. When fed, the mass increment, $g_i(t)$, depends on $X_i(t)$:

$$g_i(t) = 1\text{-}d/(1+100\exp(-X_i(t)+e)), \tag{2}$$

(i.e. the value of food decreases as nestlings approach satiation; d and e are shape parameters) and a nestling's mass in the next period (t + 1) is:

$$m_i(t + 1) = m_i(t) + g_i(t) - c. \tag{3}$$

The simulation proceeds to the next period by returning to equation (1) to calculate nestling state.

Nestling Death

We assume that the probability of nest predation, P(t), is a function of the summed begging levels of both nestlings:

$$P(t) = k \text{ (alpha level + beta level)}, \tag{4}$$

where alpha and beta levels are specified by begging strategy and nestling ability, and k is the strength of predation risk. Predation kills both nestlings.

Nestlings may also starve. Probability of starvation in period t, $S_i(t)$, is:

$$S_i(t) = [d/(1+100\exp(X_i(t)+e))]*s, \tag{5}$$

subject to $X_i(t) < 0$, where d and e are shape parameters, and s serves as an environmental starvation risk parameter. We chose the exponential form of this curve by comparing the probability of starvation with nestling condition (deviation from a regression of mass on tarsus) in a sample of 168 yellow-headed blackbird nestlings (Price 1994). Following the starvation of its nestmate, the survivor is assured of enough food.

Nestling Asymmetry

Nestling role determines the type and amount of asymmetry. In symmetrical sibling pairs, alpha and beta nestlings have similar begging abilities and long-term need. We incorporate asymmetries in competitive ability into the model by assuming that alpha nestlings are more effective beggars than beta nestlings, achieved by multiplying the alpha nestling's begging effort by 1.5.

We create asymmetries in nestling long-term need (see definitions) by giving one nestling a relatively higher risk of starvation for a given hunger state (shifting the starvation risk curve by changing parameter e). From a parent's perspective, the marginal gain of feeding the needier nestling is higher, due to the exponential shape of the starvation risk curve.

Model Output

We explored several sets of parameter values (Table 1). Figure 2 shows a sample run, illustrating begging and provisioning strategies (set randomly at generation 1) that evolve to a stable pattern after about 150 generations. The lines in the figure trace the evolution of the maximum and minimum begging effort for both the alpha (top panel) and beta (middle panel) nestlings in one run of the model, and the parent's (bottom panel) allocation strategy. For each of the five genes, we averaged the value over the last 50 (of 500) generations to express the outcome of a model run.

Table 1. Model parameters. Standard values in bold typeface; others are used in specific runs.

Description	Parameter	Range of Values
Maintenance cost	c	0.5
Steepness parameter	d	5.9
Shift parameter	e	1.0, **2.0**
Predation risk parameter (none, moderate)	k	**0.0**, 0.01
Relatedness (unrelated, full sib, identical)	r	0.0, **0.5**, 1.0
Starvation risk parameter (low, moderate, high)	s	**0.25**, 0.5, 1.0
Resource abundance (sufficient to abundant)	f	**2.0**, 2.1, 2.2
Begging ability (symmetrical or asymmetrical)	a	**1.0**, 1.5
Mutation rate	μ	**0.01**, 0.02

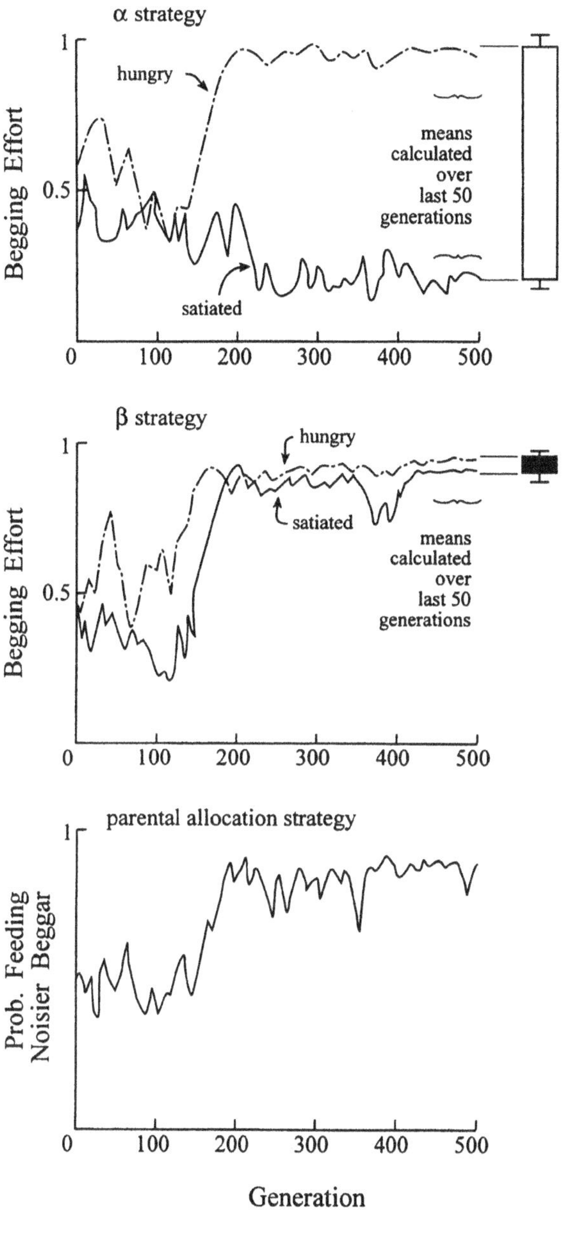

Figure 2. Sample results from one run of the model. See text for full description. Bars indicate mean ± SD of the minimum and maximum begging efforts, calculated from the last 50 generations, over 20 runs of the model. Open bar = alpha nestling; black bar = beta nestling.

We ran the model 10 or 20 times with each set of parameter values, and calculated means and standard deviations. This outcome is represented in the bar on the right hand side of the nestlings' panels in Figure 2. The top of the bar represents the mean begging effort (+ SD from 10 or 20 replicates) when hungry, and the bottom of the bar shows the mean effort (- SD) when satiated. The height of the bar reveals the state-sensitivity of begging: a long bar indicates high sensitivity. This bar is used in the figures to portray results.

In Figure 2, alpha nestlings beg at a high level when hungry, but reduce their begging greatly when satiated. In contrast, beta nestlings always beg at a high level and show little state-sensitivity. The parental provisioning strategy is portrayed in the bottom panel. In the example shown in Figure 2, the parent usually feeds the noisiest beggar. For each set of model replicates, we also calculated the proportion of runs in which stable (see above for Definitions) strategies evolved, and kept track of the average number of nestling deaths in each generation.

Note that the figures portray the strategy only. They do not show nestlings' hunger states, and hence they do not directly reveal the begging behaviour expressed. For example, a nestling might always be hungry and begging loudly, even though its strategy specifies low begging effort at low hunger levels. Though we do not show it here, it is possible to calculate the nestling states resulting from a strategy by running the evolved strategy through the nesting simulation portion of the model. In most cases, nestling states equilibrate at levels where each is fed in turn.

RESULTS

Parental Provisioning Strategy

In model runs with related symmetrical or asymmetrical nestlings and no costs, the parental strategy evolves to feed noisy nestlings half the time and to feed quiet nestlings the other half. In the former case, nestlings beg at high levels when hungry and somewhat lower levels when satiated (the classical pattern), while in the latter, nestlings beg at low levels when full and tend not to beg at all when hungry (paradoxical strategies). These alternatives evolve quickly and remain stable over 200 generations without flipping from one to the other. Note that begging signals nestling state in both cases; in the first it represents hunger, while in the second it represents satiation. Model runs with unrelated nestlings never achieve stability, and frequently flip between classical and paradoxical begging patterns. Taken together, these results show that begging can evolve to signal nestling state, even in cost-free situations, provided that nestmates are related.

When begging carries predation costs, the classical pattern evolves more frequently (~70% of runs) than the paradoxical strategy. When the latter does evolve, it occasionally flips (~10% of runs) to the classical strategy. Classical strategies, however, never flip to paradoxical, even though twice as many nestlings die when parents feed noisy rather than quiet nestlings (e.g. 60 ± 3 versus 31 ± 2 deaths per generation; standard parameters except food = 2.1, predation = 0.01; see Table 1). Evidently, the paradoxical strategy gives higher fitness, but when predation is a danger, this strategy is not stable against invasion by the classical pattern of noisy hungry nestlings.

In all the remaining results, the gene specifying parental behaviour is fixed, so that parents always feed the noisiest beggar. This approach allows us to look in detail at evolved sibling begging strategies once parents have evolved to feed beggars. Mutation rate is set at 0.02.

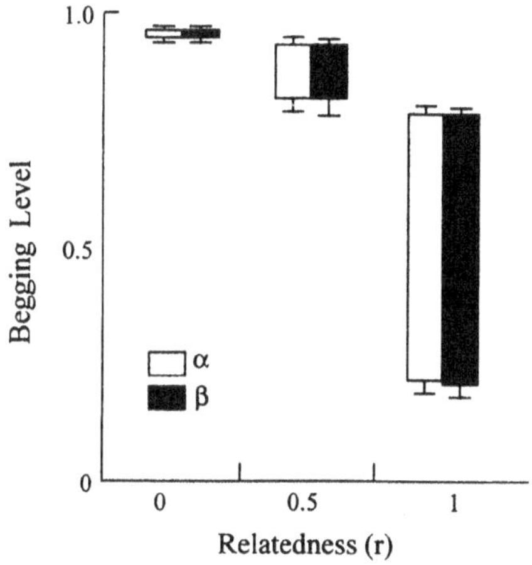

Figure 3. Evolved begging levels of unrelated, full sibling and identical pairs of symmetrical nestlings. Bars indicate mean ± SD of the minimum and maximum begging efforts over 10 runs of the model. Open bar = alpha nestling; black bar = beta nestling.

Effects of Relatedness

Evolved begging strategies vary with sibling relatedness (Figure 3). Unrelated nestlings (r = 0) always evolve escalated strategies, showing no state-sensitivity. This pattern is robust to changes in parameter values. In most families, only one nestling survives (e.g. 74 ± 1 deaths per generation; standard parameters except r = 0).

Full siblings (r = 0.5) also escalate when hungry, but show some state-sensitivity by reducing their begging intensity when full. Both nestlings exhibit similar behaviour. More families fledge two nestlings (e.g. 59 ± 2 deaths; standard parameters). Begging strategies of related nestlings are sensitive to other parameter values including the asymmetries and costs discussed below. Changes in resource abundance also affect these results: scarcity leads to escalation, while abundance increases state-sensitivity. Identical nestlings (r = 1) show even greater state-sensitivity, as well as decreased overall levels of begging; there are fewer deaths (e.g. 26 ± 1 deaths per generation; standard parameters except r = 1). These results confirm the importance of relatedness in achieving stability.

In all subsequent results, nestlings are full siblings.

Effects of Asymmetries

We consider first nestlings with equal begging ability, but unequal long-term need. Relative to symmetrical siblings (compare Figure 4a with the full siblings in Figure 3), slightly reduced state-sensitivity evolves in the nestling with higher long-term need. But if long-term need is equal and ability unequal (Figure 4b), the nestling with higher ability evolves much greater state-sensitivity, while the less able sibling retains low sensitivity. Our interpretation is that more able beggars can always outcompete their siblings, but do so only when hungry. Conversely, less able beggars (in this case the beta nestlings) risk starvation, and cannot afford to decrease begging. Beta nestlings starve almost three times as frequently as their siblings do (e.g. alpha nestlings: 7 ± 1 deaths per generation; beta nestlings: 19 ± 1 deaths per generation; standard parameters except alpha begging ability = 1.5).

Cases in which nestlings differ in both long-term need and ability are biologically more interesting, representing broods of large and small, or male and female, siblings. When alpha nestlings have higher ability and lower need (e.g. larger, older or better conditioned nestlings; Price et al. 1996), they evolve high state-sensitivity, decreasing their effort to below that of beta nestlings when satiated (Figure 4c). Conversely, if nestlings with higher need also have higher ability (e.g. the larger gender in dimorphic species; Teather 1992; Price & Ydenberg 1995), they evolve moderate state-sensitivity, and always beg more for a given state (Figure 4d). These results show that competitive ability and nestling need interact to produce different begging strategies (i.e. there are both scramble and signalling elements to begging).

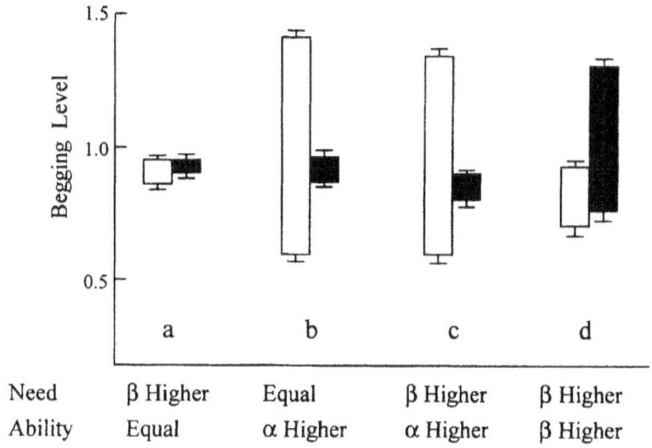

Figure 4. Evolved begging levels in pairs of nestlings with asymmetries in long-term need and ability (standard parameters except for asymmetries: for alpha nestlings, begging ability = 1.5; for needy nestlings, e = 1). The strategies differ depending on whether needy nestlings have higher or lower ability. Bars indicate mean ± SD of the minimum and maximum begging efforts over 10 runs of the model. Open bar = alpha nestling; black bar = beta nestling.

Costs of Begging

The risk of nest predation associated with begging interacts with the risk of starvation to affect evolved begging strategies. We present an example (Figure 5) with nestlings of asymmetrical ability and equal long-term need (standard parameters except alpha nestling begging ability = 1.5, starvation = 0.5 or 1, predation = 0 or 0.01).

With a moderate risk of starvation, alpha nestlings (higher begging ability) show high state-sensitivity with or without predation risk. With added predation risk, beta nestlings change their relatively state-insensitive strategy only slightly. Overall, adding predation to simulations with moderate starvation risk has little effect on relative begging strategies.

Under a harsher starvation regime, however, adding predation risk changes relative begging strategies considerably. With no predation, both nestlings beg at high levels in a state-insensitive manner. Nestlings compete intensely, and many, particularly beta nestlings, die. With the addition of predation costs, beta nestlings remain state-insensitive, but alpha nestlings greatly vary their begging level with state. These results show that increased predation costs may – or may not – influence begging strategy, depending on the risk of starvation.

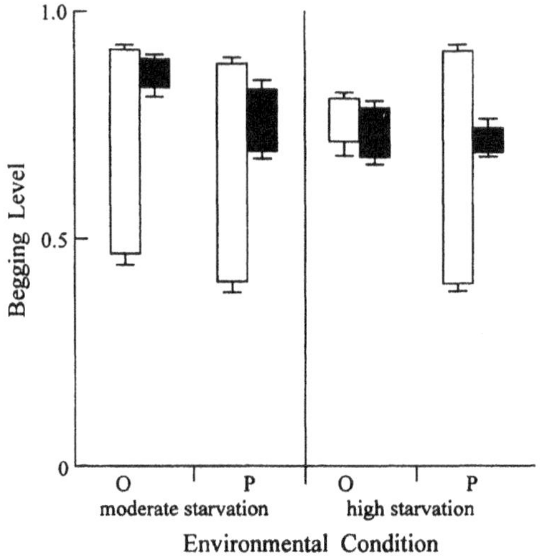

Figure 5. Evolved begging strategies under varying starvation and predation risks. O = no predation; P = predation. Note that adding predation risk affects strategies very little under moderate starvation, but considerably under high starvation by making the alpha nestling more state-sensitive. Bars indicate mean ± SD of the minimum and maximum begging efforts over 10 runs of the model. Open bar = alpha nestling; black bar = beta nestling.

DISCUSSION

The model presented here shows that nestling relatedness and predation costs influence parental provisioning strategy, and that nestling relatedness, asymmetries in competitive ability and long-term need, and predation costs all affect begging strategies. We discuss each effect in turn and compare model predictions with empirical evidence. We then compare our model with others and summarize our findings in relation to the three begging issues initially presented.

Cost-Free Begging and Relatedness

Our model demonstrates that stable patterns of begging and provisioning can evolve in the absence of costs, as long as the nestlings are related. This result counters assertions that begging costs are necessary for evolutionary stability (e.g. Stamps et al. 1978; Macnair & Parker 1979; Parker et al. 1989; Godfray 1995b; also see R.A. Johnstone & H.C.J. Godfray this volume). The dynamic aspect of our model probably accounts for the discrepancy here. Model nestlings experience a range of hunger levels and gain inclusive fitness from leniency when satiated. Our model does not suggest that begging never escalates; it merely points out that escalation does not necessarily remove information about hunger level from begging.

McCarty (1997) found little evidence for strong energetic costs of begging, and suggested that adverse impacts to kin may be sufficient to constrain begging. Our model confirms his suggestion and matches observations that begging levels increase with decreased nestmate relatedness (Briskie et al. 1994), an effect particularly obvious in brood parasites (e.g. Dearborn 1998; D.C. Dearborn & G. Lichtenstein this volume). The model further predicts that begging by brood parasites should be less state-sensitive than the begging of their hosts, as found by Redondo (1993; also see T. Redondo & J.M. Zuñiga this volume).

Maynard Smith (1991, 1994) developed an analytical model showing that cost-free signalling to a potential donor will be stable under some conditions. (In our model, nestlings 'donate' food to their siblings by reducing their begging level). Our model shows a broad range of conditions with stable cost-free begging. Non-linearities in the state-dependent risk of starvation and the possibility for continuous variation in begging (rather than the discrete options in Maynard Smith's model) increase the chances of a stable signal evolving. Our model agrees with recent extensions to the Maynard Smith model (Bergstrom & Lachmann 1998; Johnstone 1999) that the range of conditions allowing cost-free begging among competing, related nestlings may be broader than previously believed. Our model is compatible with the lack of empirical evidence for large costs to begging (M.A. Chappell & G.C. Bachman this volume; D.G. Haskell this volume).

Parker and Macnair (1979) modelled a retaliatory strategy in which parents ignore solicitation. Random feeding (the equivalent of ignoring begging) never evolved in our model. Rodríguez-Gironés et al. (1996), after discarding the strategy of feeding quiet nestlings as making no biological sense, found a non-signalling equilibrium with higher fitness. This equilibrium disappeared, however, with the introduction of sibling competition (Rodríguez-Gironés 1999). We did not detect a stable non-signalling strategy.

Asymmetries in Need and Ability

An important result of our model is that nestlings with higher ability evolve increased state-sensitivity. Conversely, nestlings with equal ability and greater long-term need have decreased state-sensitivity. When both ability and long-term need vary, we found that for a given hunger state, high ability/low need nestlings (e.g. large or early-hatched nestlings) sometimes begged less than their nestmates, whereas high ability/high need nestlings (e.g. males in dimorphic species) never begged less than their siblings. High ability/high need nestlings always used their competitive advantage to increase their survival chances. High ability/low need nestlings had good survival prospects, and were able to beg less when not hungry in order to increase the survival chances of their small, needy siblings.

From the model results, we predict that: (1) able beggars will be more state-sensitive; (2) long-term needy nestlings will be less state-sensitive; (3) in sexually dimorphic species, the faster growing gender will beg more for any given hunger state; and (4) small (or late-hatched) nestlings will beg more than their larger siblings when both are full or in good condition, but that large nestlings will always beg more when both are hungry or in poor condition. Tests of these predictions require measurements of nestlings over a range of hunger levels. Few studies have examined both short-term and long-term need. Experiments with yellow-headed blackbirds (Price et al. 1996) showed that males (high ability/high need) beg more than females for any given hunger level, and that males are more state-sensitive. Iacovides and Evans (1998) found that ring-billed gull (*Larus delawarensis*) nestlings in poor condition (greater long-term need) were less state-sensitive and begged more than individuals in good condition. Cotton et al. (1999) found that large starling (*Sturnus vulgaris*) nestlings (high ability/low need) in asynchronous broods begged less than nestlings in synchronous broods, but did not present data showing the change in begging with hunger level. Lotem (1998) found that large barn swallow (*Hirundo rustica*) nestlings begged less than small nestlings for given hunger levels when young, but not when older. Large swallow nestlings did not exhibit greater state-sensitivity in non-vocal begging components over the range of hunger levels reported. This discrepancy warrants further investigation.

Our model was designed to examine whether or not begging could signal hunger (short-term need), given nestlings with differing competitive ability. Hence, we combined information about both short-term need and competitive ability into a single begging component (level). Real world begging signals may have several components that provide different information. For example, experimental evidence shows that parents respond to both size and hunger level (e.g. Stamps et al. 1985; Kilner 1995; Price &

Ydenberg 1995). The model could be extended to consider separate begging components, but it cannot in its current form tease apart aspects of nestling state expressed through different components (e.g. jostling versus vocalizing).

Predation Risk and Starvation Risk

Although empirical evidence suggests that current predation costs of begging are low or non-existent, D.G. Haskell (this volume) suggests that begging calls may have evolved to minimize predation costs. Our model examines this evolutionary process. The addition of shared begging costs increased the probability that parents evolved to feed beggars and that hungry nestlings begged more. It makes intuitive sense that nestlings are more likely to risk predation when hungry than when satiated, even though more nestlings died following the classical as opposed to the paradoxical strategy. Unlike Godfray (1995b), we found stable strategies when predation affected all nestlings equally.

Evolved begging strategies varied with the relative risks of predation and starvation. Our results demonstrate that predation risk can influence begging behaviour (as suggested by Zahavi 1977b), particularly when nestlings must trade-off starvation risk against predation risk. We propose a scenario in which predation risk affects sibling interactions as follows. The alpha nestling reduces its begging when less hungry to improve its chances of survival. The constant begging of the beta nestling raises the predation risk felt by its sibling and is one factor in the alpha nestling's begging decision. The evolution of increased state-sensitivity by the alpha nestling moderates the predation risk felt by both nestmates. The suggestion that begging could influence the behaviour of larger siblings as well as parents is supported by findings that small yellow-headed blackbird nestlings beg when the parent is absent (Price & Ydenberg 1995). Hungry barn owl (*Tyto alba*) nestlings beg less when their sibling is hungry rather than satiated, also suggesting negotiation between siblings (A. Roulin this volume).

The model predicts that begging strategies should vary between environments with different starvation and predation regimes. Species in safe nests subject to high starvation rates (e.g. some cavity-nesters) should show relatively state-insensitive begging, while those in dangerous nests and subject to high starvation rates should show state-sensitive and size-sensitive begging levels. Nestlings with asymmetries in competitive ability, experiencing low predation and low starvation risks, should also show state- and size-sensitive begging. Testing these predictions will require interspecific comparisons of begging by nestlings with controlled hunger levels. Some, but not all, studies suggest that cavity-nesters have louder and more locatable calls than non-cavity-nesters (reviewed in D.G. Haskell this volume), but no interspecific studies have compared begging over a range of hunger levels.

The Model

Our model follows Parker et al. (1989) in examining begging strategies by asymmetric nestlings in two-nestling broods, and Harper (1986) in allowing parental strategies to evolve and in looking at the evolution of state-dependent begging levels. It fulfils two of the three criteria presented by Cotton et al. (1996) as important, allowing for dynamic variation in state and variation in competitive ability. It does not allow food availability to fluctuate within a simulation. We agree that fluctuation in food is a critical component of nestling experience and should be included in further modelling.

Our model differs from previous models primarily in using a numerical rather than analytical approach to finding stable strategies. While numerical models lose something in elegance and generality, they facilitate examination of complex problems like begging. The biggest benefit of using a numerical approach is the ability to model dynamic changes in state. The approach is flexible, and allows examination of a range of parameters.

Our model parallels Godfray's (1995b) landmark model in that both examine the evolution of state-dependent begging strategies in families with one parent and two nestmates varying in ability and long-term need. Both models start by assuming that a fixed amount of food will be distributed between two offspring. Both models find that decreased offspring condition (increased need) leads to increased begging, and that increased relatedness leads to decreased begging (or increased state-sensitivity). Godfray finds that decreased sibling condition leads to increased begging (due to competition); we find that increased long-term need decreases state-sensitivity slightly. These results, however, are not directly analogous and are thus difficult to compare.

Godfray finds that increased resources lead to increased begging; we find the reverse. This difference may be due to our simple fitness function; we assume that as long as a nestling survives, it does not benefit from increasing its food allocation. Godfray finds that costs other than indirect costs to kin are necessary for stability; we do not, although definitions of stability may differ here. Both models look at offspring asymmetries, though in somewhat different terms. Godfray finds that nestlings with lower costs of begging should beg more. We assume that larger nestlings will have lower costs of begging, and thus that this asymmetry is analogous to our asymmetry in begging ability. We find that more able beggars beg more when hungry, but less when full. We did not look at asymmetries in asymptotic fitness, and our asymmetry in long-term need is not quite analogous to Godfray's difference in attainable condition.

The Issues

Our model demonstrates that parents can evolve to respond to begging signals even if begging carries no costs – provided that nestlings are related. It shows that begging can evolve to signal nestling hunger even if nestlings have asymmetrical competitive abilities. Finally, it illustrates the limited use of terms such as 'honesty' and 'manipulation' as follows. The model proposes two conditions under which large nestlings should be more sensitive to hunger than small nestlings: (1) under low to moderate starvation risk, large siblings increase inclusive fitness by reducing the risk of their sibling's death; (2) under both a high risk of starvation and a high risk of predation, large siblings decrease the chances of predation (and death of both nestlings). The first of these hypotheses suggests that large nestlings beg less because their small siblings need food (i.e. begging carries inclusive fitness costs), and hence implies that begging carries honest information about nestling long-term need. The second suggests that large nestlings beg less because their siblings have changed the context of their begging decision (i.e. begging increases predation risk), and hence implies that small nestlings may be manipulating their nestmates' behaviour. Hence, begging can be interpreted as either an honest signal of need or manipulative, depending upon starvation and predation regimes.

FUTURE DIRECTIONS

We selected the results presented in this chapter because they relate to recent discussions in the literature and illustrate the flexibility of the genetic algorithm modelling approach. We encourage readers to copy the basic model (accessible from http://www.sfu.ca/biology/faculty/ydenberg) and try their own modifications.

Future investigations of begging should take a broad perspective. We suggest that theoretical and empirical studies incorporate nestling asymmetries in competitive ability and need, avoid classifying complexities as dichotomies, look at interactions among nestlings as well as between parents and offspring, and assess the environment (e.g. predation risks, starvation risks) in which a particular begging strategy evolved.

Some predictions from the model await testing. In particular, examination of the relationship between begging and hunger by asymmetrical nestlings under varying environmental conditions could be explored. The state-sensitivity of large (high ability/low need) and small nestlings requires further experimental testing.

Many other investigations are possible using the genetic algorithm framework; we suggest some extensions below.

Effects of energetic costs: Energetic costs of begging can easily be incorporated into the mass change calculation (equation (3)). Resource availability must be increased to cover the extra costs.

Siblicide: In simulations including predation costs, it is possible to generate siblicide by allowing a nestling to fledge without begging following the starvation death of its sibling. Under these conditions, nestlings beg more under high predation risk, resulting in death of one nestling within the first few periods.

Multiple begging components: Components correlated with role as an alpha or beta nestling could be incorporated to investigate whether provisioning strategy evolves to feed nestlings with a higher competitive ability or a higher need.

Parental strategies: Allowing parents to vary the amount of food they bring to the nest would be an interesting and important extension. This extension requires a second parental gene, however, which complicates interpretation and reduces the parameter space allowing stability. Including variation in parental investment could be followed by adding a second brood to examine between-brood interactions.

ACKNOWLEDGEMENTS

We thank Clive Welham for introducing us to genetic algorithms; Marty Leonard, Arnon Lotem, Geoff Parker, Miguel Rodríguez-Gironés, Jamie Smith, David Tait, Jon Wright, members of the Behavioural Ecology Research Group at Simon Fraser University and an anonymous reviewer for discussion and suggestions and NSERC Canada and Simon Fraser University for financial support.

REFERENCES

Bergstrom, C.T. & Lachmann, M. 1998. Signaling among relatives. III. Talk is cheap. *Proceedings of the National Academy of Sciences USA* **95**, 5100-5105.

Briskie J.V., Naugler, C.T. & Leech, S.M. 1994. Begging intensity of nestling birds varies with sibling relatedness. *Proceedings of the Royal Society of London, Series B* **258**, 73-78.

Cotton, P.A., Kacelnik, A. & Wright, J. 1996. Chick begging as a signal: are nestlings honest? *Behavioral Ecology* **7**, 178-182.

Cotton, P.A., Wright, J. & Kacelnik, A. 1999. Chick begging strategies in relation to brood hierarchies and hatching asynchrony. *American Naturalist* **153**, 412-420.

Dearborn, D.C. 1998. Begging behavior and food acquisition by brown-headed cowbird nestlings. *Behavioral Ecology and Sociobiology* **43**, 259-270.

Fiala, K.L. & Congdon, J.D. 1983. Energetic consequences of sexual size dimorphism in nestling red-winged blackbirds. *Ecology* **64**, 642-647.

Godfray, H.C.J. 1995a. Evolutionary theory of parent-offspring conflict. Nature **376**, 133-138.

Godfray, H.C.J. 1995b. Signalling of need between parents and young: parent-offspring conflict and sibling rivalry. *American Naturalist* **146**, 1-24.

Godfray, H.C.J. & Johnstone, R.A. 2000. Begging and bleating: the evolution of parent-offspring signalling. *Philosophical Transactions of the Royal Society of London, Series B* **355**, 1581-1592.

Goldberg, D.E. 1989. *Genetic Algorithms in Search, Optimization and Machine Learning.* Reading: Addison-Wesley Publishing Company.

Grafen, A. 1990. Biological signals as handicaps. *Journal of Theoretical Biology* **144**, 517-546.

Harper, A.B. 1986. The evolution of begging: sibling competition and parent-offspring conflict. *American Naturalist* **128**, 99-114.

Iacovides, S. & Evans, R.M. 1998. Begging as graded signals of need for food in young ring-billed gulls. *Animal Behaviour* **56**, 79-85.

Johnstone, R.A. 1999. Signaling of need, sibling competition, and the cost of honesty *Proceedings of the National Academy of Sciences USA* **96**, 12644-12649.

Johnstone, R.A. & Grafen, A. 1993. Dishonesty and the handicap principle. *Animal Behaviour* **46**, 759-764.

Kilner, R. 1995. When do canary parents respond to nestling signals of need? *Proceedings of the Royal Society of London, Series B* **260**, 343-348.

Kilner, R. & Johnstone, R.A. 1997. Begging the question: are offspring solicitation behaviours signals of need? *Trends in Ecology and Evolution* **12**, 11-15.

Litovich, E. & Power, H.W. 1992. Parent-offspring conflict and its resolution in the European starling. *Ornithological Monographs* **47**. Washington: American Ornithologists' Union.

Lotem, A. 1998. Differences in begging behaviour between barn swallow, *Hirundo rustica*, nestlings. *Animal Behaviour* **55**, 809-818.

Macnair, M.R. & Parker, G.A. 1979. Models of parent-offspring conflict. III. Intra-brood conflict. *Animal Behaviour* **27**, 1202-1209.

Maynard Smith, J. 1991. Honest signalling: the Philip Sidney game. *Animal Behaviour* **42**, 1034-1035.

Maynard Smith, J. 1994. Must reliable signals always be costly? *Animal Behaviour* **47**, 1115-1120.

McCarty, J.P. 1997. The role of energetic costs in the evolution of begging behavior in nestling passerines. *The Auk* **114**, 135-137.

Mock, D.W. & Forbes, L.S. 1992. Parent-offspring conflict: a case of arrested development. *Trends in Ecology and Evolution* **7**, 409-413.

Parker, G.A. & Macnair, M.R. 1979. Models of parent-offspring conflict. IV. Suppression: evolutionary retaliation by the parent. *Animal Behaviour* **27**, 1210-1235.

Parker, G.A., Mock, D.W. & Lamey, T.C. 1989. How selfish should stronger sibs be? *American Naturalist* **133**, 846-868.

Price, K. 1994. The behavioural ecology of begging by yellow-headed blackbirds. PhD thesis, Simon Fraser University.

Price, K. 1996. Begging as competition for food in yellow-headed blackbirds (*Xanthocephalus xanthocephalus*). *The Auk* **113**, 963-967.

Price, K. & Ydenberg, R. 1995. Begging and provisioning in broods of asynchronously-hatched yellow-headed blackbird nestlings. *Behavioral Ecology and Sociobiology* **37**, 201-208.

Price, K., Harvey, H. & Ydenberg, R. 1996. Begging tactics of nestling yellow-headed blackbirds (*Xanthocephalus xanthocephalus*) in relation to need. *Animal Behaviour* **51**, 421-435.

Redondo, T. 1993. Exploitation of host mechanisms for parental care by avian brood parasites. *Etologia* **3**, 235-297.

Richter, W. 1984. Nestling survival and growth in the yellow-headed blackbird *Xanthocephalus xanthocephalus*. *Ecology* **65**, 597-608.

Rodríguez-Gironés, M.A. 1999. Sibling competition stabilizes signalling resolution models of parent-offspring conflict. *Proceedings of the Royal Society of London, Series B* **266**, 2399-2402.

Rodríguez-Gironés, M.A., Cotton, P.A. & Kacelnik, A. 1996. The evolution of begging: signaling and sibling competition. *Proceedings of the National Academy of Sciences USA* **93**, 14637-14641.

Smith, H.G. & Montgomerie, R. 1991. Nestling American robins compete with siblings by begging. *Behavioral Ecology and Sociobiology* **29**, 307-312.

Stamps, J.A., Metcalf, R.A. & Krishnan, V.V. 1978. A genetic analysis of parent-offspring conflict. *Behavioral Ecology and Sociobiology* **3**, 369-392.

Stamps, J.A., Clark, A., Arrowood, P. & Kus, B. 1985. Parent-offspring conflict in budgerigars. *Behaviour* **94**, 1-40.

Sumida, B.H., Houston, A.I., McNamara, J.M. & Hamilton, W.D. 1990. Genetic algorithms in evolution. *Journal of Theoretical Biology* **147**, 59-84.

Teather, K.L. 1987. Intersexual differences in food consumption by hand-reared great-tailed grackle (*Quiscalus mexicanus*) nestlings. *The Auk* **104**, 635-639.

Teather, K.L. 1992. An experimental study of competition for food between male and female nestlings of the red-winged blackbird. *Behavioral Ecology and Sociobiology* **31**, 81-87.

Teather, K.L. & Weatherhead, P.J. 1988. Sex-specific energy requirements of great-tailed grackle (*Quiscalus mexicanus*) nestlings. *Journal of Animal Ecology* **57**, 659-668.

Zahavi, A. 1977a. The cost of honesty (further remarks on the handicap principle). *Journal of Theoretical Biology* **53**, 205-214.

Zahavi, A. 1977b. Reliability in communication systems and the evolution of altruism. In: *Evolutionary Ecology* (Ed. by B. Stonehouse & C.M. Perrins). Baltimore: University Park Press.

3.

BEGGING AND COOPERATION: AN EXPLORATORY FLIGHT

David Sloan Wilson & Anne B. Clark
Department of Biological Sciences, Binghamton University, Binghamton NY 13902, USA (dwilson@binghamton.edu; aclark@binghamton.edu)

ABSTRACT

The study of begging has been dominated by the assumption that it is primarily competitive, limited only by its inherent costs and thus often detrimental to family productivity. Dysfunctional competition supposedly exists because siblings are only partially related to each other and their parents, making conflicts of interest inevitable. We examine these issues from the perspective of multilevel selection theory, which partitions selection into within- and between-group components. Although it is certainly possible that major features of begging have evolved by within-group selection, the evidence is not strong and frequently is not presented in a way that facilitates comparison of within- versus between-group selection. Partial relatedness does not inevitably lead to within-group selection, which can be suppressed by mechanisms known to exist in other animal groups. We conclude that group-level functionality of bird begging needs to be reconsidered from an explicitly multilevel evolutionary perspective.

INTRODUCTION

Theories of bird begging have long been dominated by a number of assumptions including: (1) that the relationship among nestlings is primarily competitive; (2) that altruism exists in direct proportion to genetic relatedness; and (3) that parents follow simple decision rules such as "feed the offspring that begs the hardest". These assumptions combine to produce a picture of nestlings engaged in an evolutionary arms race, often to the detriment of all.

The purpose of this chapter is to question the standard view, including the stated and unstated assumptions upon which it rests. It is primarily an 'ideas'

J. Wright and M.L. Leonard (eds.), The Evolution of Begging, 43–64.

chapter. If we stimulate the reader to think in new ways, we have achieved our primary goal. We will make a modest effort to evaluate our conjectures empirically, but we realize that more effort is required in the future.

In addition to specific pieces of evidence (e.g. whether begging birds really attract predators to the nest), there is more general evidence suggesting that the standard picture of bird begging may need to be revised. Evolution is such a synthetic theory that information on one kind of organism and behaviour (e.g. chorusing behaviour during reproduction in frogs) can be highly relevant to other organisms and behaviour (e.g. begging behaviour in birds; see A.G. Horn & M.L. Leonard this volume). Yet, this information is often slow to disperse because the various areas of evolutionary biology are only partially connected to each other and all scientists have enough trouble keeping up with the burgeoning literatures in specific areas. Nevertheless, our impression is that the current approach to studying begging may find new directions based on developments that have been taking place in other areas of evolutionary biology. What initially seemed most plausible and parsimonious needs to be reconsidered in general terms in addition to examining the specifics.

THE CURRENT THEORETICAL LANDSCAPE

Two somewhat independent developments in evolutionary biology that are especially relevant to avian begging concern the cognitive sophistication of animals and the relationship between individuals and groups. With respect to cognitive sophistication, descriptions of animal behaviour in the 1960's have a robotic feel to them, as if nonhuman species lack the brainpower to make the nuanced decisions that we take for granted in our own species. The trend over the years has been steadily away from this robotic image toward greater cognitive sophistication. Organisms of many species know their social partners as individuals, remember their past history of interactions and behave adaptively in a highly context-sensitive fashion that can be envisioned as a network of 'if-then' rules. Against this background, a parental feeding rule such as "feed the offspring that begs the hardest" seems too simple for a complete picture, although we certainly find such rules in our data (e.g. Teather 1992; McRae et al. 1993; Kilner 1995; Leonard & Horn 1996). Such rules make no use of knowledge of individual offspring, memory of past feedings or other parental experience. What could we expect of bird begging and parental behaviour if we advanced the cognitive sophistication of both parents and even offspring up several notches to a level already known to exist in lower vertebrates such as fish?

With respect to the relationship between individuals and groups, the widespread rejection of group selection in the 1960's led evolutionary biologists to explain all adaptations as forms of self-interest (Williams 1966). Altruism was explained either as a way of increasing one's own genes in the bodies of others (kin selection), maximizing return benefits to self (reciprocity) or a coincidental by-product of selfish behaviour (by-product mutualism). According to Hamilton's rule, the glass of altruism is empty for non-relatives, full for identical twins and roughly half full for most nestlings that find themselves together in a nest. With only a partial genetic interest in each other, a degree of conflict among nestlings appears inevitable, resulting in a lower combined fitness than otherwise could be achieved. Selfish gene theory seemed to transform even individual organisms into groups of actors that frequently evolve to benefit themselves at the expense of their collective.

These ideas are so firmly established and so often interpreted as an alternative to the rejected theory of group selection that many behavioural ecologists do not even think to question them. Nevertheless, a number of developments over the past 35 years have led to a revival of multilevel selection theory on a more solid basis. In the first place, the process of group selection has become more plausible than previously thought. As theoretical biologist Joel Peck commented in a recent article on multilevel selection (Dicks 2000), "[t]here is no doubt that we were way too hasty in trashing group selection...[t]he theoretical models of the 60s and 70s were very oversimplified and should be taken with a pinch of salt". In the second place, the theoretical frameworks that seemed to replace group selection are not alternatives, but have been shown to include group selection within their own structure. For example, the concept of genes as replicators was initially regarded as an argument against group selection (Williams 1966; Dawkins 1976). Later it became clear that the group selection controversy is and always has been a question about vehicles of selection, not replicators. The common statement "group selection doesn't work because the gene is the fundamental unit of selection" is a *non-sequitur*. Similarly, Hamilton (1964) developed inclusive fitness theory as an alternative to group selection but was convinced by George Price that it included group selection within its own structure (Hamilton 1975, 1996). Thus, however valid the insights of kin selection theory, evolutionary game theory and selfish gene theory, they should not be regarded as alternatives to an erroneous way of thinking called group selection, but rather as different ways of representing selection within and among groups.

The general revival of multilevel selection theory is discussed in detail elsewhere (Wilson 1990, 1997a; Sober & Wilson 1998). Its main relevance for the subject of bird begging is as follows: during the nestling stage of the life cycle, bird populations obviously consist of a population of groups (families) in addition to a population of individuals. In principle, a gene can evolve by natural selection in three ways in such a structured population; (1) by having a high fitness relative to other genes in the same individual (between-gene, within-individual selection); (2) by causing the individual to have a high fitness relative to other individuals within the same group (between-individual, within-group selection); or (3) by causing the group to have a high fitness relative to other groups (between-group, within-population selection). If bird populations are grouped at a still larger scale, we can add a fourth level to the hierarchy (e.g. between-population, within-metapopulation selection). Multilevel selection theory seeks to determine the relative importance of these various levels of selection. The burgeoning literature on intragenomic conflict shows that genes sometimes do evolve at the expense of other genes within the same individual (Hurst et al. 1996; Pomiankowski 1999). In the language of selfish gene theory, individuals are not always the vehicles of selection. With respect to begging, the main question is whether the behavioural, anatomical and physiological traits associated with begging evolve by increasing the fitness of some individuals relative to other individuals in the same family or by increasing the fitness of families relative to other families in the population. One extreme possibility is that begging can be explained entirely as a product of between-individual, within-group selection. In this case, begging would be like intragenomic conflict frame-shifted upward, with individuals as the lower level units and groups as the higher level units. Another extreme possibility is that begging can be explained entirely as a product of between-group, within-population selection. In this case, groups would be the vehicles of selection, in just the same way that individuals are regarded as vehicles of selection. Of course, real begging systems are likely to reside between these two extremes. Our challenge is to discover the relative importance of the two levels of selection. The answer will depend upon a large number of factors; multilevel selection is a complicated process the outcome of which could vary between species and between specific traits associated with begging within a species.

At this point, some readers may be thinking that we have invented a 'new' group selection that bears no relationship to the 'old' group selection that was rejected in the 1960's. The old group selection involved big multigenerational groups like isolated demes, not little ephemeral groups like bird families. Furthermore, nobody seriously proposes that begging can be explained in its entirety as a form of intra-family conflict, but do we need a group selection explanation? This objection has two answers; one historical and the other conceptual. With respect to history, no less a figure than G.C. Williams constructed a model of family selection exactly as we have outlined above, and called it group selection (Williams & Williams 1957). No less a figure than W.D. Hamilton came to the same conclusion in 1975 (see also Hamilton 1996 for an autobiographical account). For these and most other authors who thought seriously about the subject, multilevel selection theory was a method of comparing fitnesses that could be applied to a broad range of groups, from small groups such as ant colonies and fish schools at one end of the spectrum to large groups such as demes, species and ecosystems at the other end. The question in all of these cases was whether genes could evolve by increasing the fitness of the group, despite being selectively neutral or even disadvantageous within the group. Bird families fall comfortably within this spectrum, which means that the fitness comparisons that we have outlined above are part of traditional multilevel selection theory in every sense of the word. Part of reviving multilevel selection theory involves relearning its true history.

With respect to concepts, we agree that even though students of bird begging may have rejected group selection in their own minds, they do not seriously propose that begging can be explained entirely as intra-nest conflict. Surely, at least some aspects of begging coordinate relatively efficient parental care and cause the entire family to do well as a unit (see below). History aside, the conceptually interesting question is this: when we study bird begging from an explicitly multilevel perspective, is it dysfunctional at the family level as it is often taken to be? We believe that the answer to this question is no. Among-group selection may be more important than we have previously thought. In the language of selfish gene theory, bird families may be better vehicles of selection and less driven by intra-family conflict than other perspectives have led us to believe. At the very least the proposition needs to be explicitly considered.

In the following sections we will attempt to develop this proposition by showing how bird begging, including behaviours that require sophisticated cognition, can be approached from a multilevel perspective.

TRANSLATING AMONG PERSPECTIVES

Multilevel selection theory requires fitness comparisons that are perfectly straightforward but which differ from the concept of self-interest as defined by other theoretical frameworks. A common rule of thumb for thinking about the evolution of two alternative traits (x and y) is to compare the fitness of an individual if it performs x with the fitness of the same individual if it performs y. If x delivers the highest fitness, then it is predicted to evolve by individual selection. For example, if x is 'beg for food' and y is 'remain silent', if an individual gets more food by begging than by remaining silent, it seems obvious that begging should evolve and that it evolves by individual selection.

Unfortunately, this rule of thumb, straightforward as it seems, is not sufficient to predict whether the behaviour evolves or what level of selection is responsible for its evolution. If Williams (1966) taught us anything, it is that natural selection is based on relative fitness. Knowing the effect of a behaviour on the absolute fitness of the actor does not allow us to calculate relative fitness. To evaluate relative fitness within single groups, we must compare an individual performing x with another individual in the same group performing

y. One possibility is that begging causes the parent to feed the individual who begs, in which case x is more fit than y within the same group. Another possibility is that begging causes the parent to feed all its offspring, including those who remained silent. In this case x is less fit than y within the same group, because begging provides a benefit for everyone and costs at least something in time and energy for the beggar. If begging evolves, but actually decreases the relative fitness of the beggar within groups, its evolutionary advantage must reside elsewhere. If groups with more beggars receive more food than groups with fewer beggars, we can postulate that begging evolves on the strength of relative fitness differences among groups, despite being selectively disadvantageous within groups. These crucial distinctions are lost when we know only the absolute fitness of the individual beggar.

The rule of thumb based upon absolute fitness does correctly predict what evolves under a narrow range of parameter values, such as in randomly formed groups and behaviours that are performed without regard to whether others are performing the same behaviour. It fails, however, for other parameter values that are both biologically reasonable and highly relevant to the subject of begging. As one example, Wilson (1998) considered a model of hunting versus not hunting in human groups where all food must be shared. In addition to fixed strategies such as 'hunt' and 'don't hunt', the model included more flexible strategies such as 'hunt if there are fewer than four hunters in the group; otherwise don't hunt'. The flexible strategies radically changed the dynamics of the model in ways that invalidated a rule of thumb based upon absolute fitness. A more comprehensive model that kept track of relative fitness within and between groups was required to predict the outcome and component forces. A begging model that included flexible strategies such as 'beg if no one else is begging; otherwise remain silent' would probably produce similar results.

Birds in nests are not randomly formed groups, because different broods are derived from the genes of different parents. This non-random grouping is usually modelled with inclusive fitness theory, where it appears as a modification of individual fitness. In other words, the fitness of an individual includes its effect on itself plus its effect on others, weighted by the coefficient of relatedness. Multilevel selection theory models non-random groupings more directly. Relatedness does not appear at all in the fitness comparisons within groups. If begging causes the parent to feed everyone in the nest, then beggars are less fit than non-beggars within the same group regardless of their genealogical relatedness. However, when groups are composed of relatives, there will be more variation among groups than when the groups are formed at random. The coefficient of relatedness appears as an index of variation among groups rather than as a modifier of individual fitness. As with the rule of thumb based on absolute fitness, inclusive fitness theory works only within a range of parameter values defined by its implicit assumptions, which are violated under biologically reasonable conditions, causing Hamilton's rule to break down (Wilson et al. 1992; Mitteldorf & Wilson 2000). Multilevel selection theory can explain the exceptions as well as the rule because it makes fewer assumptions about the population structure. If we know the evolutionary forces operating within single groups, plus the manner in which the groups contribute differentially to the formation of new groups, we have a relatively complete accounting procedure for tracking evolutionary change in the global population for a wide range of population structures.

The point of this discussion is not to promote multilevel selection theory over other perspectives, but to stress that care is needed when translating from one perspective to another. The rule of thumb based on absolute fitness and the inclusive fitness of an individual are often regarded as forms of individual selection, but they are not the same as between-individual, within-group selection in multilevel selection theory. With these cautions in mind, we will now attempt to translate the literature on bird begging into the language of multilevel selection theory.

WITHIN-FAMILY CONFLICT OR FAMILIES AS VEHICLES OF SELECTION?

If bird families are vehicles of selection, the total output of the nest should be close to maximum. If bird families are torn by conflict, then the total output of the nest should be far from its maximum for reasons that clearly enhance the relative fitness of some genes within the nest at the expense of others. Unfortunately, most begging studies are not explicitly designed to make these within-nest versus between-nest comparisons. Numerous features of begging, however, give the appearance of within-family conflict. First, there is the general competitive appearance of begging behaviour. Second, begging appears to be dysfunctional for the whole group by attracting predators. Third, begging appears to be dysfunctional for the whole group by requiring energy that could be spent on growth and development. Fourth, the partial genetic relatedness of nestlings makes conflict appear theoretically inevitable. Let us examine each of these possibilities in turn.

Is Begging Dysfunctional for the Group because it Is Competitive?

Genes and cells cooperate intricately with each other to make individual organisms the paradigmatic vehicle of selection. Nevertheless, if we were to become cell sized and enter the body of an individual organism, many of the interactions would look more like competition and predation than cooperation as we usually think of it. These interactions may look negative, but in a more fundamental sense they are cooperative by contributing to the adaptive design of the higher level unit. Frame-shifting upward to the level of birds in nests, we need to appreciate that seemingly negative interactions can be adaptive at the group level and are not necessarily products of within-group selection.

In some respects, researchers in avianbegging have always appreciated the group-level advantages of competition. Lack's brood reduction hypothesis envisions competition as a mechanism for adaptively regulating brood size, which maximizes the output of the brood (Lack 1947). Mock and Parker (1997) regard siblicide in ardeids as a form of brood reduction to which the parents acquiesce and from which both they and older nestlings profit on balance. O'Connor's (1978) model of nestling suicide can even be regarded as a form of group-level apoptosis. Nevertheless, it is easy to think of cooperation as what we learn in Sunday school and to casually interpret all seemingly negative interactions as dysfunctional at the nest level. Examples do exist of overt cooperation among nestlings, especially in species with highly asynchronous broods in which older siblings are developmentally capable and preen or even feed their younger siblings (Marti 1989; Stamps et al. 1989). Additionally, there are many cases in which older siblings are far less domineering than they could be (facultatively siblicidal species, Drummond 2001). That begging is largely 'honest' (Kilner & Johnstone 1997) implies that older nestlings are not competing as strongly as they could. The ultimate test of within- versus between-family selection, however, is not the appearance of niceness in anthropomorphic terms but the maximum versus sub-maximum output of the nest, achieved by any proximate mechanism. Regardless of its competitive appearance, we need to know if begging behaviour is a piece of social machinery that maximizes the combined output of the brood or if it is a form of intra-nest conflict that decreases combined output, just as intragenomic conflict decreases the fitness of individual organisms (Hurst et al. 1996). The mere appearance of competition does not provide evidence for within-family conflict.

Is Begging Dysfunctional for the Group because it Attracts Predators?

Before we attempt to answer this question for bird families, let us ask it in another context: is mate calling dysfunctional for a male frog because it attracts predators? The answer depends on the level of calling. Presumably there is an optimal level at which the advantage of attracting mates is counterbalanced by the disadvantage of attracting predators. Above this level, calling is dysfunctional because it attracts predators. Below this level, calling is functional despite the fact that it attracts predators. Returning to bird families, the mere fact that begging sounds attract predators is not sufficient to conclude that the sounds are dysfunctional at the group level. We need to know if the begging sounds reflect the optimal trade-off between attracting predators and communicating with the parents (see A.G. Horn & M.L. Leonard this volume), or if the sounds are suboptimal for the group because the risk is being increased by some offspring begging excessively to get more than their fair share. To our knowledge, no study of begging has distinguished between these two possibilities.

Experimental studies so far are few, but may offer some intriguing ways to explore this question. Clearly, loud begging played back at artificially placed nests can increase predation (Haskell 1994, D.G. Haskell this volume; Leech & Leonard 1997; Dearborn 1999; but see Halupka 1998). This cost may, however, be site specific (Haskell 1994) and it remains unclear whether predation limits calling in some cases (e.g. Halupka 1998). However, where predation risk is demonstrably higher, begging calls are less detectable (Briskie et al. 1999; Haskell 1999). Do the changes in acoustic characteristics also affect their localizability or effectiveness in getting parental attention against a background of 'normal' begging calls? If true, this possibility, testable with reciprocal transfers between species, might indicate an important role of group benefit for the evolutionary modification of begging. As part of the family, parents would be selected to modify aspects of their behaviour and responsiveness to collective signals of brood need.

Is Begging Dysfunctional for the Group because it Costs Energy?

Our answer to this question parallels our answer to the previous question. The mere fact that begging requires energy does not make it dysfunctional for the group. The question is whether the energy expenditure reflects an optimal trade-off for the group or a suboptimal trade-off that enables some individuals to get more than their fair share. Trees exist because of an evolutionary arms race to capture light. The cost of producing a tree trunk is tangible evidence for within-group, between-tree selection. It is tempting to think of bird begging as, like tree trunks, the result of an arms race by nestlings to get more than their fair share of energy, resulting in a waste of energy for all. The idea of begging as a within-group arms race, like the idea of begging attracting predators, seems so plausible that it must be true. However, when the energetic cost of begging is actually measured, it appears to add a relatively small total energetic drain at normal levels (Leech & Leonard 1996; McCarty 1996; Bachman & Chappell 1998; Soler et al. 1999; M.A. Chappell & G.C. Bachman this volume).

Experimentally induced high levels of begging over a prolonged period, however, cause a measurable decrease in growth in magpies (*Pica pica*, Rodríguez-Gironés et al. 2001) and canaries (*Serinus canaria*, Kilner 2001) but not in ring doves (*Streptopelia risoria*, Rodríguez-Gironés et al. 2001), demonstrating a possibly limiting individual fitness cost of escalation for some species. However, energetic costs, unlike predation risk, can be borne by individuals alone and will only spread to whole families if siblings escalate in kind. The influence of nestmates on begging varies markedly across species and studies (see discussion in Lichtenstein 2001; D.C. Dearborn & G. Lichtenstein this volume). Brood parasites typically show exaggerated begging (e.g. Dearborn 1998; Kilner et al. 1999), but little direct response to nestmates' begging (Lichtenstein 2001). Flexible matching could be viewed as a potentially group-beneficial tit-for-tat strategy that evolves in part due to its ability to reduce group costs commensurate with parental ability and willingness to feed (see Hussell 1988), while also providing protection against 'cheating' nestmates who would exceed their more conservative siblings. Parasites, especially those without nestmates, may benefit from a fixed strategy that matches the collective output that most host broods would produce. In any case, the variation in nestmate responsiveness offers an arena for exploring individual and group energetic costs as selective forces.

In general, models of begging evolution assume that begging must be expensive to qualify as an honest signal (Harper 1986; Grafen 1990; Godfray 1991, 1995; also see R.A. Johnstone & H.C.J. Godfray this volume). The fact that the signals are honest means that the self-serving advantages of begging described in the previous paragraph have been reduced or even eliminated. However, the fact that the signals are expensive means that energy is still being used for calling that could have been used for growth and development. By focusing explicitly on group output as a product of group-level selection, multilevel selection theory encourages us to look for systems of begging that eliminate within-group selfishness without being energetically expensive, which is something we will return to below.

Is Within-Group Conflict Theoretically Inevitable?

As we mentioned earlier, Hamilton's rule portrays the glass of altruism as only half full for birds in nests, which suggests that a degree of within-family conflict is inevitable. There is, however, more to multilevel selection theory than genetic relatedness. A variety of social control mechanisms can suppress within-group selection and promote among-group selection in groups of relatives and non-relatives alike. Even social insect colonies, which Hamilton (1964) tried to explain on the basis of their particularly high relatedness among sisters, are increasingly being explained in terms of social control mechanisms, which suppress cheating behaviours that would be expected to occur on the basis of genetic relatedness alone (e.g. Ratnieks & Visscher 1989; Seeley 1995). An example from human families is the social convention 'cut the cake' (Skyrms 1996), in which the sibling who cuts the cake is the last to choose a slice. Wonderful in its simplicity, this social convention accomplishes total fairness among individuals who would otherwise be tempted to gain at the expense of each other. Studies of bird begging should be looking for social control mechanisms similar to 'cut the cake' rather than assuming that conflict is inevitable among partially related individuals.

Are there Conflicts among Successive Broods?

A group is defined in multilevel selection theory (and indeed all theories of social behaviour) as the set of individuals who influence each other's fitness with respect to a given trait. Bird nests appear to be exceptionally tidy groups, in the sense that they are spatially discrete and begging behaviours influence the fitness of nestmates without influencing the fitness of nestlings in other nests. A variety of factors can, however, make this tidy population structure more complicated, including successive broods. When the begging behaviour of a nestling affects the fitness of future broods (through its effects on the parent), there is a sense in which the future broods are part of the same group, creating new possibilities for within-family conflict. Multilevel selection theory must attend to this possibility, just like any other theoretical perspective. One interesting possibility is that the nestlings in a given brood cooperate with each other to collectively extract more than the parent has been selected to provide. Altruism can exist within single broods leading to conflict within the multiple-brood family.

To summarize, our purpose in this section is not to claim that bird families are perfect vehicles of selection, unblemished by intra-family conflict. Instead, we make the more modest claim that the balance between levels of selection is completely unknown for bird families. Research conducted from other theoretical perspectives has contributed to the appearance of intra-family conflict, but the evidence falls apart when examined from an explicitly multilevel perspective. Bird families could be highly designed to maximize their output and to suppress within-family selection, and this conclusion would be fully consistent with the known facts of bird begging.

OTHER REASONS FOR NESTLINGS TO COOPERATE

So far we have concentrated on the standard conception of bird begging, in which there is little scope for cooperation other than to adaptively regulate brood size downward. Nestlings may, however, have other reasons to cooperate. Siblings are greatly needed for heat conservation and thermoregulation, especially at a young age (Dunn 1976, 1979; Hill & Beaver 1982; Sullivan & Weathers 1991). Not only do nestlings with siblings spend less energy in thermoregulation, but earlier and longer brood homeothermy allows parents more time for foraging (Visser 1998). In some bird species, siblings continue to associate with each other after they have fledged, acting as helpers and allies during their adult lives (see Brown 1987). The presence of siblings post-fledging can also contribute to more effective learning of social behaviour and/or foraging skills (Edwards 1989; Garnetzke-Stollmann & Franck 1991; Wanker 1996; Berger 1998). The benefits of such cooperation may not be gained during the nestling stage, and will be forestalled by the negative effects of competition before the benefits can accrue.

Parental effort involves a fixed cost that is independent of brood size (e.g. building the nest, incubating) and a variable cost that increases with brood size (e.g. amount of food that needs to be provided). It can be adaptive for a parent to neglect or abandon small broods to conserve time, energy and risk for another brood during the same season or to increase personal survival and condition for the next season (Mock & Parker 1986). Even without this, parents may decrease their feeding in proportion to brood size and it is not clear from the many studies of natural and experimental brood reduction that remaining young generally survive or grow better than those in unreduced broods (see Drummond 2001).

Partial nest predation could lead to a form of cooperation. Partial predation may be an under-reported risk sometimes confused with brood reduction (Clark & Wilson 1981), but it is observed regularly by specific predators on some species (black-crowned night herons, *Nycticorax nycticorax*, taking cattle egrets, *Bubulcus ibis*, Mock & Parker 1997; red-winged blackbirds, *Agelaius phoeniceus*, and American robins, *Turdus migratorius,* personal observation; American crows, *Corvus brachyrhynchos*, K.J. McGowan personal communication). For a nestling in a brood of four, its chances of being eaten by a predator who will remove a single nestling is 0.25. If it causes one of its siblings to starve to death, its chance of being taken by the predator has increased to 0.33. If the predator is discriminating and takes the largest nestling in the nest (observed repeatedly by one northern harrier, *Circus cyaneus*, 'harvesting' red-winged blackbird nestlings, personal observation), it should encourage its siblings to eat more than itself!

Even a basically competitive situation can be made cooperative through the behaviour of the parent. The standard picture assumes that the parent is programmed in a robotic fashion to feed the nestling that excels in some feature of begging. Let us imagine a mutant parent that is not even this sophisticated. She attends only to the combined begging signals of all her offspring and feeds them in random order. Thus, a nestling can cause its parent to bring more food to the nest by begging harder, but cannot direct the food to itself. This alteration in the parent's behaviour changes begging from an act of selfishness to an act of altruism, since the beggar expends its own energy to benefit the entire brood. If the parental rule 'feed the nestling that begs the hardest' really leads to an arms race that is dysfunctional for the whole brood (as suggested by Stamps et al. 1985), then we should expect it to be replaced by the more group-adaptive rule 'attend to combined signal and feed in random order', especially since the latter rule does not require more, and may even require less, cognitive sophistication and time to actually deliver the food while at the nest.

Davis et al. (1999) have explicitly tested various simple and compound feeding rules against random allocation with computer simulations and have shown that this ultra-simple rule was usually second best or equal to the best simple rule (feed hungriest, smallest, largest, etc.) across a range of food availability. Its relative failure in maximizing brood output was probably due to occasional long (random) lapses in feeding one model chick and its subsequent 'death'. If indeed parents used begging only to guard against such extreme lapses and otherwise fed randomly, young would be left with little to do but cooperate in requesting that more food be brought to the nest.

To be fair, no study to date shows random parental allocation of food among offspring, without regard to begging signals. Studies do, however, report differences in feeding strategies and biases of two parents (e.g. Stamps et al. 1987; Westneat et al. 1995; Kölliker et al. 1998; Krebs 2001, E.A. Krebs this volume) which may serve to decrease the predictability of being fed given a particular position or begging level. The actual level of perceived randomness a nestling would experience is seldom, if ever, measured. However, the fact that parents can be shown to attend to nestling cues does not necessarily mean that the cues can be selfishly exploited by the nestlings.

Finally, overt parental control is a theoretical possibility realized at least in some parrot species (budgerigars, *Melopsittacus undulatus*, Stamps et al. 1985; crimson rosellas, *Platycercus elegans*, Krebs 2001). E.A. Krebs (this volume) suggests a number of reasons for control in these species, including the extreme age differences among young and the existence of strong non-behavioural cues such as crop fullness. One possible result is that, although older psittacine nestlings clearly have the bill structure and strength to harm younger siblings (demonstrated toward human handlers), injurious nestling aggression is unreported in this group.

ADDING BEHAVIOURAL SOPHISTICATION

The rule "feed the nestling that begs the hardest" turns the nestlings into free agents that can operate without social control, thereby maximizing the potential for intrabrood conflict. We have just suggested that even simpler parental rules such as 'feed in random order' could impose a crude form of fairness and turn selfish beggars into altruists. We also stress, however, that animal cognition can be more complex and context sensitive than implied by either of these rules. What happens to the standard picture of bird begging and the balance between levels of selection when we increase the sophistication of both parents and their offspring?

Watching videotapes of nests quickly leads to the impression that parents are assessing the needs of their offspring and delivering food in ways that are difficult to exploit. Red-winged blackbird females frequently place food into a nestling's mouth and then take it out again to give it to someone else (pers. obs.). If the parents can directly assess the amount of food in a nestling's crop, then it becomes a back-up signal that cannot be faked. Parents also divide food among two or even three nestlings during a single visit more often, and with greater effect on intake than suggested by many reports (e.g. McRae et al. 1993). An analysis of red-winged blackbirds in progress shows that female nestlings with brothers in broods of two get as much or more food than their larger brothers during early growth, but they get it during the second feeding on a visit. An analysis of food received during first feeds would misleadingly show that males get much more of the food (J. Peet & A.B. Clark unpublished data).

Birds that breed in close proximity to each other are well known to discriminate among offspring and to direct parental care to their own young. Because cup-nesting birds are not faced with this problem (apart from brood parasitism), they are widely assumed to be unable to discriminate among offspring at least until fledging. Experiments that involve switching offspring between nests are predicated upon this assumption. There are, however, many reasons to discriminate among one's offspring other than to avoid feeding foreign young. Sophisticated social behaviours of all sorts require knowing the identity of one's social partner. Even so-called lower vertebrates such as fish have been shown to discriminate among and make decisions on the basis of past interactions with as many as six conspecifics simultaneously (Dugatkin & Wilson 1992). If we grant parent birds a similar ability, it seems likely that they could use it to their advantage. Red-winged blackbirds often show no obvious discrimination of introduced foreign nestlings (Edwards et al. 1999), but finer analyses and occasional dramatic rejections indicate that, for a short while after introduction, some combination of nestling appearance and behaviour can trigger discrimination and even rejection (A.B. Clark & K. Yasukawa unpublished data). Mild lasting effects were suggested by Teather (1992).

Another role for recognition suggested by data for redwings recently analysed in our laboratory (J. Peet & A.B. Clark unpublished data), is that parents are influenced by the intensity of begging after the nestlings have been fed and before the parent leaves for its next feeding trip and independently of relative begging on the next trip (see also Glassey 2000 for a similar result). It is more likely that only unfed nestlings will be begging as she leaves and it may be physically difficult for a recently fed nestling to beg when its crop has just been filled. However, this information is usable only if parents can discriminate among their offspring and remember which were begging when they left the nest so that these offspring can be fed upon their return. (Note that nest sites such as boxes may limit parents' ability to see and recognize their young, so we may expect species differences.)

Much has been made recently of honest communication and the need for honest signals to be expensive (e.g. Harper 1986; Grafen 1990; Godfray 1991, 1995; but see R.A. Johnstone this volume). As we previously mentioned, however, expensive signals must invariably detract from the productivity of the entire brood, which is against the interest of both parents and offspring as long as appropriate safeguards against cheating can be implemented. Some signals can be honest without being expensive, as when the parents can directly assess the crop content of the young, the colour of the gape (Kilner & Davies 1998) or remember the feeding history of each young separately. In addition, when a pair of individuals interacts many times and dishonest communication can be detected, honest and selfish communication can be modelled as an iterated prisoner's dilemma game. Parents can simply retaliate by punishing their young for begging inappropriately, presumably by withholding food, to encourage more cooperative begging behaviour in the future. Another way to make the same point is to imagine the parent shaping the behaviour of each offspring the way a scientist shapes the behaviour of an animal subject in an operant conditioning experiment. If inappropriate begging isn't rewarded, presumably it will become extinguished. Only one study so far has explicitly approached begging as a learned behaviour that can be modified by parental reinforcement schedules (Kedar et al. 2000; see also Hussell 1988; Stamps et al. 1989 for similar suggested effects).

Game theory is at least as relevant to begging behaviour as kin selection theory, since the conditions for repeated interactions required for strategies such as 'tit-for-tat' to evolve are obviously satisfied. Game theory might be useful for explaining nestling-nestling interactions in addition to parent-offspring interactions. In a paper entitled "Tit-for-tat vs. nepotism, or, why should you be nice to your rotten brother?" Wilson and Dugatkin (1991) showed that direct experience can provide better information about the behavioural propensity of one's social partner than degree of genetic relatedness. If you are an altruist and your brother is a jerk, chances are that he did not inherit the same altruistic gene that you received from one of your parents, despite the fact that his chances were 50%. Thus, there is every reason to retaliate against this sibling and associate with nicer siblings, despite their similar overall coefficient of relatedness. To our knowledge, no social interactions resembling tit-for-tat have been documented among nestling birds, but this may be because no study has ever been designed to examine begging in this way. The fact that nestlings of some species, faced with an introduced, hungry and more strongly begging nestmate, increase their own begging (Smith & Montgomerie 1991; Price & Ydenberg 1995; Leonard & Horn 1998) has been interpreted as a competitive response to increased competition. Another interpretation might be that a nestling in a brood that has not generally been so deprived could look on a new, strongly begging nestmate as a cheater and respond in tit-for-tat fashion to the sudden change. As mentioned above, the social inflexibility and high begging levels of brood-parasitic species (Lichtenstein 2001) should be of particular interest in this context.

BEGGING AS A GROUP-LEVEL COMMUNICATION SYSTEM

Earlier we stated that studies of bird begging can profit from knowledge of other subjects that superficially appear very different. One of these subjects is social insect biology. After all, a social insect colony is a family that succeeds spectacularly well at the group level. Although Hamilton's (1964) focus on the particularly high relatedness of sisters in haplodiploid organisms seemed to place social insects in a class by themselves, subsequent research has shown that single colonies are often more genetically diverse than previously thought (because of multiple queens or single queens who mate with multiple males) and that social control mechanisms are at least as important as genetic relatedness for forging the colony into a vehicle of selection.

There is a strong tendency among behavioural ecologists to think of individual organisms as self-contained strategizing units, like reincarnations of Machiavelli. Altruistic or selfish, the individual is still the decision-making unit whose behaviour can be deduced by slipping into its shoes (or hooves, or claws) and asking what you would do in the same situation and with the appropriate metric of self-interest. This way of thinking may sometimes serve as a useful heuristic for predicting the behaviours that evolve by natural selection, but it becomes counterproductive when we try to think about the cognitive mechanisms actually employed by animals, especially as members of functionally integrated groups. One of the most remarkable features of social insect colonies is the distributed nature of the cognitive mechanisms that enable the colony to function as a unit. Highly intelligent decisions are made, for example about which resource patches to exploit over a wide area. However, when the decision-making process is studied in detail, each individual insect plays such a small role that it resembles a neuron more than a cognitive system in its own right. Isolating a social insect from its group does not simply cause it to make different decisions about the best way to behave; it gives the individual a lobotomy by removing much of the information processing circuitry that always resided outside its own head. Even humans, with their vaulted individual brains, might function as part of a group-level cognitive architecture more than most of us currently realize (Wilson 1997b). What would begging look like if parents and offspring alike are components of a single cognitive system rather than separate reincarnations of Machiavelli?

The recent discovery that parasitic cuckoo offspring mimic the begging behaviour of an entire brood rather than a single nestling and thus influence parental feeding (Davies et al. 1998; Kilner et al. 1999) adds credence to the concept of begging as a group-level communication system. Of course, one obvious difference between a bird family and a social insect colony is that the adults do all the foraging. The offspring can communicate only the state of their own hunger. Nevertheless, there are numerous possibilities for offspring to merge their begging into a collective signal for communicating with the parent.

Studies on vocal behaviour in other species and contexts provide valuable information regarding how to distinguish cooperative begging from competitive begging in birds. Males calling for mates are obviously competing with each other, but in some species, including a number of arthropods and anurans, males in a local area merge their calls to collectively compete with males from other local areas (Greenfield 1994). In a very useful review of the mechanics of these signal interactions *vis-à-vis* their effects on receivers, Greenfield provides a number of ways in which calling individuals can cooperate as well as compete. Sound is a complex physical process and the combined effect of several individuals calling together is not a simple sum of their individual calls (see A.G. Horn & M.L. Leonard this volume). Hearing is also a complex psychological process that adds another set of constraints on the effectiveness of a vocal signal. Interactions between nearly simultaneous signals can be competitive or cooperative, depending on very small differences in timing and other adjustments. If nestlings have evolved to speak 'with a single voice' to their parents, this will almost certainly require a sophisticated coordination of their individual vocal behaviours. On the other hand, if they have evolved to maximally 'stand out from the crowd' to get more food for themselves, a very different kind of coordination will be required.

Exactly what group attributes will be most successful depends in part on what receivers are looking for and in part on the environmental conditions under which signals are transmitted. If parents are influenced by cues that reliably indicate brood size, as suggested by the cuckoo mimicry evidence, then group begging might occur so as to make a brood appear even larger than it is (Harrington (1989) suggests this phenomenon may occur in the howling of wolf packs). To do so would be to avoid complete synchrony, but also highly individualistic signatures. If nesting is somewhat colonial, brood signatures might also be important. One very intriguing suggestion by Greenfield (1994) allows for a specific pattern incorporating elements of both cooperation and competition based on biases in receiver psychology. The 'precedence effect', or selective attraction of receiver attention to the first one of a group of closely synchronized signals, has apparently affected calling patterns in a tettigoniid (Greenfield & Roizen 1993). The precedence effect is widespread across taxa (see Greenfield 1994). If nestlings were selected to beg together in such a way as to sound like 'large brood-all hungry', it would not preclude competition for being the first to initiate a call. Hence nestlings might be sensitive to incipient begging motions of a fellow nestling or cues of an actual parent and start to vocalize as quickly as possible, but once begging as a group, fall into a pattern selected by its group effect.

Another factor involves the environmental context of vocal signals. The standard picture portrays begging as an interaction with the parent while the parent is at the nest. Begging also, however, occurs when the parent is not at the nest and may function in this context as a form of long-distance communication especially when young are very hungry. By analogy, imagine that you are part of a small group of hikers that has become separated from the main group and decides to call for help. You will certainly merge your calls to make them carry the longest distance, although you may not know the best way to do this, other than to call at the same time. If bird broods have evolved over thousands of generations to 'call for help' as a group and over distance, they may have achieved levels of sophistication that puts a group of hikers to shame. Note that we again should expect interspecific variation in the use of such signals, depending on how far parents forage. The contrasting results of Clark and Lee (1998), Price (1998) and Burford et al. (1998) may be due to such species or population differences (Clark & Lee 1998).

With sufficient knowledge about the physics of sound production and the psychophysics of hearing, it should be possible to distinguish between these two very different hypotheses and empirically settle the issue of whether the vocal component of begging functions as a group signal or as a form of intra-nest conflict. One important first report on signal dynamics in tree swallows (*Tachycineta bicolor*) suggests that both individual and group considerations may be moulding signals (Leonard & Horn 2001; see also A.G. Horn & M.L. Leonard this volume).

Optimal feeding behaviour is highly influenced by predation risk and other factors. At some times parents may be able to make frequent feeding trips while at other times they may need to stay close to the nest to ward off potential predators. Predation risk varies at all temporal scales: minute-to-minute, day-to-day and season-to-season. We know that parental feeding behaviour is highly responsive to predation risk, but how about offspring begging behaviour? For example, when predation risk is low and nestlings are being fed every five minutes, it may make sense to become hungry and to beg after seven minutes. When risk is high and nestlings are being fed more sporadically, it may make sense to be patient. These differences in begging behaviour may exist not only between species, but facultatively within species, as long as parents can communicate predation risk to their offspring. In addition to vocal communication, the feeding schedule itself might become a form of communication. For example, if there is a series of long periods without feeding, this may act as a signal to become patient rather than to beg harder.

FUTURE DIRECTIONS

This entire chapter has been about a future direction - studying bird begging explicitly from a multilevel evolutionary perspective. The entire field of inquiry has been guided by the assumption that group selection can be ignored and that all features of begging can be explained as a form of individual selection. Yet, when we try to correctly partition selection into within- and between-group components, we discover that we know remarkably little about whether begging counts as a group-level analogue of intragenomic conflict or as a sophisticated set of adaptations that maximizes the output of the whole group. Multilevel selection theory, combined with an awareness of the relevance of other subjects that superficially appear very different from begging, will enable these fundamental questions to be answered more decisively in the future.

REFERENCES

Bachman, G.C. & Chappell, M.A. 1998. The energetic cost of begging behaviour in nestling house wrens. *Animal Behaviour* **55**, 1607-1618.

Berger, M.L. 1998. Development and function of social relationships in young budgerigars (*Melopsittacus undulatus*). M.Sc. thesis, Binghamton University.

Briskie, J.V., Martin, P.R. & Martin, T.E. 1999. Nest predation and the evolution of nestling begging calls. *Proceedings of the Royal Society of London, Series B* **266**, 2153-2159.

Brown, J.L. 1987. *Helping and Communal Breeding in Birds: Ecology and Evolution*. Princeton: Princeton University Press.

Burford, J., Friedrich, T.J. & Yasukawa, K. 1998. Responses to playback of nestling begging in the red-winged blackbird, *Agelaius phoeniceus*. *Animal Behaviour* **56**, 555-561.

Clark, A.B. & Lee, W.-H. 1998. Red-winged blackbird females fail to increase feeding in response to begging call playbacks. *Animal Behaviour* **56**, 563-570.

Clark, A.B. & Wilson, D.S. 1981. Avian breeding adaptations: hatching asynchrony, brood reduction and nest failure. *Quarterly Review of Biology* **56**, 253-277.

Davies, N.B., Kilner, R.M. & Noble, D.G. 1998. Nestling cuckoos, *Cuculus canorus*, exploit hosts with begging calls that mimic a brood. *Proceedings of the Royal Society of London, Series B* **265**, 673-678.

Davis, J.N., Todd, P.M. & Bullock, S. 1999. Environmental quality predicts parental provisioning decisions. *Proceedings of the Royal Society of London, Series B* **266**, 1791-1797.

Dawkins, R. 1976. *The Selfish Gene*. London: Oxford University Press.

Dearborn, D.C. 1998. Begging behavior and food acquisition by brown-headed cowbird nestlings. *Behavioral Ecology and Sociobiology* **43**, 259-270.

Dearborn, D.C. 1999. Brown-headed Cowbird nestling vocalizations and risk of nest predation. *The Auk* **116**, 448-457.

Dicks, L. 2000. All for one! *New Scientist* **8 July 2000**, 30-35.

Drummond, H. 2001. The control and function of agonism in avian broodmates. *Advances in the Study of Behavior* **30**, 261-301.

Dugatkin, L.A. & Wilson, D.S. 1992. The prerequisites for strategic behaviour in bluegill sunfish (*Lepomis macrochirus*). *Animal Behaviour* **44**, 223-230.

Dunn, E.H. 1976. The relationship between brood size and age of effective homeothermy in nestling house wrens. *Wilson Bulletin* **88**, 478-482.

Dunn, E.H. 1979. Age of effective endothermy in nestling tree swallows according to brood size. *Wilson Bulletin* **91**, 455-457.

Edwards, S., Messenger, E. & Yasukawa, K. 1999. Do red-winged blackbird parents and their nestlings recognize each other? *Journal of Field Ornithology* **70**, 297-309.

Edwards, T.C. Jr. 1989. Similarity in the development of foraging mechanics among sibling ospreys. *Condor* **91**, 30-36.

Garnetzke-Stollmann, K. & Franck, D. 1991. Socialisation tactics of the spectacled parrotlet (*Forpus conspicillatus*). *Behaviour* **119**, 1-29.

Glassey, B. 2000. Resource competition among nestling red-winged blackbirds. PhD thesis., University of Manitoba.

Godfray, H.C.J. 1991. Signalling of need by offspring to their parents. *Nature* **352**, 328-330.

Godfray, H.C.J. 1995. Evolutionary theory of parent-offspring conflict. *Nature* **376**, 133-138.

Grafen, A. 1990. Biological signals as handicaps. *Journal of Theoretical Biology* **144**, 517-546.

Greenfield, M.D. 1994. Cooperation and conflict in the evolution of signal interactions. *Annual Review of Ecology and Systematics* **25**, 97-126.

Greenfield, M.D. & Roizen, I. 1993. Katydid synchronous chorusing is an evolutionarily stable outcome of female choice. *Nature* **364**, 618-620.

Halupka, K. 1998. Vocal begging by nestlings and vulnerability to nest predation in meadow pipits *Anthus pratensis*; to what extent do predation costs of begging exist? *Ibis* **140**, 144-149.

Hamilton, W.D. 1964. The genetical evolution of social behaviour: I and II. *Journal of Theoretical Biology* **7**, 1-52.

Hamilton, W.D. 1975. Innate social aptitudes in man, an approach from evolutionary genetics. In: *Biosocial Anthropology* (Ed. by R. Fox). London: Malaby Press.

Hamilton, W.D. 1996. *The Narrow Roads of Gene Land*. Oxford: W.H. Freeman/Spektrum.

Harper, A.B. 1986. The evolution of begging: sibling competition and parent-offspring conflict. *American Naturalist* **128**, 99-114.

Harrington, F.H. 1989. Chorus howling in wolves: acoustic structure, pack size and the beau geste effect. *Bioacoustics* **2**, 117-136.

Haskell, D. 1994. Experimental evidence that nestling begging behaviour incurs a cost due to nest predation. *Proceedings of the Royal Society of London, Series B* **257**, 161-164.

Haskell, D. 1999. The effect of predation on begging call evolution in nestling wood warblers. *Animal Behaviour* **57**, 893-901.

Hill, R.W. & Beaver, D.L. 1982. Inertial thermostability and thermoregulation in broods of redwing blackbirds. *Physiological Zoology* **55**, 250-266.
Hurst, L.D., Atlan, A. & Bengtsson, B.O. 1996. Genetic conflicts. *The Quarterly Review of Biology* **71**, 317-364.
Hussell, D.J.T. 1988. Supply and demand in tree swallow broods: A model of parent-offspring food-provisioning interactions in birds. *American Naturalist* **131**, 175-202.
Kedar, H., Rodríguez-Gironés, M.A., Yedvab, S., Winkler, D.W. & Lotem, A. 2000. Experimental evidence for offspring learning in parent-offspring communication. *Proceedings of the Royal Society of London, Series B* **267**, 1-5.
Kilner, R. 1995. When do canary parents respond to nestling signals of need? *Proceedings of the Royal Society of London, Series B* **260**, 343-348.
Kilner, R.M. 2001. A growth cost of begging in captive canary chicks. *Proceedings of the National Academy of Sciences USA* **98**, 11394-11398.
Kilner, R. & Davies, N.B. 1998. Nestling mouth colour: ecological correlates of a begging signal. *Animal Behaviour* **56**, 705-712.
Kilner, R. & Johnstone, R.A. 1997. Begging the question: are offspring solicitation behaviours signals of need. *Trends in Ecology and Evolution* **12**, 11-15.
Kilner, R.M., Nobel, D.G. & Davies, N.B. 1999. Signals of need in parent-offspring communication and their exploitation by the common cuckoo. *Nature* **397**, 667-672.
Kölliker, M., Richner, H., Werner, I. & Heeb, P. 1998. Begging signals and biparental care: nestling choice between parental feeding locations. *Animal Behaviour* **55**, 215-222.
Krebs, E.A. 2001. Begging and food distribution in crimson rosella (*Platycercus elegans*) broods: why don't hungry chicks beg more? *Behavioral Ecology and Sociobiology* **50**, 20-30.
Lack, D. 1947. The significance of clutch size. Parts 1 and 2. *Ibis* **89**, 302-352.
Leech, S.M. & Leonard, M.L. 1996. Is there an energetic cost to begging in nestling tree swallows (*Tachycineta bicolor*)? *Proceedings of the Royal Society of London, Series B* **263**, 983-987.
Leech, S.M. & Leonard, M.L. 1997. Begging and the risk of predation in nestling birds. *Behavioral Ecology* **8**, 644-646.
Leonard, M. & Horn, A.G. 1996. Provisioning rules in tree swallows. *Behavioral Ecology and Sociobiology* **38**, 341-347.
Leonard, M.L. & Horn, A.G. 1998. Need and nestmates affect begging in tree swallows. *Behavioral Ecology and Sociobiology* **42**, 431-436.
Leonard, M.L. & Horn, A.G. in press. Dynamics of calling by tree swallow (*Tachycineta bicolor*) nestmates. *Behavioral Ecology and Sociobiology* **50**, 430-435.
Lichtenstein, G. 2001. Selfish begging by screaming cowbirds, a mimetic brood parasite of the bay-winged cowbird. *Animal Behaviour* **61**, 1151-1158.
Marti, C.D. 1989. Food sharing by sibling common barn-owls. *Wilson Bulletin* **101**, 132-134.
McCarty, J.P. 1996. The energetic cost of begging in nestling passerines. *The Auk* **113**, 178-188.
McRae, S.B., Weatherhead, P.J. & Montgomerie, R. 1993. American robin nestlings compete by jockeying for position. *Behavioral Ecology and Sociobiology* **33**, 101-106.
Mitteldorf, J.J. & Wilson, D.S. 2000. Population viscosity and the evolution of altruism. *Journal of Theoretical Biology* **204**, 481-496.
Mock, D.W. & Parker, G.A. 1986. Advantages and disadvantages of ardeid brood reduction. *Evolution* **40**, 459-470.
Mock, D.W. & Parker, G.A. 1997. *The Evolution of Sibling Rivalry*. Oxford: Oxford University Press.
O'Connor, R.J. 1978. Brood reduction in birds: selection for fratricide, infanticide and suicide? *Animal Behaviour* **26**, 79-96.
Pomiankowski, A. 1999. Intragenomic conflict. In: *Levels of Selection in Evolution* (Ed. by L. Keller). Princeton: Princeton University Press.

Price, K. 1998. Benefits of begging for yellow-headed blackbird nestlings. *Animal Behaviour* **56**, 571-577.
Price, K. & Ydenberg, R. 1995. Begging and provisioning in broods of asynchronously-hatched yellow-headed blackbird nestlings. *Behavioral Ecology and Sociobiology* **37**, 201-208.
Ratnieks, F.L. & Visscher, P.K. 1989. Worker policing in the honeybee. *Nature* **342**, 796-797.
Rodríguez-Gironés, M.A., Zuñiga, J.M. & Redondo, T. 2001. Effects of begging on growth rates of nestling chicks. *Behavioral Ecology* **12**, 269-274.
Seeley, T. 1995. *The Wisdom of the Hive*. Cambridge: Harvard University Press.
Skyrms, B. 1996. *Evolution of the Social Contract*. Cambridge: Cambridge University Press.
Smith, H.G. & Montgomerie, R. 1991. Nestling American robins compete with siblings by begging. *Behavioral Ecology and Sociobiology* **29**, 307-312.
Sober, E. & Wilson, D.S. 1998. *Unto Others: The Evolution And Psychology Of Unselfish Behaviour*. Cambridge: Harvard University Press.
Soler, M., Soler, J.J., Martinez, J.G. & Moreno, J. 1999. Begging behaviour and its energetic cost in great spotted cuckoo and magpie host chicks. *Canadian Journal of Zoology* **77**, 1794-1800.
Stamps, J.A., Clark, A.B., Arrowood, P. & Kus, B. 1985. Parent-offspring conflict in budgerigars. *Behaviour* **94**, 1-39.
Stamps, J.A., Clark, A.B., Kus, B. & Arrowood, P. 1987. The effects of parent and offspring gender on food allocation in budgerigars. *Behaviour* **101**, 177-199.
Stamps, J., Clark, A.B., Arrowood, P. & Kus, B. 1989. Begging behavior in budgerigars. *Ethology* **81**, 177-192.
Sullivan, K.A. & Weathers, W.W. 1991. Brood size and thermal environment influence field metabolism of nestling yellow-eyed juncos. *The Auk* **109**, 112-118.
Teather, K.L. 1992. An experimental study of competition for food between male and female nestlings of the red-winged blackbird. *Behavioral Ecology and Sociobiology* **31**, 81-87.
Visser, G.H. 1998. Development of temperature regulation. In: *Avian Growth and Development. Evolution within the Altricial-Precocial Spectrum*. (Ed. by J.M. Starck & R.E. Ricklefs). New York: Oxford University Press.
Wanker, R., Bernate, L.C. & Franck, D. 1996. Socialization of spectacled parrotlets *Forpus conspicillatus*: the role of parents, crèches, and sibling groups in nature. *Journal für Ornithologie* **137**, 447-461.
Westneat, D.F., Clark, A.B. & Rambo, K.C. 1995. Within-brood patterns of paternity and paternal behavior in red-winged blackbirds. *Behavioral Ecology and Sociobiology* **37**, 349-356.
Williams, G.C. 1966. *Adaptation And Natural Selection: A Critique Of Some Current Evolutionary Thought*. Princeton: Princeton University Press.
Williams, G.C. & Williams, D.C. 1957. Natural selection of individually harmful social adaptations among sibs with special reference to social insects. *Evolution* **11**, 32-39.
Wilson, D.S. 1990. Weak altruism, strong group selection. *Oikos* **59**, 135-140.
Wilson, D.S. 1997a. Altruism and organism: disentangling the themes of multilevel selection theory. *American Naturalist* **150**, S122-S134.
Wilson, D.S. 1997b. Incorporating group selection into the adaptationist program: a case study involving human decision making. In *Evolutionary Social Psychology* (Ed. by J. Simpson & D. Kendricks). Hillsdale: Lawrence Erlbaum Associates.
Wilson, D.S. 1998. Hunting, sharing and multilevel selection: the tolerated theft model revisited. *Current Anthropology* **39**, 73-97.
Wilson, D.S. & Dugatkin, L.A. 1991. TIT-FOR-TAT vs. Nepotism, or, why should you be nice to your rotten brother? *Evolutionary Ecology* **5**, 291-299.
Wilson, D.S., Pollock, G. & Dugatkin, L.A. 1992. Can altruism evolve in purely viscous populations? *Evolutionary Ecology* **6**, 331-341.

4.

PARENTAL INVESTMENT IN RELATION TO OFFSPRING SEX

Catherine M. Lessells
Netherlands Institute of Ecology, 6666ZG Heteren, The Netherlands (lessells@cto.nioo.knaw.nl)

ABSTRACT

Offspring sex may influence both the begging behaviour of nestlings and the parental response. An important first step in interpreting these strategies is understanding how parents would invest in their offspring if they had full information about their sex. Models are presented which predict the relative amounts of care received, and consequent fitness of, sons compared with daughters. These give predictions that are consistent with observed patterns of care. Models are also presented where the distribution of care is under the control of sons, and where offspring are produced in more than one 'situation', where situations differ in the fitness consequences of care for the parent or offspring. The models highlight two areas that merit future attention: first, allowing the parental investment strategy of the parent and the developmental strategy of the offspring to coevolve in an evolutionary game, and second, treating the discrimination of offspring sex by parents as a signalling problem.

INTRODUCTION

Begging behaviour is generally thought to be the costly outcome of parent-offspring conflict when the state of the offspring is not known with certainty by the parent (Godfray 1991). One aspect of offspring state that may have a large influence on the optimal division of parental investment between offspring is their sex, and empirical studies demonstrate differences between male and female offspring in their begging behaviour (e.g. Price et al. 1996). An important step in interpreting such differences is understanding how

J. Wright and M.L. Leonard (eds.), The Evolution of Begging, 65–85.

parents would optimally distribute investment between offspring if they had full information about the sex of the offspring. This chapter is about this life history decision over the relative amount that parents invest in total in offspring of each sex (Fisher 1930; Charnov 1982). In species such as birds, with separate sexes, there are two ways in which a parent can vary sex allocation: it can manipulate the primary sex ratio – the number of sons and daughters that are produced in the first place – or it can invest more on average in individual sons or daughters. The majority of the development and application of sex allocation theory has concerned primary sex ratios (Charnov 1982; West et al. 2000). This chapter seeks to redress this balance by focusing on the amount of investment made in individual sons and daughters.

SEX RATIO MODELS

Although the primary focus of this chapter is sex-biased investment in individual offspring, it is useful briefly to review sex ratio models because many of the same selection pressures apply to parental investment in each offspring. Current explanations of sex ratio in terms of selection pressures are generally based on Fisher's (1930) explanation that the total fitness of males in the population must equal the total fitness of females (the 'Fisherian constraint'). As a result, the fitness of offspring of each sex depends on the sex ratio that is being produced in the population as a whole. This 'Fisherian frequency dependence' results in selection for an equilibrium sex ratio at which parents make an equal total investment in sons and in daughters (producing a 1:1 sex ratio when the costs of producing an individual son or daughter are equal).

The most important other idea for understanding avian sex allocation was put forward in 1973, when Trivers and Willard provided an explanation for individual variation in sex ratio. They developed their theory with polygynous mammals in mind, and pointed out that when the condition of mothers has an effect on the condition of offspring that lasts into adulthood, and this in turn has a bigger effect on the reproductive success of sons than daughters, mothers in good condition are selected to produce sons and those in poor condition to produce daughters (see also Leimar 1996). In fact, Trivers and Willard's conclusion applies much more generally, whenever the expected relative fitness of sons and daughters varies predictably with the 'situation' in which offspring are produced (Charnov 1979, 1982). Situation might be defined by variation in any aspect of parental phenotype, such as maternal condition, as in Trivers and Willard's theory. Situation may also be defined by environmental variation, such as with laying date in the kestrel

(*Falco tinnunculus*) where sons have relatively high fitness compared to daughters if hatched early in the year (and relatively low fitness if hatched late) because all one-year-old females may breed, but only early-hatched one-year-old males do so (Dijkstra et al. 1990).

SEX ALLOCATION IN BIRDS

Empirical studies of sex allocation in birds have concentrated on sexually dimorphic species, because it was assumed that the larger sex of offspring would be more costly to produce. These species therefore offer an opportunity to test Fisher's equal allocation principle because the predicted sex ratios differ from the equal numbers of sons and daughters expected from random segregation of the sex chromosomes. Three kinds of information have been gathered.

First, studies have been directed at determining whether the larger sex was indeed more costly to produce, and if so, whether sex ratios were biased to give overall equal allocation. For example, the energy requirements of the larger male nestlings are higher than those of female nestlings in both red-winged blackbirds (Fiala & Congdon 1983) and great-tailed grackles (Teather & Weatherhead 1988). Food consumption in these two species is also higher by male than female nestlings (Fiala 1981; Teather 1987; Teather & Weatherhead 1988). These studies suggest over-allocation in the larger sex. In marsh harriers (*Circus aeruginosus*), however, the higher food intake of female nestlings and slightly male-biased secondary sex ratio are consistent with equal allocation (Krijgsveld et al. 1998). More species can be included in the analysis by using the extent of sexual dimorphism in adult body mass as a crude measure of the relative cost of producing a son or daughter. The most complete such analysis to date (Pen 2000) examines data on sex ratios at fledging and sexual dimorphism in 48 species. The sex ratio becomes increasingly female-biased as adult body mass of males relative to females increases, but not to the extent predicted on the basis of equal allocation judged either by body mass or food consumption predicted on the basis of body mass.

Second, differential mortality of sons and daughters has been investigated as a potential method of skewing secondary sex ratios, either to restore equal sex allocation when primary sex ratios are unbiased, or by mothers who are unable to rear successful sons (Trivers & Willard 1973). Nestling mortality is indeed biased towards males when adult body mass is larger in males, but not usually to a degree sufficient to reinstate equal allocation (Weatherhead & Teather 1991). Regarding the second possibility, Clutton-Brock et al. (1985) showed for birds that differential mortality becomes increasingly pronounced as sexual dimorphism in adult body mass increases. This is expected if more extreme dimorphism is associated with greater variance in male reproductive success and hence stronger selection for mothers in poor condition to abandon investment in sons. However, they also point out that this pattern could simply result from faster growing males being more vulnerable to resource shortage. In mammals their alternative interpretation is strengthened because there is little evidence for rejection of male offspring, and differential mortality often occurs late in the period of parental care or after it has terminated. These patterns of mortality do not suggest that the female is using a tactic of deliberately discontinuing care in order to avoid investing in sons with poor reproductive prospects.

Third, a small amount of information is available on how the growth or survival of sons relative to daughters varies with environmental conditions (Table 1). In two species in which sexual dimorphism is virtually absent, males fare worse than females in adverse conditions in one species, but not the other. In four out of five species with sexual dimorphism, it is the larger sex that fares worse in adverse conditions, but in the other species (the great tit), which has rather modest sexual dimorphism, it is the smaller females who fare worse.

Overall, the empirical evidence provides little support for equal allocation. There are two possible explanations for this. First, Fisher's prediction of equal allocation applies when there is no individual optimization of sex ratio. When sex ratio varies depending on individual circumstances as suggested by Trivers and Willard, the predicted population sex ratio no longer necessarily leads to equal allocation (Frank 1990). In general, models predict overproduction of the sex of offspring that is produced in the poorer situation, but this is not always the case (Charnov 1979, 1982; Frank 1987; Frank & Swingland 1988). Alternatively, the assumptions on which Fisher's theory is based may not hold. The remainder of this chapter focuses on models where his assumptions are reversed.

MODELS OF SEX-BIASED PARENTAL INVESTMENT

Fisher's theory assumes that the cost of producing a son or a daughter is fixed and asks what sex ratio a parent is selected to produce. A more recent body of theory addresses the opposite question (Maynard Smith 1980; Lessells 1998): if the sex ratio is fixed, how much should a parent invest in each son or daughter? In birds, recent studies of variation in sex ratio with convincing adaptive explanations (Ellegren et al. 1996; Komdeur et al.1997; Nager et al. 1999) suggest that sex ratio may be under parental control. However, some species show inconsistent patterns (Lessells et al. 1996;

Kölliker et al. 1999; Radford & Blakey 2000), and it is not yet clear to what extent parental control of sex ratio is limited to certain species or populations, or the published examples represent publication bias (Csada et al. 1996). It is therefore worth considering the theoretical consequences for sex-biased investment of sex ratio not being under parental control.

Table 1. Offspring sex differences in fitness components under varying environmental conditions

Species	Body mass dimor-phism*	Environmental variation	Sex showing greater reduction in fitness in poorer conditions: Size	Mass	Survival	Refer-ence
Blue-footed booby *Sula nebouxii*	~-30%	Hatching date			F	1.
Lesser black-backed gull *Larus fuscus*	~15%	Manipulated parental condition	=	M	M	2.
Pied flycatcher *Ficedula hypoleuca*	~0%	Nest mite density	M			3.
Collared flycatcher *Ficedula albicollis*	~0%	Manipulated brood size	=	=	=	4.
Great-tailed grackle *Quiscalus mexicanus*	~85%	Food abundance			M	5.
		Sex of competing nestling		M		
Great tit *Parus major*	~5%	Habitat, month			F	6.
		Manipulated brood size			F	7.
		Nestling mortality			F	8.
Red-winged blackbird *Agelaius phoeniceus*	~40%	Manipulated brood size			M	9.

Only species that feed their young (i.e. with a clearly depreciable form of care) are included in the table. Studies where the environmental variation is in the position in the laying sequence are excluded because these may represent a brood reduction strategy. * % dimorphism = 100(Msize - Fsize)/[(Msize - Fsize)/2]. References: 1. Torres & Drummond 1997; 2. Nager et al. 2000; 3. Potti & Merino 1996; 4. Sheldon et al. 1998; 5. Teather & Weatherhead 1989; 6. Dhondt 1970; 7. Smith et al. 1989; 8. Lessells et al. 1996; 9. Cronmiller & Thompson 1981.

In order to predict the evolutionarily stable amount of investment in sons and daughters it is necessary to know the shape of the offspring fitness curves. These are the relationships between the expected fitness accrued through a son or daughter and the amount of care that it receives. The shapes of these relationships depend on the physiology and ecology of male and female offspring, including the target growth trajectory and asymptotic size, and the usefulness of reserves for events such as dispersal. These curves are assumed to show diminishing returns, with offspring fitness reaching an asymptote at higher levels of care (e.g. Box Figure 1; Smith & Fretwell 1974). In sexually dimorphic species it is likely that the larger sex (assumed in the following to be the male) will have the lower offspring fitness curve; that is, they will have lower fitness for a given level of investment. For the sake of brevity, the sex with the lower offspring fitness curve will be referred to as being more 'needy'. The models must also include Fisherian frequency dependence: whenever there is a preponderance of one sex in the population – irrespective of whether that is brought about through the primary sex ratio or through heavy investment in individuals of that sex – that sex must suffer a reproductive disadvantage. Taken together, the offspring fitness curves and Fisherian frequency dependence allow the evolutionarily stable division of care between a fixed number of sons and daughters to be predicted (see Maynard Smith 1980 for details of an explicitly population genetic derivation and Lessells 1998 for a phenotypic derivation of the results presented in Box 1). In the simplest model, the total amount of care to be divided between the offspring is taken as fixed, but it is easy to extend the models to allow the total amount of care to be decided by the parent based on the cost of care (Box 1; Lessells 1998). The evolutionarily stable distribution of a given amount of care between the offspring does not depend on whether the amount of care is fixed or depends on the cost of care.

The most important conclusion from these models is that the evolutionarily stable division of care may not involve equal total investment in sons and daughters. In general, if males are the more needy sex, sons will receive more care than daughters but have lower fitness before Fisherian frequency dependence is taken into account (although some offspring fitness curves can be found where this is not the case; Maynard Smith 1980). For example, Box Figure 2 shows an evolutionarily stable distribution of care (represented by the filled circles) in which sons receive more care (m) than daughters (f), but have lower individual fitness (values on the y axis). As sons become increasingly needy compared with daughters, the disparities between sons and daughters in the care received and consequent fitness increase (Figure 1). These predictions are consistent with the empirical observations presented above: both allocation and mortality are biased towards males, and the differential mortality of males increases with the degree of sexual dimorphism. The model presented here agrees with Clutton-Brock et al. (1985) in interpreting the higher mortality of males as the result of their greater neediness (and can therefore also explain their additional evidence suggesting that females are not completely withdrawing investment from sons), but adds to their explanation by providing an adaptive explanation of relative levels of care for sons and daughters.

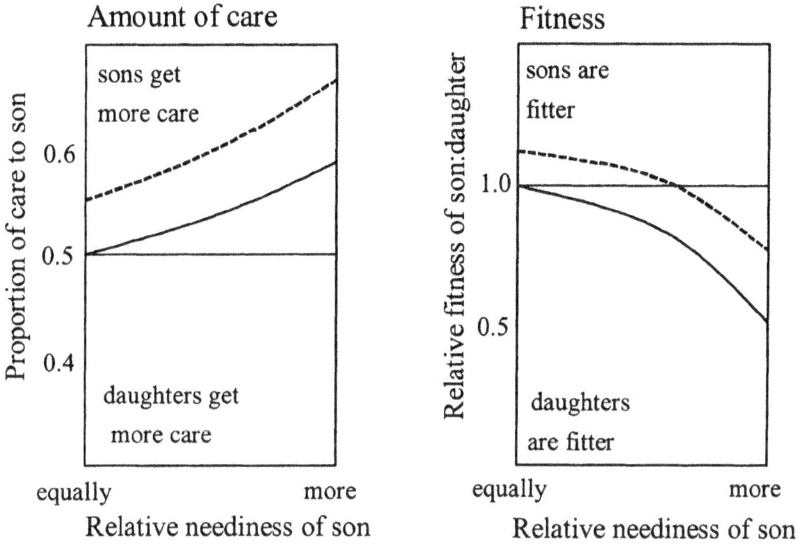

Figure 1. The predicted amount of care received and fitness of sons relative to daughters when the initial family sex ratio is fixed (at one son : one daughter). The evolutionarily stable pattern of investment was predicted assuming that the parent determines the total amount of care provided, and the distribution of care is under parental control (solid line) or control by sons (dashed line). (See Box 1, Maynard Smith 1980 and Lessells 1998 for the calculation of the evolutionarily stable distribution of care. See Appendix 1 for the fitness functions and parameters used in this numerical example.)

Box 1. Calculating the evolutionarily stable (ES) investment in sons and daughters.

Fisher's (1930) theory on sex ratios assumes that the cost of a son or a daughter is fixed and asks what sex ratio a parent is then selected to produce. An alternative body of theory addresses the opposite problem: it assumes that the sex ratio is fixed and asks how much investment a parent should make in each son or daughter (Maynard Smith 1980; Lessells 1998). In order to answer this question it is necessary to know the offspring fitness curves, $F_m(m)$ and $F_f(f)$ - the relationships between the fitness of a son or daughter, respectively, and the amount of care that each has received (m and f; Box Figure 1). Let us ignore for a moment the fact that the sons and daughters produced in the population as a whole will subsequently mate with each other to produce the next generation. Then, if the parent has a fixed

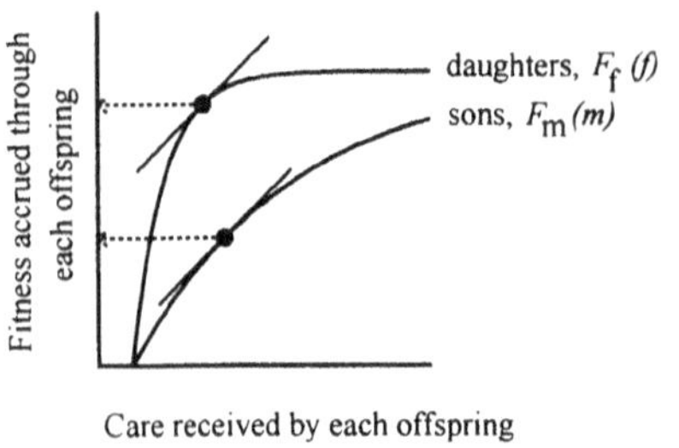

Box Figure 1. Offspring fitness curves. This figure does not show the ES distribution of care because Fisherian frequency dependence has not been taken into account.

amount of investment, c, to divide between one son and one daughter, its fitness will be equal to $F_m(m)+F_f(f)$. The value of m (and hence f, which equals $c-m$) which gives the parent highest fitness occurs when $F_m'(m) = F_f'(c-m)$ (where ' indicates the first derivative). Graphically, this is where the slopes of the $F_m(m)$ and $F_f(f)$ curves are the same, and m and f add up to c (Box Figure 1; see also Temme 1986; Haig 1990; Lessells in press). It is easy to see why the slopes must be equal: if we imagine the parent transferring a small amount of care from one offspring to the other, the rate at which it loses fitness from the first nestling is represented by the slope of the tangent to this deprived offspring's curve, and the rate at which it will gain extra fitness from the recipient is represented by the slope of the tangent to this second offspring's curve. Only when the slopes are equal will it not be possible for the parent to increase its fitness by changing the way in which it distributes care between the offspring.

This is not, however, the ES distribution of care because we have not taken into account the Fisherian constraint. In Box Figure 1 sons and daughters are initially produced in equal numbers and have different predicted fitness (the values indicated by an arrow on the y axis). This cannot be the case, and the reproductive success of one or both sexes must be constrained, for instance by failure to establish a territory or attract a mate. If Fisherian frequency dependence (FFD) acts after and independently of the effects of the amount of parental care, then the overall fitness of a son or a daughter will be the fitness in relation to the care received multiplied by a factor r_m or r_f, respectively. We can calculate the relative magnitudes of r_m and r_f by remembering the Fisherian constraint that $F_m(m^*)r_m = F_f(f^*)r_f$, so that $r_m/r_f = F_f(c-m^*)/F_m(m^*)$ (where * indicates the values of m and f for the population). We can use this to calculate that the ES division of an amount of care c between a son and a daughter occurs when $F_m'(m)r_m/r_f = F_f'(c-m)$, where $r_m/r_f = F_f(c-m)/F_m(m)$ (Lessells 1998, equation 8). Graphically this means that the ES division of care occurs when the slopes of the offspring fitness curves corrected for the Fisherian frequency dependence generated by that division of care are equal, and m and f add up to c (Box Figure 2).

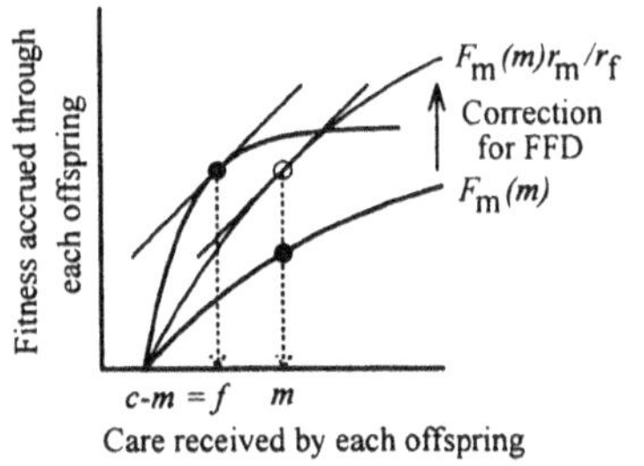

Box Figure 2. ES distribution of care between one son and one daughter of an amount, c, of care.

So far we have assumed that the parent has a fixed amount of care to divide between its offspring, but we can also ask what happens when the amount of investment is chosen by the parent depending on the cost of care. In order to do this

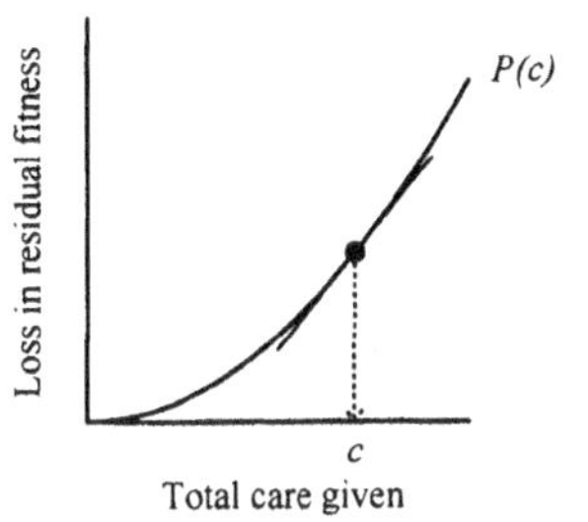

Box Figure 3. Mother's fitness curve.

we need to know the parent's (here assumed to be the mother) fitness curve, $P(c)$ - the loss in her residual fitness (what she gains through future reproductive attempts) in relation to the total amount of care that she gives (Box Figure 3). The ES amount and division of care are then given by $F_m'(m)r_m/r_f = F_f'(c\text{-}m) = P'(c)$, where $r_m/r_f = F_f(c\text{-}m)/F_m(m)$ (provided that FFD acts on the mother's future fitness in the same way as the daughter's; Lessells 1998, equations 11 and 12b). Graphically this means that not only are the slopes of the offspring fitness curves equal after correcting for FFD (Box Figure 2), but that the mother's fitness curve also has the same slope as the corrected offspring slopes (cf. Box Figures 2 and 3).

We can also ask what happens when offspring are produced in more than one situation. The offspring and mother's fitness curves are then assumed to be determined only by the situation in which the mother is reproducing. However, if the offspring disperse and mate at random in the population as a whole, the FFD by which the fitness curves should be corrected needs to be calculated taking into account the offspring produced in all situations. If situation i is referred to by the subscript i and occurs with frequency p_i, the ES amount and distribution of care in situation i are given by $F_{i,m}'(m_i)r_f/r_m = F_{i,f}'(c_i\text{-}m_i) = P_i'(c_i)$, where $r_f/r_m = \Sigma[F_{i,m}(m_i)p_i]/\Sigma[F_{i,f}(c_i\text{-}m_i)p_i]$ (for the amount of FFD acting across the habitat as a whole). Box Figure 4 shows an example in which there are two breeding situations, A (circles and solid slopes) and B (triangles and dashed slopes), that are used

Box Figure 4. ES distribution of care between one son and one daughter in two different situations (A and B) which differ in the mother's fitness curve.

equally frequently ($p_A = p_B = 0.5$). The corrected offspring slopes and mother's slope are the same only within each situation (Box Figure 4). Although the figure illustrates a case where only the mother's fitness curve differs between situations, the same equations can be used to calculate the ES amounts and distribution of care when only the offspring curves differ, or when both offspring and mother's fitness curves differ between situations.

Lastly we have assumed that the parent controls the amount and distribution of care. However, the same approach can be used to predict the ES pattern of investment when the parent controls the amount of care, but the son controls its distribution (cf. Parker et al. 1989; Forbes 1993; Lessells in press). The son's fitness (W_{son}) is calculated assuming that the fitness that he accrues through himself is twice what the parent accrues, fitness accrued through his sister the same as what the parent accrues (i.e. assuming they are full siblings), and what he loses through the decrease in his parent's residual fitness is half what the parent loses (see Lessells & Parker 1999). The ES amount and distribution of care are then found by setting $\partial W_{son}/\partial m$ and $\partial W_{mother}/\partial c$ equal to zero, which gives $2F_m'(m)r_m/r_f = F_f'(c\text{-}m) = P'(c)$, where $r_m/r_f = F_f(c\text{-}m)/F_m(m)$. The ES amount and distribution of care under control by sons of the distribution of care can be similarly found when offspring are produced in more than one situation.

ASSUMPTIONS OF THE MODELS

The models presented here make a number of simplifying assumptions. Some are made because they do not qualitatively alter the predictions. For example, that care is given by a single parent (see Lessells 1998 for the inclusion of the second parent), or that the fixed family size is two (Maynard Smith 1980; Lessells 1998). Other assumptions are made for the sake of tractability. For example, the amount and distribution of care are treated as a single investment decision, rather than a series of decisions, each affecting the state of the offspring (including satiation, condition, growth and development) and thus having knock-on effects on the fitness consequences of future investment decisions (Godfray 1995). Still other assumptions have been made in order to concentrate on one main line of argument. For example, care is assumed to be provided in some form that cannot be divided between the offspring ('depreciable care' *sensu* Clutton-Brock 1991). There is also no possibility for the parent to terminate care for some offspring (brood reduction; O'Connor 1978; Haig 1990), even after the parent has made some initial investment (Maynard Smith 1980). There are, however, three sets of assumptions which deserve further attention.

Parental Control of the Distribution of Care

Although the parent is always likely to have control over how much care it provides, one or more of the offspring may control how it is distributed. For example, when males are larger than females, sons are likely to outcompete their smaller sisters in competition over access to care. When the distribution of care is under the control of sons, the model becomes an evolutionary game between the parent, who decides the amount of care, and the son, who controls its distribution (see Box 1; Parker et al. 1989; Forbes 1993; Lessells in press). The change from parental control to control by sons increases the amount of care received by the son, decreases the amount of care received by the daughter and increases the amount of care provided overall by the parent (not shown), so that the son receives a higher proportion of the care, and has higher fitness relative to his sister (dashed lines in Figure 1). However, as the relative neediness of sons increases, the qualitative trends are the same under parental control and control by sons: the proportion of care received by the son increases, but his fitness relative to the daughter decreases. Overall, the only qualitative prediction that distinguishes parental control and control by sons is that under control by sons there are some values of relative neediness for which sons are expected to be fitter than daughters, which does not occur under parental control.

Costless Discrimination of Offspring Sex

We can use a modification of the model to ask how much care the parent should provide if it cannot discriminate offspring sex (and therefore cares equally for sons and daughters). This results in sons receiving less care, but daughters receiving more care, thus raising the evolutionary question of why daughters do not conceal their sex by mimicking their brother's characteristics. In some cases this may be impossible; for example, in European sparrowhawks (*Accipiter nisus*) sexual size dimorphism is so extreme that older nestlings can be sexed by human observers "at a glance" (Newton & Marquiss 1979). However, it is more difficult to see why this should not occur for plumage dichromatism, especially when this does not occur in adults of the same species (e.g. crown coloration in European bee-eaters, *Merops apiaster*; personal observation). One possible explanation is that the coloration that signals offspring sex is costly to produce (e.g. Hill & Montgomerie 1994), outweighing the benefit to daughters of being confused with their brothers. Alternatively, when the relative neediness of sons is sufficiently high, daughters may actually be selected to be distinguishable from their brothers (see Appendix 1 for details): although they lose in terms of their individual fitness, this is more than compensated by the inclusive fitness gain through their brother. Such costless honest signalling is possible because the daughter and son are related and the state that is being signalled (sex) has discrete values (Maynard Smith 1991; Bergstrom & Lachmann 1998). Selection for daughters to signal their sex may lead to selection for subtle changes in developmental trajectory; for example, for an initial period of rapid growth and development (Newton & Marquiss 1979; Richter 1983) before sex is 'declared' to the parent. Alternatively, the signalling of sex might be subsumed within begging behaviour that also signals other components of neediness such as hunger and condition. Begging models have yet to be developed which take into account multiple components of need.

Simultaneous Optimization of Sex Ratio and Parental Investment

The models presented here make the opposite assumptions to Fisher (1930), in that sex ratio is taken as fixed and investment per individual son or daughter allowed to respond to selection. Models which allow both sex ratio and sex-biased investment to evolve have only recently been developed (Pen 2000; Pen & Weissing in press). A parent can then be regarded as making four decisions: clutch size, sex ratio and the amount of investment in individual sons and daughters. Such models predict that Fisher's equal allocation principle holds as long as clutch size and sex ratio are optimized and the fitness of offspring is not affected by clutch size or sex ratio *per se* (but only through their indirect effects on the amount of care that an offspring receives), whether or not the investment per son or daughter is fixed or allowed to evolve (Pen 2000; Pen & Weissing in press). The empirical observations of unequal sex allocation in birds therefore appear to argue against optimization of both (but not necessarily one of) clutch size and sex ratio.

MODELS WHEN OFFSPRING ARE PRODUCED IN MORE THAN ONE SITUATION

The models presented so far assume that all families are identical: that they have the same fitness curves for offspring of each sex and the parent. This is often not the case, however, and models analogous to that of Trivers and Willard (1973) for sex ratios are then needed which predict the evolutionarily stable parental investment in sons and daughters in the different breeding situations encountered by the population. Such models have not previously been developed. In these models, the predicted amount and distribution of care in each situation will depend on the corresponding fitness curves for that situation. However, in a panmictic population, offspring produced in different situations will mate at random in the population as a whole, so the Fisherian frequency dependence is the same for all offspring, and must be calculated combining the offspring produced in all situations (Box 1; Lessells 1998). Such models often result in both the relative and absolute amounts of care received by offspring of each sex varying between situations.

Although the ways that breeding situations may differ in terms of environment or parental phenotype are multitudinous, in the model these differences boil down to differences in the offspring or parental fitness curves. Two major classes of difference will be considered here: in the first, the offspring fitness curves are identical, but the cost of care to the parent differs between the two situations considered. The consequence of this is that the parent invests more in total in the situation with the less steep parental fitness curve. An exactly equivalent effect results if both sexes of offspring suffer an equal proportional loss of fitness, independent of the amount of care that they have received, in one of the two situations (see comments on the effects of extra-pair paternity in Lessells 1998, in press). The difference in the costs of care can therefore be thought of as representing a difference in the severity of the two situations. The second major class of difference between the situations is when the offspring curves differ between the situations for one sex, but not the other. For instance, in kestrels, late-hatched (but not early-hatched) males (but not females) lose fitness because they fail to breed at one year old (Dijkstra et al. 1990). Here this difference was modelled by making the fitness curve in one situation a constant multiple of the fitness curve in the other situation for one sex only.

Under parental control, when the situations differ in the cost of care, the situation with the lower cost of care can be clearly identified as the 'better' situation for the offspring: both sons and daughters receive more care and have higher fitness there (not shown). In terms of the relative prospects of sons and daughters within situations, sons receive more than 50% of the care in the better situation, and slightly less than 50% in the poorer situation, the divergence increasing as the relative neediness of sons increases (Figure 2a). This differential investment is reflected in the fitness of sons relative to daughters in each of the situations, but sons always have lower fitness than daughters within a situation (Figure 2a). When the situations differ in the offspring fitness curves for one of the offspring sexes, each sex of offspring receives more care and has higher fitness in the situation in which it has the relatively higher fitness curve (not shown). (See Appendix 1 for the range of fitness curves for sons and daughters in the two situations investigated.) The better situation for one sex is therefore the poorer situation for the other. When sons and daughters are equally needy (before applying the constant multiple difference in one situation), sons get more than 50% of the care in the better situation for sons, and less than 50% in the other situation (Figure 2b). As the relative neediness increases, the proportion of care that sons

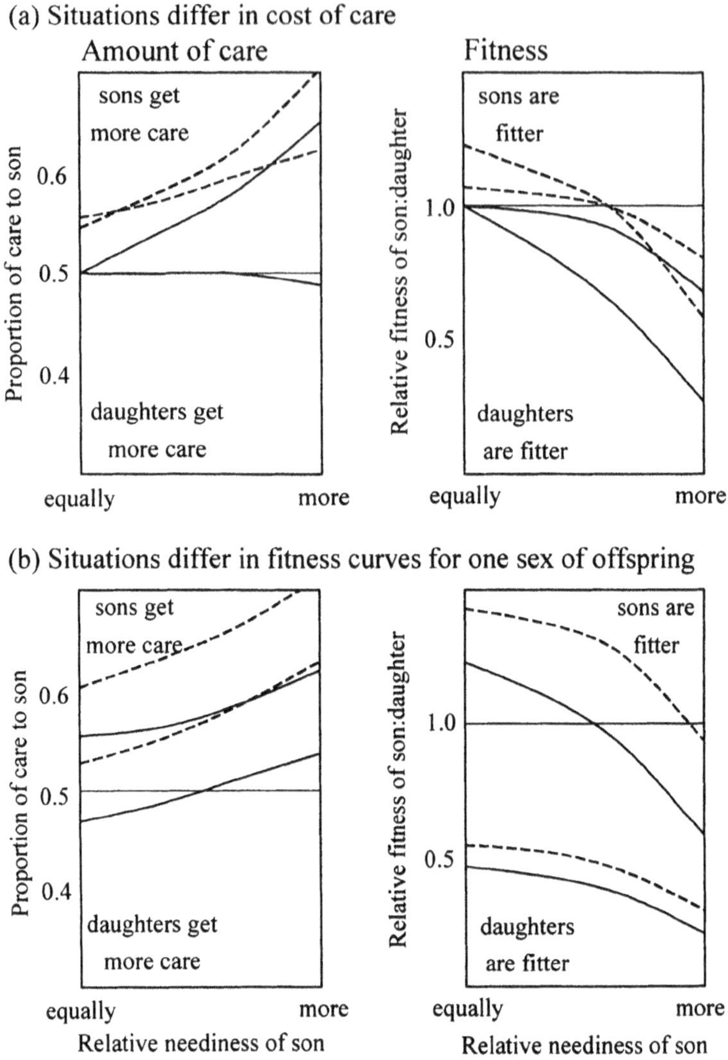

Figure 2. The predicted amount of care received and fitness of sons relative to daughters when the initial family sex ratio is fixed (at one son : one daughter) and offspring are produced in two situations which differ in (a) the cost of care or (b) the fitness of daughters receiving a given quantity of care. The higher of the pair of lines on the right hand side of the figures refers to (a) the situation with the less steep parental fitness curve (the better situation) or (b) the lower fitness curve for daughters (the better situation for sons). The evolutionarily stable pattern of investment was predicted assuming that the parent determines the total amount of care provided, and the distribution of care is under parental control (solid line) or control by sons (dashed line). (See Box 1, Maynard Smith 1980 and Lessells 1998 for the calculation of the evolutionarily stable distribution of care. See Appendix 1 for the fitness functions and parameters used in this numerical example.)

receive tends to increase in both situations so that at higher levels of neediness sons sometimes receive more than 50% of the care in both situations (counter cases not shown). When sons and daughters are equally needy (before applying the constant multiple difference in one situation), sons are fitter than daughters in the better situation for sons and less fit in the other situation (Figure 2b). As the relative neediness of sons increases, the fitness of sons relative to daughters decreases and eventually drops below one in both situations.

In summary, when sons are needier than daughters and parents control the distribution of care, sons receive more care than daughters in one or both situations, and are fitter than daughters in one or neither of the situations.

Models when offspring are produced in more than one situation can be extended to include control by sons of the distribution of care in the same way as the models with all families in identical breeding situations (see Box 1; predictions are shown as dashed lines in Figure 2). In general, this has the same qualitative effect as for a single situation: sons receive a higher proportion of care and have higher fitness relative to their sisters (Figure 2). When there is more than one situation, the only qualitative differences between control by sons and parental control are that sons only receive less than 50% of the care in any situation under parental control and only have greater fitness than daughters in both situations under their own control.

Empirical observations against which these predictions can be tested are limited (Table 1). Two species have little sexual dimorphism, so there are no clear predictions. If the relative growth and survival rates are indicative of relative fitness, four of the other species can be accommodated within models with parental control or control by sons. The great tit is a more interesting case because sons have higher fitness than daughters under less favourable conditions. It could also be argued that the environmental factors associated with different relative fitnesses represent variation in the availability of resources (i.e. in the cost of care; Figure 2a) rather than differences in the fitness functions of the offspring. If this is the case, the pattern of relative fitness is only consistent with control by sons, because only then can sons have higher fitness than daughters (proportion of males among survivors ranges from about 50% upwards; Dhondt 1970; Smith et al. 1989; Lessells et al. 1996), and the relative fitness of sons be higher in the poorer situation. It is worth noting that these predictions only occur when sons are slightly more needy than females, and sexual dimorphism in great tits is slight. However, given that the observed pattern can also be accommodated by models with parental control when one of the offspring fitness functions differs between situations, the conclusion of control by sons is provisional at best.

CONCLUSIONS

Empirical observations of sex allocation in birds lend little support for Fisher's equal allocation principle, and hence for the assumptions of his theory that the costs of a son or daughter are fixed and sex ratio free to vary evolutionarily. In contrast, the three main features of empirical observations – that males (when these are the larger sex) receive more investment, but have lower fitness, than females, and that the fitness differential increases with increasing dimorphism, match perfectly with the predictions of models (Maynard Smith 1980; Lessells 1998) based on reverse premises – suggest that sex ratio is fixed and parental investment per offspring is free to vary. Recent models (Pen 2000; Pen & Weissing in press) demonstrate that Fisher's equal allocation principle follows when sex ratio and clutch size are optimized, whether or not parental investment per offspring is also optimized. Hence, the empirical observations imply that at least one of these two variables is constrained in most natural populations.

The models presented here also predict the patterns of care expected when offspring are produced in more than one situation. Like Trivers and Willard's model (1973), which predicts that sex ratio should vary between situations, the present models predict variation in sex allocation, but in the investment per offspring. Biased secondary sex ratios could therefore be consistent with optimization of either sex ratio or investment per offspring, and primary sex ratios would need to be measured to determine which was the case. The models presented here differ, however, in predicting intermediate secondary sex ratios, while Trivers and Willard's hypothesis predicts a strategy with offspring of only one sex being produced in each situation (provided there is no cost to sex ratio control; Pen & Weissing in press).

Faced with the general empirical pattern that sons receive more investment than daughters in species where males are larger, it is tempting to speculate that this results from sons rather than parents controlling the distribution of care. However, this bias is also expected under parental control and predictions that discriminate between parental control and control by sons are more limited than naïve intuition might lead one to expect. In particular, whenever there is variation in the situation in which offspring are produced, there are some situations in which sons are predicted to receive more care and be fitter than their sisters – even under parental control of the distribution of care.

Although the models presented here seem to be successful in offering insights into the selection pressures acting on sex allocation, two caveats must be given. The first concerns the generality of conclusions. Maynard Smith pointed out in 1980 that even the qualitative predictions of models depend on the exact shape of the offspring fitness curves. A similar difficulty is encountered in models where offspring are produced in more than one situation (and also occurs in the equivalent sex ratio models; e.g. Frank 1987). This does not mean that the models are wrong or misconceived, but it does mean that there is an asymmetry in the conclusions that can be drawn using specific fitness functions. Used in this way, the models can demonstrate that a particular pattern of investment may be evolutionarily stable (e.g. unequal allocation in sons and daughters), but not that a particular pattern cannot (e.g. sons never receiving more care than daughters in all environments).

The second caveat concerns applying the models to empirical examples. Research into life histories has underlined the difficulties of empirically measuring the fitness consequences of investment decisions (Lessells 1991), and these problems apply equally to measuring the fitness curves embedded in the present models. The empirical application of the models will also be thwarted if the fitness consequences of care are not temporally separated from Fisherian frequency dependence as neatly as in the mathematical equations. For example, the care received as an offspring may have effects on the subsequent fecundity of females (Haywood & Perrins 1992). Thus, the offspring fitness curve overlaps Fisherian frequency dependence in the stages in the life cycle at which they apply. (The models will still apply provided that these effects are multiplicative; see Maynard Smith 1980 for a model where that is not the case). In such cases, predictions regarding the relative fitness of sons and daughters will be close to impossible to test.

FUTURE DIRECTIONS

Despite these caveats, these models have a role to play, not only as an initial rather simple-minded attack on a complex problem, but also in making clear their assumptions. This highlights areas where the complexities of the real world deserve further attention, two of which seem especially promising.

First, parental investment is not a single decision, but a series of decisions each of which has an effect on the consequences of subsequent decisions. If multiple components of offspring state affect the parent's optimal investment decisions, then offspring can also influence their own fitness by the way that they change state in response to investment. In other words, the offspring's growth and development strategy and the parent's investment strategy are entwined in an evolutionary game. Models exploring the consequences of regarding growth and development trajectories as the outcome of selection already exist (e.g. Ricklefs & Schew 1994), but almost none have addressed the evolutionary game confronting parental investment strategies with offspring growth and development strategies (but see Godfray & Parker 1992).

Second, male and female offspring are not instantly and costlessly discriminable from the moment that they prise themselves out of the egg. Some sex-specific characters may be impossible to hide from the parent, but others may be regarded as signals that allow the offspring the evolutionary choice of concealing or revealing its sex. Conversely, parents may face a choice of whether to use information about offspring sex which is either imperfect or is costly to acquire, or whether instead to provide care indiscriminately. Alternatively, information about sex may be transmitted by the same begging signals as communicate other components of offspring need. Theoretical models will predict the consequences of signalling multiple components of need, but empirical studies investigating the extent to which separate information about offspring sex is available to and used by the parents will be equally crucial. Integrating problems of sex allocation with the signalling system between parents and offspring that is the central theme of this book is an exhilarating challenge for the future.

ACKNOWLEDGEMENTS

I am grateful to Jon and Marty for organizing the stimulating and enjoyable workshop on which this book is based, and to them and Rüdiger Cordts, Tobias Limbourg, Ido Pen and Ben Sheldon for helpful comments on earlier versions of this chapter.

REFERENCES

Bergstrom, C.T. & Lachmann, M. 1998. Signaling among relatives. III. Talk is cheap. *Proceedings of the National Academy of Sciences USA* **95**, 5100-5105.

Charnov, E.L. 1979. The genetical evolution of patterns of sexuality: Darwinian fitness. *American Naturalist* **113**, 645-480.

Charnov, E.L. 1982. *The Theory of Sex Allocation*. Princeton: Princeton University Press.

Clutton-Brock, T.H. 1991. *The Evolution of Parental Care*. Princeton: Princeton University Press.

Clutton-Brock, T.H., Albon, S.D. & Guinness, F.E. 1985. Parental investment and sex differences in juvenile mortality in birds and mammals. *Nature* **313**, 131-133.

Cronmiller, J.R. & Thompson, C.F. 1981. Sex ratio adjustment in malnourished red-winged blackbird broods. *Journal of Field Ornithology* **52**, 65-67.

Csada, R.D., James, P.C. & Espie, R.H.M. 1996. The 'file drawer problem' of non-significant results: does it apply to biological research? *Oikos* **76**, 591-593.

Dhondt, A.A. 1970. The sex ratio of nestling great tits. *Bird Study* **17**, 282-286.

Dijkstra, C., Daan, S. & Buker, J.B. 1990. Adaptive seasonal variation in the sex ratio of kestrel broods. *Functional Ecology* **4**, 143-147.

Ellegren, H., Gustafsson, L. & Sheldon, B.C. 1996. Sex ratio adjustment in relation to paternal attractiveness in a wild bird population. *Proceedings of the National Academy of Sciences USA* **93**, 11723-11728.

Fiala, K.L. 1981. Reproductive cost and the sex ratio in red-winged blackbirds. In: *Natural Selection and Social Behaviour: Recent Research and New Theory* (Ed. by R.D.Alexander & D.W. Tinkle). New York: Chiron Press.

Fiala, K.L. & Congdon, J.D. 1983. Energetic consequences of sexual size dimorphism in nestling red-winged blackbirds. *Ecology* **64**, 642-647.

Fisher, R.A. 1930. *The Genetical Theory of Natural Selection.* Oxford: Clarendon Press.

Forbes, L.S. 1993. Avian brood reduction and parent-offspring "conflict". *American Naturalist* **142**, 82-117.

Frank, S.A. 1987. Individual and population sex allocation patterns. *Theoretical Population Biology* **31**, 47-74.

Frank, S.A. 1990. Sex allocation theory for birds and mammals. *Annual Review of Ecology and Systematics* **21**, 13-55.

Frank, S.A. & Swingland, I.R. 1988. Sex ratio under conditional sex expression. *Journal of Theoretical Biology* **135**, 415-418.

Godfray, H.C.J. 1991. The signalling of need by offspring to their parents. *Nature* **353**, 328-330.

Godfray, H.C.J. 1995. Signalling of need between parents and young: parent-offspring conflict and sibling rivalry. *American Naturalist* **146**, 1-24.

Godfray, H.C.J. & Parker, G.A. 1992. Sibling competition, parent-offspring conflict and clutch size. *Animal Behaviour* **43**, 473-490.

Haig, D. 1990. Brood reduction and optimal parental investment when offspring differ in quality. *American Naturalist* **136**, 550-566.

Haywood, S. & Perrins, C.M. 1992. Is clutch size in birds affected by environmental conditions during growth? *Proceedings of the Royal Society of London, Series B* **249**, 195-197.

Hill, G.E. & Montgomerie, R. 1994. Plumage colour signals nutritional condition in the house finch. *Proceedings of the Royal Society of London, Series B* **258**, 47-52.

Kölliker, M., Heeb, P., Werner, I., Mateman, A.C., Lessells, C.M. & Richner, H. 1999. Offspring sex ratio is related to male body size in the great tit (*Parus major*). *Behavioral Ecology* **10**, 68-72.

Komdeur, J., Daan, S., Tinbergen, J. & Mateman, C. 1997. Extreme adaptive modification in sex ratio of the Seychelles warbler. *Nature* **385**, 522-525.

Krijgsveld, K.L., Dijkstra, C., Visser, G.H. & Daan, S. 1998. Energy requirements for growth in relation to sexual size dimorphism in marsh harrier *Circus aeruginosus* nestlings. *Physiological Zoology* **71**, 693-702.

Leimar, O. 1996. Life-history analysis of the Trivers and Willard sex-ratio problem. *Behavioral Ecology* **7**, 316-325

Lessells, C.M. 1991. The evolution of life histories. In: *Behavioural Ecology*, 3rd Edition (Ed. by J.R. Krebs & N.B. Davies). Oxford: Blackwell Scientific Publications.

Lessells, C.M. 1998. A theoretical framework for sex-biased parental care. *Animal Behaviour* **56**, 395-407.

Lessells, C.M. in press. Parentally-biased favouritism: why should parents specialize in caring for different offspring? *Philosophical Transactions of the Royal Society Series B.*

Lessells, C.M. & Parker, G.A. 1999. Parent-offspring conflict: the full-sib - half-sib fallacy. *Proceedings of the Royal Society of London, Series B* **266**, 1637-1643.

Lessells, C.M., Mateman, A.C. & Visser, J. 1996. Great tit hatchling sex ratios. *Journal of Avian Biology* **27**, 135-142.

Maynard Smith, J. 1980. A new theory of sexual investment. *Behavioral Ecology and Sociobiology* **7**, 247-251.

Maynard Smith, J. 1991. Honest signalling: the Philip Sidney game. *Animal Behaviour* **42**, 1034-1035.
Nager, R.G., Monaghan, P., Griffiths, R., Houston, D.C. & Dawson, R. 1999. Experimental demonstration that offspring sex ratio varies with maternal condition. *Proceedings of the National Academy of Sciences USA* **96**, 570-573.
Nager, R.G., Monaghan, P., Houston, D.C. & Genovart, M. 2000. Parental condition, brood sex ratio and differential young survival: an experimental study in gulls (*Larus fuscus*). *Behavioral Ecology and Sociobiology* **48**, 452-457.
Newton, I. & Marquiss, M. 1979. Sex ratio among nestlings of the European sparrowhawk. *American Naturalist* **113**, 309-315.
O'Connor, R.J. 1978. Brood reduction in birds: selection for fratricide, infanticide and suicide? *Animal Behaviour* **33**, 519-533.
Parker, G.A., Mock, D.W. & Lamey, T.C. 1989. How selfish should stronger sibs be? *American Naturalist* **133**, 846-868.
Pen, I.R. 2000. *Sex Allocation in a Life History Context*. PhD thesis., Rijksuniversiteit Groningen.
Pen, I.R. & Weissing, F.J. in press. Optimal sex allocation: steps towards a mechanistic theory. In: *The Sex Ratio Handbook* (Ed. by I. Hardy). Cambridge: Cambridge University Press.
Potti, J. & Merino, S. 1996. Parasites and the ontogeny of sexual size dimorphism in a passerine bird. *Proceedings of the Royal Society of London, Series B* **263**, 9-12.
Price, K., Harvey, H. & Ydenberg, R. 1996. Begging tactics of nestling yellow-headed blackbirds, *Xanthocephalus xanthocephalus*, in relation to need. *Animal Behaviour* **51**, 421-435.
Radford, A.N. & Blakey, J.K. 2000. Is variation in brood sex ratios adaptive in the great tit (*Parus major*)? *Behavioral Ecology* **11**, 294-298.
Richter, W. 1983. Balanced sex ratios in dimorphic altricial birds: the contribution of sex specific growth dynamics. *American Naturalist* **121**, 158-171.
Ricklefs, R.E. & Schew, W.A. 1994. Foraging stochasticity and lipid accumulation by nestling petrels. *Functional Ecology* **8**, 159-170.
Sheldon, B.C., Merilä, J., Lindgren, G. & Ellegren, H. 1998. Gender and environmental sensitivity in nestling collared flycatchers. *Ecology* **79**, 1939-1948.
Smith, C.C. & Fretwell, S.D. 1974. The optimal balance between size and number of offspring. *American Naturalist* **108**, 499-506.
Smith, H.G., Källander, H. & Nilsson, J.-Å. 1989. The trade-off between offspring number and quality in the great tit *Parus major*. *Journal of Animal Ecology* **58**, 383-401.
Teather, K.L. 1987. Intersexual differences in food consumption by hand-reared great-tailed grackle (*Quiscalus mexicanus*) nestlings. *The Auk* **104**, 635-639.
Teather, K.L. & Weatherhead, P.J. 1988. Sex-specific energy requirements of great-tailed grackle (*Quiscalus mexicanus*) nestlings. *Journal of Animal Ecology* **57**, 659-668.
Teather, K.L. & Weatherhead, P.J. 1989. Sex-specific mortality in nestling great-tailed grackles. *Ecology* **70**, 1485-1493.
Temme, D.H. 1986. Seed size variability: a consequence of variable genetic quality among offspring? *Evolution* **40**, 414-417.
Torres, R. & Drummond, H. 1997. Female-biased mortality in nestlings of a bird with size dimorphism. *Journal of Animal Ecology* **66**, 859-865.
Trivers, R.L. & Willard, D.E. 1973. Natural selection of parental ability to vary the sex ratio of offspring. *Science* **179**, 90-92.
Weatherhead, P.J. & Teather, K.L. 1991. Are skewed sex ratios in sexually dimorphic birds adaptive? *American Naturalist* **138**, 1159-1172.
West, S.A., Herre, E.A. & Sheldon, B.C. 2000. The benefits of allocating sex. *Science* **290**, 288-290.

APPENDIX 1: FUNCTIONS AND PARAMETERS USED IN GRAPHICAL EXAMPLES

In all the numerical examples given in this chapter, offspring fitness curves have the form $F(x) = 1-e^{-z(1-x)}$, and parental fitness curves the form $P(c) = kc^2$, where z and k are constants. All families are assumed to consist initially of one son and one daughter. In addition the following assumptions are made:

Figure 1: z_m varies linearly from 1 to 0.2 along each x axis, $z_f = 1$, $k = 0.02$.

Figure 2: $z_{m,A} = z_{m,B}$ and varies linearly from 1 to 0.2 along each x axis, $z_{f,A} = z_{f,B} = 1$, and situations A and B equally frequent. (a) $k_A = 0.04$, $k_B = 0.01$. (b) Fitness of males in both situations, and females in situation A were reduced by multiplying the original offspring fitness curves by a factor j, so that $F(x) = j(1-e^{-z(1-x)})$, where $j_{m,A} = j_{m,B} = j_{f,A} = 0.5$ and $j_{f,B} = 1$. $k_A = k_B = 0.02$. Similar predictions of the model were also generated, but not reported in detail here, for: (i) $j_{m,A} = j_{f,A} = j_{f,B} = 1$ and $j_{m,B} = 0.5$; (ii) $j_{m,A} = j_{m,B} = j_{f,A} = 1$ and $j_{f,B} = 0.5$; (iii) $j_{m,A} = 1$ and $j_{m,B} = j_{f,A} = j_{f,B} = 0.5$.

Box Figure 1: $z_m = 0.2$, $z_f = 1$, $c = 7$.

Box Figures 2 & 3: $z_m = 0.2$, $z_f = 1$, $k = 0.0127$ (giving predicted ES value of $c = 7$).

Box Figure 4: $z_m = 0.2$, $z_f = 1$, $k_A = 0.01$, $k_B = 0.04$, situations A and B equally frequent.

Signalling of sex by the offspring was investigated in a model where $z_f = 1$, $c = 7$, and z_m varied from 1 to 0.1 (i.e. increasing neediness of sons relative to daughters). Dimorphic sons have higher fitness than monomorphic sons when the resident population is either monomorphic (when they are mutants) or dimorphic (when they are the resident strategy). Sons are therefore always selected to reveal their sex. Dimorphic daughters have higher fitness than monomorphic daughters in a monomorphic population when $z_m < 0.48$, and in a dimorphic population when $z_m < 0.44$. Thus, when z_m is less than 0.44 the ESS is for daughters to signal their sex.

BEGGING AS A SIGNAL AND THE ISSUE OF COSTS

5.

THE EVOLUTION OF COMPLEX BEGGING DISPLAYS

Rebecca M. Kilner
Department of Zoology, University of Cambridge, Cambridge CB2 3EJ, UK
(r.m.kilner@zoo.cam.ac.uk)

ABSTRACT

Passerine nestlings solicit food by performing a vigorous postural display, while revealing a brightly coloured gape and calling repetitively. In this chapter, I suggest that the individual elements of the begging display may collectively function in the resolution of parent-offspring conflict. A general conclusion from two experimental studies is that multiple elements increase the information content of the display, thereby preventing exploitation of parents by potentially manipulative offspring. I also review experimental evidence that provisioning males and females respond differently to the various elements of the begging display, and suggest three hypotheses to explain this curious observation.

INTRODUCTION

The arrival of an adult with food at the passerine nest typically provokes a frenzy of nestling begging activity. Nestlings reveal brightly coloured gapes (Pycraft 1907), and assume a range of begging postures (e.g. Redondo & Castro 1992). They also call loudly (e.g. Haskell 1999) and jostle for position in the nest (e.g. McRae et al. 1993). The evolution of such bizarrely excessive nestling begging displays has been attributed to conflicts of interest at the nest over the allocation of investment among dependent offspring (Trivers 1974; Parker & Macnair 1979; Godfray 1995a; Mock & Parker 1997). Such conflicts can arise because selection yields different optimal patterns of investment for parents and young (Hamilton 1964; Trivers 1974; Mock & Parker 1997). Within each brood, parents and young may be in conflict over the division of investment among offspring, with

J. Wright and M.L. Leonard (eds.), The Evolution of Begging, 87–106.

individual offspring seeking greater levels of investment than parents should be selected to provide (intrabrood conflict: Macnair & Parker 1979). Between broods there can be additional conflict, with offspring effectively demanding resources which parents would do best by allocating to future young (interbrood conflict: Trivers 1974; Parker & Macnair 1979; Lessells & Parker 1999). Extravagant begging may have evolved originally either to facilitate offspring attempts to extort additional resources from resistant parents (Trivers 1974) or through sibling competition for limited parental resources (Rodríguez-Gironés et al. 1996). Begging is thought to persist because it plays a key role in resolving parent-offspring conflict (Godfray 1995a), both within and between broods.

The specific function of begging in the resolution of parent-offspring conflict varies between models. In those concerned with food allocation during each nest visit, begging might be a means of scramble competition for food (Parker & Macnair 1979) or it might be a signal, honestly advertising need (Godfray 1995b). Similarly, begging may influence nest visit rate either by signalling offspring condition (Godfray 1991) or by blackmailing parents into providing food, thereby preventing further wasteful solicitation (Zahavi 1977; Parker & Macnair 1979; Eshel & Feldman 1991; Johnstone 1996a). Despite the range of possible functions, individual models typically treat begging intensity as a univariate parameter with a single function. However, even a brief glimpse at a nest crowded with nestlings gaping, posturing and calling indicates that this is likely to be an oversimplification.

This chapter suggests ways in which the complexity of nestling begging displays could generally serve to resolve parent-offspring conflict. I begin with an overview of the functions of the individual elements that make up the begging display before discussing more recent experiments looking at the collective function of the multiple elements. The chapter concludes by discussing why mothers and fathers react differently to complex begging displays. Aside from their role in resolving family conflicts, individual elements of begging displays may collectively enhance the efficacy and efficiency of communication at the nest (see A.G. Horn & M.L. Leonard this volume), but this will not be discussed further here.

WHY DO NESTLINGS HAVE COMPLEX BEGGING DISPLAYS?

Each Element Has a Discrete Function

An intuitively appealing idea is that multicomponent begging displays have evolved because each element is discretely involved in a different aspect of nestling solicitation behaviour. For example, nestling mouth colour might function to enhance nestling detectability (Pycraft 1907; Swynnerton 1916; Ingram 1920), while posturing may settle scramble competition for food at each nest visit and calling could determine nest visit rate (see Kacelnik et al. 1995). The conceptualization of the begging display in this way by empiricists has probably been reinforced by theoreticians using separate models for dealing with detectability (Johnstone 1998), intrabrood conflict (e.g. Parker & Macnair 1979; Godfray 1995b) and interbrood conflict (e.g. Parker & Macnair 1979; Godfray 1991; but see Rodríguez-Gironés et al. 2001). Given its roots in empirical observation, it is easy to find support for this idea.

Nestling Mouths and Detectability

Evidence that gapes function to enhance nestling detectability dates as far back as 1920 when Ingram first pointed out that the gapes of cavity-nesting corvids are commonly adorned with broad white fleshy borders which are usually absent from open-nesting corvids. A much later study formalized this comparison with a wider range of species, finding a continuous relationship between nest light availability and relative border width (Kilner & Davies 1998). The same study also showed that the brightness contrast between the gape and its surrounding border was at its greatest in dark nests, again presumably to enhance conspicuousness (Kilner & Davies 1998).

When it comes to the colour of the gape itself, however, there is mixed evidence that it functions to enhance nestling detectability. In general, the cruder the measurement of colour and nest light availability, the clearer the patterns that emerge, an observation which may be biologically meaningful given the potential variation in nest lighting conditions within cup-nesting species (Kilner 1999). Classifying species as either yellow or red, Ficken (1965) found that cavity-nesting species were more likely to display yellow mouths. The result was replicated in a study using a colour ranking scheme to convert literature descriptions of mouth colour into numbers (Kilner 1999). However, when measuring mouth colours directly, using hue, saturation and brightness scores of video images, no relationship was found between photographic light meter scores of nest light availability and mouth colour (Kilner & Davies 1998). Whether or not nestling mouth colour functions to enhance detectability is unlikely to be resolved without measuring reflectance spectra of nestling mouths in conjunction with ambient nest light colour.

Nestling Posture and Food Distribution

Turning to the function of nestling begging posture, virtually every study that has looked for a positive relationship between postural intensity and food allocation has found one (e.g. Redondo & Castro 1992; Kacelnik et al. 1995; Kilner 1995; Leonard & Horn 1998). These results are, of course, simply correlational. It is not clear whether parents are responding to posture *per se*, or some close correlate like nestling height. Observations of sexually dimorphic species like red-winged blackbirds (*Agelaius phoeniceus*, Teather 1992) suggest the latter is more likely, as do studies of brood parasites reared alongside host young (Lichtenstein & Sealy 1998).

It is also unclear in many studies whether posture functions to advertise offspring condition, or to enhance competitive ability or both. Work on Arabian babblers (*Turdoides squamiceps*) suggests it plays an important role in scramble competition (Ostreiher 1997), while in captive canaries (*Serinus canaria*) posture can be a signal (Kilner 1995). In great tits (*Parus major*), parental response to posture may additionally be influenced by nestling position in the nest (Kölliker et al. 1998). Variation in nestling size appears further to complicate the relationship between postural begging intensity, nestling position and the likelihood of being fed in starlings (*Sturnus vulgaris*, Cotton et al. 1999). Postural begging could therefore be just one facet of nestling competitive ability, or it could be a signal which is amplified by a suite of non-signalling nestling attributes (Lotem et al. 1999) or it could be a combination of the two.

Begging Calls and Provisioning Rate

Experiments in which the calls produced by a brood are artificially supplemented have shown that nestling calling influences nest visit rate (e.g. von Haartman 1953; Muller & Smith 1978; Bengtsson & Rydén 1981; Ottosson et al. 1997). One criticism of this technique is that it is not always clear whether parents are responding to the experimental manipulation alone or the effect it has on the begging of their brood (Muller & Smith 1978; but see Kilner et al. 1999). Although the majority of experiments augmented brood vocalizations during nest visits alone, provisioning levels could also be elevated when calls were broadcast at other times (Burford et al. 1998; Price 1998; but see Clark & Lee 1998).

Multiple Elements and Multiple Functions

Drawing the evidence together, the empirical work appears at first sight to be overwhelmingly consistent with the hypothesis that each component of the display has a discrete function. A danger, though, is that such a conclusion could be merely an artefact of the simplest experimental and theoretical techniques for understanding begging. More recent experimental data suggest that in reality complex begging displays are likely to have complex functions. The discrete elements of begging displays could have multiple functions. For example, mouth colour may function to enhance detectability, but it can also signal hunger in some finch species (Kilner 1997; Kilner & Davies 1998), and perhaps health in barn swallows (*Hirundo rustica*, Saino et al. 2000). Similarly, the calling of yellow-headed blackbird (*Xanthocephalus xanthocephalus*) young is correlated with both food allocation (Price 1996) and the rate at which parents visit the nest with food (Price 1998). Alternatively, multiple display elements could have the same discrete function. In canaries, food allocation patterns within broods are correlated with postural intensity, but can be altered by broadcasting calls from speakers placed near individual offspring (Kilner 1996).

In summary, the hypothesis that each element has a discrete function is intuitively attractive, but probably simplistic. To understand the evolution of multicomponent begging displays, and their possible role in resolving parent-offspring conflict, we should instead consider how the different elements function together.

Multiple Elements Provide More Information

General theoretical analyses provide several alternative explanations for why multicomponent displays may have evolved. One possibility is that the separate elements of the display may interact to enhance parental response to the display. Experimental psychologists have shown with a variety of species that the magnitude of response to multicomponent displays often exceeds the summed response to the individual components alone (reviewed by Rowe 1999). Here, an accessory non-informative display might increase the response to a focal display (Rowe 1999). For example, in the case of complex begging, visual cues might enhance parental responsiveness to vocal cues, providing a powerful psychological tool for nestlings attempting to extort additional investment from resistant parents. It is possible that the white papillae that adorn the mouths of great-spotted cuckoo (*Clamator glandarius*) nestlings improve their chance of being fed by functioning in this manner (Soler et al. 1995).

Three further hypotheses explain how multicomponent displays could improve parental resistance to potential offspring manipulation by providing more information about nestling condition. For instance, the multiple elements may carry 'multiple messages' about nestling quality (Johnstone 1995, 1996b). Different parts of the display might be correlated with a different aspect of nestling condition. This hypothesis differs from the idea that each element has a discrete function because there need be no one-to-one relationship between the signal and nestling quality. The individual elements of the begging display could instead be correlated with multiple aspects of nestling quality.

Alternatively, it may be that the multiple elements advertise the same aspect of nestling condition, but are differently correlated with it. Here, multicomponent displays could function to provide more accurate information about nestling quality than individual components can alone. In other words, the separate elements of the display might function as 'back-up signals' (Johnstone 1996b).

Finally, it may be that there is redundancy in the display (Partan & Marler 1999). The separate elements might identically advertise the same aspect of nestling quality, so that collectively they provide no more information than each does alone. Redundancy could persist because of perceptual errors by parents, unable to make full use of information advertised in the discrete elements of the display.

Do Individual Begging Elements Interact to Increase Provisioning Rates?

Field experiments on reed warblers (*Acrocephalus scirpaceus*) tested the merit of these various hypotheses. Reed warblers typically rear four offspring in a cup nest slung between the stems of reeds growing in shallow water. Both parents provide care by brooding and feeding young, and the sexes are monomorphic. From two days after hatching, nestling reed warblers solicit food by displaying bright yellow gapes and calling repetitively. To test whether each element of the display could influence provisioning rate by parents, Kilner et al. (1999) carried out two types of experimental manipulation, using unmarked birds. The first involved changing the brood size, so that broods contained from one to eight nestlings, and hence manipulated both the visual and vocal elements of the display. The second manipulation altered the vocal display alone by broadcasting the calls of different numbers of nestlings at different brood sizes through a small speaker fixed on the rim of the nest. Combining the two manipulations relaxed the usual tight correlation between the visual and vocal displays, so that variance in provisioning rate could be partitioned to either element.

The experiments revealed that parents responded to both aspects of the display when feeding young at the nest; together visual and vocal begging displays explained more variance in provisioning rate than either could alone. Surprisingly, however, the two elements of the begging display did not interact to influence provisioning rate, as might be expected from the psychological literature (Rowe 1999). The exact reason for this result is unclear at present. One difficulty with interpreting the data is that the parental response to begging measured in the experiments combined the behaviour of both parents. It may be that differences between mothers and fathers in their response to the begging display masked more subtle interactions between the elements upon their separate provisioning behaviours.

Do Complex Displays Contain 'Back-Up Signals' or Carry 'Multiple Messages'?

Subsequent laboratory experiments examined the information content of the display. Broods of four nestlings were temporarily borrowed from their natural nest and taken to the laboratory, either three to four days after hatching or six to seven days after hatching. There they were fed until they stopped begging and placed in a heated artificial nest. After ten minutes, and every ten minutes for the next 110 minutes, the brood was induced to beg and the visual and vocal displays were measured. At the end of the experiment, nestlings were fed and returned to their natural nests. Analysis of the experimental begging data supported the suggestion that multicomponent displays carry multiple messages. As anticipated by theory, the visual and vocal elements of the begging display did not broadcast separate information about nestling condition, but were individually correlated both with the extent of food deprivation and the age of the nestlings (Kilner et al. 1999).

The experiments also indicated that the different elements of the display could function as back-up signals. Visual and vocal displays were differently related to nestling age and food deprivation, and together they explained more variance in nestling hunger than either could alone (Kilner et al. 1999). By responding to two signals rather than one, reed warbler parents might therefore gain more accurate information about their young, as well as information about the short- and long-term needs of their brood. In terms of interbrood conflict, perhaps parents have selected complex displays, so that they can adjust their investment precisely to the needs of their brood and hence avoid being manipulated by selfish offspring.

Theoretically, two alternative mechanisms may explain how multiple signalling equilibria can persist. One requires that the costs of the individual elements of the display are strongly accelerating (Johnstone 1996b), while the alternative view is that one or more of the signalling elements could be cost-free (Bergstrom & Lachmann 1998). A defining characteristic of cost-free signals is that they form pooling equilibria: the signal appears to be a step function of nestling condition. The signals might be cheap, but they convey less information (but see Johnstone 1999; Rodríguez-Gironés 1999). In contrast, costly displays yield separating equilibria with signals changing continuously with respect to nestling quality (Bergstrom & Lachmann 1998).

Close inspection of the begging display performed by individual reed warbler broods suggests that the latter mechanism may be the more plausible for explaining how complex begging displays persist (Figure 1). Figure 1a shows the change in the number of gapes displayed by a single brood with respect to increasing food deprivation, a plot that clearly resembles a step function. Although reed warbler parents respond to continuous increases in gape area (Kilner et al. 1999), and gape area is continuously correlated with nestling hunger (Kilner & Davies 1999; Kilner et al. 1999), at each nest visit this signal can only be a step function of brood condition.

Figure 1b shows the equivalent change in call rate which, by contrast, appears to change more continuously. Thus, call rate has the characteristics of a costly signal while gaping bears the hallmark of a cost-free display. Experimental evidence from other species supports the suggestion that call rate is costly. When western bluebird (*Sialia mexicanus*) begging calls were broadcast at a fast rate from fake nests placed on the ground and baited with eggs, more eggs were taken than when calls were broadcast at a slower rate (Haskell 1994). Perhaps, generally, begging calls were selected by parents to supplement the information provided by the cost-free, yet relatively uninformative, gaping display.

Parasitism by Cuckoos – A Special Case of Redundancy in Complex Displays?

The reed warblers tested in these experiments occasionally succumbed to brood parasitism by the common cuckoo (*Cuculus canorus*, see Brooke et al. 1998 for parasitism rates at the study site during the last ten years). The cuckoo egg requires less incubation than the reed warbler's own eggs, so often hatches in advance of the host brood. Within a few hours of hatching, while still tiny, naked and blind, the cuckoo nestling sets about destroying the reed warbler's unhatched eggs and newly-hatched young. It manoeuvres them into the small of its back and then climbs backwards up the side of the nest to tip them over the rim. The cuckoo thus dispenses with sibling competition and, reared alone in the nest, becomes the sole beneficiary of parental care.

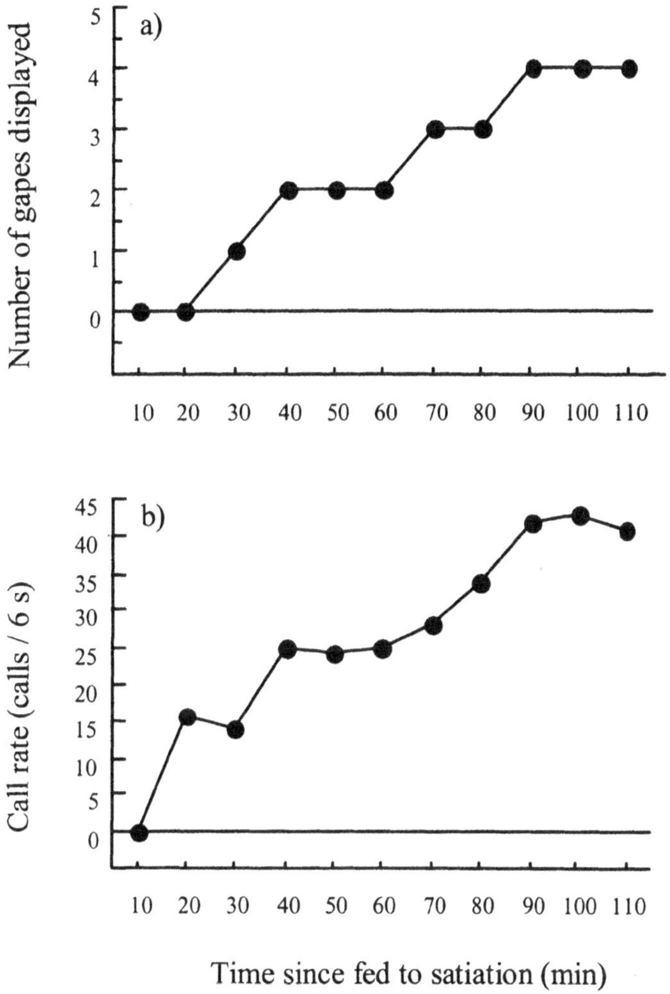

Figure 1. The effect of food deprivation on a) the visual and b) vocal begging displays of a single brood of four reed warblers, six to seven days after hatching.

Kilner et al.'s (1999) work, however, suggests that nestling evicting behaviour by cuckoo nestlings could also incur a cost. Alone in the nest, the cuckoo presents a deficient visual stimulus of a single gape. To elicit adequate care from its reed warbler hosts, the cuckoo must compensate for this visual discrepancy by calling excessively (Kilner et al. 1999). Nevertheless, despite calling at up to twice the rate of an average host brood, and even though its selfishness is unconstrained by kinship, the cuckoo extracts roughly the same level of care as a brood of host young, for the same level of need (Kilner & Davies 1999). By chance, the complex begging displays of hosts could therefore also function to limit the selfishness of brood parasites which evict and, in some circumstances, might even prevent successful parasitism.

The begging display of common cuckoo nestlings incorporates the display of a vivid orange-red gape and production of a highly repetitive begging call. Comparative analyses indicate that other cuckoo species also have redder mouths than their hosts (Kilner 1999). What then is the function of such a red gape? Kilner et al.'s (1999) work showed precisely how the gape area displayed and the rate of call production together could account for much of the variance in cuckoo nestling provisioning rate. However, further experiments (Noble et al. 1999) suggest that the colour of the cuckoo's gape is functionally redundant in soliciting care from reed warbler hosts. When the yellow gapes of reed warbler young were painted with orange-red food colouring, there was no change in the rate at which adults provisioned the brood. Since three species of non-parasitic cuckoos also display red gapes, one parsimonious interpretation of these experimental data is that cuckoo mouth colour confers no adaptive advantage to brood parasites and their redder gapes are simply the by-product of phylogeny (Kilner 1999).

A more intriguing possibility is that a red mouth may be part of the trickery used by brood parasites to dupe some, but not all, of their hosts. The common cuckoo, for example, parasitizes two other hosts in Britain: the dunnock (*Prunella modularis*) which has orange-mouthed nestlings and the meadow pipit (*Anthus pratensis*) whose nestlings display flesh-red mouths (Kilner & Davies 1998). Adult females faithfully parasitize one host species and form distinct 'races' or gens, each recognizable by egg patterning which usually mimics the eggs of the particular host (Brooke & Davies 1988). In contrast, adult males mate promiscuously across the different female races (Marchetti et al. 1998). Egg mimicry persists, nonetheless, presumably because it is inherited maternally and expressed only by females. Mimicry of host nestling appearance or behaviour, however, may be impossible (Marchetti et al. 1998).

In reed warbler nests, the cuckoo nestling elicits sufficient care from its hosts by exploiting the usual communication system between host young and adults, relying on an exaggerated call to compensate for a deficient visual stimulus of a single gape (Davies et al. 1998; Kilner et al. 1999). Different host species, however, vary in the details of their begging display (Kilner & Davies 1999). To elicit care successfully from its various hosts, perhaps the cuckoo nestling has to perform a variety of begging tricks, some of which may be essential in one host, but quite useless in others. The different provisioning rules of the cuckoo's various hosts may mean that the cuckoo nestling has to resort to vocal trickery with some, but visual trickery with others. A red mouth may be vital for duping meadow pipit hosts, for example, but irrelevant for fooling reed warbler hosts.

Whatever the species of brood parasite, provided that male and female cuckoos are reared by different hosts, there can never be perfect mimicry of host begging displays (Davies 2000), and selection for redundancy in parasitic nestling begging displays may result. It is this possibility of redundancy which might explain generally why red cuckoo nestling mouths persist even though they appear functionally useless for extracting care from some hosts (Kilner 1999).

Multiple Elements Are Required to Maintain Reliable Signalling Throughout the Nestling Period

Experiments with captive canaries provide a different explanation for the evolution of multicomponent begging displays. In this species, two signals that play a key role in influencing food allocation among the brood are begging postures and begging calls. Experiments have shown that both elements indicate nestling hunger (Kilner 1995, 1996). Parents actively select nestlings to provision and the amount of food transferred is correlated with the intensity of the nestling's postural display (Kilner 1995). Playback manipulations have further shown that food allocation is influenced by begging calls (Kilner 1996).

A Marginal Growth Cost of Postural Begging

Recent work on canaries has found that postural begging incurs a marginal growth cost (Kilner 2001). The marginal cost of begging was measured experimentally at three different stages during the nestling period. Pairs of siblings were removed from the nest, weighed and kept in an incubator. During the next six hours, nestlings were fed every 20 minutes with the same amount of food, but had to beg for markedly different amounts of time before they were rewarded, which were nevertheless within the range typically observed at the nest. (Nestlings in the low begging treatment solicited for 10 seconds while their siblings in the high begging treatment had to beg for 60 seconds.) After six hours of being fed on one of these regimes, nestlings were weighed and returned to their nest. All the food given to nestlings during the experiments was weighed, as were the faecal sacs produced during the experiment. The net difference in weight between nestlings at the start and end of the test could therefore be attributed only to mass lost through energetic expenditure. In addition, nestlings were filmed to quantify the difference in postural begging intensity experienced by siblings during the experimental manipulation. The marginal cost of begging was measured by comparing the effect of the treatments on siblings.

Three separate lines of evidence revealed a trade-off between energy devoted to growth and energy spent on postural begging. First, excessive begging slowed growth, both during the experiment and in the subsequent 24 hours, and the impact of the manipulation was greatest in nestlings with the highest potential rate of daily mass gain. Second, the greater the difference in postural begging intensity between siblings during the experiment, the greater the difference between them in the mass lost as a result of metabolic expenditure. Third, the treatment had least effect on mass gain, during the subsequent 24 hours, in older nestlings that had completed most growth (Kilner 2001).

At first sight, this result appears to stand at odds with the metabolic evidence that the energetic cost of begging is relatively small (M.A. Chappell & G.C. Bachman this volume). However, canary begging displays are typically two or three times longer than those of tree swallows (*Tachycineta bicolor*) or house wrens (*Troglodytes aedon*) and can continue for over 60 seconds. Protracted displays may additionally incur anaerobic costs as a result of sustained muscle contraction. Even so, despite their briefer displays, the energy devoted to house wren begging may still be traded off against energy for growth. Recalculating data presented in Bachman and Chappell (1998), the greater the proportion of the daily energy budget that is devoted to growth, the less that is spent on begging (Kilner 2001).

The Cost of Postural Begging Declines with Nestling Age

The growth rate of canary nestlings is correlated with a measure of fitness, namely the likelihood of survival to independence. Logistic regression analyses of the growth rates, and subsequent fates, of over 300 nestlings yielded equations relating mass and mass gain at different ages to the probability of survival to independence. By substituting the experimental data into these logistic regression equations, the decrease in the probability of survival imposed by the high begging treatment can be estimated. Translated into estimates of fitness, the excessive begging treatment proved most costly for younger nestlings, reducing the chance of survival to independence by roughly 5%. For older nestlings, prolonged begging carried virtually no equivalent cost.

The Reliability of Posturing Declines with Nestling Age

The change in the marginal fitness cost of begging with age fortuitously provided the chance to test the theoretical idea that the reliability of nestling begging should be maintained by the cost of the display (Godfray 1991, 1995b). According to this view, a decrease in the cost of signalling should precipitate a decrease in the reliability of signalling. As nestlings get older, theory predicts that their postural displays become more and more unreliable.

The reliability of begging was measured by regressing nestling hunger on postural intensity to calculate the standardized regression coefficient *R*. This statistic quantifies the strength of the relationship between hunger and begging: the weaker the relationship, the less reliable the information that begging conveys about nestling hunger. By measuring *R* at five different stages of the nestling period it was found that the reliability of the postural display declined with increasing nestling age (R.M. Kilner unpublished data). Just as predicted by theory, the decline in the cost of posturing was linked with a decline in the reliability of posturing.

If the postural display becomes increasingly unreliable as nestlings get older, parents would benefit by ignoring it in favour of a more informative signal. Detailed observations revealed that the influence of posturing on food allocation by mothers changed during the course of the nestling period. Maternal attentiveness to postural begging was measured by regressing the number of feeds transferred to nestlings on the strength of their display to calculate *R*: the less attentive mothers were to the display, the lower the measure of *R*. These analyses revealed that as nestlings grew older, mothers became less and less responsive to the postural display (R.M. Kilner unpublished data).

Begging Calls Provide an Alternative Source of Reliable Information?

Although the postural display became less reliable as nestlings grew older, mothers were not fooled by their increasingly deceptive offspring. One possibility is that they turned to a different element of the display for reliable information about the condition of their young. A likely candidate is the vocal display. As begging posture decreased in reliability with nestling age, begging calls became increasingly reliable (R.M. Kilner unpublished data).

Exactly why calling should become more reliable with nestling age remains unclear at this point. One possibility is that young nestlings are physically constrained in their ability to call loudly or rapidly, much as the frequency of toad (*Bufo bufo*) croaking is constrained by body size (Davies & Halliday 1978). Perhaps the nature of these physical constraints changes as nestlings get older. Alternatively, it may be that the cost of calling increases with age. Playback experiments using western bluebird calls have demonstrated a marginal predation cost of call rate (see above; Haskell 1994), and the more rapid calling rate of older canary nestlings (R.M. Kilner unpublished data) may therefore incur a greater risk of predation. Perhaps, then, the array of signals acquired during the nestling period successively imposes sufficient costs to maintain reliable signalling throughout this time. Whatever the mechanism enforcing reliability in begging calls, perhaps mothers have selected vocal begging to reduce their vulnerability to exploitation through visual begging by increasingly deceptive young.

WHY DO ADULT MALES AND FEMALES REACT DIFFERENTLY TO COMPLEX BEGGING DISPLAYS?

A curious feature of the canary work described above is that the paternal response to begging was quite different from that displayed by mothers (R.M. Kilner unpublished data; see also Lessells in press). Detailed work on two other species has also revealed sex differences in the response to individual elements of the begging display.

Using brood size manipulations combined with playback of different numbers of nestlings calling, the visual and vocal displays produced by broods of great tits were experimentally manipulated and the response of both parents measured (C.A. Hinde unpublished data). It was found that females integrated both elements of the display to determine their brood provisioning rates. In contrast, males responded only to the visual display in terms of the number of gapes displayed. These results replicate Kölliker et al.'s (2000) findings that, whereas female great tits elevate their feeding rates in response to begging calls broadcast at the nest, males do not.

Experiments with the cooperatively breeding superb fairy wren (*Malurus cyaneus)* have found the opposite result. In this species, single females can be assisted in nestling provisioning by one to four males. Observations show that females increase their provisioning rate in relation to brood age, while males provision at a constant rate throughout the nestling period (Dunn & Cockburn 1996). The vocal display of the brood increases in intensity as nestlings grow. Macgregor (2000) broadcast the calls of older wren broods while males or females fed nestlings in the nest, thus more than doubling the vocal display of the brood. Males responded by increasing their rate of provisioning. Females, however, showed no change at all in the rate at which they delivered food to their young, either when playback was directed exclusively at them, or when all provisioning adults were exposed to additional calls. The results of the playback experiment at first appear to conflict with the observational data. Macgregor, however, suggests that females were simply not fooled by the noisy broods created in the playback treatment. Males, by contrast, alarmed at the female's neglect of the apparently starving brood, boosted their provisioning rate. Perhaps in this species there is a strong interaction between the visual and vocal begging displays on female provisioning rate, so that females cannot be made to work harder by manipulating the begging calls alone. If males respond to vocal begging alone, they would be more easily fooled by the playback experiment.

Exactly why males and females should respond differently to complex begging displays is not easily explained by current theory. I conclude by suggesting three tentative hypotheses, which are not mutually exclusive.

Maternal Investment Is More Costly

One possibility is that the sex differences stem from differences in male and female life history strategies. Some evidence suggests that the costs of care may be greater for females than for males (e.g. Nur 1988) which may select for greater precision in the way that females allocate care between broods. Under selection to allocate investment precisely females may, in turn, utilize complex begging displays as a means of deriving more accurate information about the state of their brood. By contrast, rather than trading off current and future reproductive success, males may be selected to balance rearing young with the opportunity for mating (e.g. Magrath & Elgar 1997). Or they may adjust their level of care to minimize the penalties of investment for females, thereby retaining a fecund mate for future reproduction (Lessells & Parker 1999). Either way, males need only respond to relatively crude indicators of offspring quality, such as gaping.

Cryptic Sexual Conflict

Alternatively, it is possible that a cryptic form of sexual conflict underlies the differences in provisioning rules exhibited by the sexes. When females engage in extra-pair copulations, different paternal genomes end up competing for maternal care within each brood. Just as male mice (*Mus musculus*) attempt to manipulate females *via* the hormonal secretions of their unborn offspring (Haig & Graham 1991), so male birds might manipulate their partners *via* the signalling of their nestlings.

Superb fairy wrens exhibit the highest known rates of extra-group paternity, with around half the broods fathered entirely by extra-group males, and the remaining broods showing mixed paternity (Mulder et al. 1994). Within each brood, therefore, multiple paternal genomes typically vie for maternal attention. Perhaps female fairy wrens do not respond to vocal begging alone because offspring calls are a paternally inherited device designed to manipulate the mother into providing more care than is optimal for her. Female wrens might resist paternal manipulation by responding only to an appropriate combination of visual and vocal displays.

A Mechanism for Incomplete Compensation

Houston and Davies (1985) analysed a provisioning game in which two parents decide how hard each must work to feed the brood. They showed that an ESS can arise when one parent only partially compensates for a decrease in provisioning by the other. It is possible that sex differences in provisioning rules might provide a mechanism for resolving sexual conflict over nestling provisioning in the manner that they predict.

The various elements of the begging display can change differently with respect to nestling hunger (Kilner et al. 1999). If mothers and fathers respond to different components of the display they may receive different information about the state of the brood (see also Kölliker et al. 1998). A change in provisioning rate by one parent, and the consequent change in brood begging behaviour, might therefore provoke only a partially compensating change in provisioning by the other parent, as envisaged by the Houston and Davies (1985) model (see Wright & Cuthill 1989; MacNamara et al. 1999; Wright & Dingemanse 1999; but see Sanz et al. 2000).

For example, suppose that males adjust their provisioning in relation to the gape area displayed by the brood, whereas females respond to a mixture of visual and vocal displays, as has been discovered for the great tit. If the female independently slows her provisioning rate, the brood will become more hungry. Initially nestlings may call more rapidly, but males will be insensitive to this change in the begging display and so, at first, will not change their rate of provisioning at all, despite the decrease in maternal feeding rates. Only when the brood has become sufficiently hungry for more nestlings to gape at each nest visit will the male increase the rate at which he supplies food to the brood, even then only partly compensating for the reduction in female provisioning.

FUTURE DIRECTIONS

The emphasis of this review has been on the information content of begging displays, simply because that has been the focus of the few theoretical and experimental studies to investigate the function of complex solicitation displays. Multicomponent begging displays need not have evolved only for their signalling properties. At this stage, it is just as likely that they could enable scramble competition, or blackmail parents into providing food. For example, if the elements of the display are more costly together than they are individually, complex displays may increase the potential for nestlings to blackmail parents for resources. One challenge for future work is to develop a more realistic theoretical framework yielding experimentally testable predictions, which can distinguish between these different possibilities. Only then can we unravel the complexities of nestling begging displays.

ACKNOWLEDGEMENTS

I wrote this chapter as a Royal Society Dorothy Hodgkin Research Fellow, sponsored by the Wolfson Foundation. I thank The Royal Society, The Wolfson Foundation, The British Ecological Society, BBSRC, NERC and Magdalene College, Cambridge for financially supporting the research described here, and C.A. Hinde, R.A. Johnstone, M.L. Leonard and J. Wright for comments on the chapter. C.A. Hinde kindly allowed me to discuss her unpublished data here.

REFERENCES

Bachman, G.C. & Chappell, M.A. 1998. The energetic cost of begging behaviour in nestling house wrens. *Animal Behaviour* **55**, 1607-1618.

Bengtsson, H. & Rydén, O. 1981. Development of parent-young interaction in asynchronously hatched broods of altricial birds. *Zeitschrift für Tierpsychologie* **56**, 255-272.

Bergstrom, C.T. & Lachmann, M. 1998. Signaling among relatives III. Talk is cheap. *Proceedings of the National Academy of Sciences USA* **95**, 5100-5105.

Brooke, M. de L. & Davies, N.B. 1988. Egg mimicry by cuckoos *Cuculus canorus* in relation to discrimination by hosts. *Nature* **335**, 630-632.

Brooke, M. de L., Davies, N.B. & Noble, D.G. 1998. Rapid decline of host defences in response to reduced cuckoo parasitism: behavioural flexibility of reed warblers in a changing world. *Proceedings of the Royal Society of London, Series B* **265**, 1277-1282.

Burford, J.E., Friedrich, T.J. & Yasukawa, K. 1998. Response to playback of nestling begging in the red-winged blackbird, *Agelaius phoeniceus*. *Animal Behaviour* **56**, 555-561.

Clark, A.B. & Lee, W.-H. 1998. Red-winged blackbird females fail to increase feeding in response to begging call playbacks. *Animal Behaviour* **56**, 563-570.

Cotton, P.A., Wright, J. & Kacelnik, A. 1999. Chick begging strategies in relation to brood size hierarchies and hatching asynchrony. *American Naturalist* **153**, 412-420.

Davies, N.B. 2000. *Cuckoos, Cowbirds and Other Cheats*. London: T & AD Poyser.

Davies, N.B. & Halliday, T.R. 1978. Deep croaks and fighting assessment in toads *Bufo bufo*. *Nature* **274**, 145-169.

Davies, N.B., Kilner, R.M. & Noble, D.G. 1998. Nestling cuckoos, *Cuculus canorus*, exploit hosts with begging calls that mimic a brood. *Proceedings of the Royal Society of London, Series B* **265**, 673-678.

Dunn, P.O. & Cockburn, A. 1996. Evolution of male parental care in a bird with almost complete cuckoldry. *Evolution* **50**, 2542-2548.

Eshel, I. & Feldman, M.W. 1991. The handicap principle in parent-offspring conflict: comparison of optimality and population genetic analysis. *American Naturalist* **137**, 167-185.

Ficken, M.S. 1965. Mouth color of nestling passerines and its use in taxonomy. *Wilson Bulletin* **77**, 71-75.

Godfray, H.C.J. 1991. Signalling of need by offspring to their parents. *Nature* **352**, 328-330.

Godfray, H.C.J. 1995a. Evolutionary theory of parent-offspring conflict. *Nature* **376**, 133-138.

Godfray, H.C.J. 1995b. Signaling of need between parents and young: parent-offspring conflict and sibling rivalry. *American Naturalist* **146**, 1-24.

Haig, D. & Graham, C. 1991. Genomic imprinting and the strange case of the insulin-like growth factor. *Cell* **64**, 1045-1046.

Hamilton, W.D. 1964. The genetical evolution of social behaviour. *Journal of Theoretical Biology* **7**, 1-52.

Haskell, D.G. 1994. Experimental evidence that nestling begging behaviour incurs a cost due to nest predation. *Proceedings of the Royal Society of London, Series B* **257**, 161-164.

Haskell, D.G. 1999. The effect of predation on begging-call evolution in nestling wood warblers. *Animal Behaviour* **57**, 893-901.

Houston, A.I. & Davies, N.B. 1985. The evolution of cooperation and life history in the dunnock *Prunella modularis*. In: *Behavioural Ecology: Ecological Consequences of Adaptive Behaviour* (Ed. by R.M. Sibly & R.H. Smith). Oxford: Blackwell Scientific Publications.

Ingram, C. 1920. A contribution to the study of nestling birds. *Ibis* **2**, 856-880.

Johnstone, R.A. 1995. Honest advertisement of multiple qualities using multiple signals. *Journal of Theoretical Biology* **177**, 87-94.

Johnstone, R.A. 1996a. Begging signals and parent-offspring conflict: do parents always win? *Proceedings of the Royal Society of London, Series B* **263**, 1677-1681.

Johnstone, R.A. 1996b. Multiple displays in animal communication: 'backup signals' and 'multiple messages'. *Philosophical Transactions of the Royal Society of London, Series B* **351**, 329-338.

Johnstone, R.A. 1998. Efficacy and honesty in communication between relatives. *American Naturalist* **152**, 45-48.

Johnstone, R.A. 1999. Signaling of need, sibling competition and the cost of honesty. *Proceedings of the National Academy of Sciences USA* **96**, 12644-12649.

Kacelnik, A., Cotton, P.A., Stirling, L. & Wright, J. 1995. Food allocation among nestling starlings: sibling competition and the scope of parental choice. *Proceedings of the Royal Society of London, Series B* **259**, 259-263.

Kilner, R. 1995. When do canary parents respond to nestling signals of need? *Proceedings of the Royal Society of London, Series B* **260**, 343-348.

Kilner, R.M. 1996. Parental investment in canaries and zebra finches. PhD thesis, University of Cambridge.

Kilner, R. 1997. Mouth colour is a reliable signal of need in begging canary nestlings. *Proceedings of the Royal Society of London, Series B* **264**, 963-968.
Kilner, R.M. 1999. Family conflicts and the evolution of nestling mouth colour. *Behaviour* **136**, 779-804.
Kilner, R.M. 2001. A growth cost of begging in captive canary chicks. *Proceedings of the National Academy of Sciences USA* 98, 11394-11398.
Kilner, R.M. & Davies, N.B. 1998. Nestling mouth colour: ecological correlates of a begging signal. *Animal Behaviour* **56**, 705-712.
Kilner, R.M. & Davies, N.B. 1999. How selfish is a cuckoo chick? *Animal Behaviour* **58**, 797-808.
Kilner, R.M., Noble, D.G. & Davies, N.B. 1999. Signals of need in parent-offspring communication and their exploitation by the cuckoo. *Nature* **397**, 667-672.
Kölliker, M., Richner, H., Werner, I. & Heeb, P. 1998. Begging signals and biparental care: nestling choice between parental feeding locations. *Animal Behaviour* **55**, 215-222.
Kölliker, M., Brinkhof, M.W.G., Heeb, P., Fitze, P.S. & Richner, H. 2000. The quantitative genetic basis of offspring solicitation and parental response in a passerine bird with biparental care. *Proceedings of the Royal Society of London, Series B* **267**, 2127-2132.
Leonard, M.L. & Horn, A.G. 1998. Need and nestmates affect begging in tree swallows. *Behavioral Ecology and Sociobiology* **42**, 431-436.
Lessells, C.M. in press. Parentally-biased favouritism: why should parents specialize in caring for different offspring? *Philosphical Transactions of the Royal Society of London, Series B.*
Lessells, C.M. & Parker, G.A. 1999. Parent-offspring conflict: the full-sib - half-sib fallacy. *Proceedings of the Royal Society of London, Series B* **266**, 1637-1643.
Lichtenstein, G. & Sealy, S.G. 1998. Nestling competition, rather than supernormal stimulus, explains the success of parasitic brown-headed cowbird chicks in yellow warbler nests. *Proceedings of the Royal Society of London, Series B* **265**, 249-254.
Lotem, A., Wagner, R.H. & Balshine-Earn, S. 1999. The overlooked signaling component of nonsignaling behavior. *Behavioral Ecology* **10**, 209-212.
Macgregor, N.A. 2000. Nestling begging and parental response in the superb fairy wren *Malurus cyaneus*. Honours thesis, Australian National University.
Macnair, M. & Parker, G.A. 1979 Models of parent-offspring conflict. III. Intrabrood conflict. *Animal Behaviour* **27**, 1202-1209.
MacNamara, J.M., Gasson, C.E., & Houston, A.I. 1999. Incorporating rules for responding into evolutionary games. *Nature* **401**, 368-371.
Magrath, M.J.L. & Elgar, M.A. 1997. Paternal care declines with increased opportunity for extra-pair matings in fairy martins. *Proceedings of the Royal Society of London, Series B* **264**, 1731-1736.
Marchetti, K., Nakamura, H. & Gibbs, H.L. 1998. Host-race formation in the common cuckoo. *Science* **282**, 471-472.
McRae, S.B., Weatherhead, P.J. & Montgomerie, R. 1993. American robin nestlings compete by jockeying for position. *Behavioral Ecology and Sociobiology* **33**, 101-106.
Mock, D.W. & Parker, G.A. 1997. *The Evolution of Sibling Rivalry*. Oxford: Oxford University Press.
Mulder, R.A., Dunn, P.O., Cockburn, A., Lazenby-Cohen, K.A. & Howell, M.J. 1994. Helpers liberate female fairy-wrens from constraints on extra-pair mate choice. *Proceedings of the Royal Society of London, Series B* **255**, 223-229.
Muller, R.E. & Smith, D.G. 1978. Parent-offspring interactions in zebra finches. *The Auk* **95**, 485-495.
Noble, D.G., Davies, N.B., Hartley, I.R. & McRae, S.B. 1999. The red gape of the nestling cuckoo (*Cuculus canorus*) is not a supernormal stimulus for three common cuckoo hosts. *Behaviour* **136**, 759-777.

Nur, N. 1988. The consequences of brood size for breeding blue tits. III. Measuring the cost of reproduction: survival, future fecundity and differential dispersal. *Evolution* **42**, 351-362.

Ostreiher, R. 1997. Food division in the Arabian babbler nest: adult choice or nestling competition? *Behavioral Ecology* **8**, 233-238.

Ottosson, U., Backman, J. & Smith, H.G. 1997. Begging affects parental effort in the pied flycatcher, *Ficedula hypoleuca*. *Behavioral Ecology and Sociobiology* **41**, 381-384.

Parker, G.A. & Macnair, M.R. 1979. Models of parent-offspring conflict IV. Suppression: evolutionary retaliation by the parent. *Animal Behaviour* **27**, 1210-1235.

Partan, S. & Marler, P. 1999. Behavior - communication goes multimodal. *Science* **283**, 1272-1273.

Price, K. 1996. Begging as competition for food in yellow-headed blackbirds. *The Auk* **113**, 963-968.

Price, K. 1998. Benefits of begging for yellow-headed blackbird nestlings. *Animal Behaviour* **56**, 571-577.

Pycraft, W.P. 1907. Nestling birds, and some of the problems they present II. *British Birds* **1**, 129-132.

Redondo, T. & Castro, F. 1992. Signalling of nutritional need by magpie nestlings. *Ethology* **92**, 193-204.

Rodríguez-Gironés, M.A. 1999. Sibling competition stabilizes signalling resolution models of parent-offspring conflict. *Proceedings of the Royal Society of London, Series B* **266**, 2399-2402.

Rodríguez-Gironés, M.A., Cotton, P.A. & Kacelnik, A. 1996. The evolution of begging: signaling and sibling competition. *Proceedings of the National Academy of Sciences USA* **93**, 14637-14641.

Rodríguez-Gironés, M.A., Enquist, M. & Lachmann, M. 2001. Role of begging and sibling competition in foraging strategies of nestlings. *Animal Behaviour* **61**, 733-745.

Rowe, C. 1999. Receiver psychology and the evolution of multicomponent signals. *Animal Behaviour* **58**, 921-931.

Saino, N., Ninni, P., Calza, S., Martinelli, R., de Benardi, F. & Møller, A.P. 2000. Better red than dead: carotenoid-based mouth coloration reveals infection in barn swallow nestlings. *Proceedings of the Royal Society of London, Series B* **267**, 57-61.

Sanz, J.J., Krananberg, S. & Tinbergen, J.M. 2000. Differential response by males and females to manipulation of partner contribution in the great tit (*Parus major*). *Journal of Animal Ecology* **69**, 74-84.

Soler, M., Martinez, J.G., Soler, J.J. & Møller, A.P. 1995. Preferential allocation of food by magpies *Pica pica* to great spotted cuckoo *Clamator glandarius* chicks. *Behavioral Ecology and Sociobiology* **37**, 7-13.

Swynnerton, C.F.M. 1916. On the coloration of the mouths and eggs of birds. I. The mouths of birds. *Ibis* **4**, 264-294.

Teather, K.L. 1992. An experimental study of competition for food between male and female nestlings of the red-winged blackbird. *Behavioral Ecology and Sociobiology* **31**, 81-87.

Trivers, R. 1974. Parent-offspring conflict. *American Zoologist* **14**, 249-264.

von Haartman, L. 1953. Was reizt den Trauerfliegenschnapper (*Muscicapa hypoleuca*) zu füttern? *Vogelwarte* **16**, 157-164.

Wright, J. & Cuthill, I. 1989. Manipulation of sex differences in parental care. *Behavioral Ecology and Sociobiology* **25**, 171-181.

Wright, J. & Dingemanse, N.J. 1999. Parents and helpers compensate for experimental changes in the provisioning effort of others in the Arabian Babbler. *Animal Behaviour* **58**, 345-350.

Zahavi, A. 1977. Reliability in communication systems and the evolution of altruism. In: *Evolutionary Ecology* (Ed. by B. Stonehouse & C. Perrins). London: MacMillan.

6.

THE SIBLING NEGOTIATION HYPOTHESIS

Alexandre Roulin
Department of Zoology, University of Cambridge, Cambridge CB2 3EJ, UK
(ra241@hermes.cam.ac.uk)

ABSTRACT

I propose that when siblings strongly differ in need, in the absence of parents they signal to each other their willingness to compete for non-divisible food, provided upon the parent's return. A needy individual signals to its siblings that it will vigorously contest the impending food resources in order to deter siblings from competing when parents return to the nest, thus ensuring that it will be fed without having to beg too intensely. In contrast, since less needy siblings have little chance of being fed, they may expect little reward from investment in sibling competition. They should refrain from signalling to siblings, therefore indicating that they will retreat from sibling competition. This would allow them to avoid wasting energy in negotiation, competitive behaviour and begging. Sibling negotiation should allow nestlings to invest effort optimally in competitive begging when food resources are non-divisible.

INTRODUCTION

The observation that offspring communicate with their parents using apparently costly begging signals has prompted numerous studies on the begging behaviour of nestling birds. Offspring are supposed to produce expensive begging displays because parents require from them an honest signal of their food requirements. Such signals may allow parents to adjust feeding rate in relation to short-term variation in offspring need, but also to allocate food to the offspring with the greatest fitness return (Godfray 1991, 1995). Given that parents preferentially feed those offspring that beg most intensely (Mondloch 1995; Hofstetter & Ritchison 1998), each nestling is

J. Wright and M.L. Leonard (eds.), The Evolution of Begging, 107–126.

expected to compete and escalate begging to convince parents that it is the neediest and to monopolize a larger share of the resources (Godfray 1995).

Although tremendously valuable, theoretical tests of the hypothesis that begging is an evolutionarily stable signal of need suffer from the drawback of analysing parent-offspring interactions within a single parental feeding visit (Godfray 1991, 1995; but see Johnstone 1996). The omission of parent-offspring, as well as sib-sib, interactions over an expanded period of time can have profound effects on the way nestlings are actually predicted to adjust begging effort, but also on the way parents are predicted to respond to offspring solicitations (Johnstone 1996). In situations where parents and offspring are allowed to interact only once, the value of food resources that parents bring to the nest is high. For this reason, nestlings should escalate begging when competing with each other. However, under the more realistic assumption that family members interact on several occasions, and that successive interactions are not independent of each other, nestlings may behave differently. For instance, when sibling competition is intense and entails a significant cost, some nestlings within a brood may achieve higher fitness by retreating from sibling contests over non-divisible resources until the probability of being fed increases (Roulin et al. 2000). Since broods are composed of nestlings that differ in hunger, motivation to compete intensely should differ between individuals, but also vary between feeding visits.

In this context, nestlings may assess the likelihood of monopolizing a non-divisible food item and thereby adjust investment in competitive behaviour at the appropriate level. Given a particular food requirement and probability of succeeding in sibling competition, a nestling may decide at which parental visit it should enter into or retreat from that competition. When the net benefit of delaying feeding time is greater than the net benefit of entering into sibling competition, individuals that are momentarily outcompeted would achieve higher fitness if they retreat from the current contest. The energy saved may be used in a subsequent contest once the benefit of competing increases. Therefore, individuals may temporarily refrain from competing until the probability of succeeding in sibling competition becomes non-negligible. In such a situation, adjustment of competitive begging (referred to as signals aimed at influencing within-brood parental food distribution, but not parental feeding rate) becomes strategic and implies that nestlings have the capacity to monitor variation in the probability of winning a contest. Empirical evidence for such ability has been provided for the barn owl (*Tyto alba*), in which nestlings vocalize all night in the absence of parents to inform siblings that they will contest the next food delivery (Roulin et al. 2000).

In this chapter, I develop the hypothesis that nestlings not only signal need to parents, but also to siblings, as a means of settling sibling contests over impending food. This hypothesis is referred to as the 'Sibling

Negotiation Hypothesis' (Roulin et al. 2000). The first part of the chapter is devoted to how and when sibling negotiation may take place, the second part to observational and experimental tests using the barn owl.

DEVELOPING THE SIBLING NEGOTIATION HYPOTHESIS

Modes of Sibling Competition and Definition of Sibling Negotiation

To date, three types of within-family interactions have been proposed.

Pure honest signalling defines the situation where nestlings adjust their begging level to their need and not to that of their siblings, implying that a high level of sibling competition does not escalate begging (Cotton et al. 1996). Since begging is assumed to signal need honestly, parents are expected to preferentially allocate food to the offspring that beg most intensely (Godfray 1995; Kilner 1995; Ostreiher 1997). Parents are also expected to adjust feeding rate in relation to the degree to which offspring beg, because the fitness return from feeding a given nestling will be correlated with its begging level (Godfray 1991; Burford et al. 1998; Davies et al. 1998; Price 1998; but see Clark & Lee 1998). In this model, nestling behaviour is clearly directed at parents and not at siblings.

Scramble competition defines the situation where siblings compete non-aggressively by escalating begging. A target nestling is predicted to beg more intensely in the presence of needy rivals (Smith & Montgomerie 1991; Leonard & Horn 1998; but see Kacelnik et al. 1995; Cotton et al. 1996) to divert parental attention to itself and thereby increase its likelihood of being fed. Since only needy individuals may be willing to escalate competitive begging, the level of signalling both before and during escalation may honestly reflect need (Godfray 1995). In this model, nestlings are fed with divisible resources in direct proportion to their begging levels. Although siblings assess each other's begging signals to adjust their own behaviour, begging is solely directed at parents (Godfray 1995). Competitive begging is assumed to influence the way parents allocate food within the brood, but not the way they adjust feeding rate (Macnair & Parker 1979).

Dominance hierarchy defines the situation where a nestling A dominates its sibling B throughout the rearing period and can control, to some extent, within-brood food distribution. The amount of food nestling B consumes is the amount nestling A is willing to concede. Nestling A can impose its dominant role in sibling competition through aggressive acts against its younger nestmate (Mock & Parker 1997; H. Drummond this volume).

Aggression may not always be required to establish or maintain dominance hierarchies, since asymmetries in body size may be sufficient to determine which nestling has priority of access to food resources (Price & Ydenberg 1995; Ostreiher 1997; Lichtenstein & Sealy 1998; Cotton et al. 1999). Signals used to maintain a dominance hierarchy are primarily directed at siblings, but may also be used by parents as a criterion to favour a particular offspring (e.g. some female passerines preferentially feed smaller nestlings; Slagsvold 1997). Since within-brood food allocation may not be entirely under the control of the most dominant nestling, offspring are expected to beg both to influence within-brood food allocation and parental feeding rate. In the case where dominance relationships are controlled through aggressive behaviour, subordinates have been shown to refrain from begging to avoid aggression from dominants (Nuechterlein 1981; Braun & Hunt 1983; Drummond & Chavelas 1989; H. Drummond this volume). As predicted by scramble competition, where dominance hierarchies depend upon a size hierarchy and not on aggression, subordinates escalate competitive begging to increase the chance of being fed (Price et al. 1996; Cotton et al. 1999) .

Under the Sibling Negotiation Hypothesis, siblings vary in their need for food and, in the absence of parents, they signal to each other in relation to how needy they are. When parents arrive, nestlings beg at a level that depends upon their own need and on the results of sib-sib communication (i.e. negotiation) that took place before the parents returned to the nest. Under appropriate circumstances, selection will favour investment in negotiation, because negotiation by a needy nestling will lead its less needy nestmates to reduce their begging. This should lead to an increase in the probability of monopolizing a non-divisible food item without having to compete too intensely. Since the settlement of conflict may require both time and successive challenges of each other's motivation to compete, sibling negotiation should mainly occur in the prolonged absence of parents. Siblings are assumed to negotiate resources if three conditions are met. (1) Nestlings are able to monitor each other's willingness or motivation to enter into competition for the non-divisible food that parents will next deliver. (2) Nestlings that are willing to contest a food item are able to signal this decision thereby causing less motivated siblings to retreat from this contest. This may delay feeding time and occasionally result in a missed feeding if parents do not quickly return to the nest. (3) Nestlings that refrain from competition enter into subsequent contests, but only once rivals have consumed a food item and indicate that they now retreat from the competition. Therefore, in the case where siblings negotiate priority of access to food resources, momentarily out-competed nestlings are expected to refrain from displaying signals aimed at influencing parental allocation decisions, but they should not necessarily stop signalling to parents in order to increase or maintain feeding rate.

How Can Siblings Negotiate?

For negotiation processes to take place, siblings must be able to monitor each other's willingness to compete for the next food item. Willingness to compete depends on need for food that siblings can detect only if it is explicitly signalled. Therefore, before the arrival of a parent, an individual may display various signals to indicate the effort it is willing to invest in sibling competition over the impending food, just in case siblings may contest the same resources. Nestlings may frequently move and push siblings to assess their competitiveness. A challenged individual that is highly motivated to get fed may resist and enter into confrontation with its challenger to keep its position or even to move to a more favourable position. In contrast, when a challenged individual is outcompeted, it may move to an unfavourable nest position where the probability of being subsequently fed is lower (McRae et al. 1993). Nestlings can also vocally signal to siblings their willingness to compete, as in the barn owl (Roulin et al. 2000). Ritualized and intimidating behaviour might also play a role with needy and less needy nestlings displaying dominant and submissive behaviour, respectively (Braun & Hunt 1983). In summary, nestlings may indicate their willingness to compete for non-divisible food items by behaving in the same way as they do when begging for food from parents, but in a less intense manner (Figure 1). Signals may include begging vocalizations (Roulin et al. 2000), jostling for nest position, body movements, wing flapping and aggression. The idea is that the display of such behaviours in the absence of parents is a means of siblings negotiating future access to food. Negotiation therefore serves as a means by which nestlings save energy otherwise spent in the acquisition of food items provided by parents. This occurs because sibling competition and its associated costs are reduced as part of the negotiation processes (Figure 1).

When Should Siblings Negotiate Resources?

Provided that parents will bring other food items subsequently, a nestling should retreat from a contest when delaying feeding entails fewer costs than does engaging in intense competition. In other words, an individual should retreat from sibling competition when the costs of entering into it are greater than the benefits (benefits being the value of the contested food resources times the probability of monopolizing them). This means that sibling negotiation should take place whenever negotiation can reduce the costs of sibling competition to a larger extent than the costs imposed by negotiating and eventually delaying feeding time. Sibling negotiation would therefore allow each individual to assess the probability of monopolizing a food item, and given this knowledge, to invest an optimal amount of effort into sibling competition. This implies that 'retreat from sibling competition' does not necessarily mean a nestling completely gives up, but rather that the optimal level of investment is reduced. For example, if the probability of securing a food item is small, it may not be optimal to compete intensely. Instead, the optimal investment should correspond to a level that sometimes returns a food item whose value is greater than the sum of similar investments in sibling competition that did not result in the monopolization of an item.

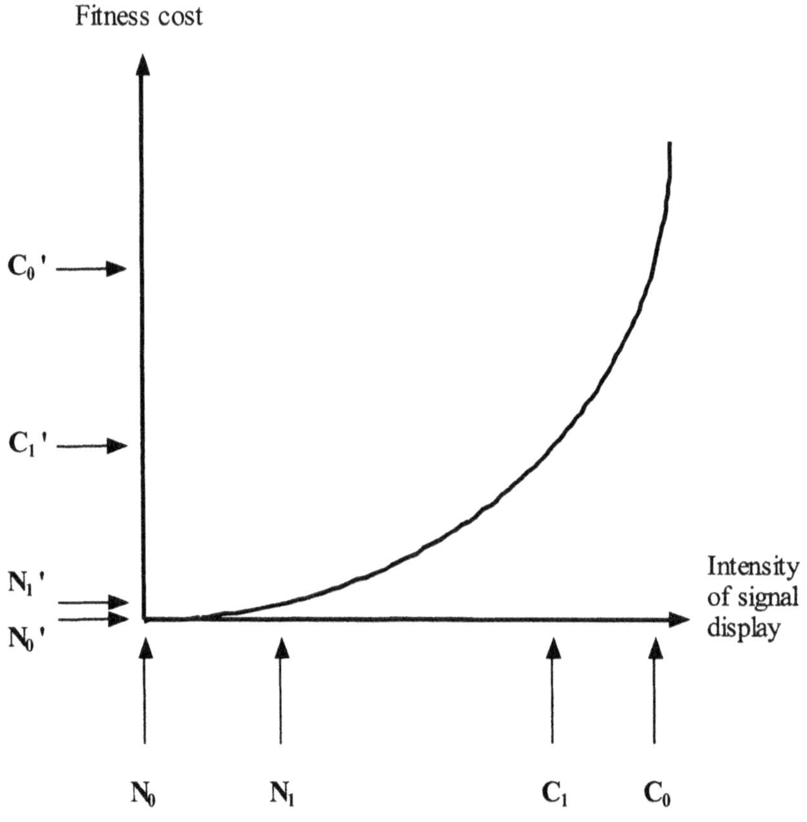

Figure 1. Hypothetical cost function of signalling during sibling negotiation and begging competition both in the case where siblings negotiate priority of access to non-divisible food resources and do not negotiate. In this example siblings use the same signal to negotiate and compete over food resources provided by parents. N refers to signalling level performed during negotiation and C during competition. 0 refers to the level when siblings do not negotiate and 1 when they negotiate. Negotiation should evolve whenever for a fixed amount of food resources the reduction in the cost of sibling competition given by C_0'-C_1' is greater than the costs involved in sibling negotiation given by N_1'-N_0'.

The Sibling Negotiation Hypothesis postulates that sibling negotiation allows nestlings to optimize investment in sibling competition in a way that would not otherwise be possible. This contrasts with Godfray's (1995) model in which nestlings are fed in direct proportion to begging so that it pays to escalate competitive begging. Under the Sibling Negotiation Hypothesis there is no such benefit because food items are not divisible, or because the share of divisible food resources among siblings is not proportional to begging levels. The notion of food-divisibility is crucial because when food cannot be shared among siblings only one individual is fed and all the investment made in sibling competition by its nestmates does not provide any fitness return.

Following the Sibling Negotiation Hypothesis, siblings negotiate which one of them will have priority of access to the non-divisible food item next delivered by parents. This implicitly means that the effort each individual invests in sibling competition is negotiated. Such a situation may arise when the asymmetry between siblings in their food requirements is large. For an illustration, one can consider a two-nestling brood with, all else being equal, one of the two nestlings being needier. Following the logic of the Sibling Negotiation Hypothesis, the needy individual signals its willingness to compete, whereas its less needy sibling signals that it retreats from the contest. In this way, the two nestlings predefine the roles that they will play in sibling competition once the parents return to the nest, thereby allowing them to invest effort in sibling competition at the appropriate (i.e. the most cost-effective) level. (Note that if outcompeted nestlings temporarily retreat from competitive begging, but they may still signal to parents at some level in order to keep feeding rates up.) Such negotiation may be adaptive when the outcome of sibling competition is predictable – i.e. when only the needy nestling would have been fed. A less needy nestling should not compete intensely for resources it is unlikely to acquire and a needy sibling should confirm its willingness to fight for the resources, if contested.

To better understand the potential fitness value of sibling negotiation, imagine the situation in which the needy and less needy nestlings do not negotiate priority of access to non-divisible food resources. In this case, the less needy nestling has no indication about the need of its sibling, and hence when a parent arrives at the nest both individuals may escalate competitive begging in order to attract the attention of the parent. If it was predictable that only the most needy nestling would be fed and this outcome was known in advance, then begging escalation could have been avoided and the less needy nestling could also have refrained from competitive begging.

When Should Siblings Not Negotiate Resources?

Siblings may be willing to negotiate resources if by doing so they substantially reduce the costs involved in sibling competition, whilst only slightly altering the benefits (they delay feeding times). In contrast, sibling negotiation should not occur if it would reduce the benefits of sibling competition to a larger extent than the costs. Three examples illustrate this view. Firstly, consider a species in which a fixed dominance hierarchy is established among siblings due to a pronounced hatching asynchrony. Since a first-hatched nestling imposes its dominance over its younger nestmate, it can decide the amount of resources it consumes. In the hypothetical case in which a first-hatched nestling negotiates resources with its younger nestmate, it would get less food, but only slightly reduce the cost of sibling competition compared to the situation in which sibling negotiation does not occur. This is because its larger size allows it to monopolize food relatively easily. For this reason, a first-hatched nestling may be less willing to negotiate resources if the inclusive fitness benefit to its nestmate does not compensate for the reduction in its own condition. Secondly, when food supply is poor, the cost of waiting may be extremely high, and in such situations each nestling should escalate begging in order to monopolize non-divisible food items. Thirdly, if parents distribute food resources to all offspring in direct proportion to their begging levels, it may pay to escalate competitive begging (Godfray 1995) rather than to negotiate.

Sibling Negotiation versus Sibling Competition

Sibling negotiation can reduce the level of sibling competition compared with the situation in which siblings have no information regarding each other's intention to compete. It cannot, however, nullify sibling competition. Firstly, there may be an optimal level of investment in sibling competition if each nestling has some probability of monopolizing a non-divisible food item. Secondly, the outcome of sibling negotiation can be contested. That is, a needy nestling may request more resources than it can actually receive. Even though priority of access may be settled before the arrival of a parent, less needy nestlings may nevertheless try to attract the attention of parents by begging relatively intensely. This is because, although in the absence of parents a nestling can show that it is the most willing to enter into competition, it cannot simultaneously attract the attention of a parent and prevent the parent's access to nestmates. This new conflict may decrease the benefit of sibling negotiation.

In this context, a promising issue to be addressed is the extent to which sibling negotiation and sibling competition determines within-brood food allocation. To this end, I recently explored the relative effect of food supply on sibling negotiation and sibling competition in the barn owl (Roulin 2001a). On two successive days I provided broods with dead laboratory mice (treatment: 'food-added') or I removed prey remains (treatment 'food-removed'), and hence could compare vocalization behaviour in relation to food supply. In the absence of parents, nestlings vocalized at a significantly greater rate when food was removed than when food was added, but in their presence they vocalized at a similar rate. This suggests that sibling negotiation is more related to food supply than sibling competition. Two interpretations can explain this result. Firstly, the extra energy invested in sibling negotiation by food-removed nestlings may have decreased the intensity of sibling competition resulting in similar levels in the two treatments. Secondly, when the whole brood is offered mice, siblings may not differ in their short-term need for food, and hence the outcome of sibling competition may not be predictable, implying that any negotiation may be contested upon the parents' arrival. In this case, nestlings may reduce investment in sibling negotiation and increase investment in sibling competition, and hence may compete at a similar level in the two treatments. This high investment in sibling competition may be adaptive because nestlings often monopolize prey items by sitting on them and then consuming them at a later time after a previous meal has been digested. This study exemplifies the need to further investigate the relationship between sibling negotiation and competition.

Evolution of Sibling Negotiation

Following the arguments above regarding when we should expect sibling negotiation to occur, it appears that (1) a high cost of sibling competition associated with (2) a high level of short-term asymmetry in food requirement among siblings and (3) non-divisible food resources brought by parents may favour the evolution of sibling negotiation. Low levels of extra-pair fertilizations may also contribute to a more rapid evolution, since a target nestling may be more prone to retreat from a contest when its nestmate is more closely related (Johnstone 1999). Indeed, the inclusive fitness benefits obtained from allowing its nestmate to be fed first are larger when relatedness is high (Mock & Parker 1997). Furthermore, because the major aim of sibling negotiation is to reduce the overall level of sibling competition, there is no *a priori* reason to believe that the evolution of sibling negotiation should depend upon whether parents or offspring have control over within-brood food distribution. In both cases, investment in sibling competition is a key factor that determines this outcome, and therefore in the two situations nestlings may benefit from negotiating investment in sibling competition.

How Can Sibling Negotiation Be Honest?

Sibling negotiation will persist only if it proves to be reliable to siblings. Reliability would be achieved if three conditions are satisfied. Firstly, a nestling that signals high willingness to enter into sibling competition should be truly motivated to do so. To put it another way, investment in sibling negotiation should accurately reflect prospective investment in sibling competition. In order to be honest, signals of the willingness to enter into sibling competition should themselves be costly to produce so that intense signalling provides net benefits only when individuals are needy (Godfray 1991, 1995). Costs of signalling would prevent cheaters misrepresenting their true motivation to compete for food once a parent arrives at the nest. Signals may be energetically expensive to produce or the honesty of signalling may be socially controlled. Nestmates may challenge signallers to assess whether signalling level corresponds to motivation, and eventually punish cheaters that back down during negotiation, but which compete when the parent arrives. Punishment in this case could simply take the form of stealing a recently acquired food item (the dishonest nestling can be considered to be punished because any investment in sibling negotiation is thereby lost). Note that if cheating becomes too widespread the reliability of signalling breaks down and sibling negotiation would disappear.

Secondly, a nestling that is motivated to compete should signal this decision. Imagine the hypothetical case of a motivated nestling that does not signal and hence does not negotiate resources. Given that siblings negotiate priority of access to food in order to reduce the cost of competition, any reduction in sibling negotiation would imply that this individual has now to invest more effort in sibling competition (Figure 1). If the costs of acquiring a food item are greater without previous negotiations, then an individual would have much to lose by not previously signalling its motivation to enter into sibling competition. This mechanism may allow the evolution of costly displays that signal motivation (although Maynard Smith (1982) questions whether such evolution could take place).

Finally, as mentioned above, the primary function of sibling negotiation is to reduce the cost of sibling competition. Siblings may negotiate by displaying the same behaviours as the ones used to beg for food from parents. As far as the honesty of sibling negotiation is concerned, investment in sibling negotiation must be lower than the decrease in investment in sibling competition due to negotiation (Figure 1). Since sibling negotiation and competition may not necessarily involve the same currencies, care should be taken to consider all the behaviours involved in the costs and benefits of negotiating and competing. For instance, in the barn owl sibling negotiation involves begging vocalizations (Roulin et al. 2000), whereas sibling competition involves begging vocalizations, but also pushing, moving, wing flapping, aggressive interactions and stealing of recently acquired prey items (Roulin 2001a; A. Roulin unpublished data).

A STUDY IN THE BARN OWL

Is the Barn Owl a Good Model Organism for the Study of Sibling Negotiation?

I proposed three major factors that appear to favour the evolution of sibling negotiation: (1) costly sibling competition; (2) large asymmetries between siblings in need; and (3) non-divisible food resources. As we can see below, the barn owl fulfils these three conditions, and thus represents a good model organism to study sibling negotiation.

Sibling competition seems to entail costs for barn owls given that brood sizes can be large (mean = 4, range = 1-9 offspring). When a parent arrives at the nest with a food item, offspring usually vocalize intensely, push each other to be closer to the source of the food and in a tumult try to take the prey from their parent's bill (A. Roulin personal observation) and from each other (Roulin 2001a). The finding that the youngest nestling sometimes falls out of nests located in houses, churches and nestboxes suggests that they incur a risk during sibling contests by being positioned too close to the nest entrance (Bunn & Warburton 1977; Baudvin 1986; A. Roulin personal observation). When food becomes scarce then aggression among siblings, brood reduction and cannibalism occur (Baudvin 1986). As a tangible sign of intense competition, siblings frequently monopolize prey items by sitting on them. This behaviour may have evolved because siblings steal each other's food in 9% of parental feeding visits (Roulin 2001a). Furthermore, parents did not appear to adjust feeding rate to either brood size (Roulin et al. 1999) or brood hunger level (Roulin et al. 2000) suggesting that they do not behave in a way that would reduce the overall level of sibling competition when food supply is poor.

The asymmetry between siblings in hunger level is large. Parents deliver a single small mammal approximately once an hour, which is consumed by a single offspring. Given that each offspring eats three to four items per 24 hours (Roulin 2001a), the consumption of a single item is likely to create large asymmetries in hunger among-siblings.

How Can Sibling Barn Owls Negotiate?

Another reason why the barn owl is particularly suited to an investigation of the Sibling Negotiation Hypothesis is that nestlings vocalize all night long, in the absence of parents (Bühler & Epple 1980; Roulin et al. 2000). Since parents are not always in the vicinity of the nest (territory sizes range between 90 and 392 ha, Brandt & Seebass 1994; Taylor 1994), vocalizations may be primarily directed towards siblings. Even if calling in the absence of parents evolved as an offspring-parent communication system, selection should favour siblings that assess each other's calling behaviour and use this information to adjust their own vocalization level and related competitive behaviour. Alternatively, if such calling evolved as sib-sib communication, parents may also have been selected to monitor these calls. However, I will not consider this possibility here, and consider calling behaviour in the absence of parents only in the context of sibling negotiation. Before presenting tests of predictions arising from the Sibling Negotiation Hypothesis, I test three assumptions.

Test of Three Assumptions

The Sibling Negotiation Hypothesis relies on three fundamental assumptions. Firstly, if vocalizing in the absence of parents signals the willingness to compete for the next prey item delivered, needy individuals should vocalize more intensely than less needy ones. To test this assumption, I selected 26 broods and at 09:00 hours either added twice as many dead mice as the number of offspring (treatment: 'food-added') or removed prey remains from the nest (treatment: 'food-removed'). From 21:30 to 05:30 hours, I filmed parental feeding visits with an infrared sensitive camera and with a microphone recorded offspring vocalization behaviour in the absence of parents. On the films, I counted the number of calls produced during 30-second samples every 15 minutes. Mean calling rate was significantly greater when broods had food removed than when food was added (paired t-test on log transformed data, $t_{25} = 5.69$, $P < 0.001$). This result confirms a previous study showing that in two-nestling broods the food-supplemented nestling vocalized less intensely than its food-deprived sibling (Roulin et al. 2000). Food-supplemented broods still produced a relatively large number of calls (Figure 2), probably because satiated nestlings often monopolize prey items by sitting on them for later consumption (Roulin 2001a), implying that food supplements cannot nullify the competitive inclination of nestlings. Siblings may indeed compete for food items to be eaten after having digested laboratory mice.

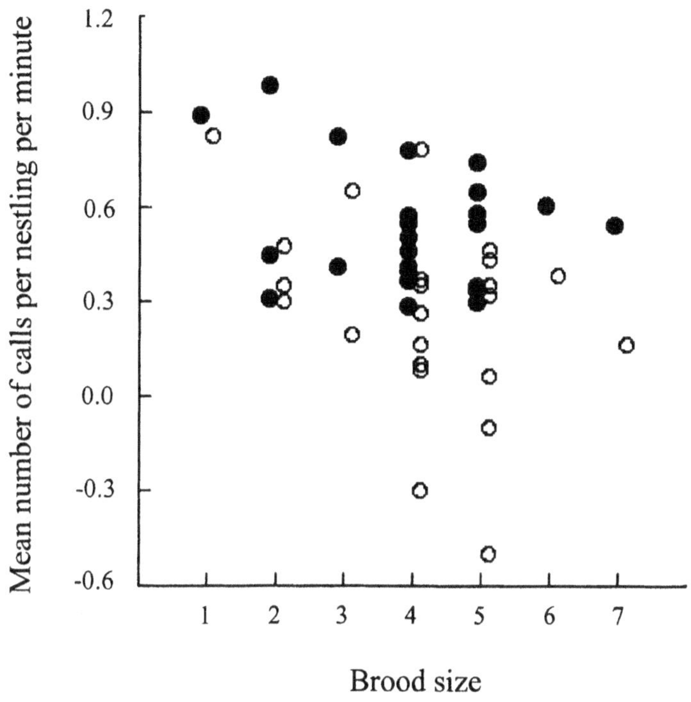

Figure 2. Mean number of calls per nestling per minute (log transformed) in relation to brood size when nestlings had food added (open circles; mice were deposited in the nest in the morning) or food removed (filled circles; prey remains in the nest were removed). Calls were counted in the absence of parents during 30-second samples every 15 minutes. For each nest and food treatment, a mean value is calculated.

The second assumption states that if calling in the absence of parents signals willingness to compete for the impending food resource, the nestling that is fed first should have been more vocal than its siblings. To test this assumption, I created 65 two-nestling broods over the course of three years (1997, 1998 and 2000). Two nestlings were randomly chosen and their siblings removed from the nest from 21:30 to 23:30 hours. During this interval, I recorded the begging behaviour of the two siblings in the absence of parents. I could determine which nestling consumed the first prey item of the night in a sample of 53 nests. This nestling had produced an average of 7.7 calls per minute and its nestmate produced 3.9 (Roulin 2001a). Therefore, vocalizations in the absence of parents appear to signal need.

The final assumption proposes that vocalizations produced in the absence of parents entail fitness costs, thereby ensuring that calls truly reflect siblings' motivation to compete once the parents return to the nest. The assessment of the fitness cost of begging is not a trivial exercise (Roulin 2001b), and so far I have been unable to measure it directly. However, two pieces of evidence suggest that vocalizations may be expensive to produce. Firstly, each call contains substantial noise energy given that a single call lasts 0.3 to 0.9 seconds, with a frequency range of 6 to 10 kHz (Bühler & Epple 1980). Also, in a single night each 36-day-old nestling produced 1786 ± 181 calls (mean ± S.E.; range = 922-4608) when food was removed and 1031 ± 143 calls (range = 144-3168) when food was added (calculated from the 26 broods above), suggesting that investment in sibling negotiation in the absence of parents may not be negligible. Secondly, if investment in sibling negotiation is costly, it should be traded off against other activities, such as preening. To test this assumption, I used data from artificially created two-nestling broods, and measured preening activity as the proportion of time a nestling spent preening fifteen minutes before the first nocturnal nest visit of a parent. Note that in some cases nestlings were not always visible on the videos, and hence I could not record preening behaviour in all nests. When a nestling had food added it vocalized less intensely in the absence of parents than its food-deprived sibling (Roulin et al. 2000), but spent more time preening (food-added nestlings, median = 13.2% of the time; food-removed nestlings = 1.7%; $z = 2.07$, $n = 19$, $P = 0.039$). Similar findings in unmanipulated two-nestling broods showed that the nestling that vocalized more intensely also preened less frequently compared to its less vocal sibling (5.1% versus 7.9%, $z = 2.14$, $n = 48$, $P = 0.033$). Note that calling does not seem to incur a strong predation cost. For instance, in French churches stone martens (*Martes foina*) predated only two of 1031 broods (Baudvin 1986).

Test of Predictions

In this section, I propose a non-exhaustive list of four predictions of the Sibling Negotiation Hypothesis tested on a Swiss population of barn owls.

Prediction 1: Sibling negotiation starts before the first parental feeding visit of the night.

When visiting nestboxes in the evening or afternoon, I often heard nestling begging calls before it was dark enough for parents to fly. Unfortunately, I rarely recorded the exact time at which I heard such calls, and hence I cannot present quantitative data (note the observation of nestlings calling two hours before the night). It can nevertheless be taken for granted that nestlings start to vocalize well before the first possible parental feeding visit.

Prediction 2: Sibling negotiation takes place between parental feeding visits.

To examine this prediction I recorded vocalizations in 26 broods from 21:30 to 05:30 hours. Call number was counted in the absence of parents during 30-second samples every 15 minutes. Then, for each hour of the night I calculated the mean calling rate per minute per nestling. Based on square root transformed data, calling rate significantly declined from 22:00 to 05:00 ($F_{1,310} = 26.37$, $P < 0.001$) after controlling for food treatment (food-added versus food-removed: $F_{1,310} = 17.93$, $P < 0.001$) and nest ($F_{1,310} = 2.66$, $P < 0.001$). There was no significant interaction between time of the night and treatment ($F_{1,310} = 0.003$, $P = 0.95$) and nest ($F_{1,310} = 1.31$, $P = 0.15$). Given that mean calling rate per nestling in the first hour of the night was 12.0 for the food-added treatment and 24.1 for the food-removed treatment, and in the last hour 3.6 and 11.9 respectively, sibling negotiation takes place in the absence of parents throughout the night.

Prediction 3: When food resources are not scarce (i.e. when an individual will not be penalised if food is delayed), some nestlings signal their willingness to contest the next food item delivered, whereas others retreat from the contest. As a consequence, mean calling rate in the absence of parents should be lower in smaller broods where competitors are less numerous (i.e. fewer individuals retreat per contest).

Using observations made in the 26 broods above, nestlings appeared to vocalize more frequently in smaller broods (mean number of calls per nestling with food-added versus food-removed as the repeated measure on log transformed data, and brood size as a first covariate: $F_{1,23} = 7.62$, $P = 0.011$; Figure 2). In this analysis, I statistically controlled for mean nestling body condition, defined as 'residuals from the regression of mean nestling wing length on mean nestling body mass', because nestlings in poorer condition vocalized more intensely (second covariate: $F_{1,23} = 9.88$, $P = 0.005$). This confirms results from brood size manipulation experiments where calling rate per nestling was lower in enlarged as compared to reduced broods (Roulin et al. 2000). Two non-mutually exclusive interpretations can account for this result. Firstly, the number of competitors in a nest tends to inhibit calling behaviour because when a target nestling perceives that the probability of being fed is low, as is the case in large broods, it tends to retreat from the contest. If true, then in artificially created two-nestling broods calling rate should not be correlated with original brood size. This was the case ($r = -0.10$, $n = 65$, $P = 0.43$; using the same sample of broods used in testing the second assumption). Secondly, if only one or a few nestlings vocalize in the absence of parents (Bühler & Epple 1980), the total number of calls produced in large and small broods may be similar and therefore calling rate per nestling should be smaller in large broods. Empirical data have still to be collected to test this interpretation.

Prediction 4: Investment in sibling negotiation by a target nestling depends on the extent to which it values a food item, as compared with its nestmates. When the likelihood of acquiring a food item is low, a nestling vocally signals that it retreats from the contest and, once the probability of being subsequently fed increases, this individual increases its begging level to signal that it is now willing to compete.

I used a sample of 23 artificially created two-nestling broods where I could monitor vocalizations in the absence of parents before and just after the first nocturnal feeding visit of a parent. I entered square root transformed calling rate before and after the first parental visit as repeated measures and the time lapse separating these two periods as a covariate (range: 1-17 minutes). The nestling that consumed the first prey item of the night significantly reduced its calling rate after that feed (repeated measures: $F_{1,21} = 6.03$, $P = 0.023$; time lapse: $F_{1,21} = 4.15$, $P = 0.054$; Figure 3). In contrast, the individual that did not get the prey item significantly increased calling rate after its sibling consumed the item (repeated measures: $F_{1,21} = 5.20$, $P = 0.033$; time lapse: $F_{1,21} = 2.18$, $P = 0.16$; Figure 3). This shows that outcompeted nestlings increase their investment in sibling negotiation once a rival has consumed a prey item. Given that the time lapse between the recording of vocalizations produced before and after the rival was fed by a parent was short and controlled statistically, the increase in vocalization level cannot be explained by a sudden increase in food need. This result was already demonstrated in an experiment where two siblings had been randomly assigned to one of the two roles (to be fed by a parent or not) by keeping them during daylight hours with food (food-added nestling) or without food (food-removed nestling). After a parent fed the food-removed nestling, the food-added nestmate immediately increased its vocalization level in the absence of parents (Roulin et al. 2000).

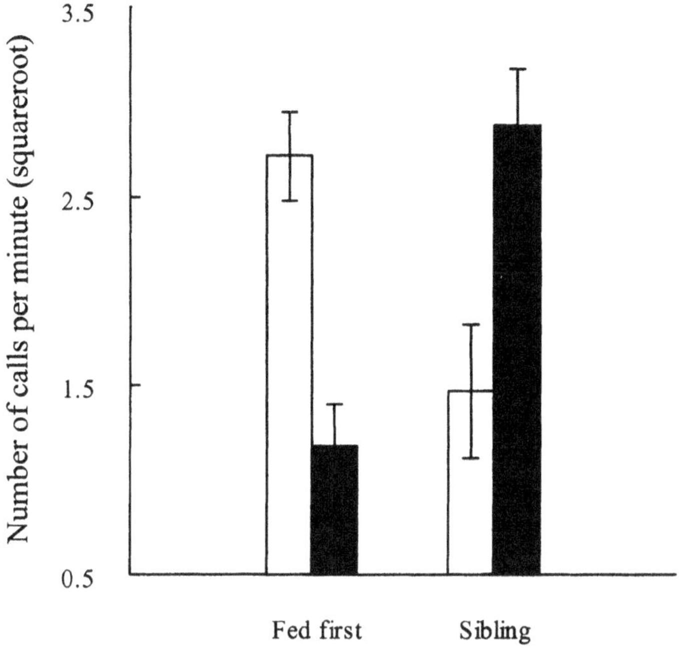

Figure 3. Mean (± SE) number of calls before (white bars) and after (dark bars) the first parental feeding visit of the night. Calls were counted 15 minutes before the arrival of a parent and 15 minutes after one of the two nestlings consumed the delivered prey item.

FUTURE DIRECTIONS

From a theoretical point of view, the Sibling Negotiation Hypothesis is of interest because it emphasizes the importance of the dynamics of interactions between family members, and hence the importance of past and future interactions on food allocation and sibling competition at a given interaction. Furthermore, within a family it is usually assumed that offspring signal to parents their need (Godfray 1991, 1995), whereas the Sibling Negotiation Hypothesis also explores the importance of sib-sib communication in contests over parental food distribution. In other words, the level with which a nestling communicates its need to its parents is not only a function of need and sibling competition (Godfray 1995), but also of previous sibling interactions. Signalling models are now required to formally investigate the conditions under which sibling negotiation can evolve and be stable, and to establish what factors govern the intensity of negotiation. Among the factors that may be of importance, I suggest: the number of offspring the parents can feed per visit, the level of hatching asynchrony, the level of asymmetry in need between siblings and the cost of sibling competition. Furthermore, based on the Sibling Negotiation Hypothesis, any extension of signalling models of begging should consider that a begging signal can have two different functions. A begging display can simultaneously signal to parents to increase feeding rate and influence parental decisions over within-brood food distribution. This is of importance because sibling negotiation may inhibit begging escalation due to sibling competition, but not necessarily the transfer of information that parents use to maintain an appropriate feeding rate. Finally, the major message of the Sibling Negotiation Hypothesis is that sibling interactions taking place in the absence of parents may allow nestlings to adjust the optimal level of competitive begging effort. Such adjustment has to be made in relation to need, but also according to the probability of successfully monopolizing a non-divisible food item.

From an empirical point of view, I see two main aspects of this idea that still have to be tackled. Firstly, the fitness value of sibling communication taking place in the absence of parents has to be investigated. Experiments should test the prediction that sibling negotiation reduces the level of sibling competition, and hence allows nestlings to optimize competitive begging. Secondly, since sibling negotiation has been postulated to allow nestlings to adjust their optimal level of competitive behaviour and reduce sibling competition, there is no *a priori* reason to believe that sibling negotiation should be exclusively restricted to barn owls. Researchers should therefore start to monitor nestling behaviour displayed in the absence of parents in other avian species (e.g. Budden & Wright 2001; Leonard & Horn 2001) and perform specific experiments to explore the fitness consequences of these behaviours (see also Roulin 2001b). This may require the quantification of vocalizations, movements, jostling and aggression, but also the asymmetry in hunger level between siblings and the number of nestlings parents feed within a single visit. Ultimately, these studies allow comparative analyses to detect in which species or ecological contexts sibling negotiation is particularly prevalent. In any case, the verbal arguments developed here may stimulate further investigations of nestling begging behaviour, since it appears that in the context of parent-offspring conflict both parent-offspring and sib-sib communication may be relevant.

ACKNOWLEDGEMENTS

This chapter is dedicated to the memory of Jean-Charles Daiz, a naturalist who devoted his life to protect birds. For the empirical part of this work, I benefited from the technical assistance of the late Jean-Charles Daiz and Martin Epars and of Laurent Hirt. Martin Epars, Henri Etter and Anne-Lyse

Ducrest helped me in the field. They are all warmly thanked. Experiments were conducted under authorization of the 'Service vétérinaire du canton de Vaud' n° 1145. I am also grateful to Anne-Lyse Ducrest, Rufus Johnstone, Marty Leonard, Jon Wright and an anonymous reviewer for their useful comments on a previous version of this chapter. I reiterate my congratulations to Jon Wright and Marty Leonard for the quality of the 'Begging Workshop' they organized at Gregynog, Wales. This chapter was written during a post-doctorate financed by the Swiss National Science Foundation (N° 81-59899).

REFERENCES

Baudvin, H. 1986. La reproduction chez la chouette effraie (*Tyto alba*). *Le Jean le Blanc* **25**, 1-25.

Brandt, T. & Seebass, C. 1994. *Die Schleiereule*. Wiesbaden: AULA Verlag.

Braun, B.M. & Hunt, G.L. 1983. Brood reduction in black-legged kittiwakes. *The Auk* **100**, 469-476.

Budden, A.E. & Wright, J. 2001. Falling on deaf ears: the adaptive significance of begging in the absence of parents. *Behavioral Ecology and Sociobiology* **49**, 474-481.

Bühler, P. & Epple, W. 1980. The vocalizations of the barn owl. *Journal für Ornithologie* **121**, 36-70.

Bunn, D.S. & Warburton, A.B. 1977. Observations on breeding barn owls. *British Birds* **70**, 246-256.

Burford, J.E., Friedrich, T.J. & Yasukawa, K. 1998. Response to playback of nestling begging in the red-winged blackbird, *Agelaius phoeniceus*. *Animal Behaviour* **56**, 555-561.

Clark, A.B. & Lee, W.-H. 1998. Red-winged blackbird females fail to increase feeding in response to begging call playbacks. *Animal Behaviour* **56**, 563-570.

Cotton, P.A., Kacelnik, A. & Wright, J. 1996. Chick begging as a signal: are nestlings honest? *Behavioral Ecology* **7**, 178-182.

Cotton, P.A., Wright, J. & Kacelnik, A. 1999. Chick begging strategies in relation to brood hierarchies and hatching asynchrony. *American Naturalist* **153**, 412-420.

Davies, N.B., Kilner, R.M. & Noble, D.G. 1998. Nestling cuckoos, *Cuculus canorus*, exploit hosts with begging calls that mimic a brood. *Proceedings of the Royal Society of London, Series B* **265**, 673-678.

Drummond, H. & Chavelas, C.G. 1989. Food shortage influences sibling aggression in the blue-footed booby. *Animal Behaviour* **37**, 806-819.

Godfray, H.C.J. 1991. Signalling of need by offspring to their parents. *Nature* **352**, 328-330.

Godfray, H.C.J. 1995. Signaling of need between parents and young: parent-offspring conflict and sibling rivalry. *American Naturalist* **146**, 1-24.

Hofstetter, S.H. & Ritchison, G. 1998. The begging behavior of nestling eastern screech-owls. *Wilson Bulletin* **10**, 86-92.

Johnstone, R.A. 1996. Begging signals and parent-offspring conflict: do parents always win? *Proceedings of the Royal Society of London, Series B* **263**, 1677-1681.

Johnstone, R.A. 1999. Signaling of need, sibling competition, and the cost of honesty. *Proceedings of the National Academy of Sciences USA* **96**, 12644-12649.

Kacelnik, A., Cotton, P.A., Stirling, L. & Wright, J. 1995. Food allocation among nestling starlings: sibling competition and the scope of parental choice. *Proceedings of the Royal Society of London, Series B* **259**, 259-263.

Kilner, R. 1995. When do canary parents respond to nestling signals of need? *Proceedings of the Royal Society of London, Series B* **260**, 343-348.

Leonard, M.L. & Horn, A.G. 1998. Need and nestmates affect begging in tree swallows. *Behavioral Ecology and Sociobiology* **42**, 431-436.

Leonard, M.L. & Horn, A.G. 2001. Begging in the absence of parents by nestling tree swallows. *Behavioral Ecology* **12**, 501-505.

Lichtenstein, G. & Sealy, S.G. 1998. Nestling competition, rather than supernormal stimulus, explains the success of parasitic brown-headed cowbird chicks in yellow warbler nests. *Proceedings of the Royal Society of London, Series B* **265**, 249-254.

Macnair, M.R. & Parker, G.A. 1979. Models of parent-offspring conflict. III. Intra-brood conflict. *Animal Behaviour* **27**, 1202-1209.

Maynard Smith, J. 1982. *Evolution and the Theory of Games*. Cambridge: Cambridge University Press.

McRae, S.B., Weatherhead, P.J. & Montgomerie, R. 1993. American robin nestlings compete by jockeying for position. *Behavioral Ecology and Sociobiology* **33**, 101-106.

Mock, D.W. & Parker, G.A. 1997. *The Evolution of Sibling Rivalry*. Oxford: Oxford University Press.

Mondloch, C.J. 1995. Chick hunger and begging affect parental allocation of feedings in pigeons. *Animal Behaviour* **49**, 601-613.

Nuechterlein, G.L. 1981. Asynchronous hatching and sibling competition in western grebes. *Canadian Journal of Zoology* **59**, 994-998.

Ostreiher, R. 1997. Food division in the Arabian babbler nest: adult choice or nestling competition? *Behavioral Ecology* **8**, 233-238.

Price, K. 1998. Benefits of begging for yellow-headed blackbird nestlings. *Animal Behaviour* **56**, 571-577.

Price, K. & Ydenberg, R. 1995. Begging and provisioning in broods of asynchronously-hatched yellow-headed blackbird nestlings. *Behavioral Ecology and Sociobiology* **37**, 201-208.

Price, K., Harvey, H. & Ydenberg, R. 1996. Begging tactics of nestling yellow-headed blackbirds, *Xanthocephalus xanthocephalus*, in relation to need. *Animal Behaviour* **51**, 421-435.

Roulin, A. 2001a. Food supply differentially affects sibling negotiation and competition in the barn owl (*Tyto alba*). *Behavioral Ecology and Sociobiology* **49**, 514-519.

Roulin, A. 2001b. On the cost of begging: implications of vigilance. *Behavioral Ecology* **12**, 506-515.

Roulin, A., Ducrest, A.-L. & Dijkstra, C. 1999. Effect of brood size manipulations on parents and offspring in the barn owl *Tyto alba*. *Ardea* **87**, 91-100.

Roulin, A., Kölliker, M. & Richner, H. 2000. Barn owl (*Tyto alba*) siblings vocally negotiate resources. *Proceedings of the Royal Society of London, Series B* **267**, 459-463.

Slagsvold, T. 1997. Brood division in birds in relation to offspring size: sibling rivalry and parental control. *Animal Behaviour* **54**, 1357-1368.

Smith, H.G. & Montgomerie, R. 1991. Nestling American robins compete with siblings by begging. *Behavioral Ecology and Sociobiology* **29**, 307-312.

Taylor, I.R. 1994. *Barn Owls: Predator-Prey Relationships*. Cambridge: Cambridge University Press.

7.

EFFICACY AND THE DESIGN OF BEGGING SIGNALS

Andrew G. Horn & Marty L. Leonard
Department of Biology, Dalhousie University, Halifax NS B3H 4J1, Canada
(aghorn@is.dal.ca; mleonard@is.dal.ca)

ABSTRACT

Begging displays not only must encode information on the needs of nestlings, but also must effectively convey this information to parents. We discuss some problems in transmission and reception that begging signals must overcome to successfully convey information, and suggest ways these problems may have affected the design of begging signals, particularly begging calls. Acoustic effects within nests, especially reflection of sound from nest walls, may select for begging calls that circumvent or exploit these sound effects. Similarly, acoustic interference from nestmates may select for calls that enhance locatability and discriminability. We suggest that the apparent exaggeration of begging displays may not only be a result of evolutionary conflicts of interest between parents and young, but could also be needed to convey the signal from offspring to parents.

INTRODUCTION

Animal signals come in an amazing diversity of forms, from the subtle whispers of mated pairs to the spectacular displays of courting males. Signals on the showier end of the spectrum have particularly fascinated evolutionary biologists, because they seem excessively conspicuous. Indeed, understanding the selective pressures that favour conspicuous signals has been the focus of countless studies in the fields of animal communication and sexual selection.

One unifying concept in this research is that animal signals, subtle and spectacular alike, must be designed to encode information and to convey this information to receivers. Design features that serve each of these two

J. Wright and M.L. Leonard (eds.), The Evolution of Begging, 127–141.

functions, which we will call 'content' and 'efficacy' (Dawkins & Guilford 1997), must be contained in every signal. For example, consider why many bird songs are loud and complex. On one hand, these features advertise the singer's vigour and experience, while on the other, they are also necessary to ensure that the sound reaches distant listeners and stands out from background noise (Kroodsma & Miller 1996). As the copious literature on bird song suggests, both content and efficacy must be considered to understand how selection has shaped the design of animal signals.

Studies on the evolution of nestling begging displays have concentrated mainly on the information content of the signal. More specifically, they have focused on determining what information is encoded in the visual and acoustic components of the display and on measuring the response of parents to this information (reviewed in Budden & Wright 2001). Few studies have examined how begging is designed for effective transmission and reception. One reason for this bias is that the delivery of a begging signal from a nestling to a parent only a few centimetres away seems to raise few, if any, transmission or reception problems. Indeed, the initial interest in nestling begging arose partly because the display seemed more vigorous than necessary for conveying information over the relatively short distances within a nest.

In this chapter, we suggest that the transmission and reception of begging signals might be more difficult than expected for such a short range signal and that some of the seemingly unnecessary elaboration of begging may be a result of selection to overcome these difficulties (see also Dawkins & Guilford 1997). After some brief comments on the information that begging conveys, we discuss two main problems that begging signals must overcome: transmission noise and interference from nestmates. Throughout most of the chapter, we focus on the acoustic components of begging (i.e. begging calls) because the efficacy of acoustic signals is better understood than that of visual signals, thanks largely to a huge literature on the transmission and reception of insect, frog, bird and mammal sounds (Gerhardt 1994; Hauser 1996; Kroodsma & Miller 1996; Hoy et al. 1998).

The purpose of this chapter is twofold. We wish to encourage more research on the efficacy of begging because, as we will hopefully show, the transmission and reception problems faced by nestlings are every bit as complex and interesting as those faced by frogs and adult birds, yet they are poorly understood. We also wish to illustrate how the apparent intensity and complexity of the begging signal could be shaped not only by strategic conflicts between parents and offspring (see R.M. Kilner this volume), but also by the need to effectively convey information to parents.

INFORMATION CONTENT AND NESTLING BEGGING CALLS

In order to understand how begging calls are designed to effectively convey information, we must first know what information the calls encode. A number of studies have shown that call features such as rate, duration and harmonic structure encode information on nestling size, gender, hunger and thermal condition (Redondo & Castro 1992; Price et al. 1996; Leonard & Horn 2001a) and that parents respond to increases in the rate and duration of calls emanating from the brood by increasing their feeding rate (Kilner & Johnstone 1997; Budden & Wright 2001). Additional studies have also shown that parents respond similarly to playback of begging calls at nests (Kilner & Johnstone 1997; Budden & Wright 2001; but see Clark & Lee 1998). Thus, the evidence suggests that begging calls contain information on nestling need and condition and that the calls are used by parents to adjust the feeding rate to the brood.

The information content of begging calls and how it is used by parents to adjust feeding rates to broods has been well established. Relatively less is known, however, about whether parents also use the information in the calls to make feeding decisions about individual nestlings within broods. Indeed, only a handful of studies have included the experiments necessary to separate the visual and vocal components of the display by temporarily devocalizing nestlings (B. Glassey personal communication) or by playing back manipulated calls to parents (Leonard & Horn 2001b; R.M. Kilner this volume).

A brief example from our own work illustrates how we approached the problem and suggests how parents might use these calls when making feeding decisions (Leonard & Horn 2001b). We conducted a paired-choice test to determine if parent tree swallows (*Tachycineta bicolor*) discriminated between the begging calls of hungry nestlings and those that had been recently fed. We placed miniature speakers on opposite walls of tree swallow nestboxes and placed either a model nestling or a live nestling fed to satiation (i.e. so that it would not beg) below each speaker. When parents returned to the box to feed, we played pairs of playback tapes that gave them a choice between calls or features of calls associated with high and low levels of hunger. We found that parents preferentially responded to the calls associated with longer periods of food deprivation. Although our experimental paradigm was crude, it showed that parents attended to differences in the calls of individual nestlings and directed their feedings accordingly.

To summarize, most work on begging has shown that it advertises the needs of the brood and of individual nestlings and that parents respond to changes in the intensity of the display (Budden & Wright 2001). As the above work on begging calls illustrates, however, we cannot completely understand the function of the various components of this display without more experimental presentations of manipulated begging signals. In particular, we cannot tell whether begging is designed to effectively convey cues bearing information about need without knowing what cues are being used by parents. Experiments that reveal these cues will be hard to perfect, but are essential for understanding the role of efficacy in the design of the begging display.

PROBLEMS IN TRANSMISSION AND RECEPTION OF BEGGING CALLS

If parents use begging calls strictly for regulating feeding rates to the nest, then assessing changes in brood-level calling rates in relation to hunger, for instance, would be a relatively easy task. As the results of our playback study illustrate, however, parents also use the calls to distribute feedings to individuals within nests, which is probably more difficult. Indeed, at least two factors might influence the transmission and reception of calls. First, during transmission from nestlings to parents, the signal may interact with the structure of the nest in complex ways that may absorb or amplify certain features of the call. Secondly, as they are received by parents, the calls of individual nestlings may interfere with those of calling nestmates. In the following sections, we examine these transmission and reception problems in detail and discuss possible mechanisms for reducing their impact on signal efficacy.

Transmission

As we mentioned above, transmission of the begging signal from offspring to parents seems to raise few issues in signal design because signaller and receiver are centimetres apart. Indeed, begging calls do not encounter the acoustic filters that, for instance, bird songs encounter as they travel through forests or open fields. The nest does, however, present an obstruction that may change the signal, even over the short distances between parents and young. It may act as a muffler or an amplifier, favouring certain frequencies and filtering others, depending on its shape, size and composition. These acoustic effects may, in turn, select for calls that have design features that overcome or exploit these sound effects for more efficient reception by parents.

Predicting exactly what happens to sound in enclosures like nests is difficult. Most acoustic effects, however, occur when sounds encounter obstructions that are comparable in size to their wavelength (Fletcher 1992). Begging calls typically have frequencies between 1 and 8 kHz (Popp & Ficken 1991; Briskie et al. 1999) and therefore wavelengths of between 34 and 4 cm. Nests of most altricial species form enclosures around nestlings that are similar in their linear dimensions to the wavelengths of begging calls, and thus will probably affect call transmission. At the same time, variation in both call wavelength and nest dimensions over this size range should yield considerable variation across species in how nest acoustics affect begging calls.

Reflection of sound waves from nest walls probably affects calls the most, so we will use it to illustrate how nest acoustics could influence call design. We restrict ourselves to some general effects, recognizing that the specifics of each situation would require more complex models and measurements *in situ* (Hodgson & Warnock 1992; Nefske & Sung 1992; Fletcher & Rossing 1998; also see Handel 1989 for a highly readable intuitive introduction to these topics).

Sound waves will reflect from nest walls if the walls are sufficiently hard, non-porous and extensive relative to the wavelength of the sound. The problem for the parent is that the reflected sound will interfere with the calls coming directly from the nestlings. Such reverberations within the nest may obscure information encoded in the fine structure of the call, especially in rapid amplitude modulations. Reverberation should vary considerably with nest type, being higher in nests that provide broad reflective surfaces like wood, mud or leaves, than in nests made of varied surfaces that scatter sound, like sticks or grass. Thick nest linings of feathers and other soft materials reduce reverberation, by absorbing sound energy that otherwise would have been reflected.

Design features that deal with reverberation are also likely to vary among nest types. Specifically, reverberant nests such as cavity nests may select for calls that lack rapid amplitude modulation, are shorter, have a broader frequency range and are more directional. Rapid amplitude modulations may be selected against, because the quiet portions of the modulations will be blurred by the continuing reverberations of the modulation peaks that precede them. Short calls, on the other hand, may be favoured, because they have less time to produce reverberations and to be obscured by them. Similarly, calls with a broad frequency range may be advantageous because they clearly mark the onset and offset of calls smeared by reverberation. Finally, increasing the directionality of calls (i.e. radiating calls in a narrow beam from the nestling to the parent) may be favoured because it would strengthen the sound coming from the nestling, relative to sound reflected from the cavity walls. This strategy would work best in nests with low to moderate reverberation, because in highly reverberant nests the reflections may be too intense to overcome. Also, directionality is probably only achieved when calls have wavelengths that are small relative to the nestling's head and body size (Larsen & Dabelsteen 1990). This would generally select against lower frequency calls, whose longer wavelengths would tend to wrap around the caller's head and propagate in all directions.

So far, no one has directly tested these relationships between nest structure and call design. Supportive evidence, however, comes from studies relating call features and nest type to predation risk. Specifically, calls with broader frequency ranges occurred more commonly in cavity nests, which should tend to be highly reverberant (Redondo & Arias de Reyna 1988; Briskie et al. 1999, Table 1). Also, higher frequency calls, which are generally more directional (see above), occurred in species with nests that should have low to moderate reverberation (i.e. non-cavity nests; Redondo & Arias de Reyna 1988; but see Briskie et al. 1999, Table 1). These observations are intriguing, because they suggest a link between nest type and call structure. Another study failed to show these trends, however (Popp & Ficken 1991). Thus, they clearly need to be re-examined more directly, by measuring the impact of nests on calls *in situ*.

An alternative to circumventing reverberations, especially in cavity nests where they are probably inescapable, is to exploit sound reflections in ways that amplify rather than distort signals. This could be done by calling from particular locations in the nest or at particular frequencies. For instance, by calling within a half wavelength of a reflective wall, a nestling may increase the loudness of its calls. This is because the reflection of the sound from the wall combines with the sound coming directly from the caller (Hodgson & Warnock 1992). Calling at the modal frequencies of the nest may also be advantageous. The modal frequencies of an enclosure are those for which constructive and destructive interference between reflections produce standing waves of sound, with pronounced peaks of intensity at certain locations in the enclosure (Nefske & Sung 1992). A nestling calling at one of these modal frequencies may thus present a more intense signal to the parent than a nestmate calling equally intensely, but not at a modal frequency.

The amplification that nestlings could achieve by exploiting reflections and modal frequencies is hard to predict, because it will be strongly affected by variation in the shape or composition of the nest walls. In a hard-walled nest, however, amplification could be substantial. For instance, some species of crickets and frogs increase the loudness of their calls by 10 and 6 dB, respectively, by calling against concave leaf surfaces (Prozesky-Shulze et al. 1975; Wells & Schwartz 1984). Others amplify their calls by 18 dB and 33 dB, respectively, by calling at the resonant frequency of their burrows (Bailey & Roberts 1981; Bennet-Clark 1987). Similar levels of amplification would be a significant source of variation in the case of nestling calls, when one considers that variation in begging sound pressure levels is in the order of 40 dB between and 10 dB within species (Briskie et al. 1999; Leonard & Horn 2001a).

The examples discussed above should be sufficient to illustrate that nest acoustics are likely to have a strong effect on sound even across the short distance between parents and young and that these effects may in turn select for particular design features in begging calls. The specific nature of these features are hard to predict, though, because small variations in nests can have large acoustic implications and because nestlings might either circumvent or exploit these acoustic effects.

Reception: Jamming by Nestmates

Perhaps the most obvious source of noise in communication between nestlings and their parents is interference from competing signals. Loud concurrent calls from several begging nestlings might mask one another and obscure call characteristics that parents could use when making feeding decisions. Such signal jamming should select for signals that are detectable against this background noise. Note that selection for more detectable signals is expected whether parents are discriminating amongst competing calls or simply feeding the first nestling to capture their attention.

Either way, selection should favour acoustic signals with features that make them locatable and easier to discriminate from other signals (Klump 1996). We will discuss each of these design requirements in turn.

Call Location

For acoustic signals, locatability is the ease with which a listener can determine the location of a sound source. Some researchers (e.g. Choi & Bakken 1990) have suggested that locatability of begging calls within the nest is a lost cause for parents because nestlings are packed so tightly together that their sounds intermingle. The task does seem difficult at first glance, but the few bird species that have been tested can localize individual sound sources that are as little as 20 to 30 degrees apart (Dooling 1992). This would be equivalent to a parent bird discriminating between calls that are 3 to 5 cm apart when it is positioned 10 cm above the nestlings. The ability to localize sounds can be improved further, depending on the nature of the sound and the task involved. For example, performance is better for broadband sounds and for differentiating the relative, as opposed to the absolute, position of two sound sources (Dooling 1992).

Given that parents can potentially localize sounds within the nest, there should be selection for design features that further enhance the locatability of begging calls. These features could counteract the masking effect of noisy nestmates by helping parents to attend to particular nestlings and to associate overlapping calls with individual callers (Klump 1996). The design features that aid localization by birds are poorly understood (Dooling 1992), but may include use of frequencies that parents can easily perceive (Klump et al. 1986) and use of broadband sounds (Dooling 1992).

To date, no study has directly examined whether the structure of begging calls enhances locatability by parents. A few studies have tested whether the features that make calls more locatable have been suppressed in species with heavier predation (D.G. Haskell this volume); a different but related question. For instance, species with relatively low predation rates tend to have call features thought to aid localization, such as wide frequency ranges, low frequencies and rapid modulation, although which features show the predicted patterns varies among different studies (Redondo & Arias de Reyna 1988; Briskie et al. 1999; Haskell 1999; but see Popp & Ficken 1991), and the effect of these features on locatability in birds is still largely unconfirmed (Dooling 1992). Perhaps if predation is low, features that enhance locatability will be favoured because they help nestlings attract the attention of parents within the nest. It is worth noting, however, that call characteristics that help a parent locate a particular nestling may be different from those that help a predator find a nest.

Clearly, direct tests are needed to examine whether calls with the characteristics described above actually enhance locatability within the nest. In particular, it will be important to determine how reverberation affects locatability. As it stands, the current empirical evidence is equivocal, with reverberation enhancing locatability in some studies and degrading it in others (Brown & May 1990; Nelson & Stoddard 1998).

Call Discrimination

Another way that nestlings might overcome acoustic interference from their nestmates is by making their calls easier to discriminate. Studies on how insects, frogs and adult birds make themselves heard when other individuals are calling provide hints on how nestling birds might do the same (Gerhardt 1994; Greenfield 1994; Todt & Naguib 2000). First, nestlings might simply increase their call output, by giving longer or louder calls or by delivering their calls at a higher rate. Second, they might change the call's frequency or overall form (i.e. its shape on the spectrogram) to make their calls less similar to those of their nestmates. Third, they might adjust the timing of their calls to avoid acoustic overlap with their nestmates' calls.

Different strategies might be better suited to particular situations. For example, in tree swallows, parents usually select a nestling to feed within a second or two of their arrival. Given that the calls must be transmitted in a short time period, producing a quick succession of many calls may be more effective than, for instance, adjusting the timing of fewer calls. This reasoning may partly explain why nestling tree swallows increased their call rate in response to the calls of nestmates, but otherwise did not show some of the subtler acoustic responses listed above (Leonard & Horn 2001c).

To date, few researchers have examined how nestlings change their signals to overcome nestmate interference (but see Leonard & Horn 2001c). Instead, most have simply documented changes in begging intensity when nestmates are present (e.g. Leonard & Horn 1998; Leonard et al. 2000). The situation could be rectified relatively easily using two approaches. First, dynamic responses of nestlings to competition (or competition simulated by playback) could be tested as in previous experiments, but with the inclusion of more detail on which specific signal features change (Leonard & Horn 2001c). Second, relatively static features of begging, like individual differences in call structure, could be examined to see whether they enhance the discriminability of begging displays within nests. Conceptual tools already developed for studies of parent-offspring recognition will be useful here, as they include quantitative measures of discriminability and how they change with relevant variables such as brood size (Beecher 1991; Medvin et al. 1993).

To summarize, the structure and delivery of begging calls should show adaptations for overcoming the interfering signals of nestmates, including design features that enhance locatability and discriminability. Some studies have examined begging calls for features that enhance locatability to predators, although those features need to be re-examined in terms of locatability to parents within the confined space of the nest. Features that enhance discriminability have received little attention, yet relatively minor refinements of studies of nestling competition would be required to examine these features.

Determining the specific strategies that nestlings take in the face of acoustic interference is directly relevant to the evolution of 'conspicuous' begging. If nestlings can circumvent interference, for example by adjustments in call timing or structure, escalation of overall begging intensity is less likely to occur, than if nestlings increase the rate, loudness and length of their calls to scramble above the din of their nestmates. Note that these considerations are simply consequences of sharing a communication channel and apply whether or not nestlings are engaged in some sort of conflict with nestmates or parents over the division of resources. Given the intricacies of adaptations to counteract signal interference in animal choruses (Gerhardt 1994; Greenfield 1994; Todt &

Naguib 2000), we should not be surprised if signal interference also influences the form of displays given in crowded nests.

The Contribution of Visual Components

We focused this chapter on acoustic signalling, because the efficacy of acoustic signals are well understood. We should stress, however, that factors similar to those discussed for acoustic components of begging might also explain the design of the visual components, such as gaping, neck stretching and wing flapping. Here, we briefly relate some of the factors discussed above to the visual aspects of the signal.

First, transmission and reception requirements of visual signals may select for particular signal features. For example, across species, the conspicuousness of the fleshy flanges surrounding gapes increases with decreasing illumination, presumably giving parents clearer targets when feeding in dim light. The colour of the gapes themselves did not show this pattern, but might be expected to correlate with the spectral composition of ambient light instead (Kilner & Davies 1998).

For effective reception by parents, visual components of begging must not only be illuminated, but must also be conspicuous against the visual background of the nest and nestmates. Gapes, for instance, may be coloured so that they stand out against nesting material. Other visual features of begging may be designed to be conspicuous against a background of begging nestmates. Humans pick out visual targets more easily if the targets move, have sudden onsets and appear to be at a different depth from distracting stimuli (van der Heijden 1996). Similarly, nestlings might attract the parent's attention by waving their wings, begging suddenly and stretching their necks. Indeed, the benefits of capturing parental attention quickly may have selected for the most distinctive features of the begging display (Dawkins & Guilford 1997).

Another point concerning the visual components of begging is that they may well interact with begging calls in ways that increase the efficacy of the begging display as a whole. For example, acoustic and visual components that start simultaneously and issue from the same place should help parents to localize and attend to individual nestlings. Moreover, combining cues from different modalities may be particularly effective. In a variety of psychological tasks, combinations of cues from different sensory modalities, such as sound and movement, enhance performance more than combinations of cues from single modalities, such as call loudness and length or gape colour and shape (Rowe 1999). This advantage may be one reason why begging is a multimodal display. Certainly the many components of begging, taken in combination, probably enhance the efficacy of the entire display. How they all fit together to do so has nevertheless been neglected and deserves more attention (see also R.M. Kilner this volume).

BALANCING THE COSTS OF EFFICACY

Up to this point, we have predicted that begging should have design features that enhance its efficacy. These same features, however, may also make nestlings more obvious to predators. Signal design should thus be constrained to some degree by the impact of inappropriate audiences (Wiley 1994). In fact, studies like those on call locatability discussed above suggest that some variation in begging call structure across species may be the result of differences in susceptibility to predation.

We would expect design features to vary in relation to predation risk within species, as well. Nestlings might be expected to modulate the efficacy of their begging not only according to the likelihood that the parent will respond, but also according to the perceived risk of predation. If so, some of the variation in begging intensity that has been attributed to reliable signalling of need, manipulation or sibling competition might instead be related to the nestlings' attempts to be conspicuous to appropriate receivers (parents), yet cryptic to inappropriate audiences (e.g. predators; see also Johnstone 1998).

We have some experimental evidence consistent with this suggestion (M.L. Leonard & A.G. Horn unpublished data). We presented nestling tree swallows under standardized periods of food deprivation with three acoustic stimuli: (1) the sound of an adult tree swallow landing on a nestbox giving a call known to stimulate begging; (2) the same sound, but without the call; and (3) the sound of a nest predator, the common grackle (*Quiscalus versicolor*), landing on the box. We found that the nestlings begged to all three stimuli, but gave a more intense begging display in response to the tree swallow sound with the call than to the other two stimuli.

This increased begging intensity was presumably not because of greater motivation (i.e. nestlings were equally hungry), but rather because nestlings were more certain that the stimuli they heard was the parent. In effect, the parental call is the nestlings' insurance that the bump on the box is a parent and not a gust of wind or, worse, a predator. It ensures that the benefits of conspicuousness to parents outweigh the costs of eavesdropping by predators. Conversely, when nestlings are less sure whether the parent has arrived, but too hungry not to beg, they might give a less conspicuous signal to hedge their bets. Some other experiments have shown that nestlings do not beg or beg less after they hear parental alarm calls (Greig-Smith 1980; Halupka 1998; but see Maurer 2001). Further tests of intermediate levels of threat and their effect on aspects of begging displays may prove interesting.

IS BEGGING CONSPICUOUS?

We have suggested that at least some features of begging that seem excessively conspicuous might in fact be needed for effective signalling. If so, then the display should show design features we would expect based on transmission and reception requirements. Such a relationship on its own, however, does not eliminate alternative explanations. For instance, signals might be conspicuous because nestlings are exaggerating their needs to extract extra resources from parents or because conspicuous signals have costs necessary for enforcing signal honesty. Indeed, if nestlings are to manipulate parents, they should produce signals that are easily received. Similarly, if they are to incur a cost by signalling, they may as well do it by producing a signal that parents will notice.

So how do we tell whether begging is just conspicuous enough to get the signal across, and no more? Tests of parental responses to the various features of begging are an obvious first step. In principle, these tests can be designed to measure the minimal visual and acoustic stimulation that parents can detect and the minimal variation in begging displays that they can discriminate. Transmission and reception noise characteristics of nest sites can then be added, to see how much nestlings must intensify their begging displays to ensure the signals are detected by parents (Klump 1996). If actual begging displays exceed this requirement, then they are more conspicuous than they need to be. Presumably the extra conspicuousness is required to elicit a response from parents that can perceive the signal, but are reluctant to respond.

The sales resistance of parents could also take the form of a perceptual insensitivity to begging signals, rather than a reluctance to respond to signals once they are perceived. Even though the begging display is just effective enough to be perceived by the parent, it is nevertheless more conspicuous than would be needed if parents were more perceptually acute. Some laboratory results suggest that parents may perceive begging calls differently to other vocalizations (Beecher & Stoddard 1990; Dooling et al. 1990; Loesche et al. 1992) suggesting that they may have evolved perceptual mechanisms specifically for dealing with begging displays. Further tests could examine whether parents are particularly insensitive to features of begging that signal offspring demands (e.g. Loesche et al. 1991).

FUTURE DIRECTIONS

We began this chapter with a reminder that the initial interest in begging stemmed from its apparently superfluous intensity and complexity. We have shown, however, that problems in the transmission of begging signals could favour features that increase their intensity and complexity. In the future, efficacy should be considered alongside information content and conflicts of interest, as a possible explanation for the form and function of begging signals.

Further research on the efficacy of begging signals will have four main questions to tackle: (1) How do parents extract information from begging displays? (2) What are the acoustic and optical characteristics of nests and how do they affect the design of begging signals and their reception by parents? (3) How do dynamic and static features of the begging display help nestlings deal with signal jamming by nestmates? (4) How does the risk of eavesdropping by predators influence the efficacy of begging displays?

The unique design features that enable nestlings to cope with transmission and reception problems should contribute new insights into our understanding of signals in general. In particular, since the solutions are likely to vary considerably among different ages and species of nestling, begging displays offer unique and barely tapped opportunities for developmental and comparative studies of the design of animal signals.

ACKNOWLEDGEMENTS

Our particular thanks to the many students that have contributed to our work on the tree swallows, and to the Coldwell, Hines and Minor families for permission to work on their land. We are grateful to the participants of the 2000 Begging Workshop in Gregynog, Wales for stimulating and useful discussion, to Rob Magrath and Mike Crisp of the Australian National University, for intellectual and logistic support during the writing of this chapter and to Peter McGregor and Jon Wright, for invaluable comments on an earlier draft. This work is supported by an NSERC research grant.

REFERENCES

Bailey, W.J. & Roberts, J.D. 1981. The bioacoustics of the burrowing frog *Heleioporus* (Leptodactylidae). *Journal of Natural History* **15**, 693-702.

Beecher, M.D. 1991. Successes and failures of kin recognition in animals. In: *Kin Recognition* (Ed. by P.G. Hepper). Cambridge: Cambridge University Press.

Beecher, M.D. & Stoddard, P.K. 1990. The role of bird song and calls in individual recognition: contrasting field and laboratory perspectives. In: *Comparative Perception. Vol. 2. Complex Signals* (Ed. by M.A. Berkley & W.C. Stebbins). New York: John Wiley & Sons.

Bennet-Clark, H.C. 1987. The tuned singing burrow of mole crickets. *Journal of Experimental Biology* **128**, 383-409.

Briskie, J.V., Martin, P.R. & Martin, T.E. 1999. Nest predation and the evolution of nestling begging calls. *Proceedings of the Royal Society of London, Series B* **266**, 2153-2159.

Brown, C.H. & May, B.J. 1990. Sound localization and binaural processes. In: *Comparative Perception. Vol. 1. Basic Mechanisms* (Ed. by M.A. Berkley & W.C. Stebbins). New York: John Wiley & Sons.

Budden, A.E. & Wright, J. 2001. Begging in nestling birds. *Current Ornithology* **16**, 83-118.

Choi, I.-H. & Bakken, G.S. 1990. Begging response in nestling red-winged blackbirds (*Agelaius phoeniceus*): effect of body temperature. *Physiological Zoology* **63**, 965-986.

Clark, A.B. & Lee, W.-H. 1998. Red-winged blackbird females fail to increase feeding in response to begging call playbacks. *Animal Behaviour* **56**, 563-570.

Dawkins, M.S. & Guilford, T. 1997. Conspicuousness and diversity in animal signals. *Perspectives in Ethology* **12**, 55-72.

Dooling, R.J. 1992. Hearing in birds. In: *The Evolutionary Biology of Hearing* (Ed. by D.B. Webster, R.R. Fay & A.N. Popper). New York: Springer-Verlag.

Dooling, R.J., Brown, S.D., Park, T.J. & Okanoya, K. 1990. Natural perceptual categories for vocal signals in budgerigars (*Melopsittacus undulatus*). In: *Comparative Perception. Vol. 2. Complex Signals* (Ed. by M.A. Berkley & W.C. Stebbins). New York: John Wiley & Sons.

Fletcher, N.H. 1992. *Acoustic Systems in Biology*. Oxford: Oxford University Press.

Fletcher, N.H. & Rossing, T.D. 1998. *The Physics of Musical Instruments*, 2nd Edn. New York: Springer-Verlag.

Gerhardt, H.C. 1994. The evolution of vocalization in frogs and toads. *Annual Review of Ecology and Systematics* **25**, 293-324.

Greenfield, M.D. 1994. Cooperation and conflict in the evolution of signal interactions. *Annual Review of Ecology and Systematics* **25**, 97-126.

Greig-Smith, P.W. 1980. Parental investment in nest defence by stonechats (*Saxicola torquata*). *Animal Behaviour* **28**, 604-619.

Halupka, K. 1998. Nest predation in meadow pipits *Anthus pratensis* nesting in natural conditions. *Ornis Fennica* **75**, 139-143.

Handel, S. 1989. *Listening: An Introduction to the Perception of Auditory Events*. Cambridge: MIT Press.

Haskell, D.G. 1999. The effect of predation on begging-call evolution in nestling wood warblers. *Animal Behaviour* **57**, 893-901.

Hauser, M. 1996. *The Evolution of Communication*. Cambridge: MIT Press.

Hodgson, M. & Warnock, A.C.C. 1992. Noise in rooms. In: *Noise and Vibration Control Engineering: Principles and Applications* (Ed. by L.L. Beranek & I.L. Vér). New York: John Wiley & Sons.

Hoy, R.R., Popper, A.N. & Fay, R.R. 1998. *Comparative Hearing: Insects*. New York: Springer-Verlag.

Johnstone, R.A. 1998. Conspiratorial whispers and conspicuous displays: games of signal detection. *Evolution* **52**, 1554-1563.

Kilner, R. & Davies, N.B. 1998. Nestling mouth colour: ecological correlates of a begging signal. *Animal Behaviour* **56**, 705-712.

Kilner, R. & Johnstone, R.A. 1997. Begging the question: are offspring solicitation behaviours signals of need? *Trends in Ecology and Evolution* **12**, 11-15.

Klump, G.M. 1996. Communication in the noisy world. In: *Ecology and Evolution of Acoustic Communication in Birds* (Ed. by D.E. Kroodsma & E.H. Miller). Ithaca: Cornell University Press.

Klump, G.M., Windt, W. & Curio, E. 1986. The great tit's (*Parus major*) auditory resolution in azimuth. *Journal of Comparative Physiology A* **158**, 383-390.

Kroodsma, D.E. & Miller, T.E. 1996. *Ecology and Evolution of Acoustic Communication in Birds*. Ithaca: Cornell University Press.

Larsen, O.N. & Dabelsteen, T. 1990. Directionality of blackbird vocalization - implications for vocal communication and its further study. *Ornis Scandinavica* **21**, 37-45.

Leonard, M.L. & Horn, A.G. 1998. Need and nestmates affect begging in tree swallows. *Behavioral Ecology and Sociobiology* **42**, 431-436.

Leonard, M.L. & Horn, A.G. 2001a. Acoustic signalling of hunger and thermal state by nestling tree swallows. *Animal Behaviour* **61**, 87-93.

Leonard, M.L. & Horn, A.G. 2001b. Begging calls and parental feeding decisions in tree swallows (*Tachycineta bicolor*). *Behavioral Ecology and Sociobiology* **49**, 170-175.

Leonard, M.L. and Horn, A.G. 2001c. Dynamics of calling by tree swallow (*Tachycineta bicolor*) nestmates. *Behavioral Ecology and Sociobiology* **50**, 430-435.

Leonard, M.L., Horn, A.G., Gozna, A. & Ramen, S. 2000. Brood size and begging intensity in nestling tree swallows. *Behavioral Ecology* **11**, 196-201.

Loesche, P., Stoddard, P.K., Higgins, B.J. & Beecher, M.D. 1991. Signature versus perceptual adaptations for individual recognition in swallows. *Behaviour* **118**, 15-25.

Loesche, P., Beecher, M.D. & Stoddard, P.K. 1992. Perception of cliff swallow calls by birds (*Hirundo pyrrhonota* and *Sturnus vulgaris*) and humans (*Homo sapiens*). *Journal of Comparative Psychology* **106**, 239-247.

Maurer, G. 2001. Nestling begging behaviour and the impact of parental alarm-calls in white-browed scrubwrens, *Sericornis frontinalis* (Vigors & Horsfield). Diplom-Biologen Diplomarbeit, Rheinischen Friedrich-Wilhelms-Universität.

Medvin, M.B., Stoddard, P.K. & Beecher, M.D. 1993. Signals for parent-offspring recognition: a comparative analysis of the begging calls of cliff swallows and barn swallows. *Animal Behaviour* **45**, 841-850.

Nefske, D.J. & Sung, S.H. 1992. Sound in small enclosures. In: *Noise and Vibration Control Engineering: Principles and Applications* (Ed. by L.L. Beranek & I.L. Vér). New York: John Wiley & Sons.

Nelson, B.S. & Stoddard, P.K. 1998. Accuracy of auditory distance and azimuth perception by a passerine bird in natural habitat. *Animal Behaviour* **56**, 467-477.

Popp, J. & Ficken, M.S. 1991. Comparative analysis of acoustic structure of passerine and woodpecker nestling calls. *Bioacoustics* **3**, 255-274.

Price, K., Harvey, H. & Ydenberg, R. 1996. Begging tactics of nestling yellow-headed blackbirds, *Xanthocephalus xanthocephalus*, in relation to need. *Animal Behaviour* **51**, 421-435.

Prozesky-Schulze, L., Prozesky, O.P.M., Anderson, F. & van der Merve, G.J.J. 1975. Use of a self-made sound baffle by a tree cricket. *Nature* **255**, 142-143.

Redondo, T. & Arias de Reyna, L. 1988. Locatability of begging calls in nestling altricial birds. *Animal Behaviour* **36**, 653-661.

Redondo, T. & Castro, F. 1992. Signalling of nutritional need by magpie nestlings. *Ethology* **92**, 193-204.

Rowe, C. 1999. Receiver psychology and the evolution of multicomponent signals. *Animal Behaviour* **58**, 921-931.

Todt, D. & Naguib, M. 2000. Vocal interactions in birds: the use of song as a model in communication. *Advances in the Study of Behavior* **29**, 247-296.

van der Heijden, A.H.C. 1996. Visual attention. In: *Handbook of Perception and Action. Vol. 3. Attention* (Ed. by O. Neumann & A.F. Sanders). London: Academic Press.

Wells, K.D. & Schwartz, J.J. 1984. Vocal communication in a neotropical treefrog, *Hyla ebraccata*: aggressive calls. *Behaviour* **91**, 128-145.

Wiley, H.R. 1994. Errors, exaggeration, and deception in animal communication. In: *Behavioral Mechanisms in Behavioral Ecology* (Ed. by L.A. Real). Chicago: University of Chicago Press.

8.

ENERGETIC COSTS OF BEGGING BEHAVIOUR

Mark A. Chappell[1] & Gwendolyn C. Bachman[2]
[1]*Department of Biology, University of California, Riverside CA 92521, USA (chappell@citrus.ucr.edu)*
[2]*School of Biological Sciences, University of Nebraska, Lincoln NE 68588-0118, USA (gbachman@unlserve.unl.edu)*

ABSTRACT

Energy expenditure during begging by nestling birds is of interest because of its potential role in the social dynamic between provisioning parents and their young. In this chapter we discuss methods of measuring the energy cost of begging and argue that continuous-flow respirometry is the method of choice. Evaluating the impact of begging on nestling energetics and fitness requires an estimate of nestling energy budgets. To date, studies show that the energy cost of begging comprises a very small fraction of the nestling's total energy budget. With existing estimates of the time budget and energy cost of begging, we calculate that even large increases in begging activity are likely to have little negative impact on nestling fitness. We suggest that additional work on begging energetics should broaden the range of behaviours and social groupings that are examined in order to explore the limits to begging activity in nestling birds.

INTRODUCTION

Altricial nestlings and a parent bringing food to the nest may face conflicts of interest over the amount of food delivered and how it is allocated among the brood (Godfray & Parker 1992). If parents rely on nestling begging signals to indicate the need for food, the potential exists for nestlings to be dishonest about their needs despite the potential fitness detriment to the parent or siblings. For example, individual nestlings could attempt to obtain a disproportionate fraction of the food delivered to the nest (thereby harming siblings and possibly parents) or coerce parents into delivering food at rates

J. Wright and M.L. Leonard (eds.), The Evolution of Begging, 143–162.

that benefit the nestling, but are harmful to parental fitness. Theoretical analyses suggest that evolutionarily stable communication (i.e. honest signalling) in such a situation requires that signals used for begging have a cost that reduces fitness (e.g. Johnstone & Grafen 1992; Godfray 1995; Kilner & Johnstone 1997; R.A. Johnstone & H.C.J. Godfray this volume). Two of the more obvious ways in which begging could reduce nestling fitness are (1) increasing predation risk by making nests easier to locate (Haskell 1994; D.G. Haskell this volume) and (2) increasing energy expenditure. If begging has a substantial energy cost, it can divert energy from growth, development or maintenance and thereby decrease nestling or fledgling survival. High energy costs of begging have been assumed by some authors (e.g. 3.5 times resting metabolism; Beauchamp et al. 1991), but until recently there were no empirical data on the energetics of begging.

In this chapter we consider whether the energy cost of begging could play a role in the evolution of an honest begging signal. We describe methodologies for quantifying the energy cost of begging, discuss the ecological context within which these costs should be considered and summarize current data on the energetics of begging. We conclude with an analysis of the key factors that determine the energy cost of begging and discuss the likelihood that this cost is an important selective factor in the evolution of begging behaviour.

MEASURING THE ENERGY COST OF BEGGING

Begging is very brief compared to most behavioural or physiological events that affect energy metabolism and always occurs against a background of substantial, continuous and often variable energy utilization. Therefore, accurate measurements of the energy cost of begging pose a considerable technical challenge. More importantly, the various methods used for measuring begging energetics yield data of highly variable quality and it is crucial to understand the strengths and weaknesses of different techniques when designing experiments or interpreting results.

Most physiologists rely on respirometry (measurement of gas exchange) to quantify energy metabolism and it is the method of choice for studying the energetics of short-term events like begging. Rates of gas exchange such as oxygen consumption (VO_2) or carbon dioxide production (VCO_2), obtained from respirometry are converted into units of energy by the appropriate scaling factor. That factor varies with the metabolic substrate being oxidized (carbohydrate, lipid or protein), and the effect of substrate on energy equivalence - and hence the uncertainty of conversion - is much greater for VCO_2 (20.9-28.2 joules/ml) than for VO_2 (19.5-20.9 joules/ml).

Good CO_2 analysers, however, generally have substantially higher resolution and are much more stable than most O_2 analysers, which confers a considerable advantage when attempting to measure the low metabolic rates typical of small nestlings.

For studies of VO_2 or VCO_2, there are two respirometry methods: closed-system and open-system. In both, nestlings are enclosed in a metabolism chamber, begging is elicited through various stimuli (e.g. tactile, hand movement, light fluctuation, sound) and rates of gas exchange determined.

Closed-System Techniques

In closed-system (i.e. constant volume) respirometry, the animal is sealed inside a gas-tight chamber for the period of measurement and gas exchange over that interval is determined from the difference between initial and final gas concentrations. This method has several advantages. It does not require a stable and accurately determined flow of air and thus is mechanically and logistically simpler than open-system methods. Also, a closed system may be more effective than an open system for animals with low metabolic rates (e.g. small altricial hatchlings). This is because animals with low metabolism will generate correspondingly small changes in gas concentration in open-flow systems and these minimal deflections may not be resolvable with many gas analysers, especially O_2 analysers. The same animals will, however, eventually produce a comfortably measurable change in gas concentration in a closed system if the measurement interval is sufficiently long. For determinations of VO_2 it is even possible to dispense with gas analysers by using manometric methods; water vapour and CO_2 in the chamber are absorbed chemically and the change in air volume or pressure resulting from O_2 uptake is monitored.

For studies of begging energetics, the main limitation of closed-system respirometry is that it has poor resolution of short-term metabolic events. In its basic form, each measurement provides a single datum that is the average metabolism over the complete measurement interval. In order to separate the metabolic effects of begging or any other activity from background metabolism, it is necessary to make multiple measurements that incorporate different levels of the activity of interest. With a range of data in hand, regression techniques are employed to resolve the influences of the various components of total metabolism (Figure 1). A notable weakness of this approach is that it necessarily assumes the different components of metabolism (e.g. rates of resting and begging metabolism) are constant. That assumption is frequently invalid. Begging intensity, and hence metabolic cost, varies considerably and continuously, as do other types of activity such as movement and preening. Even resting metabolism may show substantial changes over relatively short periods because of changes in nestling body temperature (and the resulting Q10 effect on metabolism) or because of the energy consumption associated with food processing.

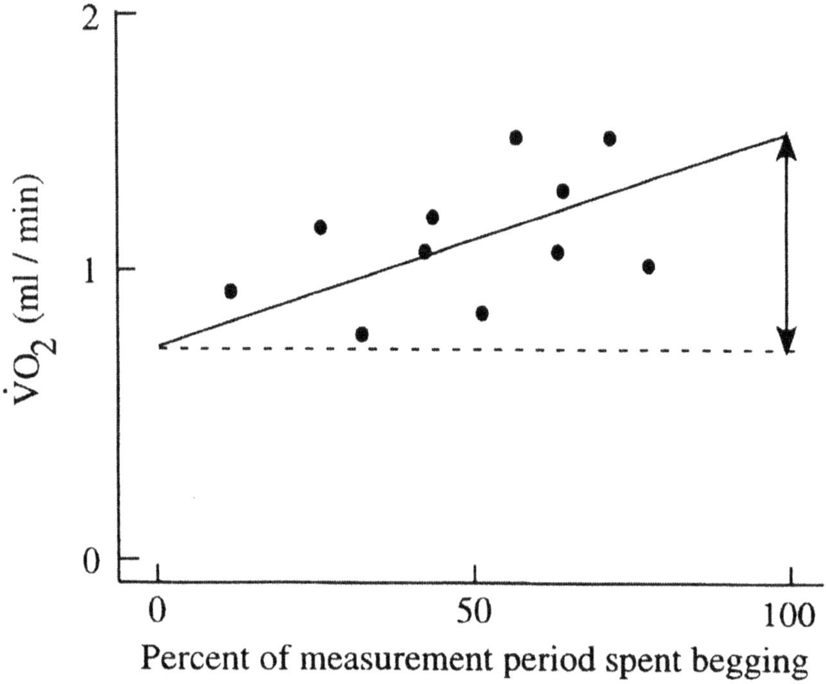

Figure 1. Estimation of begging costs from closed-system measurements of oxygen consumption. Here the assumption is that nestlings are either begging or resting, so the correlation between oxygen consumption and the percent of measurement time spent begging (i.e. the duty factor) yields an estimate of begging costs - the increase in metabolic rate during begging - as the difference between resting metabolism (dashed line) and the estimated total metabolism at a begging duty factor of 100% (vertical arrow).

It is theoretically possible to avoid these problems by continuously recording gas concentrations in a closed system and then taking the derivative of the resulting curve to determine instantaneous rates of gas exchange. However, this approach (which as far as we know has not been tried with birds) requires excellent gas mixing in the metabolism chamber and the same high resolution of the gas analysers as for open-system respirometry. Other considerations important for closed systems include avoiding excessive CO_2 build-up and selecting a chamber size that allows sufficient oxygen availability for the duration of measurements without being so large that the system is not able to resolve short bouts of behaviour.

Studies of begging energetics based on closed-system techniques include McCarty (1996) and Leech and Leonard (1996).

Open-System Techniques

In open-system (i.e. continuous flow) respirometry, the metabolism chamber is perfused with a constant flow of air or other respiratory gas mixtures and gas exchange is determined from flow rate and the difference in O_2 or CO_2 concentration between incurrent and excurrent gas. Open-system respirometry requires more equipment for air handling than closed-system methods and demands a high resolution gas analyser when used to study small animals with low metabolic rates. It provides, however, a continuous record of metabolism and therefore has much better resolution of short-term metabolic changes, such as those caused by begging, than is possible in closed-system respirometry. This makes it considerably easier to separate the energy cost of begging from that of other activities, even if the background resting metabolism varies substantially during measurements (Figure 2).

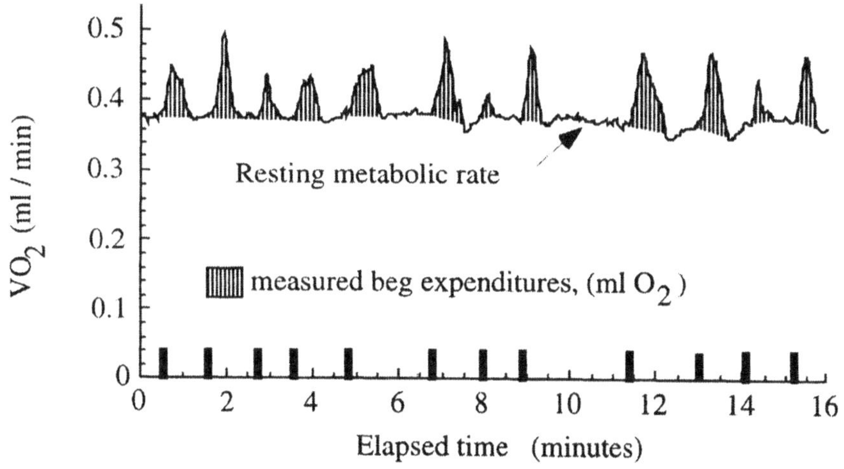

Figure 2. Oxygen consumption rates (VO_2) of a house wren nestling obtained with open-flow respirometry, showing peaks associated with begging events (black bars at the bottom of the figure; the width of the bars indicates the duration of begging events). The shaded areas indicate the integrated amount of oxygen consumed during each begging event.

Despite these clear advantages, open-system respirometry has pitfalls, some of which are poorly understood, that can seriously compromise results. A well-known issue is the capacitance-like effect that results from the mixing characteristics of all open-flow systems. Because the incurrent gas flow does not instantly exchange all of the gas in the respirometry chamber, there is mixing of new and old gas. As a result, rapid shifts in metabolism are buffered by mixing and read by the gas analyser as a gradual approach of gas concentration to a new steady-state (Figure 3a). Importantly, if the metabolic change is a brief transient, as in a begging event, there may be insufficient time to establish a new steady-state gas concentration before metabolism reverts to resting levels. Therefore, the calculated rate of gas exchange during the transient event can be considerably lower than the actual metabolic change and the apparent duration of the event will be longer than the actual duration (Figure 3b).

There are two ways to ameliorate mixing problems. One is to maximize the flow rate and minimize chamber volume, both of which reduce the washout time of the system and increase its temporal resolution. Unfortunately, high flow rates reduce the deflection in gas concentration caused by metabolism and hence may exceed the resolution limits of the gas analyser, while small chamber volumes may constrain movement unacceptably. A system that samples respiratory gas *via* a facemask instead of enclosing the animal in a chamber is ideal for temporal resolution, but is clearly inappropriate for studying begging. A popular alternative is to use a mathematical approach to back-calculate actual metabolic kinetics from the raw data, using the known mixing characteristics of the chamber (Bartholomew et al. 1981). This frequently used method is often termed the Z or instantaneous correction (Figure 3c). It can be quite effective, but it requires that the gas in the metabolism chamber be very well mixed. Mixing is facilitated by a fan in the chamber or by a reasonable match between flow rate and chamber volume. A good rule of thumb is to restrict the chamber volume in ml to no greater than two to three times the flow rate in ml/minute. Unless mixing is excellent, use of the instantaneous correction with a combination of low flow rate and high chamber volume is likely to generate substantial error.

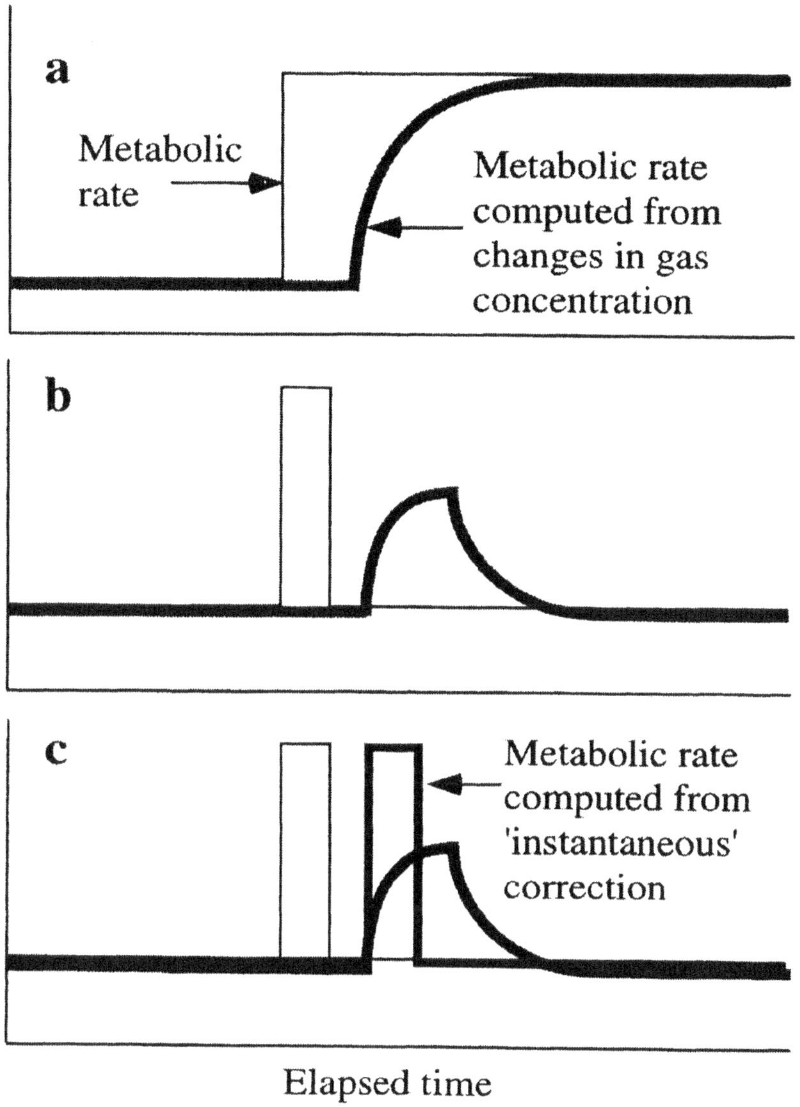

Figure 3. The effects of gas mixing in open-flow respirometry. The narrow lines represent actual metabolic rate; thick lines represent the metabolic rate computed from measured deflections in gas concentration. Note that this figure assumes perfect mixing. (a) Step change in metabolism, showing mixing artefacts and time lag. (b) Transient step change in metabolism. Note that as a result of mixing artefacts the metabolic change is recorded as a change in gas concentration that is longer and lower than the real metabolic change and does not attain steady state. (c) Transient change in metabolism with metabolic rate calculated using the instantaneous conversion (labelled thick line). Note that the mixing artefacts have been removed by the conversion, but the time lag remains.

Movement artefacts are another, more poorly appreciated difficulty with open-system respirometry. If the animal moves within the chamber relative to the inlet and outlet ports for gas flow, this will produce transient enrichments (if movement is towards the outlet port) or rarefactions (if movement is away from the outlet port) of exhaled gas in the excurrent air stream (Berrigan & Lighton 1993; Bachman & Chappell 1998). These movement effects, which have nothing to do with changes in metabolism, create spurious artefacts in calculated metabolic rates. The artefacts are particularly severe when an instantaneous correction is applied (Figure 4), as the conversion algorithm assumes a temporally variable but spatially fixed source of gas exchange. Since nearly all avian begging is characterized by extensive movements, especially of the head (i.e. the source of exhaled air), the problem of movement artefacts is very difficult to avoid.

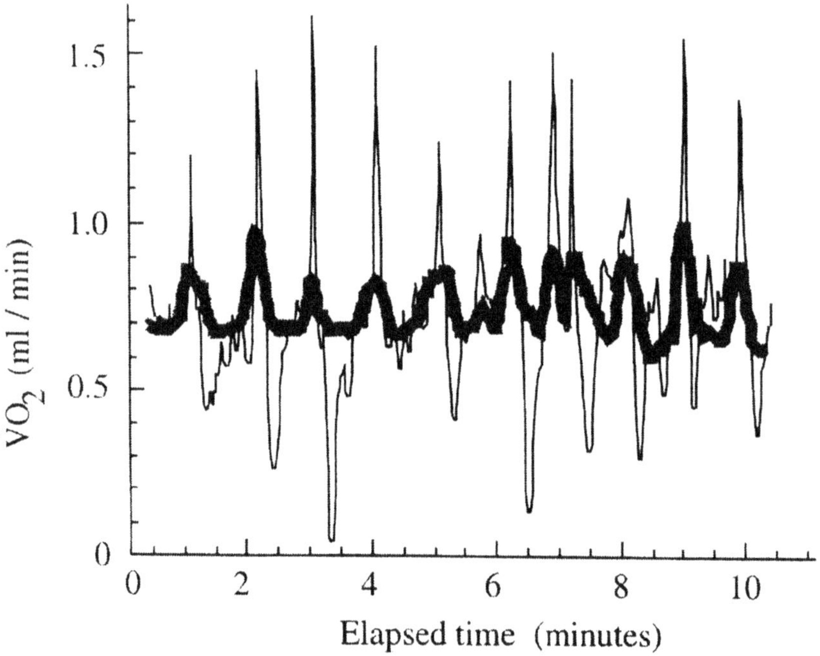

Figure 4. Effects of position changes on calculated oxygen consumption in a begging house wren nestling. The thick line is the uncorrected oxygen consumption rate and the thin line is oxygen consumption calculated with an instantaneous correction. Note that the head movements during begging cause artefacts, such as high peaks and low valleys, in the output of the instantaneous conversion.

To minimize such complications during our studies of begging energetics in house wren (*Troglodytes aedon*) nestlings (Bachman & Chappell 1998), we first subtracted baseline resting metabolism (measured as oxygen consumption in ml/minute, VO_2) from the metabolic rate during a beg and then integrated the remaining VO_2 over time (Figure 2). This yielded the cumulative total gas exchange associated with a begging event (in ml O_2). To calculate the average rate of energy consumption during a begging event, we divided the total gas exchange by the duration of the begging event (determined from simultaneous videotape records) and multiplied the result by the heat equivalent for oxygen consumption, which is about 20 joules/ml O_2. The limitation of this approach is that it does not reveal instantaneous rates of gas exchange during a begging event (e.g. the peak power output). Unfortunately, with current technology and the inherent physical constraints of open-flow respirometry, it is difficult to make measurements of sufficient precision to yield true instantaneous data.

It should be emphasized that the respirometric methods described above measure only the aerobic components of energy metabolism. For short-term, relatively intense activities such as begging, it is possible that anaerobic pathways provide some of the necessary ATP (Weathers et al. 1997). A rough indication of anaerobic metabolism can be gained from analysis of blood lactate levels or from the presence of a post-exercise oxygen debt (high rates of O_2 uptake persisting after the completion of exercise). We are, however, aware of no methods that effectively measure anaerobic metabolism on a continuous basis; indeed, an accurate determination generally requires whole-body lactate analysis. Data we have obtained from house wren nestlings of various ages (Bachman & Chappell 1998; Chappell & Bachman 1998) suggest that anaerobic contributions to begging energetics or to short bouts of forced exercise are minimal.

Doubly-Labelled Water

The doubly-labelled water (DLW) technique measures an animal's integrated metabolic expenditures over a fairly long period. It is often used in examinations of daily energy expenditure in free-living animals, has occasionally been employed in examinations of signal costs (e.g. Speakman et al. 1989; Vehrencamp et al. 1989) and in at least one study it was applied to a study of begging energetics (Soler et al. 1999). Briefly, the DLW method uses the differential rates of loss of isotopically labelled oxygen and hydrogen atoms, usually injected as water, to estimate energy metabolism (see Nagy 1983 or Speakman 1997 for detailed descriptions). Oxygen is lost as water and CO_2, whereas hydrogen is lost only as water. Accordingly, the difference in loss rates is a measure of CO_2 production. For studies of begging energetics, the DLW method has a serious disadvantage; it requires substantial turnover of both isotopes, which necessitates long measurement periods (this is dependent on overall metabolic rates as well as rates of water turnover). As discussed above for closed-system respirometry, long measurement intervals make it difficult to separate the energy cost of begging from other components of the total metabolic rate. The DLW technique is also intrinsically less accurate than respirometry for measuring metabolic rates and requires considerable handling and sampling, which may induce errors due to stress and struggling.

INTEGRATED BEGGING COSTS

Appropriately used, respirometry can yield reliable data on the energy costs of individual begging events or bouts of begging. With carefully designed experiments and sufficient data on a range of begging behaviour, various statistical methods (e.g. multiple regression) can quantify the energetic effects of individual variation in begging such as the intensity or duration of begging events, the possible influence of parental or sibling behaviours or other factors such as age, size, body temperature, nutritional state or presence of parasites.

With this information in hand, there remains the issue of calculating ecologically relevant begging energy costs. As discussed in the next section, a key factor in evaluating begging energetics in a useful context is determining how frequently and how long nestlings beg in typical nest environments. Begging rate may vary substantially during the course of a day and is also dependent on relatively unpredictable parameters such as weather and food availability. It is important to account for such changes in begging cost estimates. Time budget studies are probably the most useful approach. For our house wren work we used video monitoring of nests containing individually marked nestlings to quantify begging behaviour over substantial periods (Bachman & Chappell 1998). Briefly, we mounted video cameras in the rooves of nestboxes and taped behaviour in two-hour blocks throughout the day. Activities in respirometry chambers were scored in the same way as activities in natural nests. We scored each nestling's begging activity and other behaviours for frequency of occurrence, duration and intensity. For integrative studies of begging energetics, the most useful breakdown of begging activity is usually in terms of duty factor, which is the fraction of total time occupied by begging. These or similar time budget analyses are applicable to most avian species.

THE ENERGETIC CONTEXT OF BEGGING COSTS

To consider the significance of any metabolic costs of begging in terms of selection and fitness, they must be evaluated in an appropriate energetic context. That context is a nestling's overall budget for chemical potential energy, of which begging costs are one of many components. A brief overview of the major parts of a nestling's energy budget, in addition to begging costs, is appropriate:

Basal or minimal metabolism. The basal metabolic rate is the minimum rate of energy consumption necessary to sustain life in the absence of all other additive metabolic costs, such as the energy costs of activity, endothermic thermoregulation and so on. In the strictest sense, a growing nestling does not have a metabolic rate that satisfies the criteria for basal metabolism since the energy necessary for growth (see below) is an additive cost. Accordingly, resting metabolic rate (RMR, the metabolic rate in the absence of energy expenditure on activity, food processing and thermoregulation) is commonly used as the index of minimum metabolism. RMR must be measured in standardized thermal conditions even for nestlings that have not attained endothermy, as it is strongly dependent on body and hence ambient temperature.

Thermoregulation. In endothermic nestlings exposed to low ambient temperature, energy is used for thermostatic heat production (shivering) to maintain constant body temperature (e.g. Chappell et al. 1997). Expenditure for thermoregulation is highly variable depending on factors such as ambient conditions or parental brooding, but can be a substantial fraction of the energy budget.

Food processing costs. These costs, known to physiologists as specific dynamic action or the heat increment of feeding (HIF), are necessary for the digestion, absorption and biochemical processing of food. They are highest for protein diets and are equivalent to 5-15% or more of assimilated energy in nestlings fed arthropods (Janes & Chappell 1995; Chappell et al. 1997). Since all the chemical energy used by an animal comes from assimilated food, these processing costs account for an equivalent fraction of the overall energy budget.

Activity metabolism. This is the energy cost of muscle contraction associated with movement. Like thermoregulatory costs, activity costs are highly variable; unlike thermoregulatory costs, they are dependent on behaviour and hence are largely under the nestling's control.

Growth. These costs include the chemical potential energy contained in new tissue, as well as the synthesis cost necessary to deposit that tissue. Although a fairly small fraction of the overall energy budget, this is probably the most important component in terms of the potential fitness impact of begging costs (e.g. Verhulst & Wiersma 1997). Growth rates and efficiencies may be affected by many factors, such as energy intake, the relative balance of nutrients in the diet and temperature.

The role of begging in an energy budget is complex. It contributes to activity costs, may impact thermoregulatory costs by increasing or decreasing thermal needs and is necessary to obtain the energy needed for growth. To specifically evaluate the impact of the activity cost of begging relative to the other components of the energy budget, all components must be converted to a similar currency. This is difficult to achieve in an acceptably rigorous manner and is often subjective. Besides knowing the size of the energy budget (e.g. Weathers 1992; Weathers & Siegel 1995), there are several major questions that need to be addressed.

First, one needs to determine if the nestling's energy budget is strongly constrained. That is, increases in one component necessarily require compensatory decreases in other components. In some species, parents regulate provisioning in response to changing food requirements or solicitation (e.g. Henderson 1975; Stamps et al. 1985; Hussell 1988). In other species the rate of food delivery by parents is largely or completely unresponsive to variation in the nutritional needs of nestlings (Ricklefs 1992). In the former scenario an increase in expenditure for a particular behaviour, such as begging, does not invariably necessitate reduced expenditure elsewhere in the energy budget, but the latter situation is one in which energetic trade-offs are difficult to avoid.

Second, if there are energetic trade-offs, it is important to know which components of the energy budget are affected, as some are relatively fixed (e.g. RMR) while others are plastic (e.g. activity metabolism or the rate of new tissue deposition). Depending on how energy is allocated, a particular begging energy cost could be interpreted as having a fitness impact ranging from inconsequential to substantial.

Third, if trade-offs exist, there may be important differences in the impacts of energetic constraints on the tempo and trajectory of growth. For example, nestlings of some species are apparently constrained to a fixed developmental schedule: they fledge at the normal time, but at subnormal mass if subjected to food restriction (Schew & Ricklefs 1998; M.A. Chappell unpublished data on Adélie penguins, *Pygoscelis adeliae*). In contrast, other species appear to be constrained to a certain developmental state or size: when food-restricted they can slow or halt growth and maturation and delay fledging, but eventually fledge at approximately normal mass (Schew & Ricklefs 1998; Starck & Ricklefs 1998). Both strategies are likely to affect fitness, but in different ways.

CURRENT DATA ON ENERGY COSTS OF BEGGING

At the time of writing there are few published studies on the energetics of avian begging, so it is risky to attempt general statements. Nevertheless, data to date strongly suggest that the energy cost of begging is a very small component of the energy budget of nestlings and therefore is arguably insignificant in terms of fitness costs.

McCarty (1996) measured the energy costs of begging in nestlings of several North American passerine species. Metabolic rates during begging were 1.05 times RMR in European starlings (*Sturnus vulgaris*) and 1.27 times RMR in tree swallows (*Tachycineta bicolor*), with comparable results for another five species. Use of a closed-system respirometer limited the temporal resolution of begging events, but another study using the same method (Leech & Leonard 1996) produced almost identical results (1.28 times RMR) for five- and ten-day old tree swallows. The authors calculated that begging would increase daily energy expenditure by 12.9% and therefore concluded that begging costs might be important to fitness. However, a change of daily energy consumption of that magnitude would require a duty factor of 46% (almost half of a 24-hour day spent begging), which seems unlikely.

Somewhat higher begging costs were reported for embryonic white pelicans (*Pelecanus erythrorhynchos*) using vocal signals to solicit brooding (Abraham & Evans 1999). Increases of 20-80% above resting metabolism occur during begging, although the higher values were for embryos exposed to low temperatures and result more from a temperature induced decrease in RMR than an increase in absolute begging costs (which changed relatively little). Begging costs at normal body temperatures (35-37.8 °C) were similar to those reported by McCarty (1996) and Leech and Leonard (1996). Abraham and Evans (1999) interpreted their results as indicating a potentially significant energy cost of begging, particularly in view of the fixed energy supply in an avian egg, but they did not provide data on the duty factor for begging in natural situations.

Soler et al. (1999) attempted to use the doubly-labelled water approach to estimate begging energy costs in magpie (*Pica pica*) nestlings and their brood parasites, great spotted cuckoo (*Clamator glandarius*) nestlings. They experimentally increased or decreased the frequency of begging by selectively withholding food or feeding to satiation, but found no between-group differences in daily energy expenditure. The authors concluded that any energy costs of begging must have been low or a difference would have been observed. Their conclusion of low cost is probably correct, but the data were compromised by inadequate correction for HIF (i.e. they did not quantify the amount of food given to nestlings), the poor temporal resolution of the DLW method (i.e. they used a 24-hour measurement period) and its inherent accuracy limit of about ± 7% of daily energy expenditure (Nagy 1983; Speakman 1997), which is much larger than any begging energy cost measured to date.

In the most comprehensive study to date of energy costs in an integrated ecological and energetic context, we found that the contribution of begging costs to the energy budget of house wrens was extremely small and arguably of no significance to fitness (Bachman & Chappell 1998). Our data from open-circuit respirometry show that metabolic rate increased an average of 27% above RMR during begging; remarkably similar to the results of McCarty (1996) and Leech and Leonard (1996), with occasional begging events attaining twice that cost. When calculated as a fraction of the total daily energy budget (DEB), begging costs ranged from 0.02-0.22% over an age range of three to ten days, respectively. When calculated on the basis of the energy content of new tissue per day, begging costs were somewhat greater (0.05-2.3%), but the higher figure mainly reflects a decrease in growth rate in older nestlings instead of an increase in absolute begging costs. In contrast, non-begging activity accounted for 2-9% of DEB and was equivalent to 5-97% of the energy sequestered in new tissue in three- and ten-day old nestlings, respectively.

In a subsequent paper (Chappell & Bachman 1998) we showed that house wren nestlings probably have the physiological ability to beg with higher intensity and cost than they normally exhibit. In three- to ten-day old nestlings, metabolic rates during forced exercise averaged twice the mean metabolic rates during begging. This indicates that behaviour, not exercise capacity, limits the average intensity and metabolic cost of begging, although the maximum power output during begging is quite similar to maximum power output during forced exercise and hence may be physiologically constrained. Interestingly, rates of energy use during thermostatic heat production in older nestlings and during digestion in younger nestlings, were both considerably higher than maximum power output during begging or exercise.

THE ENERGY COST OF BEGGING AND NESTLING FITNESS

In our papers on house wren begging, we argued that the energy cost of begging was so minor (averaging 1/500 or less of the total energy budget) as to be irrelevant to fitness. There are two reasons that begging costs are low in house wrens; both are germane to predictions of begging costs for other species. First, absolute increases in metabolic rate during begging are rather small in house wrens, despite what appears subjectively to be fairly vigorous locomotor activity (see also Chappell et al. 1994). Second, the duty factor for begging is tiny. Begging consumes at most 3% (mean of 0.3-1% in nestlings of different ages) of a 24-hour day, even though the number of parental visits to the nest may be impressively large (up to 350/day).

Given our results for house wren nestlings and other published data, is it reasonable to expect that the energy costs of begging in other species might be large enough to substantially impact fitness? Of course, one can postulate that any energy cost, no matter how small, will affect fitness to some degree, but at very low costs the question assumes more rhetorical than biological importance. If we arbitrarily assume a threshold begging cost equal to 5% of the daily energy budget as being significant, some quantitative predictions can be made. Given that threshold, we believe a meaningfully large begging cost is possible, but only if certain conditions, which may be unusual in birds, are satisfied. One such condition is an absolute cost per begging event that is considerably greater than has been reported in the species studied to date. For example, we have the subjective impression, but no empirical data, that in some species such as American robins (*Turdus migratorius,* see also Smith & Montgomerie 1991), the vigour and extent of movement during begging is considerably higher than for house wrens. Therefore, it seems likely that the proportional metabolic increase above RMR is also greater. When considered as a single factor, however, the enhancement of begging effort needed to attain a substantial cumulative begging cost is dauntingly large. We estimated that ten-day-old house wren nestlings would have to increase their per-beg energy cost almost four-fold (i.e. three times their measured maximal exercise capacity) in order to achieve cumulative begging costs of 5% of the DEB - and that value was calculated using the highest begging rates we observed. The increase in effort necessary to achieve the same 5% elevation in DEB at the average begging rates is closer to 20-fold and is even greater in younger nestlings.

A second condition that would increase the net cost of begging is a high duty factor, which could be attained if the duration of each begging event was long, the frequency of begging events was high, or both. As for begging intensity, we believe that begging duty factors for some bird species are higher than for house wrens. Nevertheless, simple calculations suggest that duty factors must be considerable for begging costs to become an important fraction of the DEB. For example, if the metabolic rate during begging is twice RMR, which is higher than has been measured in any species, then a begging-derived increase of 5% in the DEB requires a begging duty factor of 5% (1.2 hours/day begging). If parents bring food to the nest 300 times per day, a schedule similar to that observed in house wrens, and all nestlings beg during every feeding visit (which is twice as often as occurred in our house wren studies), then each begging event would have to last almost 15 seconds (i.e. three to five times the average duration of house wren begging events) to generate the necessary duty factor. At the average begging frequencies and metabolic costs we measured in house wrens, the required duty factor is 18% or 4.4 hours/day of begging, with each begging event lasting almost two minutes. These figures may be conservative, since non-begging activity such as thermoregulatory costs will increase the 24-hour average metabolic rate above RMR, which means that additional begging would be needed to produce the threshold 5% increase in DEB.

As mentioned previously, the component of the DEB that is probably most important to nestling fitness is the energy used to form new tissues (McCarty 1997; Verhulst & Wiersma 1997). If we assume that all of the energy expended upon begging is directly and exclusively traded from the energy otherwise available for growth, considerably less begging time and effort is needed to achieve a cost of 5% of the growth budget than to achieve the same fraction of the DEB. As a rough approximation, growth absorbs about 20% of a house wren nestling's total energy budget from age zero-ten days, although this varies considerably with the age of the nestling. Extending the example from the previous paragraph, house wren nestlings would need to beg for 53 minutes/day to reduce growth by 5%, instead of the 4.4 hours/day of begging needed to expend 5% of the DEB. That is still more than three times the average daily begging time of ten-day old nestlings and 13 times that of three-day old nestlings. If increased begging resulted in slight increases in food intake, a negative impact on growth would be even more unlikely.

In conclusion, current data on energy expenditure during begging indicate that in the species studied to date it is a very small and arguably trivial portion of the energy budget. These small costs suggest that reductions or even increases in begging behaviour are not likely to have an impact on fitness through the energy cost of the begging activity itself. Only in a fairly restrictive and thus far unverified set of conditions, such as a combination of high metabolic cost per begging event and high duty factors for begging, is it reasonable to expect that begging energetics will be subject to selection. Hence, there is little evidence to date that energy cost can be viewed as an important factor affecting the evolution of signal honesty in begging and as pointed out by other authors (McCarty 1996, 1997), claims of high begging energy costs should not be accepted unless backed up by robust and comprehensive experimental data.

As a caveat, we note two recent reports showing that artificially augmented begging regimens can reduce nestling growth rates. Experimental groups of canaries (*Serinus canaria*, Kilner 2001, this volume) and magpies (Rodríguez-Gironés et al. 2001) that were induced to beg for substantial durations (70-120 seconds) had significantly reduced growth rates. However, no effect was found in ring doves (*Streptopelia risoria*) induced to beg for 70 minutes/day. These interesting results implicate energetic constraints, but they fail to demonstrate clearly that begging energy expenditure *per se* was the direct cause of growth rate changes.

Viewed from a slightly different perspective, begging appears to be an extremely energetically efficient way for nestlings to communicate with their parents, at least in terms of the energy expended during begging itself. Nevertheless, there may be more subtle aspects of the energetics of this signal system and we focus on some of these in our suggestions for future research.

FUTURE DIRECTIONS

We make four suggestions for further studies of the energy cost of begging. First, the conclusions in this chapter are based on a very small number of species. While it would be tedious and possibly redundant to drastically increase the number of species for which we have data on begging energetics, we think there is a need to increase the behavioural and phylogenetic diversity of the data set by targeting species of special interest. In particular, it would be useful to measure begging costs in species that are known to beg with unusual intensity and/or for long periods. This would help establish an upper limit to begging costs. For example, brood parasites such as the brown-headed cowbird (*Molothrus ater*) beg with greater intensity than non-parasitic species (Briskie et al. 1994; D.C. Dearborn & G. Lichtenstein this volume) and hence may show correspondingly greater energy expenditures.

Second, we strongly urge that future measurements of begging energy costs, especially in terms of possible fitness effects, be firmly anchored within the context of the nestling's entire energy budget. In addition to information on begging energy costs *per se,* this requires time budget studies of begging behaviour in natural nests to provide the crucial data on begging duty factors. It also requires information on daily energy budgets (allometric estimates for these are available in the literature; e.g. Weathers 1992; Weathers & Siegel 1995). This degree of rigour is necessary to confirm or falsify our suggestion that begging may be a universally small and evolutionarily inconsequential component of the energy budget, and may provide insights into possible trade-offs.

Third, most studies of begging energetics have focused on single nestlings. This facilitates the measurement of an individual's metabolic rate, but data from solitary nestlings may misrepresent begging costs in a real nest, where group dynamics are likely to be important. Potentially, the presence of siblings could increase begging energy costs by requiring greater effort per begging event (e.g. Price et al. 1996; Leonard & Horn 1998; but see Cotton et al. 1996) or by requiring a nestling to beg in a greater proportion of parental visits. As we have seen in house wren nests, siblings may increase the energetic cost of begging by interfering with begging in ways that make it more difficult or conversely, siblings may provide physical support for each other. This may be especially true for young nestlings. Thermoregulation, which is important for energy utilization as well as affecting the ability to perform activities such as begging, may be facilitated in groups. In short, the effect of siblings on the begging activity of a given nestling needs to be clarified, along with any energetic consequence of the social environment.

Finally, our understanding of the energetics of nestling competition and signalling may be enhanced if future studies look beyond begging behaviour. To date, the majority of work on the energy cost of begging focuses on the well-known response of nestlings to the arrival of a parent at a nest. This is not the only way in which young birds compete for food. For example, there are suggestions that nestlings may compete for proximity to a provisioning parent (McRae et al. 1993; Kilner 1995). Though all species may not show this effect clearly (e.g. Teather 1992), in crowded nest cups nestlings do appear to jostle for sufficient vertical height or to get out from under siblings, thereby amplifying effects of nestling reach in determining the outcome of begging. This jostling for position can contribute to energy expenditure through activity and possibly thermoregulatory costs. If these behaviours enhance a nestling's ability to acquire food, then their energy costs (which have not been measured) should be added to that of begging *per se*, keeping in mind that a primary reason for studying begging in the first place is to understand how nestlings communicate need to parents.

REFERENCES

Abraham, C.L. & Evans, R.M. 1999. Metabolic costs of heat solicitation calls in relation to thermal need in embryos of American white pelicans. *Animal Behaviour* **57**, 967-975.

Bachman, G.C. & Chappell, M.A. 1998. The energetic cost of begging behaviour in nestling house wrens. *Animal Behaviour* **55**, 1607-1618.

Bartholomew, G.A., Vleck, D. & Vleck, C.M. 1981. Instantaneous measurements of oxygen consumption during pre-flight warm-up and post-flight cooling in sphingid and saturnid moths. *Journal of Experimental Biology* **90**, 17-32.

Beauchamp, G., Ens, B.J. & Kacelnik, A. 1991. A dynamic model of food allocation to starling (*Sturnus vulgaris*) nestlings. *Behavioral Ecology* **2**, 21-37.

Berrigan, D. & Lighton, J.R.B. 1993. Bioenergetic and kinematic consequences of limblessness in larval Diptera. *Journal of Experimental Biology* **179**, 245-259.

Briskie, J.V., Naugler, C.T. & Leech, S.M. 1994. Begging intensity of nestling birds varies with sibling relatedness. *Proceedings of the Royal Society of London, Series B* **258**, 73-78.

Chappell, M.A. & Bachman, G.C. 1998. The exercise capacity of House Wren nestlings: begging chicks are not working as hard as they can. *The Auk* **115**, 863-870.

Chappell, M.A., Zuk, M., Kwan, T.H. & Johnsen, T.S. 1994. Energy cost of an avian vocal display: crowing in red jungle fowl. *Animal Behaviour* **49**, 255-257.

Chappell, M.A., Bachman, G.C. & Hammond, K.A. 1997. The heat increment of feeding in house wren chicks: magnitude, duration, and substitution for thermostatic costs. *Journal of Comparative Physiology B* **167**, 313-318.

Cotton, P.A., Kacelnik, A. & Wright, J. 1996. Chick begging as a signal: are nestlings honest? *Behavioral Ecology* **7**, 178-182.

Godfray, H.C.J. 1995. Signaling of need between parents and young: parent-offspring conflict and sibling rivalry. *American Naturalist* **146**, 1-24.

Godfray, H.C.J. & Parker, G.A. 1992. Sibling competition, parent-offspring conflict and clutch size. *Animal Behaviour* **43**, 473-490.

Haskell, D. 1994. Experimental evidence that nestling begging behaviour incurs a cost due to nest predation. *Proceedings of the Royal Societyof London, Series B* **257**, 161-164.

Henderson, B.A. 1975. Role of the chicks' begging behavior in the regulation of parental feeding behavior of *Larus glaucenscens*. *Condor 77*, 488-492.

Hussell, D.J.T. 1988. Supply and demand in tree swallow broods: a model of parent-offspring food-provisioning interactions in birds. *American Naturalist* **131**, 175-202.

Janes, D.N. & Chappell, M.A. 1995. The effect of ration size and body size on specific dynamic action in Adélie penguin chicks, *Pygoscelis adeliae*. *Physiological Zoology* **68**, 1029-1044.

Johnstone, R.A. & Grafen, A. 1992. The continuous Sir Philip Sidney game: a simple model of biological signalling. *Journal of Theoretical Biology* **156**, 215-234.

Kilner, R. 1995. When do canary parents respond to nestling signals of need? *Proceedings of the Royal Society of London, Series B* **260**, 343-348.

Kilner, R.M. 2001. A growth cost of begging in captive canary chicks. *Proceedings of the National Academy of Sciences USA* **98**, 11394-11398.

Kilner, R. & Johnstone, R.A. 1997. Begging the question: are offspring solicitation behaviours signals of need? *Trends in Ecology and Evolution* **12**, 11-15.

Leech, S.M. & Leonard, M.L. 1996. Is there an energetic cost to begging in nestling tree swallows (*Tachycineta bicolor*)? *Proceedings of the Royal Society of London, Series B* **263**, 983-987.

Leonard, M.L. & Horn, A.G. 1998. Need and nestmates affect begging in tree swallows. *Behavioral Ecology and Sociobiology* **42**, 431-436.

McCarty, J.P. 1996. The energetic cost of begging in nestling passerines. *The Auk* **113**, 178-188.

McCarty, J.P. 1997. The role of energetic costs in the evolution of begging behavior of nestling passerines. *The Auk* **114**, 135-137.

McRae, S.B., Weatherhead, P.J. & Montgomerie, R. 1993. American robin nestlings compete by jockeying for position. *Behavioral Ecology and Sociobiology* **33**, 101-106.

Nagy, K.A. 1983. *The Doubly Labeled Water ($^{3}HH^{18}O$) Method: A Guide To Its Use*. Los Angeles: UCLA Publication.

Price, K., Harvey, H. & Ydenberg, R. 1996. Begging tactics of nestling yellow-headed blackbirds, *Xanthocephalus xanthocephalus*, in relation to need. *Animal Behaviour* **51**, 421-435.

Ricklefs, R.E. 1992. The roles of parent and chick in determining feeding rates in Leach's storm-petrel. *Animal Behaviour* **43**, 895-906.

Rodríguez-Gironés, M.A., Zuñiga, J.M. & Redondo, T. 2001 Effects of begging on growth rates of nestling chicks. *Behavioral Ecology* **12**, 269-274.

Schew, W.A. & Ricklefs, R.E. 1998. Developmental plasticity. In: *Avian Growth and Developmen: Evolution Within the Altricial-Precocial Spectrum* (Ed. by J.M. Starck & R.E. Ricklefs). New York: Oxford University Press.

Smith, H.G. & Montgomerie, R. 1991. Nestling American robins compete with siblings by begging. *Behavioral Ecology and Sociobiology* **29**, 307-312.

Soler, M., Soler, J.J., Martinez, J.G. & Moreno, J. 1999. Begging behaviour and its energetic cost in great spotted cuckoo and magpie host chicks. *Canadian Journal of Zoology* **77**, 1794-1800.

Speakman, J.R. 1997. *Doubly Labelled Water: Theory and Practice.* London: Chapman and Hall.

Speakman, J.R., Anderson, M.E. & Racey, P.A. 1989. The energy-cost of echolocation in pipistrelle bats (*Pipistrellus pipistrellus*). *Journal of Comparative Physiology A* **165**, 679-685.

Stamps, J., Clark, A., Arrowood, P. & Kus, B. 1985. Parent-offspring conflict in budgerigars. *Behaviour 94*, 1-40.

Starck, J.M. & Ricklefs, R.E. 1997. *Avian Growth and Development: Evolution within the Altricial-Precocial Spectrum.* New York: Oxford University Press.

Teather, K.L. 1992. An experimental study of competition for food between male and female nestlings of the red-winged blackbird. *Behavioral Ecology and Sociobiology* **31**, 81-87.

Vehrencamp, S.L., Bradbury, J.W. & Gibson, R.M. 1989. The energetic cost of display in male sage grouse. *Animal Behaviour* **38**, 885-896.

Verhulst, S. & Wiersma, P. 1997. Is begging cheap? *The Auk* **114**, 134.

Weathers, W.W. 1992. Scaling nestling energy requirements. *Ibis* **134**, 142-153.

Weathers, W.W. & Siegel, R.B. 1995. Body size establishes the scaling of avian postnatal metabolic rate - an interspecific analysis using phylogenetically independent contrasts. *Ibis* **137**, 532-542.

Weathers, W.W., Hodum, P.J. & Anderson, D.J. 1997. Is the energy cost of begging by nestling passerines surprisingly low? *The Auk* **114**, 133.

9.

BEGGING BEHAVIOUR AND NEST PREDATION

David G. Haskell
Department of Biology, University of the South, Sewanee TN 37383, USA (dhaskell@sewanee.edu)

ABSTRACT

This chapter reviews the literature on the effects of predation on the evolution of begging behaviour. I conclude that there is little evidence for the existence of a present-day predation cost of begging for calls emanating from the nest site in which the calls evolved. Begging calls can, however, incur a predation cost when they are played back from foreign nest sites (e.g. calls of tree-nesters played back from ground nests). This suggests that begging calls show the imprint of "the ghost of predation past". This might also explain some of the interspecific diversity in the acoustic structure of begging calls. The lack of predation costs underscores the importance of theoretical investigations of low cost begging. I also assess the strengths and weaknesses of different approaches to studying the costs of begging and argue that our understanding has been hampered by reliance on correlations and by a paucity of manipulative experiments.

INTRODUCTION

Why do nestling birds call loudly when their parents deliver food? This question attracts evolutionary biologists because it offers us the hope of resolving a paradox - how natural selection produces a behaviour that, at first glance, seems to be unnecessarily costly. In addressing this paradox researchers have focused on uncovering the benefits of these extravagant displays. Attention to these benefits has, however, often been accompanied by inattention to the source of our paradox; the details of the costs themselves. This is unfortunate because costs are central to many models of

J. Wright and M.L. Leonard (eds.), The Evolution of Begging, 163–172.

the evolution of begging (see R.A. Johnstone & H.C.J. Godfray this volume) and because the scant literature on costs reveals some interesting patterns. Begging has been assumed to be costly in two ways: begging calls might attract predators (the predation cost) and begging may use energy (the metabolic cost; M.A. Chappell & G.C. Bachman this volume). In this chapter I review the literature on the predation cost of begging and ask: (1) whether there is empirical evidence for the existence of a predation cost of begging; (2) whether predation costs vary in ways that might explain the acoustic diversity of begging behaviour; and (3) how empirical studies of predation costs might broaden our understanding of the evolution of begging.

TESTS OF THE PREDATION COSTS OF BEGGING

Correlational Tests

I define a correlational study as one in which an investigator tests whether natural intraspecific variation in begging behaviour is correlated with some measure of the risk of nest predation. We might expect a positive correlation between these variables if predators use begging calls to detect and locate nests. Two types of intraspecific correlational studies have been conducted. In the first, the rate of predation on nests in the incubation stage is compared with that on nests with nestlings. The second type of study examines variation within the nestling stage only, comparing predation rates across a range of noisiness.

The simplest type of correlational investigation of the predation costs of begging involves comparing the rate of predation on nests in the incubation stage with the rate of predation in the nestling stage. Several authors have reported increased rates of predation on nestlings (e.g. Slagsvold 1982; Roper & Goldstein 1997). As these authors point out, however, there are many other potentially confounding variables in comparisons of this type. For example, after the eggs hatch, the parent birds may visit the nest more often, the nestlings may emit odours and, depending on the fastidiousness of parental nest sanitation, the nest may become soiled with faecal matter. All these factors might also attract predators to nests, so comparisons of this type are a poor test of the predation cost of begging.

Comparisons made among nests in the nestling stage control for some of the differences between nests in the incubation and nestling stage and therefore offer a more focused approach to testing the predation cost of begging. Both Redondo and Castro (1992) and Halupka (1998) assayed the noisiness of broods in their studies by quantifying the nestlings' response to human investigators. Halupka found no relationship between this measure of begging activity and the rate of predation on nests of meadow pipits (*Anthus pratensis*), even when he controlled for variation in nest location. Redondo and Castro found that for magpie (*Pica pica*) nests that were preyed upon there was a negative correlation between survival time and the number of gaping nestlings, but not the number of vocalizing nestlings. Both these studies are hard to interpret because it is unclear whether the response of nestlings to humans is a good measure of the noisiness of natural broods. These responses may reflect differences in fear, hunger or neophobia, rather than natural levels of begging.

Dearborn's (1999) study took a different approach and compared the rate of predation on nests of indigo buntings (*Passerina cyanea*) with and without brood-parasitic brown-headed cowbird (*Molothrus ater*) nestlings. Begging calls of cowbird nestlings are louder and more frequent than those of their hosts so, if begging calls attract predators, we might expect parasitized nests to experience a higher rate of nest predation than unparasitized nests. Dearborn did not, however, find this predicted relationship during the nestling stage.

None of these correlational studies could control for the rate of parental visitation, variation in nest site quality or other potentially confounding factors. These factors make the correlational method a weak approach to investigating the predation costs of begging.

Experimental Tests

Experimental manipulations of begging calls offer a more direct way of testing whether begging calls attract predators to nests. To date, all these experimental studies have used playbacks from artificial nests to measure changes in the rate of nest predation. This is a powerful tool, allowing us to focus on the predation cost of specific types of begging calls, but its narrow focus does have limitations.

When I played back the nestling begging vocalizations of western bluebirds (*Sialia mexicanus*) from artificial ground and tree nests baited with quail (*Coturnix coturnix*) eggs, I found that begging vocalizations significantly increased the rate of nest predation for ground nests, but not for tree nests (Haskell 1994). I also found that, for ground nests, nests with high rates of begging vocalizations had higher rates of predation than did nests with lower rates. Leech and Leonard (1997) also used artificial nests baited with quail eggs and found that tree swallow (*Tachycineta bicolor*) begging calls increased rates of predation for both ground nests and for nests placed on platforms below nestboxes. Their study included more spatial replication than Haskell (1994).

In a study focused on wood warbler (Parulidae) begging calls, I used nests baited with small clay eggs to examine whether interspecific variation in begging call structure is related to the predation cost of begging (Haskell 1999). I found that the calls of both tree- and ground-nesting warblers incurred statistically nonsignificant increases in the rate of predation when played back from artificial nests in the natural location for each species. When I played back the calls of the tree-nesting species from the ground, however, I measured a large increase in the rate of predation. I observed no such increase when ground-nesting calls were 'transplanted' into tree nests. Dearborn (1999) also used artificial nests baited with clay eggs to conduct an experimental investigation of the predation costs of begging that paralleled his correlational study (see above). He found that playbacks of begging calls from parasitized indigo bunting nests increased the rate of predation on artificial nests, but that playbacks of begging calls from unparasitized nests had no significant effect on the rate of predation.

Before I address the implications of these studies for our thinking about the evolution of begging, I will outline some important limitations implicit in the experimental studies that have been conducted to date. These limitations stem from the experimenters' reliance on artificial nests. Major and Kendal (1996) have reviewed all the biases of artificial nest studies; I will discuss the two biases that offer particular problems for experimental studies of begging. The first bias arises from the very focused nature of the experiments. The experimental scalpel allows us to carefully excise information from our study systems, but only at the cost of severing connections to the rest of the species' natural history. For example, Halupka (1998) points out that many parent birds give alarm calls when they see a predator. Nestlings may respond by becoming silent, thus reducing the actual predation cost of begging. Playback experiments cut natural parental behaviour out of the picture, and thus may produce over-inflated estimates of the predation cost of begging. Of course, interactions between begging and other aspects of bird behaviour are amenable to future experimental studies, but until we conduct such studies our measures of the predation costs of begging will be tentative.

The second bias arises from the type of bait used in the artificial nests. Quail eggs, in particular, are too large for some predators to eat (Roper 1992; Haskell 1995a, b; DeGraaf & Maier 1996). Thus, experiments that use quail eggs (Haskell 1994; Leech & Leonard 1997) potentially exclude all small-mouthed predators. If these predators use sound to locate nests, quail egg experiments will produce underestimates of the increase in predation caused by begging. In experiments in which I used clay eggs (Haskell 1999),

I found that chipmunks (*Tamias striatus*) were responsible for most of this increase. Chipmunks cannot eat quail eggs, so a quail egg experiment would have reported no predation cost of begging. Artificial nests, quail eggs and clay eggs also presumably have unnatural odours, so predators that are either neophobic or that are attracted to human scents will be either repelled or attracted to the nests. Thus, although playback experiments have been able to tease out the predation cost of begging, their results should be interpreted in light of the biases associated with the details of the experimental set-up.

IMPLICATIONS FOR UNDERSTANDING BEGGING

Implications for Models of Signalling

What do these previous studies tell us about the magnitude of the predation costs of begging? While begging calls can, in some circumstances, attract predators to nests, there is scant evidence for the existence of a present-day predation cost of begging for begging calls emanating from the nest site in which the calls evolved. Both Haskell (1994) and Leech and Leonard (1997) played back begging calls from unnatural nest sites - western bluebirds and tree swallows both nest in cavities and the eggs used as bait in both these experiments were placed outside cavities (although the sound playback did occur from within the cavity in Leech and Leonard's study). Haskell (1999) found that predation costs of warbler begging calls were only detectable when begging calls were transplanted out of their natural nest site. Likewise, Dearborn (1999) found no significant predation cost of begging for indigo bunting begging calls, although he did report increased predation on nests with brown-headed cowbird begging calls. Perhaps we are observing the "ghost of predation past" (*sensu* Connell's (1980) ghost of competition past). That is, begging calls may have evolved to minimize predation costs in their natural setting. Of course, we have only four experimental studies with which to assess this hypothesis, so there is clearly a need for studies of more species across a wider range of geographic areas. The information at hand, however, does not look encouraging for models that assume that begging incurs a large predation cost. Thus, because the metabolic cost of begging also appears to be low (M.A. Chappell & G.C. Bachman this volume), the empirical data seem to point towards a need for further theoretical exploration of low cost or cost-free signalling (see R.A. Johnstone & H.C.J. Godfray this volume), or a search for new costs that might underpin existing models.

Even if future studies do find that natural begging behaviour does attract predators, there is still more theoretical work to be done. Because all the nestlings in a nest are often eaten when a predator visits a nest, the predation cost of begging falls not just on the beggar, but also on its unfortunate nestmates. This shared predation cost of begging contrasts with metabolic costs of begging, which are borne individually. The distinction between shared and individual costs has fallen by the wayside in recent theoretical studies (Rodríguez-Gironés et al. 1996, 1998; Johnstone 1999).

Implications for Understanding Acoustic Structure

Nestling begging calls vary greatly from species to species. Redondo and Arias de Reyna (1988), Popp and Ficken (1991), Briskie et al. (1999) and Haskell (1999) all report remarkable variation in the frequency, amplitude and degree of modulation of begging calls. Could variation in the predation costs of begging be responsible for part of this diversity?

Redondo and Arias de Reyna (1988), Popp and Ficken (1991) and Briskie et al. (1999) attempted to answer this question by looking across species and testing for a correlation between the rate of nest predation and acoustic attributes of begging calls. I believe that this approach is flawed because there is no reason to expect that the rate of nest predation should be positively correlated with the predation cost of begging behaviour or the cost of any other trait (Haskell 1996). The predation cost of begging is the increase in the rate of predation caused by begging behaviour, not the overall rate of predation. The conflation of absolute rates of predation and predation costs is particularly worrisome in the case of nest predation. Many animals prey on bird nests, and these animals vary in the extent to which they use sound to locate nests. Thus, predators that strike at night (as do many mammals) or that are essentially deaf (as are many snakes) will increase the rate of nest predation, but not the predation cost of begging (which is the increase in the rate of predation caused by begging calls). Consider also the finding (Haskell 1999) that playbacks of begging calls from different species from the same nest site (e.g. begging calls of tree-nesters and ground-nesters played back from ground nests) can result in radically different predation costs of begging, depending upon the acoustic structure of the begging call. We therefore have two different predation costs of begging for nests with only one 'background' (*sensu* Briskie et al. 1999) rate of predation, something that should be logically impossible if we equate the rate of predation with the cost of begging. We can carry this empirical example one step further and look at all published experimental measures of the predation cost of begging and ask whether any correlation exists between the rate of predation on silent nests (an estimate of the rate of predation on incubating birds, or the background rate) and the measured predation cost of begging.

No such correlation exists (Figure 1). We cannot, therefore, use the rate of predation as a surrogate for actual measures of the predation cost of begging.

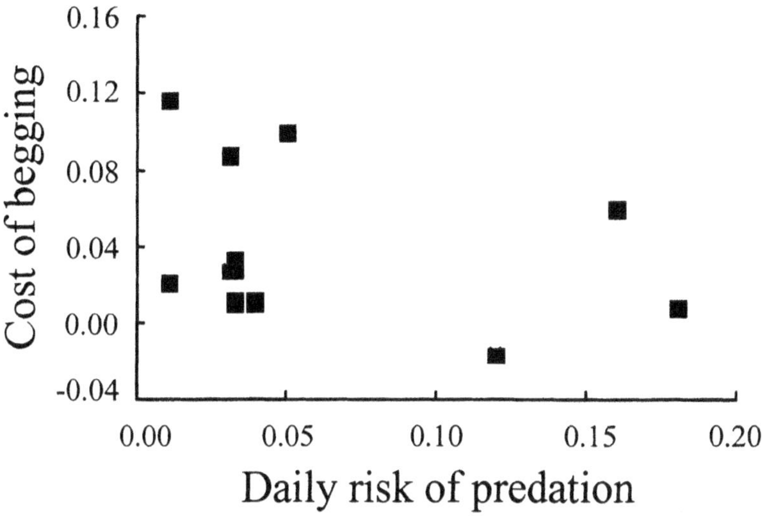

Figure 1. The relationship between the daily risk of predation (calculated using Mayfield's (1975) method) and the predation cost of begging (the difference between the daily risk of predation for nests with begging calls and for silent nests). The daily risk of predation is a measure of predation risk that allows standardized comparisons to be made across studies. I converted proportional data into daily rates of predation by assuming that within each study the risk of predation did not vary through time. For example, an experimental treatment in which 30% of nests were preyed upon over seven days would produce a daily risk of predation of 0.05 ($[1 - \text{daily risk}]^{\text{number of days}}$ = [proportion of nests surviving]). As expected on theoretical grounds, there is no significant relationship (linear regression: $F_{1,13} = 0.68$, $P = 0.42$). Data are from Haskell (1994, 1999), Leech and Leonard (1997) and Dearborn (1999).

Despite this conceptual handicap, the correlational studies do indicate some interesting patterns that would be fruitful avenues for further study. For example, Redondo and Arias de Reyna's (1988) conclusion that cavity-nesting species have louder and more locatable calls suggests that conspicuous begging calls do not increase the risk of nest predation for cavity-nesters (or perhaps there is some extra benefit for locatable begging calls in cavity nests). Their conclusion seems to be upheld by Briskie et al.'s (1999, Table 1) study of 24 species nesting in central Arizona, but not by Popp and Ficken's (1991) study which included both North American and Eurasian birds. These conflicting results suggest that the magnitude of the predation cost of begging for cavity-nesters might vary depending on geographical area or nest site. Given that the spectrum of nest predators changes from one habitat to another, this potential variation in the predation costs of begging for cavity-nesters is perhaps to be expected.

Another approach to investigating the effects of predation on the evolution of acoustic structure is to combine comparative studies of call structure with measurements of the predation costs of begging. I used artificial nests with playbacks to make these measurements for the begging calls of two wood warblers (ovenbirds, *Seiurus aurocapillus*, and black-throated blue warblers, *Dendroica caerulescens*) and found that the cost of making conspicuous begging calls was higher for the ground-nesting ovenbird than it was for the tree-nesting black-throated blue warbler (Haskell 1999). Tooth imprints on clay eggs indicated that these differences were due to predation by one species, the chipmunk. My parallel comparative study of the calls of these two species, and other warblers, showed that ground-nesters did seem to have more acoustically cryptic (but not quieter) calls. Thus, the species with the higher cost of begging had the most cryptic begging calls. As I discussed earlier, artificial nest studies are powerful because they allow us to quantify the predation costs of begging and, in this case, to transplant begging calls and measure how costs vary among nest sites. However, my reliance on artificial nests and eggs may have excluded some predators from the experiment, and definitely excluded the important three-way interactions between parent birds, nestlings and predators.

Although variation in the predation costs of begging might explain some of the interspecific differences in begging behaviour, much of this variation is still unexplained. The most grievous gap in our knowledge concerns the costs and benefits of many differences in acoustic structure. Most theoretical studies of begging have modelled the evolution of one character, the intensity of begging. A nestling might, however, have a number of ways of escalating intensity - increase the amplitude, decrease the pitch, increase the rate of modulation or even push and gape without vocalizing. Depending on the responses of parents and predators, and on the physiology of the nestlings, these different tactics may result in different costs and benefits. Tactics that produce loud, easily locatable sounds might attract predators and therefore incur shared costs for the whole brood, whereas tactics that produce little sound, or hard to locate sounds, would incur solely individual costs. Thus, a challenge for future studies will be to grapple with this multidimensional nature of begging behaviour.

FUTURE DIRECTIONS

In summary, four areas for future research seem to hold the most promise for deepening our knowledge of this topic: (1) experimental measurements of the predation cost of begging in a wider variety of species and nest sites, coupled with comparative studies of variation in begging call acoustic structure; (2) investigations of the theoretical implications of low cost signalling, and of signalling where costs are shared by the whole brood; (3) experimental and theoretical investigations of the three-way interaction between the behaviour of nestlings, predators and parents; and (4) studies of the costs and benefits of variation in the acoustic structure of begging calls.

ACKNOWLEDGEMENTS

I thank S.A. Vance, D.C. Dearborn, M.L. Leonard and J. Wright for comments on the manuscript. M.L. Leonard kindly shared data to help me construct Figure 1. My work on begging was funded by the U.S. National Science Foundation, Cornell University, the American Ornithologists' Union, the Animal Behavior Society, the Chapman Fund, Sigma Xi, the Benning Fund and the Wilson Ornithological Society.

REFERENCES

Briskie, J.V., Martin, P.R. & Martin, T.E. 1999. Nest predation and the evolution of nestling begging calls. *Proceedings of the Royal Society of London, Series B* **266**, 2153-2159.

Connell, J.H. 1980. Diversity and the co-evolution of competitors, or the ghost of competition past. *Oikos* **35**, 131-138.

Dearborn, D.C. 1999. Brown-headed cowbird nestling vocalizations and risk of nest predation. *The Auk* **116**, 448-457.

DeGraaf, R.M. & Maier, T.J. 1996. Effect of egg size on predation by white-footed mice. *Wilson Bulletin* **108**, 535-539.

Halupka, K. 1998. Vocal begging by nestlings and vulnerability to nest predation in meadow pipits *Anthus pratensis*; to what extent do the predation costs of begging exist? *Ibis* **140**, 144-149.

Haskell, D.G. 1994. Experimental evidence that nestling begging behaviour incurs a cost due to nest predation. *Proceedings of the Royal Society of London, Series B* **257**, 161-164.

Haskell, D.G. 1995a. Forest fragmentation and nest-predation: Are experiments with Japanese quail eggs misleading? *The Auk* **112**, 767-770.

Haskell, D.G. 1995b. A reevaluation of the effects of forest fragmentation on rates of bird-nest predation. *Conservation Biology* **9**, 1316-1318.

Haskell, D.G. 1996. Do the bright plumages of birds incur a cost due to nest predation? *Evolutionary Ecology* **10**, 285-288.

Haskell, D.G. 1999. The effect of predation on begging-call evolution in nestling wood warblers. *Animal Behaviour* **57**, 893-901.

Johnstone, R.A. 1999. Signaling of need, sibling competition, and the cost of honesty. *Proceedings of the National Academy of Sciences USA* **96**, 12644-12649.

Leech, S.M. & Leonard, M.L. 1997. Begging and the risk of predation in nestling birds. *Behavioral Ecology* **8**, 644-646.

Major, R.E. & Kendal, C.E. 1996. The contribution of artificial nest experiments to understanding avian reproductive success - a review of methods and conclusions. *Ibis* **138**, 298-307.

Mayfield, H. 1975. Suggestions for calculating nest success. *Wilson Bulletin* **87**, 456-466.

Popp, J. & Ficken, M.S. 1991. Comparative analysis of acoustic structure of passerine and woodpecker nestling calls. *Bioacoustics* **3**, 255-274.

Redondo, T. & Arias de Reyna, L. 1988. Locatability of begging calls in nestling altricial birds. *Animal Behaviour* **36**, 653-661.

Redondo, T. & Castro, F. 1992. The increase in risk of predation with begging activity of Magpies, *Pica pica*. *Ibis* **134**, 180-187.

Rodríguez-Gironés, M.A., Cotton, P.A. & Kacelnik, A. 1996. The evolution of begging: signaling and sibling competition. *Proceedings of the National Academy of Sciences USA* **93**, 14637-14641.

Rodríguez-Gironés, M.A., Enquist, M. & Cotton, P.A. 1998. Instability of signaling resolution models of parent-offspring conflict. *Proceedings of the National Academy of Sciences USA* **95**, 4453-4457.

Roper, J.J. 1992. Nest predation experiments with quail eggs: too much to swallow? *Oikos* **65**, 528-530.

Roper, J.J. & Goldstein, R.R. 1997. A test of the Skutch hypothesis: does activity at nests increase nest predation risk? *Journal of Avian Biology* **28**, 111-116.

Slagsvold, T. 1982. Clutch size variation in passerine birds: the nest predation hypothesis. *Oecologia* **54**, 159-169.

NESTLING PHYSIOLOGY

10.

APPETITE AND THE SUBJECTIVITY OF NESTLING HUNGER

Anne B. Clark
Department of Biological Sciences, Binghamton University, Binghamton NY 13902, USA (aclark@binghamton.edu)

ABSTRACT

Begging is important to behavioural ecologists as a signal between parents and offspring that may, or may not, have evolved to carry honest information on nestling need. Defining and measuring the need underlying begging is both critical and problematic. The literature on the regulation of foraging and feeding provides insights into this problem. In particular, the concept of appetite, which combines physiological hunger with other cost-benefit weightings concerning an individual's motivation to feed, makes good sense as a regulator of adaptive foraging behaviour and, it is argued, adaptive begging behaviour. Physiological, maturational and experiential differences among nestlings can be compared with characteristic differences in begging to show that adaptive variation in appetite can help explain begging patterns of individuals, broods and species. An appreciation of underlying mechanisms provides a realistic basis for studies of the signal aspects of begging and for assessment of its honesty.

INTRODUCTION

The begging behaviour of nestling birds has achieved its broad interest because it provides an ecologically relevant and tractable testing ground for theories on signal evolution and conflict resolution between related individuals. One prominent question continues to be "Has nestling begging behaviour evolved as an honest representation of offspring need?" - an answer to which requires some definition of need. Need has been measured

J. Wright and M.L. Leonard (eds.), The Evolution of Begging, 173–198.

empirically over both the short- and long-term. To a large extent, differences in begging behaviour within a brood do correlate with one short-term measure; differences in the amount of food that brood members have received in the recent past (Kilner & Johnstone 1997; Mock & Parker 1997; Budden & Wright 2001a). A number of studies demonstrate that growth or the amount of food received over a longer term (a day or more) do not predict relative begging effort within broods (e.g. Stamps et al. 1989; Redondo & Castro 1992; but see Price et al. 1996; Lotem 1998a). On the other hand, begging intensity is reported to vary consistently within broods according to hatching order, relative age and sometimes sex (Khayutin & Dmitrieva 1977; Rydén & Bengtsson 1980; Fujioka 1985; Grieg-Smith 1985; Stamps et al. 1989; Teather 1992; Lotem 1998a, b), even when apparently controlling for recent feeding history (Price et al. 1996). Nestlings in some species alter their begging behaviour depending upon the begging behaviour, number and even relative size of begging nestmates (Smith & Montgomerie 1991; Price 1994; Price & Ydenberg 1995; Lotem 1998b; Leonard et al. 2000; but see Cotton et al. 1996). Between populations of a single species, very different average levels of begging can yield similar levels of parental provisioning (Hussell 1988). Thus, either signal honesty is in question or need, and related descriptors such as hunger and condition, are more subjective, cryptic and relative to individual characteristics than we often assume. In either case, the interpretation of begging as a signal of need is made more complex for parents and scientists.

So what is it that makes a nestling hungry, and how is that hunger is linked to a nestling's motivation to beg? In his classic paper entitled "What is it like to be a bat?", Nagel (1974) argued that mental experience, being subjective, can never be fully described in terms of neurophysiological events, a position criticized by neuroscientists who counter that a sufficiently detailed physiological description may in fact give a real and useful understanding of that experience (Churchland 1986). As evolutionary biologists, we might approach the issue of subjective experience by defining the problem facing an organism and predicting what kinds of information should logically and functionally be included in such experience in order for the organism to use it to solve its problem. In the case of begging nestlings, the problem is ultimately one of getting an optimal amount to eat. Begging is their primary way of doing so. Thus, we should explore the information and experiences, external and internal, that might lead to begging as an adaptive foraging behaviour, rather than simply a signal (Budden & Wright 2001a). In this chapter I try to connect the concepts of need and hunger in nestlings with the literature concerned with the physiology of hunger and its relationship to the motivation to forage and eat. We can then compare begging more directly to the trade-offs and factors moulding any animal's foraging decisions and see if this approach helps to explain differences in begging behaviour within and between individuals, broods and species. This should give us a more realistic basis for investigating the specialization of begging as a signal.

DEFINING HUNGER, NEED AND CONDITION

In the literature on begging, signalling and feeding, the terms hunger and need are continually intertwined, sometimes being used as one and the same thing. Need is often used to describe a current state of nutrition (Godfray 1991; W.H. Karasov & J. Wright this volume), but also sometimes in reference to future growth requirements. Hunger is our description of what the nestling experiences as a result of its nutritional state and is implicitly equated with its motivation to beg. Continuous variation in degree of hunger could thus produce a range of begging intensities across nestlings. To control or manipulate hunger, we generally alter the amount of food recently received. In species where nestlings not only beg, but also aggressively control siblings' access to food, hunger is said to be a factor in how aggressive they are toward siblings (e.g. O'Conner 1984; Drummond & Garcia-Chavelas 1989; Drummond et al. 1991; Machmer & Ydenberg 1998; Cook et al. 2000). When nestlings alter begging or aggression in relation to brood size, it is often suggested that brood size is a predictor of future hunger because parents do not adjust the amount of food in direct proportion to the number of young (Mock & Lamey 1991; Leonard et al. 2000).

From a parental point of view, variation in offspring condition rather than hunger may be the primary concern, but condition may often be cryptic or otherwise inaccessible to parents (Mock & Parker 1997). Condition can be viewed as a measure of both the need and the reproductive value of the nestling and should be important for decisions regarding the adaptive allocation of care. Begging is important to parents if it is a window on condition; hunger will be important to the extent that it stimulates the begging that represents a nestling's condition.

In summary, the current begging literature provides us with a set of working definitions:

Hunger: The experience of a current state of nutrition, which motivates a nestling to beg and which can be changed with feeding.

Need: Either (1) operationally equivalent to hunger (short-term) or (2) an amount of food still required to reach a future goal, such as fledging (long-term).

Condition: Either (1) a nestling's accrued growth and reserves compared to that normal for its age or (2) a nestling's long-term need (above), plus its resulting reproductive value.

But what is hunger? How is it controlled mechanistically? How does it vary across individuals and contexts? Are there simple rules determining when hunger is experienced? Does it really vary with condition? Given the physiological events underlying hunger in any animal, what in fact could hunger tell the individual about itself? For instance, if we find that differences in recent food intake do not predict variation in begging, would this mean that hunger does not predict begging or rather that hunger in this species or situation varies less with recent feeding than with some other aspect of need? Thus, we need to know more about the regulation of hunger and if hunger is a sufficient explanation of the nestling's motivation to beg and acquire food. To approach these questions, I turn to the literature on hunger, feeding and its motivation in humans and domestic or laboratory animals.

THE PSYCHO–PHYSIOLOGY OF FORAGING AND FEEDING

The Hunger-Appetite System

Researchers studying feeding behaviour in humans and model vertebrates (rats, hamsters) have their own terminology in which hunger is distinguished from the concept of appetite. Appetite in reference to human behaviour describes an internal state of motivation to get food or reach some other psycho-physiological goal (Legg 1994). Le Magnen (1992) distinguishes human hunger (the result of systemic stimulation) from human appetite, the latter representing the combination of hunger with external sensory stimulation that results in a person seeking a particular category of food. Because we are trying to understand both the internal state and overt behavioural motivation of nestlings, both hunger and appetite seem appropriate concepts to explore. As a working definition, hunger is the internal physiological state that contributes directly to appetite. Appetite, in turn, is the psycho-physiological state of motivation to get food that results from hunger, plus other factors as weighted by the organism. To grasp the distinction intuitively, consider one's appetite for lunch at 11:30 am when one is grading several hundred multiple choice tests, in contrast to when one is analysing a particularly interesting data set that is giving strong support to pet hypotheses!

How might the hunger-appetite system be constructed to produce adaptive behaviour? The modern view of appetite has its roots in the "central motive states" of Morgan (1943, 1960; cited in Legg 1994). Animals are presumed to have "pre-disposition(s) to act or react, rather than a mechanical linkage between body and action..." and can "anticipate what will satisfy their needs and... select behaviour in anticipation of the outcome achieving that end" (Legg 1994, pp.6-7). A systems approach provides a useful framework for understanding the common control features of various appetites (Legg 1994). Paraphrasing Legg (1994, p.7), an adaptive appetitive system requires information about internal states (provided by blood chemistry and autonomic nervous system), about both incentive stimuli in the external environment (such as food or a mate) and learned stimuli associated with incentive or internal stimuli, and about the nature of the associations between these classes of stimuli. Finally, it requires "decision rules for translating the information into behaviour". Foraging and feeding is controlled by one such adaptive appetitive system.

Hunger and Appetite in Mammals

Feeding has been studied extensively in mammals, especially in primates and rodents (e.g. Chafetz 1990; Le Magnen 1992; Legg & Booth 1994). Hunger and its subsequent contribution to the regulation of feeding decisions results from a mix of physical, chemical and neural cues. These may usefully be divided into cephalic cues such as visual, olfactory and taste information about potential food that is still external, and internal digestive, hepatic and central nervous system (CNS) signals about the current nutritional state of the animal (Mei 1994). The size of a meal is further regulated by satiety factors including a variety of peptides originating in the gut during a meal and acting on the CNS (Woods et al. 1998). Finally, total food intake and metabolic rate are regulated over the longer term by hormones including insulin and, of recent interest, leptins produced in mammals by adipose tissue (Woods et al. 1998).

The complex of internal cues that results in the initial experience of hunger in mammals includes those from the digestive system, liver and CNS (Mei 1994). Gastric and duodenal distension suppress hunger while contractions of the digestive tract excite it. Other sensory information from receptors in the digestive tract includes information on nutrients being absorbed, hormones and peptides present in the tract and even pain, all relayed primarily *via* the vagus nerve. Glucoreceptors in the liver fire when hepatic glucose levels drop and the metabolic changes in cells resulting in changed glucose or insulin levels in the blood are detected in the hypothalamus (Le Magnen 1992; Mei 1994; Rolls 1994). This collective information on internal nutritional status is one element controlling the actual food-seeking or foraging behaviour. To the extent that it is actively experienced, we could term this hunger as we know it, but the organism need not be conscious of this experience in the way that we sometimes are.

Interestingly, the evaluation of potential food is modified by this state of hunger, but the perception itself is not controlled by it (Rolls 1994). In the lateral hypothalamus of primates, neurons respond to incoming sensory information on food (appearance, taste, smell, or conditioned stimuli associated with food) regardless of how sated an animal is, but the information is interpreted in relation to hunger (Rolls 1994). Thus, an individual that has recently eaten heavily may see a large pork chop as food, but not find the sight rewarding. The degree of reward from this sensory input to the hypothalamus may be relayed to other areas to determine ultimately if food capture or eating will occur - i.e. if it is 'worth it', given the assessment of the other costs of the required behaviour (Rolls 1994). Sensory information can also affect the experience of hunger itself, through conditioned feedback loops. Events predicting food and registered in the lateral hypothalamus may help produce the autonomic responses (e.g. salivation, insulin release) that prepare an animal for digestion. One of these, insulin release, then feeds back to lower current blood glucose levels, stimulating stronger immediate sensations of hunger. The response to food sensations can be partly food specific and the animal may continue to respond to foods it has not eaten while ceasing to respond to those on which it recently sated itself. If a great variety of foods is presented to a rat or primate, more food may be eaten, unrelated to actual nutrient content *per se*. Thus, appetite may be influenced externally as well as internally, driven by broadly adaptive rules for gaining an adequate range of nutrients (Rolls 1994).

Appetite has learned elements as well (Mei 1994; Rolls 1994). An animal must recognize an item as food in order to respond appetitively to it. Organisms learn about the palatability and nutritive value effects of digesting a specific food (Mei 1994). Animals also readily become conditioned to cues associated with food and respond appetitively without food being present. They can then show appetite for specific foods; a hungry monkey trained to expect a piece of fruit may reject an inferior lettuce leaf (Rolls 1994). The behaviours we associate with 'being hungry' are therefore dependent both on being hungry in a nutritional sense and on having an appropriate expectation that food is available and desirable. At the same time, appetite is structured such that animals do not stop attending to and assessing food just because they are nutritionally well off at the moment.

As a design to serve the tasks laid out by optimal foraging theory, one could hardly come closer than this model. It also reminds us of classical ethological studies of predatory behaviour which showed that the early searching and catching phases are hard to extinguish by feeding and persist even when the well-fed animal no longer carries on to kill and consume its prey.

Feeding History and Appetite

A more complicated question is how feeding histories influence internal hunger cues and appetitive behaviour. How might hunger represent an animal's condition with respect to its feeding history, its reserves or its achieved growth? How do animals change their appetitive behaviour with different feeding histories and nutritional condition?

Here both individual experience and predictable, species-specific challenges enter the picture. The experience of hunger and ingestive behaviour in humans and rats varies with more than circulating levels of glucose, amino acids and other nutrients from recent meals (e.g. Woods et al. 1998). In a number of ways, animals clearly adjust current feeding to accumulated deficits and future or predicted feeding problems. Laboratory rats show a provisional appetite if they are put on a feeding schedule that includes a very long deprivation period. They quickly begin to eat very heavily preceding this period, but not preceding non-feeding periods that are predictably shorter. In other words, rats can anticipate and prepare for deprivation by eating more than would normally satisfy them. After very long periods of severe food restriction, they eat to regain former body weight by eating large meals as soon as they are offered and continuing until they make up their lost weight (Le Magnen 1992). At the same time, most hamsters do not respond to deprivation with such hyperphagia (Bartness 1990). Thus, mechanisms can be species specific and presumably selected in the context of life history and ecology (Bartness 1990).

Interestingly, true fasting leads to different effects. Rats offered food after being starved to 75% of their body weight do not immediately overeat and require a longer time to regain their weight, during which time they are not using food as efficiently (Le Magnen 1992). Physiologically, fasting rats and mice are losing intestinal mass and length and, with their lowered digestive efficiency, a return to large meals might not be immediately possible (Karasov 1990). Although we cannot assess the rats' hunger in this situation, people report that the feelings of hunger disappear after about three days of fasting, while they persist very painfully during food restriction (Le Magnen 1992).

Hunger and Feeding in Birds

The control of hunger in birds is less well studied than hunger in mammals, except in the context of feeding and raising domestic fowl. Regulation of feeding in these birds, like mammals, is a complex function of direct digestive tract fullness, circulating levels of nutrients and hormones and CNS controls (Denbow 1994, 1999; Kuenzel et al. 1999; Savory 1999). As in mammals, leptins are involved in the long-term regulation of body weight and fat reserves, but they are produced not only by adipose cells, but also in the liver and in embryonic yolk (Ashwell et al. 1999). Although domestic precocial birds are self-feeding and on unnatural diets, some studies touch upon begging-related questions, such as the suppression of aggression and oral behaviours (Savory et al. 1996) and more general social effects on feeding (Nielsen 1999). There is a more extensive literature on avian digestion and food-induced changes in digestive morphology and efficiency (Karasov 1990, 1996), but surprisingly little on digestion during development. The important role that nestling digestive physiology may play in hunger and begging is reviewed by W.H. Karasov and J. Wright (this volume). Here I touch only on a few important points in order to develop the concept of appetite in relation to actual food intake.

Patterns of food intake, including meal size and digestion rate vary dramatically across phases of bird life histories. Feeding and digestion rates are both very high during, for example, pre-migratory fattening, low winter temperatures and postnatal growth (Karasov 1996). During pre-migratory fattening, at least, the intestinal tract enlarges to accommodate more food. This raises the possibility of delayed sensations of satiety from a full gut, while increased digestive rates might provide for more rapid return to hunger states after feeding. Like rats after starvation, migrating birds are slow to put on weight during stopovers, possibly due to low digestive efficiency of atrophied digestive tracts (Karasov 1996; Karasov & Pinshow 2000). Again, feedback from the gut could reduce the motivation to eat more food. Gut flexibility can also hide changes in caloric requirements; for example, greater digestive efficiency during periods of stress such as feather growth (Wijnandts 1984) should also reduce the difference between energetically challenging and less challenging periods in caloric intake observed.

Young birds during early postnatal growth probably have low assimilation efficiencies compared with adults. The reason for this is not well understood and we lack data on digestion in altricial nestlings, making generalities premature (Karasov 1990; W.H. Karasov & J. Wright this volume). Long-term selection on domestic hens (*Gallus domesticus*) for early meat production has produced a 2.5-fold increase in age-specific body weight with only a 22.5% increase in digestive efficiency, implying that greater growth is achieved largely by greater intake (Denbow 1999). In fact, these broilers seem to have evolved less responsiveness to the feeding inhibiting effects of leptins (Denbow et al. 2000), suggesting a possible mechanism for sex- or species-specific differences in growth. In the shorter term, birds change their feeding patterns with current feeding conditions. Feeding experiments on domestic chicks show that both digestive organs and behaviour respond to changes in meal patterns. Where chicks were offered food only on alternate days, the chicks ate enormously on these days and their livers (glycogen storage capacity) enlarged (Nir & Nitsan 1979). In another study, chicks fed diets containing large amounts of non-nutritive filler were able to maintain their growth by eating more bulk (Savory 1980, 1984), demonstrating responsiveness to blood nutrient levels as well as digestive tract fullness. It is also possible that decreasing leptin levels with diminishing fat stores changed the CNS response to signals of crop fullness. Prospective feeding and a change in the role of fullness is suggested by an evening increase in the degree to which Japanese quail (*Coturnix japonica*) fill their crops before ceasing to eat (Savory 1985), as well as by a seasonal adjustment to length of the night fast (reviewed in Savory 1999).

Social cues can also affect appetite. Keeling and Hurnik (1996) showed that well-fed domestic chicks housed with voraciously eating, hungry chicks did increase their food intake somewhat, but they increased their pecking far more. Apparently, the opportunity and social context stimulated interest in food, but ingestion was limited by internal hunger. As for rats and humans (above), hunger may control ingestion more than it controls the propensity to evaluate or forage for food. Why some animals, such as domestic chicks, rats and humans, are sensitive to social cues when foraging is an evolutionary question that I will raise below (see also D.S. Wilson & A.B. Clark this volume, for other considerations). The chick responses are reminiscent of situations in which nestlings join in the begging of hungrier siblings. If begging is viewed as foraging behaviour rather than as a request to be fed, such a socially facilitated response is well documented outside the nest (Budden & Wright 2001a).

An Adaptationist View of Hunger and Appetite

The studies of appetite and hunger in humans, other primates and rodents are not explicitly evolutionary or functional in outlook (but see Chafetz 1990; Rolls 1994). It is not, however, hard to envision what hunger and appetite should do for an animal and how it should be functionally structured. Hunger might be viewed as an internally generated cue like pain that gives the individual information about nutritional risks, threats and current or upcoming problems. In conjunction with information on other priorities and risks, it also motivates the individual to act to alleviate those problems or to take advantage of opportunities to offset future problems. Hunger functions adaptively both because it provides this information more or less continuously, before the organism is in a truly critical situation with no more time to act, and because it helps stimulate or motivate specific behaviour that is likely to reduce the risk; lack of food in this case. (As with pain in some life-threatening situations, it may become imperceptible when the organism is in fact at extreme risk of dying from lack of nutrients. But perhaps that is pushing functionality too far, unless we suppose that this frees the individual to take actions that would be hindered by terrible pangs of hunger.) Some of the triggers for hunger such as blood insulin are themselves subject to external information (the sight of food, conditioned expectations of food from temporal or other stimuli), so that hunger results from both current deficits and opportunities to offset future deficits. Finally, all these triggers are received in and/or transmitted to the central nervous system and can be integrated into an organism's weighting of priorities. Thus, more extreme hunger will help motivate an animal to incur increasing risks of predation or interference in order to get food (see overviews on birds: Maurer 1990, 1996; Stephens 1990).

ADAPTIVE APPETITE AND THE BEGGING OF NESTLINGS

What does the foregoing framework suggest that we need to know in order to interpret nestling begging behaviour as a foraging behaviour? To begin with, we might consider (1) what information should in principle be provided to a nestling by hunger and thus be able to affect appetite and (2) what internal or external situations might be good cues or sources of information. We need to look for empirical evidence that variation in this information, as well as variation in related risks, are reflected in begging patterns and differences within and between species. Finally, we must ask if treating begging as a foraging behaviour makes any difference to our expectations of begging as a signal.

While the ultimate issue for the nestling is getting the appropriate amount of energy and nutrients to maximize both survival and future reproduction, we are concerned here with more proximate causes of a nestling's behavioural decisions. How proximate is also one of the issues: is begging behaviour somehow sensitive *a posteriori* to long-term situations like a history of nutrient restriction or prospectively sensitive to future situations? If so, can hunger and appetite as we understand them be responsible for the relationship between begging and nutritional need on these temporal scales? Logically, we expect an adaptively constructed appetite to vary in response to: (1) the current state of nutrition, signalled by hunger; (2) the risks to that individual if food is not forthcoming; (3) other risks of foraging in that context; and (4) any prospective information about the probability of feeding in future.

The Current Nutritional and Energy State of the Nestling

A current lack of nutrients is the most likely stimulus for hunger as commonly understood. Nutritional state may include both amount of energy and the balance of nutrients, and birds may shift how much they eat of certain foods in search of specific nutrients (Denbow 1994), such as protein (Boa-Amponsem et al. 1991a; Steinruck & Kirchgessner 1992, 1993) or calcium (Lee 2000). For the purposes of this chapter, nutritional state will be described in terms of available energy. If a nestling has fed on a given amount of food and waited without feeding for a set amount of time, its hunger should vary with factors that determine how rapidly and completely the food calories have been used up. These include the current growth rate of the nestling, the reserves it possesses, and how much food it needs in absolute terms, as well as its digestive efficiency (W.H. Karasov & J. Wright this volume).

Current Growth Rate and Hunger

Faster growth rates in birds entail increased costs per unit time because of greater rates of tissue production, greater costs of synthesis and probably also higher basal metabolic rates (Weathers 1996). Basically, a given amount of food will be metabolized more rapidly, and the metabolites removed more rapidly, with faster growth. Growth rate varies with species, age and in some cases, sex. Differences in growth rate among species may result from selection for faster growth and earlier fledging under high predation risks to nests (Bosque & Bosque 1995) or for winning in sibling competition (Ricklefs 1993). Alternatively, there may be constraints on growth rate because of precociality (Ricklefs et al. 1998) or a limited food supply (Case 1978; see Ricklefs et al. 1998 for evaluation of all factors). (With respect to food supply, however, Weathers (1992) shows that increased growth rates can reduce the total energy required by the end of growth and slower growth increases it.) Selection for rapid growth in domestic chicks results in changes in the degree of gut fullness associated with satiation, as well as changes in leptin-based satiety mechanisms (Burkhart et al. 1983, cited in Boa-Amponsem et al. 1991b; Denbow et al. 2000). Thus, species differences in growth rate might well be evident in feeding responses to a given degree of hunger. Intraspecifically, growth curves usually show newly-hatched and near-fledging nestlings growing less quickly (lower absolute gain per unit time) than those midway through the postnatal period (Ricklefs 1984; Konarzewski et al. 1998). On the other hand, the youngest nestlings are growing at the highest mass-specific rate (% mass increase per day).

We thus expect inter- and intraspecific variation in begging responsiveness and levels with differences in growth rate. Within species, the most rapidly growing nestlings are much smaller than adults and may be limited in what they can ingest at one time in comparison with near-fledging nestlings. Furthermore, immediate post-meal satiation may relate closely to fullness in the digestive tract (or crop; Savory 1985), as well as to levels of circulating nutrients. Thus, after the maximal meal size that such a nestling can hold, a strong appetite and readiness to beg intensely would return more quickly in younger than in older nestlings. The former have eaten somewhat less and they are metabolizing it and using the metabolites more quickly. As noted above, it is common to find that younger siblings beg more readily and intensely than older ones in the brood even when their growth is not inhibited. One factor may be that, during the feeding trip intervals, they do become physiologically hungrier. If so, this is a confound for studies that try to equalize hunger by, for example, feeding all nestlings to satiation and waiting one hour. We need studies of the begging intensity relative to time and amount eaten for nestlings throughout ontogeny.

Sexually dimorphic growth patterns offer a more complex problem. The overall growth rate constant (K of the logistic curve) may differ little between sexes (Richner 1991), but the maximum growth rate and its timing may do (Gebhardt-Henrich & Richner 1998). Often the smaller sex has an earlier growth maximum, while it also matures more quickly. Thus, the energetic costs of maturation must be added to understand the actual difference in need between the two sexes. The energetic needs of the smaller sex may be relatively larger (calories per gram nestling) while still being absolutely smaller (Teather & Weatherhead 1988). Female red-winged blackbirds (*Agelaius phoeniceus*) are about 2/3 the size of males at fledging, but usually achieve their peak weights and feather out sooner (Teather 1993; Clark 1995). In our ongoing analyses of feeding to red-winged blackbird broods of one male and one female across nestling ontogeny, females are fed as much as (nonsignificantly more than) their brothers early in the nestling period, but males are fed significantly more than their sisters later on, at days 7-9 (J. Peet & A.B. Clark unpublished data). Teather (1992) reported that males were consistently fed more food, but his trials all used older nestlings. The sex differences in size can be dramatic, such that the smaller sex can ingest or hold in its gut much less, possibly resulting in more frequent hunger. In order to predict how hunger should compare, the time course of energetic needs as well as average meal size and digestive efficiency for both sexes must be characterized.

Fast-growing nestlings should also beg sooner or more intensely after a feeding because, prospectively, they face reduced survival earlier. During their fastest growth phases, nestlings might be most susceptible to irreparable growth deficits resulting from short-term food deprivation, such as might be experienced on a rainy, windy day of low insect activity. Supporting this idea that faster growing nestlings really do experience nutritional deficits more quickly and with more serious consequences, faster growth was linked to poorer survival of yellow-eyed penguin (*Megadyptes antipodes*) chicks (van Heezik & Davis 1990), and faster growing male lesser black-backed gulls (*Larus fuscus*) were more susceptible to starvation than slower growing females (Griffiths 1992). Such nestlings might be selected to beg at lower levels of physiological hunger or beg more intensely for a given level of hunger, much as adult birds on restricted energy budgets might forage to minimize the risk of falling below some critical minimum intake (Stephens 1990).

Absolute Need

Over much of the growth curve, larger nestlings will need an absolutely larger amount of energy per day to cover current maintenance and growth costs. Within a size class, faster growing nestlings have a larger absolute requirement. If food becomes limited below that required to satisfy the total needs of the brood and parents feed all equally, these needier nestlings may become hungrier because the same amount of food leaves them with an absolutely greater caloric deficit. Larger nestlings may also feel this deficit even after being fed, if the food items are smaller and insufficient to trigger satiety. This should lead to more consistent and intense begging by such nestlings based on their physiological information about their state.

Reserves

Both fat and, in extreme situations, other metabolizable tissue can buffer a nestling against growth or developmental harm during food shortages. While some species may reduce their metabolic rates when food shortages set in, others do not have such selected plasticity (Schew & Ricklefs 1998). Starlings (*Sturnus vulgaris*), for instance, responded to restricted food levels (adjusted for no weight gain) with continued structural growth of wings and tarsus at 80% of normal for two days until yolk and fat reserves were used up (Schew 1995, cited in Schew & Ricklefs 1998). Obviously, if nestlings are going to maintain close to normal structural growth against short-term deprivation, those with reserves should be less worried about missing a meal than those without reserves. In general, larger nestlings will have absolutely larger amounts of reserves in terms of fat. Older nestlings will also have a higher percentage accrued dry weight of tissue and thus, for their size, will have more reserves. Small or young nestlings should behave proactively to avoid the possibility of missed meals and to build up their reserves by being very responsive to any opportunity to feed and sensitive to very small changes in their nutritional state.

Exactly how nestlings might register that they have few reserves, before they have to use them, is not known, but studies of leptins in poultry (e.g. Denbow et al. 2000) suggest a role for such hormones. Leptin synthesis occurs in embryonic yolk cells as well as liver and adipose tissue of precocial birds (Ashwell et al. 1999). Young nestlings up to the age of peak absolute growth rates will be losing their yolk sacs, but will only have small fat reserves, appropriate to their body size. These fat reserves fluctuate on short time scales. Fat reserves of young red-winged blackbird nestlings can disappear within one day of restricted feeding and also return quickly (personal observation). If meal size at satiation is closely related to lipid reserves, one might expect decreasing satiety signals as yolk is lost and fat levels rapidly adjust up and down in small nestlings. For a given percent decrease in food across the brood, younger, fast-growing nestlings may then use up more of their smaller reserves. Thus, after meals restricted in part by crop and/or gut size, younger nestlings could feel less satiated and be more ready to beg again. Since leptins have been studied in the context of intake regulation and not prospective foraging effort, the direct effect on begging readiness or intensity is a completely unexplored but intriguing possibility.

Digestive Efficiency

Although far too little is known about digestive ontogeny in altricial nestlings, data on a few species indicate that assimilation efficiencies begin relatively low and increase over the nestling period (Karasov 1990). The lower assimilation efficiency of younger nestlings will compound their problems of lower intake per meal, increasing growth rates and relatively poor fat reserves (although cushioned by yolk at the beginning). This will be one more factor making it likely that younger nestlings will feel real and potentially life-threatening hunger more quickly following satiation than would an older nestling. Given that yolk may cushion them in the first few days, hunger ought to be more immediate and the begging response to that hunger more intense, at the critical point when yolk reserves are gone.

Digestive efficiency may also vary with the way food is fed, rather than just the amount. In our recent experimental study of the effects of meal schedules on male and female nestling red-winged blackbirds (Rutigliano 1999; A.B. Clark, J. Rutigliano & H. Chekanovsky unpublished data), we found that nestlings offered food in small frequent meals over an entire day grew more and more efficiently (i.e. greater gains in weight, tarsus and primaries per gram food over 24 hours) than did those offered the same amount of food in fewer larger meals. This study used nestlings during their fifth day after hatching, a period of near maximal growth for both sexes. The growth effects were greatest for the females, the smaller sex. More importantly, the begging responsiveness and intensity was significantly and dramatically reduced in the frequently fed nestlings of both sexes, a decrease made all the more striking because they actually consumed less food over the day. Thus, begging and presumably hunger were clearly not simple functions of the amount of food fed recently, but also of the pattern by which it was fed. One contributing factor may have been a greater efficiency in digesting smaller meals (McWilliams et al. 1999). In this study, total faeces (wet and dry weights) and protein content of faeces did not differ between meal schedules, but carbohydrate and fat absorption were not measured.

The Risks of Going Hungry

In a number of ways, nestlings within a brood could differ adaptively in how they evaluate their situation when food deprived. A certain level of physiological hunger might have very different risk implications for different nestlings, based on their developmental age and potential for adaptive growth variation.

Current Developmental Priorities

Nestlings are not only increasing in mass but also maturing organs and other structural components of their bodies. Some components, e.g. neural tissue, will always have priority, but others have critical periods of growth and differentiation. The risk attending a caloric and/or nutrient deficit will depend on how it affects the specific body parts currently emphasized in the maturational process. While it is possible that there could be some direct feedback to hunger-inducing mechanisms from these specific tissues, begging that is sensitive to this risk could be managed by an appetite that varies ontogenetically in its relationship to hunger *per se*. When feather development is at risk, nestlings might respond more strongly to a given level of hunger or respond at earlier signs of low food, than those much younger or older, for example. The implications are twofold: first, growth rate will not be the only predictor of quantitative differences in begging responsiveness, and second, age differences in begging intensity within broods are expected even if feeding is egalitarian on the surface. Sex differences could also appear (and disappear) during ontogeny if the timing of critical developmental events differs according to sex.

Developmental Plasticity and Compensatory Growth

Lack of food generally registers as reduced growth, but in some cases birds have evolved adaptive changes in growth patterns in the face of changes in food supply (Schew & Ricklefs 1998). Differentiating reduced growth as a cost of food restriction from reduced growth as an adaptive, protective response to food restriction can be difficult (Schew & Ricklefs 1998). Species capable of adaptively slowed growth will, however, probably not be at as great a risk as those experiencing slow growth determined directly by food levels. Even when slowed growth and development offset actual starvation, there remains the question of compensatory growth later on. The potential for compensatory growth may vary with the particular structure impacted or with the timing of the growth insult. Species differ in being more or less developmentally plastic, with more precocial birds and those subject to more frequent, predictable food shortages of intermediate length having the clearer mechanisms (see Schew & Ricklefs 1998 for a particularly helpful discussion of plasticity and compensatory growth). Typically, these birds will reduce both the growth of specific body parts (digestive tract, liver) and maturational processes. They may lower both body temperature and oxygen consumption. At least a few altricial species (e.g. white-fronted bee-eaters, *Merops bullockoides,* Emlen et al. 1991; budgerigars, *Melopsittacus undulatus*, personal observation) apparently adaptively reduce growth and development in the face of food deprivation while others (e.g. starlings; Schew 1995, cited in Schew & Ricklefs 1998) do not. Where nest predation risk is strong, slowing growth may seldom be adaptive and seems an unlikely strategy for cup-nesting passerines. Lowering metabolic rate and growth rate should slow the onset of actual physiological signals of hunger after eating; whether changes in the digestive tract itself could play a role here is not known. Without such adaptive reductions in the rate of energy demand, real and sudden food restriction should be experienced as intense hunger. Furthermore, nestlings without growth reduction mechanisms should be selected to respond strongly and consistently to hunger signals, as the only hope is to shorten the time of food restriction. Thus, we would expect marked species and perhaps age differences in the relationship between begging intensity and degree of starvation, with some species decreasing begging (foraging effort) to conserve energy while others maintain or even increase it.

Adaptive Responses Based on Experience

Nestlings have the potential to alter their behaviour with feeding experience both physiologically and behaviourally, as long as the environment provides useful cues of future problems and contingencies. In an important recent study of nestling learning, Kedar et al. (2000) demonstrated that nestlings would alter their typical begging intensity depending on the intensity that was reinforced by food. This is perhaps the clearest way that nestlings might come to differ within or between broods. But beyond this, nestlings experience many patterns with respect to food and hunger to which they might respond adaptively with changes in apparent appetite and begging behaviour.

Patterns of Feeding

The pattern of feeding in terms of visit rates to the nest, feeding rates to that nestling and typical meal size all contribute to an expectation of the feeding situation in the next period of time. If intervals between feeds are typically long, then the nestling should always be prepared to take whatever it is offered, even if large, this time. Also, it should beg hungrily on any arrival because it is likely to be a long time before an opportunity arises again. Thus, one might expect nestlings that typically go long periods between feedings to respond strongly to any parental visit, even if the visit interval is unusually short and the nestling relatively full. By analogy with rat hunger, this might be mediated by strong feedback stimulation from the cues of parental presence (which signals the presence of food) to internal pre-digestive events that themselves create sensations of hunger. That would be one interpretation of the behaviour of our experimental nestlings fed at thirty minute versus ten minute intervals (Rutigliano 1999; A.B. Clark, J. Rutigliano & H. Chekanovsky unpublished data). Short-interval nestlings were not as motivated to take a feeding for a given degree of mild hunger because they had an expectation that another opportunity would come soon. Long interval nestlings were always preparing for longer periods of food deprivation and stuffing themselves. (Another way to view it selectively is to ask, given the feeding pattern, what would be the effect of one missed feeding upon a nestling's nutritional state?) The domestic chicks fed every other day ate much more on food days than chicks on a normal *ad lib.* schedule (Nir & Nitsan 1979; Nitsan et al. 1984). Their motivation to eat might have increased because crop or gut changes altered signals of satiation after normal sized meals, but they might also have altered their priorities from eating meals that allow them mobility to eating meals that keep them from being hungry for longer. Whether altricial nestlings can adapt physiologically using gut or crop size changes to deal with the range of normal variation in feeding patterns is unknown (W.H. Karasov & J. Wright this volume).

This raises the question of uncertainty as a motivator. If intervals between feedings or amounts per feeding swing wildly, the nestlings may never have a chance to adapt physiologically or behaviourally in terms of begging. Smaller, younger siblings appear to have the most uncertain intervals. Older siblings experience one or two post-hatch days with few competitors and feedings that follow begging cues while younger siblings, from hatching onward, will experience many parental arrivals when they do not get fed. Thus, during the initial learning phase, they experience a greater uncertainty of reward than first-hatched nestlings. To the extent that larger, older young need more food, younger ones will continue to be fed on a smaller fraction of the parental visits. This problem should increase with increasing brood size and age differences. Unpredictable contingencies typically stimulate animals to work harder for food, and begging might increase just as bar pressing does in a laboratory rat.

Social Conditions and Begging

Social cues within the brood should also affect begging in a number of ways directly related to risks and predictability of food. First, if a sibling begs suddenly or over a period of visits begins to beg more strongly, why would other nestlings respond, especially if less hungry (Price et al. 1996; but see Leonard & Horn 1998)? The behaviour of siblings may be an excellent cue both to the presence of the parent (and hence food) and also to changing resource conditions. Young nestlings, in particular, are not very discriminating about the cues of parents and often beg in response to other environmental noise or, conversely, miss the parent's arrival (A.B. Clark & C. Mostello unpublished data; Budden & Wright 2001b; Leonard & Horn 2001). An older sibling offers a better filter of environmental noise and acts as an indicator of a parent. Thus, one might expect younger siblings, when young, to profit if they quickly join in the response of a more discriminating sib, just as foragers might join individuals that appear to be finding food. Moreover, a sibling becoming consistently more hungry indicates that the parents are not meeting the needs of the nest and the actor could be next to feel the pinch. At this point, an increased motivation to beg should indeed be selected for its utility in getting the parents to bring more food and to work harder. Nestlings will profit if they join hungrier siblings to influence parental behaviour before the former also go hungry. The social effect upon signalling is however mirrored in socially mediated increases in simple foraging activity by chicks (Keeling & Hurnik 1996), and may not differ motivationally.

Secondly, if group begging has a signal value all its own, but begging attracts predators, nestlings would do well to beg together but less frequently. Thus, they might evaluate the risk of begging as lower when siblings are also begging, and respond most readily in concert with siblings (see also D.S. Wilson & A.B. Clark this volume).

These cases call for appetite to be changed by an evaluation of risk as much as by actual physiological hunger. However, given the many feedback loops discovered so far in mammals, this should not be an unlikely scenario. While begging would not then co-vary with hunger in the narrow sense, it would tell us as well as the parents much about the nestling's assessment of its risk of deprivation down the road.

Faster Growth: Not an Unambiguous Benefit

One often suggested goal of begging is to get more food than parents are selected to allocate. It is not clear that if a nestling wins in this regard it would receive enough food to support a higher growth rate than typical. There is plenty of evidence suggesting both short-term and lifetime costs of growth-related smaller body size at fledging. But more food and concomitantly faster growth and/or more fat reserves may themselves carry costs. There is accruing theoretical and empirical evidence from many different groups of organisms that fast growth and rapid gains in body size may come at the expense of other important structural adaptations (Arendt 1997). Predicted or demonstrated trade-offs of more rapid growth include increased risks of body asymmetry or other developmental errors (Sibly & Calow 1986), weaker bones due to limited rates of calcium deposition, poorly developed immune systems and shorter life spans (see discussion in Arendt 1997). While some of these trade-offs follow from selected changes in average growth, others may become evident within populations with abnormally nutritious or superabundant food, such as offered to young precocial birds in captivity. They may also contribute to the poorer prospects for nestlings that have undergone food restriction and then caught up with rapid growth at an atypical age, when developmental priorities may compete even more than with normal growth (see above; Schew & Ricklefs 1998). Thus, begging for food might have one set of rules, while accepting food according to need and taking no more than an adequate share may be not only good for one's siblings, but also healthy and wise.

Begging: Foraging or Signalling?

I have suggested that it is useful to consider begging as a foraging behaviour in order to see how well the model of appetite and foraging might predict its patterns. As I have tried to indicate, if begging really were a foraging behaviour that we could score, it might vary between species, contexts and individuals much as begging does, without reference to questions of honesty. It may be going too far to say that begging evolved primarily as a foraging behaviour, rather than a competitive signal, but it is at least interesting to ask if it would make any difference to our expectations for begging patterns. In one important way, it might. As appetite is apparently structured, the would-be forager is always noticing and evaluating food if presented and only then rejecting it if the risks and costs of capture and eating outweigh the benefits. A nestling cannot really evaluate food without begging for it. If begging evolved as a foraging behaviour, we would in fact expect that even nearly sated nestlings might respond at low levels to the arrival of a parent. The willingness of nestlings to beg at a low level over a range of fullness, when they are unlikely to eat or at least do not need the food, would reduce the information in low intensity begging signals.

FUTURE DIRECTIONS

In our studies of nestling begging, I think it is clear that we cannot expect to equalize hunger or appetite among nestlings with simple regimes of short-term feeding, at least not if comparing nestlings differing in such aspects as size, age, sex, experiential history and, of course, species. Appetite or the degree to which food is actually sought will be a function both of physiological hunger and of weightings of the risks and benefits of seeking food now versus later. Both hunger and appetite should vary *a priori* among nestling types, without for the most part considering competition or interactions with parents.

Clearly, we need to work more closely with physiological ecologists and developmental physiologists, to encourage comparative studies of digestion and assimilation during avian ontogeny and to investigate the relationships between the nestling's nutritional needs and its current risks, external and internal, from the individual nestling's point of view. We should be particularly interested in the recent work on leptins, now known to be important in regulating growth patterns in poultry breeds (Denbow et al. 2000) and quite possibly also involved in nestling responses to changing fat stores. Their production in yolk (Ashwell et al. 1999) gives them a potential role in very early differences in begging and feeding rate between and within broods, but to date, there are no studies of leptins in altricial birds. Such studies should include investigations of the relationship of leptin levels to begging attributes, not just food intake.

It should also be heuristically valuable to consider begging as a foraging behaviour as well as a signalling behaviour because, in a number of aspects ranging from intensity to context, begging may be subject to selection as both (see also Budden 2001; Budden & Wright 2001a). If nothing else, this approach will lead us to investigate these questions of how hunger and appetite vary between individuals and species as a basis for asking if begging evolves as an honest signal.

ACKNOWLEDGEMENTS

I thank the Gregynog Begging Workshop organizers and participants for a very stimulating three days, and T. Amundsen, in particular, for questioning the significance of begging evolving as a foraging behaviour. W.H. Karasov, M. Leonard and J. Wright provided very helpful comments on earlier versions of this chapter. TEW deserves credit for patience pre- and post-workshop. Supported in part by NSF IBN-9306938.

REFERENCES

Arendt, J.D. 1997. Adaptive intrinsic growth rates: An integration across taxa. *The Quarterly Review of Biology* **72**, 149-177.

Ashwell, C.M., Czerwinski, S.M., Brocht, D.M. & McMurty, J.P. 1999. Hormonal regulation of leptin expression in broiler chickens. *American Journal of Physiology* **276**, R228-R232.

Bartness, T.J. 1990. Species-specific changes in the metabolic control of food-intake - integrating the animal with its environment. *International Journal of Obesity* **14**, 115-124.

Boa-Amponsem, K., Dunnington, E.A. & Siegel, P.B. 1991a. Genotype, feeding regime and diet interactions in meat chickens. 2. Feeding behaviour. *Poultry Science* **70**, 689-696.

Boa-Amponsem, K., O'Sullivan, N.P., Gross, W.B., Dunnington, E.A. & Siegel, P.B. 1991b. Genotype, feeding regime, and diet interactions in meat chickens. 3. General fitness. *Poultry Science* **70**, 697-701.

Bosque, C. & Bosque, M.T. 1995. Nest predation as a selective factor in the evolution of growth rates in altricial birds. *American Naturalist* **145**, 234-260.

Budden, A.E. 2001. Nestling begging in Southern Grey Shrikes. PhD thesis, University of Wales, Bangor.

Budden, A.E. & Wright, J. 2001a. Begging in nestling birds. *Current Ornithology* **16**, 83-118.

Budden, A.E. & Wright, J. 2001b. Falling on deaf ears: the adaptive significance of begging in the absence of a parent. *Behavioral Ecology and Sociobiology* **49**, 474-481.

Case, T.J. 1978. On the evolution and adaptive significance of postnatal growth rates in the terrestrial vertebrates. *The Quarterly Review of Biology* **53**, 243-282.

Chafetz, M.D. 1990. *Nutrition and Neurotransmitters: The Nutrient Bases of Behavior.* Englewood Cliffs: Prentice Hall.

Churchland, P.S. 1986. *Neurophilosophy: Toward a Unified Science of the Mind/Brain.* Cambridge: MIT Press.

Clark, A.B. 1995. Gapes of sexually dimorphic blackbird nestlings do not show sexually dimorphic growth. *The Auk* **112**, 364-374.

Cook, M.I., Monaghan, P. & Burns, M.D. 2000. Effects of short-term hunger and competitive asymmetry on facultative aggression in nestling black guillemots *Cepphus grylle*. *Behavioral Ecology* **11**, 282-287.

Cotton, P.A., Kacelnik, A. & Wright, J. 1996. Chick begging as a signal: are nestlings honest? *Behavioral Ecology* **7**, 178-182.

Denbow, D.M. 1994. Appetite and its control. *Poultry Science Review* **5**, 209-229.

Denbow, D.M. 1999. Food intake regulation in birds. *Journal of Experimental Zoology* **283**, 333-338.

Denbow, D.M., Meade, S., Robertson, A., McMurty, J.P., Richards, M. & Ashwell, C. 2000. Leptin-induced decrease in food intake in chickens. *Physiology and Behavior* **69**, 359-362.

Drummond, H. & Garcia Chavelas, C. 1989. Food shortage influences sibling aggression in the blue-footed booby. *Animal Behaviour* **37**, 806-819.

Drummond, H., Osorno, J.L., Torres, R., Garcia Chavelas, C., & Merchant, H. 1991. Sexual size dimorphism and sibling competition: implications for avian sex ratios. *American Naturalist* **138**, 623-641.

Emlen, S.T., Wrege, P.H., Demong, N.J. & Hegner, R.E. 1991. Flexible growth rates in nestling white-fronted bee-eaters: a possible adaptation to short-term food shortage. *Condor* **93**, 591-597.

Fujioka, M. 1985. Sibling competition and siblicide in asynchronously-hatching broods of the cattle egret *Bubulcus ibis*. *Animal Behaviour* **33**, 1228-1242.

Gebhardt-Henrich, S. & Richner, H. 1998. Causes of growth variation and its consequences for fitness. In: *Avian Growth and Development* (Ed. by J.M. Stark & R.E. Ricklefs). Oxford: Oxford University Press.

Godfray, H.C.J. 1991. Signalling of need by offspring to their parents. *Nature* **352**, 328-330.

Greig-Smith, P. 1985. Weight differences, brood reduction, and sibling competition among nestling stonechats *Saxicola torquata* (Aves: Turdidae). *Journal of Zoology* **205**, 453-465.

Griffiths, R. 1992. Sex biased mortality in the lesser black-backed gull *Larus fuscus* during the nestling stage. *Ibis* **134**, 237-244.

Heezik van, Y. & Davis, L. 1990. Effects of food variability on growth rates, fledgling sizes and reproductive success in the yellow-eyed penguin *Megadyptes antipodes*. *Ibis* **132**, 354-365.

Hussell, D.J.T. 1988. Supply and demand in tree swallow broods: a model of parent-offspring food-provisioning interactions in birds. *American Naturalist* **131**, 175-202.

Karasov, W.H. 1990. Digestion in birds: chemical and physiological determinants and ecological implications. *Studies in Avian Biology* **13**, 391-415.

Karasov, W.H. 1996. Digestive plasticity in avian energetics and feeding ecology. In: *Avian Energetics and Nutritional Ecology* (Ed. by C. Carey). New York: Chapman and Hall.

Karasov, W.H. & Pinshow, B. 2000. Test for physiological limitation to nutrient assimilation in a long-distance passerine migrant at a springtime stopover site. *Physiological and Biochemical Zoology* **73**, 335-343.

Kedar, H., Rodríguez-Gironés, M.A., Yedvab, S., Winkler, D.W. & Lotem, A. 2000. Experimental evidence for offspring learning in parent-offspring communication. *Proceedings of the Royal Society of London, Series B* **267**, 1-5.

Keeling, L.J. & Hurnik, J.F. 1996. Social facilitation acts more on the appetitive than the consummatory phase of feeding behaviour in domestic fowl. *Animal Behaviour* **52**, 11-15.

Khayutin, S.N. & Dmitrieva, L.P. 1977. Behavioural mechanisms synchronizing the development of nestlings in the same clutch under their non-simultaneous hatching. *Zoologische Zhurnal* **56**, 1195-1206.

Kilner, R & Johnstone, R.A. 1997. Begging the question: are offspring solicitation behaviours signals of need? *Trends in Ecology and Evolution* **12**, 11-15.

Konarzewski, M., Kooijman, S.A.L.M. & Ricklefs, R.E. 1998. Models for avian growth and development. In: *Avian Growth and Development* (Ed. by J.M. Stark & R.E Ricklefs). New York: Oxford University Press.

Kuenzel, W.J., Beck, M.M. & Teruyama, R. 1999. Neural sites and pathways regulating food intake in birds: a comparative analysis of mammalian systems. *Journal of Experimental Zoology* **283**, 348-364.

Lee, K.H. 2000. Application of dietary self-selection by pullets and layers in practice. *Asian-Australasian Journal of Animal Sciences* **13**, 55-65.

Legg, C.R. 1994. Appetite - a psychological concept. In: *Appetite: Neural and Behavioural Bases* (Ed. by C.R. Legg & D.A. Booth). Oxford: Oxford University Press.

Legg, C.R. & Booth, D.A. 1994. *Appetite: Neural and Behavioural Bases*. Oxford: Oxford University Press.

Le Magnen, J. 1992. *Neurobiology of Feeding and Nutrition*. New York: Academic Press.

Leonard, M.L. & Horn, A.G. 1998. Need and nestmates affect begging in tree swallows. *Behavioral Ecology and Sociobiology* **42**, 431-436.

Leonard, M.L. & Horn, A.G. 2001. Begging in the absence of parents by nestling tree swallows. *Behavioral Ecology* **12**, 501-505.

Leonard, M.L., Horn, A.G., Gozna, A. & Ramen, S. 2000. Brood size and begging intensity in nestling birds. *Behavioral Ecology* **11**, 196-201.

Lotem, A. 1998a. Differences in begging behaviour between barn swallow, *Hirundo rustica*, nestlings. *Animal Behaviour* **55**, 809-818.

Lotem, A. 1998b. Higher levels of begging behaviour by small nestlings: a case of a negatively correlated handicap. *Israel Journal of Zoology* **44**, 29-45.

Machmer, M.M. & Ydenberg, R.C. 1998. The relative roles of hunger and size asymmetry in sibling aggression between nestling ospreys, *Pandion haliaetus*. *Canadian Journal of Zoology* **76**, 181-186.

Maurer, B.A. 1990. Extensions of optimal foraging theory for insectivorous birds: implications for community structure. *Studies in Avian Biology* **13**, 455-461.

Maurer, B.A. 1996. Energetics of avian foraging. In: *Avian Energetics and Nutritional Ecology* (Ed. by C. Carey). New York: Chapman and Hall.

McWilliams, S.R., Caviedes-Vidal, E. & Karasov, W.H. 1999. Digestive adjustments in cedar waxwings to high feeding rate. *Journal of Experimental Zoology* **283**, 394-407.

Mei, N. 1994. Role of digestive afferents in food intake regulation. In: *Appetite. Neural and Behavioural Bases* (Ed. by C.R. Legg & D.A. Booth). Oxford: Oxford University Press.

Mock, D.W. & Lamey, T.C. 1991. The role of brood size in regulating egret sibling aggression. *American Naturalist* **138**, 1015-1026.

Mock, D.W. & Parker, G.A. 1997. *The Evolution of Sibling Rivalry*. Oxford: Oxford University Press.

Nagel, T. 1974. What is it like to be a bat? *Philosophical Review* **83**, 435-450.

Nielsen, B.L. 1999. On the interpretation of feeding behaviour measures and the use of feeding rate as an indicator of social constraint. *Applied Animal Behaviour Science* **63**, 79-91.

Nir, I. & Nitsan, Z. 1979. Metabolic and anatomical adaptation of light-bodied chicks to intermittent feeding. *British Journal of Poultry Science* **20**, 61-71.

Nitsan, Z., Nir, I. & Petihi, I. 1984. The effect of meal feeding and food restriction on body composition, food utilization and intestinal adaptation in light breed chicks. *British Journal of Nutrition* **51**, 101-109.

O'Connor, R.J. 1984. *The Growth and Development of Birds*. New York: John Wiley & Sons.

Price, K. 1994. The behavioural ecology of begging behaviour in yellow-headed blackbirds. PhD thesis, Simon Fraser University.

Price, K. & Ydenberg, R. 1995. Begging and provisioning in broods of asynchronously-hatched yellow-headed blackbird nestlings. *Behavioral Ecology and Sociobiology* **37**, 201-208.

Price, K., Harvey, H. & Ydenberg, R. 1996. Begging tactics of nestling yellow-headed blackbirds, *Xanthocephalus xanthocephalus*, in relation to need. *Animal Behaviour* **51**, 421-435.

Redondo, T. & Castro, F. 1992. Signalling of nutritional need by magpie nestlings. *Ethology* **92**, 193-204.

Richner, H. 1991. The growth dynamics of sexually dimorphic birds and Fisher's sex ratio hypothesis: does sex-specific growth contribute to balanced sex ratios? *Functional Ecology* **5**, 19-28.

Ricklefs, R.E. 1984. The optimization of growth rate in altricial birds. *Ecology* **65**, 1602-1616.

Ricklefs, R.E. 1993. Sibling competition, hatching asynchrony, incubation period and lifespan in altricial birds. *Current Ornithology* **11**, 199-276.

Ricklefs, R.E., Starck, J.M. & Konarzewski, M. 1998. Internal constraints on growth in birds. In: *Avian Growth and Development* (Ed. by J.M. Stark & R.E Ricklefs). New York: Oxford University Press.

Rolls, E.T. 1994. Neural processing related to feeding in primates. In: *Appetite: Neural and Behavioural Bases* (Ed. by C.R. Legg & D.A. Booth). Oxford: Oxford University Press.

Rutigliano, J. 1999. The effects of meal size on nestling growth. Honors thesis, Binghamton University.

Rydén, O. & Bengtsson, H. 1980. Differential begging and locomotory behaviour by early and late hatched nestlings affecting the distribution of food in asynchronously hatched broods of altricial birds. *Zeitschrift für Tierpsychologie* **53**, 209-224.

Savory, C.J. 1980. Meal occurrences in Japanese quail in relation to particle size and nutrient density. *Animal Behaviour* **28**, 160-171.

Savory, C.J. 1984. Regulation of food intake by brown leghorn cockerels in response to dietary dilution with kaolin. *British Poultry Science* **25**, 253-258.

Savory, C.J. 1985. An investigation of the role of the crop in control of feeding in Japanese quail and domestic fowl. *Physiology and Behavior* **35**, 917-918.

Savory, C.J. 1999. Temporal control of feeding behaviour and its association with gastrointestinal function. *Journal of Experimental Zoology* **282**, 339-347.

Savory, C.J., Hocking, P.M., Mann, J.S. & Maxwell, M.H. 1996. Is broiler breeder welfare improved by using qualitative rather than quantitative food restriction to limit growth rate? *Animal Welfare* **5**, 105-127.

Schew, W.A. & Ricklefs, R.E. 1998. Developmental plasticity. In: *Avian Growth and Development* (Ed. by J.M. Stark & R.E Ricklefs). Oxford: Oxford University Press.

Sibly, R. & Calow, P. 1986. *Physiological Ecology of Animals: An Evolutionary Approach.* Boston: Blackwell Scientific Publications.

Smith, H.G. & Montgomerie, R. 1991. Nestling American robins compete with siblings by begging. *Behavioral Ecology and Sociobiology* **29**, 307-312.

Stamps, J., Clark, A.B., Arrowood, P. & Kus, B. 1989. Begging behaviour in budgerigars. *Ethology* **81**, 177-192.

Steinruck, U. & Kirchgessner, M. 1992. The role of nutritive factors during an experience period in the development of a specific lysine hunger in layers. *Journal of Animal Physiology and Animal Nutrition* **68**, 43-52.

Steinruck, U. & Krichgessner, M. 1993. The origin of the specific protein hunger of layers by investigating their responses in dietary self-selection. *Archiv für Geflugelkunde* **57**, 42-47.
Stephens, D.W. 1990. Foraging theory: up, down and sideways. *Studies in Avian Biology* **13**, 444-454.
Teather, K.L. 1992. An experimental study of competition for food between male and female nestlings of the red-winged blackbird. *Behavioral Ecology and Sociobiology* **31**, 81-87.
Teather, K.L. 1993. Behavioral development of male and female red-winged blackbirds. *Wilson Bulletin* **105**, 159-166.
Teather, K.L. & Weatherhead, P.J. 1988. Sex-specific energy requirements of great-tailed grackle (*Quiscalus mexicanus*) nestlings. *Journal of Animal Ecology* **57**, 659-668.
Weathers, W.W. 1992. Scaling nestling energy requirements. *Ibis* **134**, 142-153.
Weathers, W.W. 1996. Energetics of postnatal growth. In: *Avian Energetics and Nutritional Ecology* (Ed. by C. Carey). New York: Chapman and Hall.
Wijnandts, H. 1984. Ecological energetics of the long-eared owl. *Ardea* **72**, 1-92.
Woods, S.C., Seeley, R.J., Porte, D. Jr. & Schwartz, M.W. 1998. Signals that regulate food intake and energy homeostasis. *Science* **280**, 1378-1383.

11.

NESTLING DIGESTIVE PHYSIOLOGY AND BEGGING

William H. Karasov[1] & Jonathan Wright[2]
[1]Department of Wildlife Ecology, University of Wisconsin, Madison WI 53717, USA (wkarasov@facstaff.wisc.edu)
[2]School of Biological Sciences, University of Wales, Bangor, Gwynedd LL57 2UW, UK (j.wright@bangor.ac.uk)

ABSTRACT

As birds grow, increases in food intake rate are matched by increases in mass of the alimentary system and/or mass-specific enzyme activity. In overfed nestlings, digesta retention time seems to decline and food is digested less efficiently. These observations are consistent with the idea that digestive capacity limits food intake and thus growth. When this is so, successful information transfer between begging nestlings and parents reduces parental costs and increases family economy. Nestling begging signals and parental provisioning responses therefore represent a system with the potential to optimize nestling digestive efficiency. However, little is known about how begging signals reflect changes in food quality or increased nestling metabolic expenditures, and how parents might respond. Nestling digestion and the physiological mechanisms that control begging provide an important link between begging effort, nestling nutritional state and the information regarding offspring 'need' that parents have been selected to acquire from begging signals.

INTRODUCTION

Students of avian begging and parental care must have a general interest in nestling digestive physiology and its possible influence(s) on avian growth, even if for no other reason than a desire to understand comprehensively their study system. In fact, many studies of begging and parental care either

J. Wright and M.L. Leonard (eds.), The Evolution of Begging, 199–219.

require specific information or are based on assumptions about nestling digestion. For example, evolutionary explanations behind patterns of parental behaviour have been explored with a diversity of models, which have all made simplified assumptions about the development of nestling physiology (e.g. Winkler & Adler 1996). Developmental changes in many aspects of nestling physiology have been reviewed (O'Connor 1984; Winkler & Adler 1996; Starck & Ricklefs 1998), but apparently not the development of digestive physiology, especially in non-domesticated birds. Hence, our goals in this chapter are to review current knowledge about development of avian digestive physiology and to illuminate links to nestling ecology and evolutionary patterns of growth, parental care and nestling begging.

A central issue in the evolution of honest solicitation signals between offspring and their parents is the definition of offspring 'need'. So far, need has been considered to mean nutritional state, simply in terms of immediate fitness consequences for fed offspring (e.g. Godfray 1991). We might therefore expect the proximate control of begging signals in nestling birds to be directly connected to their energetic intake, for example *via* gut distension due to food recently received. The physiological mechanisms involved in begging, by definition, provide us with a link between the nutritional history of the offspring and the aspect of offspring need that is of interest to parents when provisioning their young in the nest.

The ecological implications of digestion revolve around issues such as its influence on rate of food demand (i.e. hunger; A.B. Clark this volume) and how that might affect begging rate or growth rate, and digestive flexibility and how it might influence the diversity of foods accepted from parents. Consider two illustrative examples. For about a half a dozen wild species, including altricial species, there is evidence that apparent digestive efficiency (the proportion of ingested energy not eliminated in excreta) improves with age early in life (Karasov 1990), and this pattern is apparent in poultry as well (e.g. Sell et al. 1986). Parents feeding nestlings with lower digestive efficiency must deliver food at a faster rate and/or for more hours per day, perhaps in response to more vigorous or prolonged begging, or the nestlings will exhibit slower growth than otherwise possible were digestive efficiency higher. As a second example, nestlings of parasitic brown-headed cowbirds (*Molothrus ater*) thrive in nests of many passerine species that feed their young arthropods, but typically fail in nests of birds with different nestling diets, such as some granivores (A.M. Kilpatrick unpublished data). Possibly, cowbird nestlings are somewhat biochemically inflexible and lack key digestive machinery which some nestlings possess that permits them to survive and grow on foods markedly different from insects.

What anatomical and biochemical features constitute this digestive machinery? Models from chemical reactor theory predict that enzyme activity and nutrient absorption rates in relation to digesta retention time determine the rate and efficiency of digestion (Penry & Jumars 1987; Jumars & Martinez del Rio 1999; Levey & Martinez del Rio 1999). In addition, the total biochemical capacity of the digestive system depends upon the quantity of digestive organ. Accordingly, this review focuses on changes in these suborganismal features during post-hatch development, with links where possible to whole-animal function at the physiological, behavioural and ecological level.

POSTNATAL GROWTH OF THE ALIMENTARY SYSTEM

Comparative analysis of postnatal growth of the avian alimentary system began more than half a century ago (Portmann 1942) and has continued steadily (Neff 1973; Lilja 1983; Konarzewski et al. 1990; Starck 1996, 1998; Gille et al. 1999). This activity has been spurred by interest in whether differences in digestive capacity might explain the faster growth rate of birds exhibiting the altricial rather than precocial mode of development (Ricklefs et al. 1998). These analyses have shown that, as a general rule, birds hatch with proportionally larger intestines than adults (as a percentage of their body mass), with no major difference according to developmental mode (Gille et al. 1999), and for a time the intestine shows an accelerated growth as compared to the rest of the body. Thus, the digestive system reaches a high proportion of body mass early in development and then decreases as the nestling approaches adult size. Both the liver, which is important in post-absorptive processing of nutrients, and the pancreas, which is important in this as well as digestion, also tend to exhibit an early accelerated growth (Lilja 1983; Gille et al. 1999). There seems to be considerable variation in when this phase of accelerated growth of the alimentary system slows. After this transition point some birds exhibit essentially no additional alimentary growth, whereas others continue to exhibit growth, albeit more slowly. Researchers have tried to study this systematically by plotting organ mass (or its natural log) against body mass (or its natural log) and analysing the resulting slopes and inflection points (Neff 1973; Lilja 1983; Konarzewski et al. 1990; Gille et al. 1999). When analysed this way, birds exhibiting altricial versus precocial modes of development do not separate cleanly (Starck 1998; Gille et al. 1999), perhaps partly because these kinds of analyses are confounded when growth is measured in nature at a time when food is limiting (Konarzewski et al. 1990) or when the period of investigation is too short (Gille et al. 1999).

Histological study of the intestine during growth of an altricial (starling, *Sturnus vulgaris*) and precocial (Japanese quail, *Coturnix japonica*) species did not reveal any notable differences (Starck 1996). Intestinal crypts where cell proliferation occurs, and mucosal epithelium where hydrolysis and absorption occurs, were topographically separated from each other in both species, and this arrangement did not change from hatchling to adult. (Recall that enterocytes are 'born' in the crypts and mature and express more function as they migrate up the villus to the tip where they are sloughed off.) The rate of cell proliferation, indexed by the length of the S-phase (phase of DNA replication, measured by labelling *in vivo*) did not differ markedly by age or species, leading Starck (1996) to propose that variation in intestinal growth probably relates to the number of intestinal crypts (or cells per crypt), i.e. variation in growth is related more to differences in the neonate size of the gut than to differences in cytokinetics. In addition, given a rather invariant S-phase (average six hours), the intestinal turnover time of small birds (the replacement time of intestinal cells) will be 2-3 days compared with 8-12 days in larger birds.

The absence of any major difference according to developmental mode in the size or growth of the alimentary tract would seem to cast doubt on the notion that the gut limits growth rate. This assumes, however, that gut size indicates function, a critical assumption that has been little tested. Thus, in the next sections we consider whether gut function can change independently of gut mass.

DEVELOPMENT OF BIOCHEMICAL DIGESTIVE CAPACITY

If changes in alimentary tract mass were an accurate index of changes in digestive capacity, then it might be straightforward in many species to reject the notion that the gut limits growth rate. For example, house sparrow (*Passer domesticus*) nestlings captive since day 3 of life and fed a formulated diet (Lepczyk et al. 1998) achieved full intestinal mass by day 6 but doubled food intake rate between days 6 and 10 (Lepczyk et al. 1998; Figure 1). If the gut were limiting and its digestive capacity were properly indexed by gut mass, then further increases in intake would not be possible once gut mass reached its maximum at day 6.

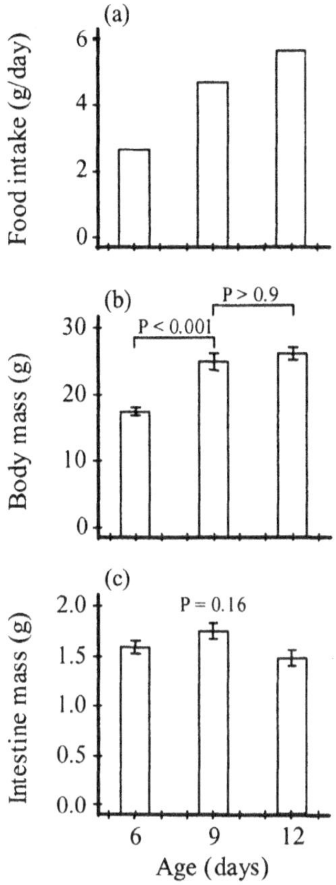

Figure 1. The effect of nestling age on: (a) food intake, (b) body mass and (c) intestine masses of captive nestling house sparrows. Bars designate means (± SE) for birds in the laboratory aged 6, 9 and 12 days (sample sizes, respectively, were 14, 10 and 12 nestlings). There are no SE bars for food intake because this graph shows the age-specific feeding schedule that yielded normal daily mass changes and 100% survival (Lepczyk & Karasov 2000). The nestlings were originally collected from wild nests on day 3. For each variable, log-transformed values from the groups were compared using a Tukey test, and the level of significance for differences in paired comparisons is indicated. Adapted from Caviedes-Vidal and Karasov (2001).

Recent biochemical measures have shown that digestive capacity does change independently of gut mass (Figure 2). Tissue-specific activity (activity per mg intestine) of maltase, which is important in starch digestion, continues to increase after day 6, and the intestine's total maltasic capacity (the product of intestinal mass and tissue-specific activity) consequently did not reach its maximum until at least days 9-12, coinciding with the asymptote in food intake rate and body mass. Sucrase activity showed a pattern similar to that of maltase. In contrast, tissue-specific activity of aminopeptidase, which is important in protein digestion, was independent of age over this time interval. Hence, focusing on gut mass alone or aminopeptidase activity might lead one to reject the hypothesis of gut limitation of feeding and growth, whereas one cannot reject the hypothesis upon consideration of the carbohydrases.

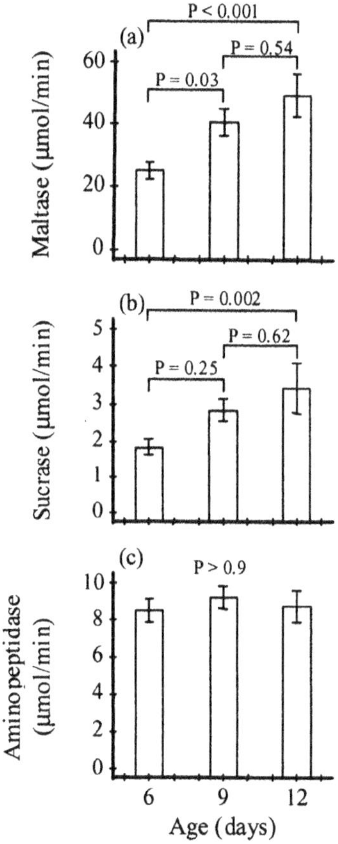

Figure 2. The effect of nestling age on intestinal summed enzyme activities (the product of tissue-specific activity and intestine mass): (a) maltase, (b) sucrase and (c) aminopeptase of captive nestling house sparrows as a function of age. Bars designate means (± SE) for birds in the laboratory aged 6, 9 and 12 days. Sample sizes are the same as in Figure 1. The birds were originally collected from wild nests on day 3. For each variable, log-transformed values from the groups were compared using a Tukey test, and the level of significance for differences in paired comparisons is indicated. Adapted from Caviedes-Vidal and Karasov (2001).

These are the only data to our knowledge on intestinal enzyme activities of an altricial species during development. Precocial, mainly domesticated, species have been studied more extensively. In poultry, after about two weeks of age the specific activity of intestinal digestive enzymes is generally constant or declines during development (Nir et al. 1978; Sell et al. 1989; Biviano et al. 1993; Jackson & Diamond 1995), although in white Pekin ducks (*Anas platyrhynchos domesticus*) a doubling of specific sucrase activity was reported during the first seven weeks of age (King et al. 2000). However, in these species continued increase in intestinal mass during development leads to increases in total enzymatic capacity that keep pace with increases in food intake during growth (Biviano et al. 1993), and so with these data one also cannot reject the hypothesis of gut limitation of feeding and growth.

The data available regarding pancreatic digestive enzymes indicate patterns somewhat similar to those for intestinal enzymes. In house sparrows, pancreatic mass, specific activity of carbohydrate-digesting amylase and protein-digesting trypsin, and total pancreatic activity (the product of specific activity and pancreatic mass) continue to increase with age (Caviedes-Vidal & Karasov 2001). In poultry, pancreatic mass, but not specific activities of the pancreatic enzymes, increase with age (Escribano et al. 1988; Krogdahl & Sell 1989).

It would be revealing to compare enzymatic activities in altricial and precocial species, which are distinguished by large differences in their growth rates. However, there are no data yet collected under uniform procedures to make such a comparison. The data available on enzymatic changes during growth and development are consistent with the notion that digestive capacity limits feeding and growth rate.

DEVELOPMENT OF ABSORPTIVE CAPACITY

There are no published studies of nutrient absorption during development of any altricial species. Eleven-day old house sparrows had *in vitro* absorption rates for the amino acid L-leucine (ca. 8 pmol mg^{-1} min^{-1} at a concentration of 0.01 mM; Lepczyk et al. 1998) comparable to those measured in a study examining the modulation of absorption in adult house sparrows (ca. 3-6 pmol mg^{-1} min^{-1}; Caviedes-Vidal & Karasov 1996).

Poultry have been studied more extensively. Overall, intestinal tissue-specific glucose and fructose *in vitro* absorption rates were similar in male white Cornish-Rock cross chicks just after hatching and at 12 weeks of age whereas tissue-specific absorption of the amino acid proline showed a decline with age (Obst & Diamond 1992). In red jungle fowl (*Gallus gallus*) tissue-specific absorption rates of both nutrients declined between weeks one and nine (Jackson & Diamond 1995). In both these studies, and as was the case for intestinal enzymes, continued increase in intestinal mass during development led to increases in total absorptive capacity for sugars and amino acids that mostly kept pace with increases in food intake during growth (Jackson & Diamond 1996). The domestic chicken was reported to exhibit pronounced spikes in tissue-specific uptake of glucose at week two and of proline at week six that were lacking in the jungle fowl and in domestic fowl studied using *in vivo* methodology (Gonzalez & Vinardell 1996). Although these spikes were interpreted as adaptive responses to specific times of energy stress, a cautious alternative explanation might be unexplained methodological differences in a chronological study lacking careful blocking for effects of time (see Levey & Karasov 1992). The caecum can also be an important site of absorption (Obst & Diamond 1989), where tissue-specific absorption rates of sugar and amino acid tend to be constant or decline with age (Planas et al. 1986; Moreto et al. 1991). Maximum intestinal L-threonine absorption in white Pekin ducks also appeared to decline with age (King et al. 2000).

Absorption rates measured *in vitro* have been compared with dietary nutrient loads in several studies (Obst & Diamond 1992; Jackson & Diamond 1995), with the conclusion that the intestine's absorptive capacity keeps pace with increasing nutrient intake as the chick grows. Such comparisons need to be undertaken cautiously because of many uncertainties, including those about tissue viability (Starck et al. 2000), luminal nutrient concentrations which affect rates *in vivo* (Cant et al. 1996), and a possible passive paracellular absorptive pathway not typically measured (Karasov 1996). A conservative conclusion for poultry, and the white Pekin duck (King et al. 2000), for which intestinal mass during development increases whereas tissue-specific absorption rate generally does not, is that one cannot reject the hypothesis that absorptive capacity limits feeding and growth rate.

DEVELOPMENT OF DIGESTA RETENTION

Digesta retention, an index of how long it takes to digest a meal, is measured by feeding birds non-absorbable markers and measuring the time course of their excretion. Its measurement could be quite important for predicting maximum food intake rate or for understanding whether begging intensity changes as a function of the proportion of the previous meal digested. Mean retention time is expected to increase with increasing gut size or decrease with increasing food intake rate (Karasov 1996), but what should be expected in nestling birds in which both gut size and intake increase with age as nestlings increase in size?

We might predict a complex pattern of change in mean retention time. In house sparrows, for example, the volumetric capacity of the gut increases rapidly to a maximum by day 6 (above), whereas food intake increases more slowly and continues to increase beyond day 6 (Blem 1975; Figure 3). Taking the quotient of gut size (in grams) and intake rate (in grams/day) as the index of retention time (in days), we predict that retention time should increase with age up to day 6, then decrease subsequently and finally level off (Figure 3). Lepczyk et al. (1998) fed captive nestling house sparrows a formulated diet with the non-absorbable marker polyethylene glycol (PEG) and measured times of first appearance (transit time, TT) and mean retention time (MRT). Though the measures were not made over the entire growth period, both measures changed in the predicted pattern (Figure 3). The initial increase in MRT corresponded to the period in which gut size increased faster than food intake rate. Once gut size levelled off, the decline in MRT correlated with increased food intake rate. This inverse relation suggests that the nestlings had limited spare volumetric capacity, because this is the pattern expected for a tube of constant dimensions in which material inflow is increased (Karasov 1996). Once gut size and food intake rate have reached asymptotic levels, no further change in MRT is expected.

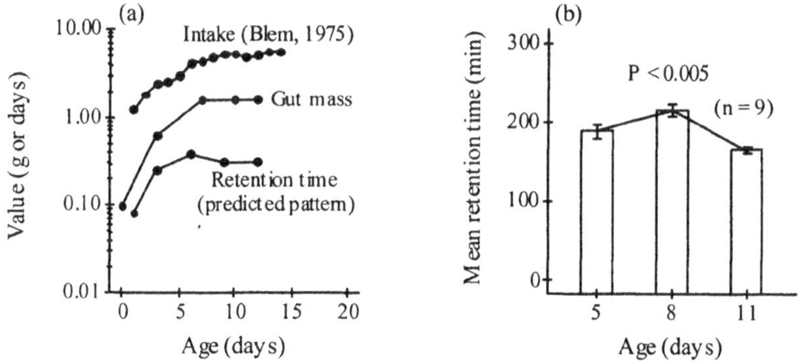

Figure 3. Predicted and measured patterns of digesta retention time in nestling house sparrows as a function of age. Graph (a) uses the quotient of gut size (in g; from Figure 1 and Neff 1973) and intake rate (in grams/day; from Blem 1975) as the index of retention time (in days). The prediction is that retention time should increase with age up to day 6, then decrease subsequently and finally level off (left-hand figure). Graph (b) shows the mean ($\pm$ SE) retention time in house sparrows measured between ages 5-11 days (Lepczyk et al. 1998), which hints at some of the predicted patterns.

We do not know of any other published studies of digesta retention during development in any avian species, but we would expect these patterns to hold in altricial species for the following reason. The alimentary tract grows rapidly, as discussed above, and early in the nestling period its size scales with nestling mass$^{>1.0}$ (see Gille et al. 1999 for review), and nestling energy needs probably scale with mass$^{0.85}$ (Weathers 1996). At some point during development (e.g. day 6 in house sparrows), intestine mass ceases to increase whereas both body mass and food intake continue to increase with age (Figures 1 & 3). At this point, the allometry changes, with gut size now scaling with mass$^{\sim 0}$, but intake still scaling with mass$^{0.85}$. Therefore, the scaling of MRT, which is proportional to gut size/intake, changes from positive (mass$^{>1.0}$/ mass$^{0.85}$ α mass$^{>0.15}$) to negative (mass^{0}/ mass$^{0.85}$ α mass$^{-0.85}$).

How will these changes in MRT of digesta interact with changing biochemical rates of digestion to influence digestive efficiency? In house sparrows, both the biochemical rates and the MRT increase up through day 6, so one might expect a notable rise in digestive efficiency. The decline in MRT after day 6 might be compensated for by the continued increase in biochemical rates of digestion, so efficiency might not decline after day 6. In accord with the first expectation, digestive efficiency of house sparrow nestlings rises steadily from 55% on day 1 to 69% on day 6 (Blem 1975). Between day 6 and day 12 both Blem (1975) and Lepczyk et al. (1998) reported slight declines in digestive efficiency (3 and 5%, respectively). Perhaps age-related changes in enzymes are not entirely compensatory for the decline in MRT, or perhaps there are no compensatory changes in absorption rates, but these have not yet been studied. We describe below how retention time and digestive efficiency decline even more when nestlings are force-fed. Overall, it is interesting that the various parameters measured in the three studies (Blem 1975; Lepczyk et al. 1998; Caviedes-Vidal & Karasov 2001) appear to fit together in logical fashion.

Mean retention time also changes in response to changing food quality. In adult birds, MRT correlates positively with food richness (digestible energy per gram; Karasov 1996). We might expect a similar pattern of change in nestlings whose parents deliver foods varying in energy density.

How might these putative changes in MRT and digestive efficiency during development, or with changing diet, relate to begging behaviour? In the case of house sparrows, perhaps nestlings with the shortest retention times and lowest digestive efficiencies, such as those under six days of age (Figure 3), will receive less net gain per gram of food than older nestlings. If there were any correspondence between begging intensity and a nestling's rate of net energy gain, we might expect begging intensity to be highest in the youngest nestlings. In addition, as A.B. Clark (this volume) points out, at the age at which nestlings are growing fastest their possibly higher mass-specific metabolic rates may catabolize absorbed nutrients faster than older nestlings with slower growth. Both of these mechanisms could lead to more intensive begging in younger nestlings, and there is evidence for this scenario. Younger and/or smaller nestlings appear to become hungry more quickly, and therefore beg more following a given level of food intake, compared with older nestlings (Lotem 1998a; Cotton et al. 1999).

As another possible example of the interplay between behaviour and digestive physiology, consider different sized nestlings at nearly the same age in an asynchronous brood. Smaller nestlings seem to receive less food reward for a given level of begging as compared with their larger nestmates (e.g. Price et al. 1996; Cotton et al. 1999). If retention time and digestive efficiency decline as nestlings approach near maximal food intake rates (as discussed below), then larger nestlings will not achieve increases in digestible energy and growth that are proportionate to their higher intake. Clearly, in order to understand the relationships between nestling begging and nutritional need, we need more information concerning the predicted patterns of change in MRT and digestive efficiency in relation to food intake during nestling development, and their possible behavioural correlates.

IS DIGESTION LIMITING DURING GROWTH AND DEVELOPMENT?

This short review of changes in alimentary tract anatomical and biochemical features offers no strong evidence to reject the notion that digestion might limit food intake rate and growth. Strong evidence to reject might be represented by a finding of large differences in food intake rate despite similar volumetric and/or biochemical digestive capacity, indicating therefore that the individual with lower food intake must have considerable spare digestive capacity. Although some species exhibit rising food intake rates even after their alimentary tract has reached an asymptotic mass, biochemical measures show that their digestive capacity is in fact still increasing (Figures 1 & 2). Ricklefs et al. (1998) presented an apparent paradox regarding the regulation of energy flow by digestive capacity, because despite similar-sized alimentary tracts, precocial species have higher food intake rates than altricial species. However, careful comparisons of biochemical digestive capacity have not been made between altricial and precocial species. Furthermore, according to Weathers (1996), peak daily energy demand of young is not related to developmental mode. It is therefore conceivable that both altricial and precocial species digest similar amounts of food as rapidly as they can and that altricials achieve faster growth rates by allocating less energy to activity and thermogenesis and more to production. Future comparisons of energy budgets based upon measurements of field metabolic rates and either feeding or growth rates will clarify this point.

There are a number of observations that are consistent with the idea that birds are feeding maximally during their growth phase. First, in force-feeding studies the nestlings of altricial species appear to be eating as much as they can accommodate and they appear to reach a ceiling in digestion rate (grams apparently digested/hour) as food intake increases. When song thrushes (*Turdus philomelos*) and house sparrows were removed from their nests and overfed to just below the point at which they regurgitated, their hourly food intake rates were higher by only 14% (Lepczyk et al. 1998) and 18% (Konarzewski et al. 1996) as compared with controls (i.e. nestlings fed amounts that yielded a growth rate similar to that of wild nestlings). Furthermore, their modest increases in food intake rate were offset by significant (Lepczyk et al. 1998) or near-significant (Konarzewski et al. 1996) declines in digestive efficiency as compared with controls (Figure 4). A detailed evaluation of what happened was possible in the study of house sparrows in which many digestive features were measured (Lepczyk et al. 1998). The picture that emerged was that the nestlings had no spare volumetric capacity, so that when food intake and digesta flow into the intestine increased, retention time declined. Because there was little spare enzymatic and absorptive capacity, the reduced contact time between digesta and hydrolases and transporters resulted in lower digestive efficiency.

Second, the apparent inability of altricial species to digest food much faster than they would normally do suggests that they have little unused growth potential that they fail to express under natural conditions (Konarzewski et al. 1996). Hence, despite overfeeding, body mass increments of overfed captives did not differ significantly from those of wild nestlings for either song thrushes (Konarzewski et al. 1996) or house sparrows (Lepczyk & Karasov 2000). Supplementation of food in the wild also failed to increase body mass accretion rates in Leach's storm-petrels (*Oceanodroma leucorhoa*, Ricklefs 1987) and red-tailed tropicbirds (*Phaethon rubricauda*, Schreiber 1996), although rate of wing growth increased in the latter case.

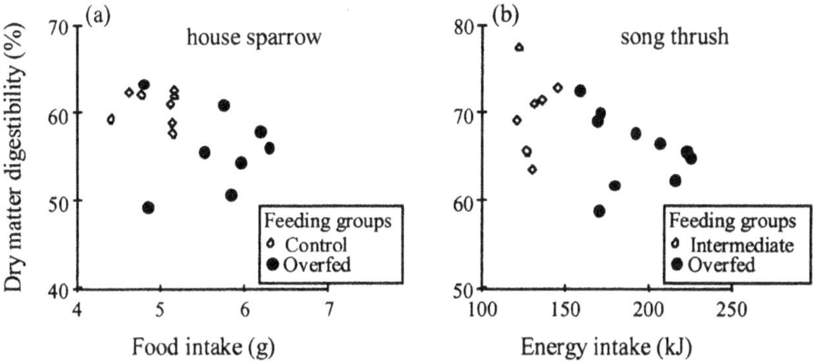

Figure 4. The apparent depression of digestive efficiency in overfed nestlings. When (a) house sparrows (Lepczyk et al. 1998) and (b) song thrushes (Konarzewski et al. 1996) were removed from their nests and overfed by only 14% and 18%, respectively, compared with controls (controls = nestlings fed amounts that yielded a growth rate similar to that of wild nestlings), their modest increases in food intake rate were offset by declines in digestive efficiency (indexed by apparent dry matter digestibility = 1 - [g excreta/g food]).

Third, experiments with force-feeding in poultry indicate that their ability to utilize extra food is associated with hypertrophy of the alimentary tract (Nir et al. 1978; Nir & Nitsan 1979): a 43% increase in intake caused a 50% increase in mass of the tract. Konarzewski et al. (1989) inferred from these kinds of studies that digestive organs in growing birds usually function at near-maximum rate, and that a substantial increase in ingestion can only be accomplished by enlargement of the digestive tract. Indeed, in both *Coturnix* (Lilja et al. 1985) and *Gallus* (Jackson & Diamond 1996), the main digestive change obtained as a result of artificial selection for more rapid growth was an increase in the relative size of the digestive organs.

The relationships between food intake, gut size and growth in birds have led some people to suggest that gut size limits food intake and hence growth rate (Kirkwood & Prescott 1984; Lilja 1997). Ricklefs et al. (1998) suspect that the gut does not set a stringent limit on avian growth, because at any one time the gut may have an upper digestive capacity which can be increased if necessary. Even though altricial nestlings may reduce gut size when underfed (Konarzewski & Starck 2000), there is no solid evidence that they can increase gut size when overfed, as do quail and chickens. The only two morphometric studies of overfed altricial nestlings both involved nestlings that were either previously food-restricted (Lepczyk et al. 1998), or the overfed nestlings were compared with food-restricted nestlings rather than unmanipulated controls (Konarzewski et al. 1996). Is the cost of building and maintaining a larger gut prohibitive as compared with the gains realized from increased demand by peripheral (growing or functioning) tissues? For passerine adults the answer is no, because in all species for whom total energy demand was increased by increased thermoregulatory and/or activity energy demand the intestinal mass and biochemical digestive capacity increased (Karasov 1996). But what about nestlings?

It is clear that some avian species do not respond to overfeeding by increasing growth rate (see above), but this may not address the issue of limits to growth if some other critical feature of physiology, or growth itself, is genetically determined and not amenable to phenotypic change (Winkler & Adler 1996; Ricklefs et al. 1998). As summarized by Ricklefs et al. (1998): "Testing a hypothesis about a growth-rate function is exceedingly difficult... because several tissues may assume symmorphic relationships to a single most limiting tissue, several tissues may constrain growth and development simultaneously, and limiting tissues may differ with age or between different developmental types".

THE IMPLICATIONS OF PHYSIOLOGICAL LIMITS FOR PARENT-NESTLING INTERACTIONS

It may not be currently possible to determine whether digestion *per se* limits avian growth, but there does seem to be evidence of rather stringent physiological limits in some species. Otherwise, how can one explain how sometimes, despite overfeeding, body mass increments of overfed nestlings do not differ significantly from those of controls (Ricklefs 1987; Konarzewski et al. 1996; Schreiber 1996; Lepczyk & Karasov 2000)? Certainly food availability proximally limits growth in some instances (Martin 1987), but there are also some demonstrations that it does not, especially where parents demonstrate spare capacity in their provisioning effort. Chief among these are: (1) experimental manipulations in which nestlings are added to clutches (e.g. Finke et al. 1987; for review see Magrath 1991; Wright et al. 1998), or where parental mates are removed (e.g. Wolf et al. 1988; Bart & Tornes 1989; Markman et al. 1996), with no subsequent significant decline in growth rate of nestlings; and (2) nearly all hosts whose nests include a parasitic cowbird are able to provision nestling cowbirds with enough food to grow at near their maximal rates (A.M. Kilpatrick unpublished data).

In cases where food is plentiful and nestling physiology apparently constrains growth, what is the nature of the parent-offspring interaction? If there were no communication between nestling and parent, and parents simply continually delivered food, then young would grow normally and there would probably be a lot of unused food at the nest. Of course, this would be wasteful, and potentially risky for the parent. In nature parents do appear to modulate their activity to some extent, in relation either to relative changes in brood demand and/or variation in environmental availability of food (and so parental costs of provisioning). Information transfer between begging nestlings and parents regarding relative brood demand thus reduces parental provisioning costs and increases family economy. Ideally, from the perspective of nestling digestion and development, nestling begging and parental provisioning responses should achieve a perfectly regulated system in which small alterations in brood demand due to varying food quality or increased metabolic expenditures (e.g. in response to changing environmental conditions) are perfectly compensated for by adjustments in parental energy delivery. Even larger alterations in demand, due to extra nestlings (either additional progeny or brood parasites) would elicit perfect compensation. Such compensatory responses (not necessarily perfect) are sometimes witnessed, as in the aforementioned studies of responses to experimental manipulation of brood size, number of carers and brood parasitism. Compensatory responses by parents to changes in nestling demands are, however, rarely perfect in natural provisioning systems (e.g. because of limits in the environmental availability of food). Nestlings therefore have to cope with varying rates of individual food intake, and for very good evolutionary reasons concerning parental provisioning costs and adaptive life history trade-offs (see Houston & Davies 1985; Wright & Cuthill 1989, 1990; Kacelnik & Cuthill 1990; Wright et al. 1998).

Despite their ultimate evolutionary complexity, most parent-nestling interactions are proximately regulated for the most part by physiologically determined food demands of the nestlings. The major research questions are how features of begging relate to nestling metabolic rate and/or physiological status, and how parents respond to these features of begging. How accurately does nestling begging reflect subtle differences in nestling state?

THEORETICAL INTEGRATION OF NESTLING DIGESTIVE PHYSIOLOGY AND BEGGING

Parental care is normally provided as a series of interactions (e.g. feeding events) within the context of progressive offspring growth and development, and with fitness pay-offs derived only after the termination of care and offspring independence (i.e. fledging). Dynamic state-dependent modelling provides a strong and appropriate theoretical framework for investigating such repeated parent-offspring and sib-sib (i.e. between nestmate) interactions (Beauchamp et al. 1991; Winkler & Adler 1996). However, at present we lack detailed knowledge concerning actual offspring 'state' variables, and we are unable to accurately define exactly what we mean by nutritional need. In addition, offspring growth and development involve multiple states, such as skeletal body size, muscle mass, energetic reserves, digestive efficiency and even immune response (see N. Saino & A.P. Møller this volume). Body size represents a tightly constrained offspring state (Ricklefs 1979) and is critical for successful competition for food in the nest (Kacelnik et al. 1995; Kilner 1995; Cotton et al. 1996). In contrast, the digestive system is a more flexible state (see Karasov 1996), which is theoretically very interesting because it influences the amount of resources available for investment in all other states. Therefore, the optimum developmental strategy for nestlings should be to invest an appropriate proportion of food resources in each component state at each stage in the nestling period, so as to maximize fitness upon fledging. Begging signals represent an important aspect of offspring investment given that they have to elicit the appropriate quantity and quality of nutrition required at every developmental stage.

Nestling birds of different ages and with contrasting developmental histories will understandably have differing nutritional needs. Thus, begging signals have been shown to reflect not just recent feeding history, or short-term need, but also long-term requirements in terms of the stage of nestling growth attained (Price et al. 1996; Lotem 1998b). Smaller or younger nestlings appear to get hungrier more quickly, and therefore beg more following a given level of food intake or satiation, compared to larger or older nestlings (Lotem 1998a; Cotton et al. 1999). Therefore, in order to qualify what we mean by 'honest' begging, it is vital that we discover whether such effects are due to differences in gut capacity and/or digestive efficiencies between nestlings of different sizes, or whether they genuinely indicate contrasting nutritional requirements that reflect differences in post-absorptive processing of energy and nutrients. It also remains to be seen whether such differences reflect alternative competitive begging strategies, because smaller nestlings often have to beg at a greater rate for a given level of food reward when in the nest (Price et al. 1996; Cotton et al. 1999). Although easy to measure in the field, long-term growth and short-term feeding history alone have proved insufficient to explain the patterns of begging observed. We must now look at nestling digestive physiology as it relates to the evolution of adaptive begging signals in the context of nestling growth and development.

FUTURE DIRECTIONS

Our knowledge of avian digestive physiology during development, especially its function in non-domesticated species, is based on very few studies. Regarding the basic question of whether digestive capacity might limit growth, current data do not permit one to reject the notion and, in fact, in many cases are consistent with it. This review reveals numerous areas for future research on development of avian digestive physiology, and we conclude with a list.

(1) Do young altricial and precocial species, which have different growth rates, have different sized alimentary systems, when studied for adequate periods of time under conditions of unlimited food, and when properly analysed with regard to the effect of body size and phylogeny?

(2) Will the finding in house sparrows of increased tissue-specific enzyme activity with age be borne out in future studies with altricial species, and are there differences between young of altricial and precocial species in enzyme activities when studied under similar conditions?

(3) Will tissue-specific nutrient absorption activity, which has barely been studied in altricial species, change in a correlated fashion with enzyme activities during growth and development?

(4) In other species, does an increase in retention time and digestive enzyme capacity early in nestling life lead to rising digestive efficiency, as apparently occurs in house sparrow nestlings?

(5) Can digestive features such as gut size, enzymes, nutrient transporters, and digesta retention time be modulated when nestlings are raised on different kinds of diets, which might occur naturally in some omnivorous species or in nest parasites?

(6) Does nestling begging contain information concerning not only the quantity of food received, but also the quality of that food and the relative quantities of different nutritional elements required, and do parents adjust their provisioning behaviour according to such begging cues?

(7) Will future studies of overfeeding in altricial species demonstrate hypertrophic response by the intestine and faster growth, as occurs for precocial species?
(8) Is it possible to identify the most limiting organ or process when several tissues might constrain growth and development (Ricklefs et al. 1998)?
(9) How do the features of begging (e.g. intensity) relate to digestive features such as extent of digestion of most recent meal, gut fill, etc.?

ACKNOWLEDGEMENTS

We thank Marm Kilpatrick and Enrique Caviedes-Vidal for sharing with us their unpublished work, and Marty Leonard, Robert Ricklefs and David Houston for comments on an earlier version of the manuscript. Supported by grants from the National Science Foundation to W.H. Karasov (IBN-9318675 and 9723793), and the A.W. Schorger Fund and the Max McGraw Wildlife Foundation.

REFERENCES

Bart, J. & Tornes, A. 1989. Importance of monogamous male birds in determining reproductive success: evidence for house wrens and a review of male-removal studies. *Behavioral Ecology and Sociobiology* **24**, 109-116.

Beauchamp, G., Ens, B.J. & Kacelnik, A. 1991. A dynamic model of allocation of food to nestlings. *Behavioral Ecology* **2**, 21-37.

Biviano, A.B., Del Rio, C.M. & Phillips, D.L. 1993. Ontogenesis of intestine morphology and intestinal disaccharidases in chickens (*Gallus gallus*) fed contrasting purified diets. *Journal of Comparative Physiology B* **163**, 508-518.

Blem, C.R. 1975. Energetics of nestling house sparrows *Passer domesticus*. *Comparative Biochemistry and Physiology* **52A**, 305-312.

Cant, J.P., McBride, B.W. & Croom, W.J. Jr. 1996. The regulation of intestinal metabolism and its impact on whole animal energetics. *Journal of Animal Science* **74**, 2541-2553.

Caviedes-Vidal, E. & Karasov, W.H. 1996. Glucose and amino acid absorption in house sparrow intestine and its dietary modulation. *American Journal of Physiology* **271**, R561-R568.

Caviedes-Vidal, E. & Karasov, W.H. 2001. Developmental changes in digestive physiology of nestling house sparrows, *Passer domesticus*. *Physiological and Biochemical Zoology* **74**, 769-782.

Cotton, P.A., Kacelnik, A. & Wright, J. 1996. Chick begging as a signal: are nestlings honest? *Behavioral Ecology* **7**, 178-182.

Cotton, P.A., Kacelnik, A. & Wright, J. 1999. Chick begging strategies in relation to brood hierarchies and hatching asynchrony. *American Naturalist* **153**, 412-420.

Escribano, F., Rahn, B.I. & Sell, J.L. 1988. Development of lipase activity in yolk membrane and pancreas of young turkeys. *Poultry Science* **67**, 1089-1097.

Finke, M.A., Milinkovich, D.J. & Thompson, C.F. 1987. Evolution of clutch size: an experimental test in the house wren (*Troglodytes aedon*). *Journal of Animal Ecology* **56**, 99-114.

Gille, U., Salomon, F.-V. & Rönnert, J. 1999. Growth of the digestive organs in ducks with considerations on their growth in birds in general. *British Poultry Science* **40**, 194-202.

Godfray, H.C.J. 1991. Signalling of need by offspring to their parents. *Nature* **352**, 328-330.

Gonzalez, E. & Vinardell, M.P. 1996. Ontogenetic development of proline intestinal transport in the domestic fowl. *British Poultry Science* **37**, 383-394.

Houston, A.I. & Davies, D.B. 1985. The evolution of cooperation and life history in the dunnock, *Prunella modularis*. In: *Behavioural Ecology: Ecological Consequences of Adaptive Behaviour* (Ed. by R.M. Sibly & R. Smith). Oxford: Blackwell Scientific Publications.

Jackson, S. & Diamond, J. 1995. Ontogenetic development of gut function, growth, and metabolism in a wild bird, the Red Jungle Fowl. *American Journal of Physiology* **269**, R1163-R1173.

Jackson, S. & Diamond, J. 1996. Metabolic and digestive responses to artificial selection in chickens. *Evolution* **50**, 1638-1650.

Jumars, P.A. & Martinez del Rio, C. 1999. The tau of continuous feeding on simple foods. *Physiological and Biochemical Zoology* **72**, 633-641.

Kacelnik, A. & Cuthill, I. 1990. Central place foraging in starlings (*Sturnus vulgaris*). II. Food allocation to chicks. *Journal of Animal Ecology* **59**, 655-674.

Kacelnik, A., Cotton, P.A., Stirling, L. & Wright, J. 1995. Food allocation among nestling starlings: sibling competition and the scope of parental choice. *Proceedings of the Royal Society of London, Series B* **259**, 259-263.

Karasov, W.H. 1990. Digestion in birds: chemical and physiological determinants and ecological implications. *Studies in Avian Biology* **13**, 391-415.

Karasov, W.H. 1996. Digestive plasticity in avian energetics and feeding ecology. In: *Avian Energetics and Nutritional Ecology* (Ed. by C. Carey). New York: Chapman and Hall.

Kilner, R. 1995. When do canary parents respond to nestling signals of need? *Proceedings of the Royal Society of London, Series B* **260**, 343-348.

King, D.E., Asem, E.K. & Adeola, O. 2000. Ontogenetic development of intestinal digestive functions in white Pekin ducks. *Journal of Nutrition* **130**, 57-62.

Kirkwood, J.K. & Prescott, N.J. 1984. Growth rate and pattern of gut development in mammals and birds. *Livestock Production Science* **11**, 461-474.

Konarzewski, M. & Starck, J.M. 2000. Effects of food shortage and oversupply on energy utilization, histology, and function of the gut in nestling song thrushes (*Turdus philomelos*). *Physiological and Biochemical Zoology* **73**, 416-427.

Konarzewski, M., Kozlowski, J. & Ziolko, M. 1989. Optimal allocation of energy to growth of the alimentary tract in birds. *Functional Ecology* **3**, 589-596.

Konarzewski, M., Lilja, C., Kozlowski, J. & Lewonczuk, B. 1990. On the optimal growth of the alimentary tract in avian postembryonic development. *Journal of Zoology* **222**, 89-101.

Konarzewski, M., Koyama, S., Swierubska, T. & Lewonczuk, B. 1996. Effect of short-term feed restriction, realimentation and overfeeding on growth of Song Thrush (*Turdus philomelos*) nestlings. *Functional Ecology* **10**, 97-105.

Krogdahl, A. & Sell, J.L. 1989. Influence of age on lipase, amylase, and protease activities in pancreatic tissue and intestinal contents of young turkeys. *Poultry Science* **68**, 1561-1568.

Lepczyk, C.A. & Karasov, W.H. 2000. The effect of ephemeral food restrictions on the growth of house sparrows (*Passer domesticus*). *The Auk* **117**, 164-174.

Lepczyk, C.A., Caviedes-Vidal, E. & Karasov, W.H. 1998. Digestive responses during food restriction and realimentation in nestling house sparrows (*Passer domesticus*). *Physiological Zoology* **71**, 561-573.

Levey, D.J. & Karasov, W.H. 1992. Digestive modulation in a seasonal frugivore, the American robin (*Turdus migratorius*). *American Journal of Physiology* **262**, G711-G718.

Levey, D.J. & Martinez del Rio, C. 1999. Test, rejection, and reformulation of a chemical reactor-based model of gut function in a fruit-eating bird. *Physiological and Biochemical Zoology* **72**, 369-383.

Lilja, C. 1983. A comparative study of postnatal growth and organ development in some species of birds. *Growth* **47**, 317-339.

Lilja, C. 1997. On the pattern of organ growth in the common tern (*Sterna hirundo*). *Growth, Development & Aging* **61**, 11-18.

Lilja, C., Sperber, I. & Marks, H.L. 1985. Postnatal growth and organ development in Japanese quail selected for high growth rate. *Growth* **49**, 51-62.

Lotem, A. 1998a. High levels of begging behaviour by small nestlings: a case of a negatively correlated handicap. *Israel Journal of Zoology* **44**, 29-45.

Lotem, A. 1998b. Differences in begging behaviour between barn swallow, *Hirundo rustica*, nestlings. *Animal Behaviour* **55**, 809-818.

Magrath, R.D. 1991. Nestling weight and juvenile survival in the blackbird, *Turdus merula*. *Journal of Animal Ecology* **60**, 335-351.

Markman, S., Yom-Tov, Y. & Wright, J. 1996. The effect of male removal on female parental care in the orange-tufted sunbird. *Animal Behaviour* **52**, 437-444.

Martin, T.E. 1987. Food as a limit on breeding birds: a life-history perspective. *Annual Review of Ecology and Systematics* **18**, 453-487.

Moreto, M., Amat, C., Puchal, A., Buddington, R.K. & Planas, J.M. 1991. Transport of L-proline and alpha-methyl-D-glucoside by chicken proximal cecum during development. *American Journal of Physiology* **260**, G457-G463.

Neff, M. 1973. Untersuchungen über das embryonale und postembryonale organwachstum bei vogelarten mit verschiedenem ontogenesemodus. *Revue Suisse de Zoologie* **79**, 1471-1597.

Nir, I. & Nitsan, Z. 1979. Metabolic and anatomical adaptations of light-bodied chicks to intermittent feeding. *British Poultry Science* **20**, 61-71.

Nir, I., Nitsan, Z., Dror, Y. & Shapira, N. 1978. Influence of overfeeding on growth, obesity and intestinal tract in young chicks of light and heavy breeds. *British Journal of Nutrition* **39**, 27-35.

Obst, B.S. & Diamond, J.M. 1989. Interspecific variation in sugar and amino acid transport by the avian cecum. *Journal of Experimental Zoology Supplement* **3**, 117-126.

Obst, B.S. & Diamond, J.M. 1992. Ontogenesis of intestinal nutrient transport in domestic chickens (*Gallus gallus*) and its relation to growth. *The Auk* **109**, 451-464.

O'Connor, R. 1984. *The Growth and Development of Birds*. Chichester: John Wiley & Sons.

Penry, D.L. & Jumars, P.A. 1987. Modeling animal guts as chemical reactors. *American Naturalist* **129**, 69-96.

Planas, J.M., Villa, M.C., Ferrer, R. & Moreto, M. 1986. Hexose transport by chicken cecum during development. *Pfluegers Archiv European Journal of Physiology* **407**, 216-220.

Portmann, A. 1942. Die ontogenese und das problem der morphologischen entwicklungszustände von verschiedener wertigkeit bei vögeln und säugern. *Revue Suisse de Zoologie* **49**, 169-185.

Price, K., Harvey, H. & Ydenberg, R. 1996. Begging tactics of nestling yellow-headed blackbirds, *Xanthocephalus xanthocephalus*, in relation to need. *Animal Behaviour* **51**, 421-425.

Ricklefs, R.E. 1979. Adaptation, constraint, and compromise in avian postnatal development. *Biological Reviews* **54**, 269-290.

Ricklefs, R.E. 1987. Response of adult Leach's storm-petrels to increased food demand at the nest. *The Auk* **104**, 750-756.

Ricklefs, R.E., Starck, J.M. & Konarzewski, M. 1998. Internal constraints on growth in birds. In: *Avian Growth and Development: Evolution Within the Altricial-Precocial Spectrum* (Ed. by J.M. Starck & R.E. Ricklefs). New York: Oxford University Press.

Schreiber, E.A. 1996. Experimental manipulation of feeding in red-tailed tropicbird chicks. *Colonial Waterbirds* **19**, 45-55.

Sell, J.L., Krogdahl, A. & Hanyu, N. 1986. Influence of age on utilization of supplemental fats by young turkeys. *Poultry Science* **65**, 546-554.

Sell, J.L., Koldovsky, O. & Reid, B.L. 1989. Intestinal disaccharidases of young turkeys: temporal development and influence of diet composition. *Poultry Science* **68**, 265-277.

Starck, J.M. 1996. Intestinal growth in the altricial European starling (*Sturnus vulgaris*) and the precocial Japanese quail (*Coturnix coturnix japonica*). A morphometric and cytokinetic study. *Acta Anatomica (Basel)* **156**, 289-306.

Starck, J.M. 1998. Structural variants and invariants in avian embryonic and postnatal development. In: *Avian Growth and Development: Evolution Within the Altricial-Precocial Spectrum* (Ed. by J.M. Starck & R.E. Ricklefs). New York: Oxford University Press.

Starck, J.M. & Ricklefs, R.E. 1998. *Avian Growth and Development: Evolution Within the Altricial-Precocial Spectrum.* New York: Oxford University Press.

Starck, J.M., Karasov, W.H. & Afik, D. 2000. Intestinal nutrient uptake measurements and tissue damage: Validating the everted sleeves method. *Physiological and Biochemical Zoology* **73**, 454-460.

Weathers, W.W. 1996. Energetics of postnatal growth. In: *Avian Energetics and Nutritional Ecology* (Ed. by C. Carey). New York: Chapman and Hall.

Winkler, D.W. & Adler, F.R. 1996. Dynamic state variable models for parental care. I. A submodel for the growth of the chicks of passerine birds. *Journal of Avian Biology* **27**, 343-353.

Wolf, L., Ketterson, E.D. & Nolan, V. Jr. 1988. Behavioural response of female dark-eyed juncos to the experimental removal of their mates: implications for the evolution of male parental care. *Animal Behaviour* **39**, 125-134.

Wright, J. & Cuthill, I. 1989. Manipulation of sex differences in parental care. *Behavioral Ecology and Sociobiology* **25**, 171-181.

Wright, J. & Cuthill, I. 1990. Biparental care: short-term manipulations of partner contribution and brood size in the starling, *Sturnus vulgaris*. *Behavioral Ecology* **1**, 116-124.

Wright, J., Both, C., Cotton, P.A. & Bryant, D. 1998. Quality versus quantity: energetic and nutritional trade-offs in parental provisioning strategies. *Journal of Animal Ecology* **67**, 620-634.

12.

HORMONAL REGULATION OF BEGGING BEHAVIOUR

Hubert Schwabl & Joseph Lipar
School of Biological Sciences and Center for Reproductive Biology, Washington State University, Pullman WA 99163-4236, USA (huschwabl@wsu.edu; jlipar@wsu.edu)

ABSTRACT

Begging is the first coordinated behaviour altricial birds perform after hatching. As the neuromuscular substrates, brain, sensory organs and endocrine systems mature, this simple reflex develops rapidly into a more complex behaviour that is influenced by external stimuli and by internal signals such as hormones. In this chapter we discuss the maturation of endocrine systems and the role of hormones in the regulation of begging in altricial birds and describe the development of physiological systems that may influence begging performance. First, we elaborate on the neuromuscular and sensory substrates that may be involved in the begging response. Then, we describe the development and maturation of relevant endocrine regulatory systems that may influence intensity, frequency and duration of begging. We also discuss effects of endogenous and maternally-derived hormones on the development, regulation and performance of begging. Finally, we suggest several approaches to the study of begging that may be useful for testing evolutionary theories such as parent-offspring conflict, sibling rivalry and parental favouritism.

INTRODUCTION

Begging, the solicitation of food from parents by postures and vocalizations, is the first coordinated behaviour a truly altricial bird performs after hatching. It seems that begging is initially a rather simple and stereotyped behavioural response. As the neuromuscular control circuits, brain, sensory

J. Wright and M.L. Leonard (eds.), The Evolution of Begging, 221–244.

organs and regulatory endocrine systems of the nestling mature, this simple reflex develops rapidly into a more complex behaviour that is influenced by external signals from feeding parents and nestmates and by internal signals such as the hormonal environment. Further, experience and learning seem to influence begging performance (Kedar et al. 2000). Begging is a more complex behavioural pattern in semi-altricial and semi-precocial birds because young of these species hatch with better developed sensory and motor control systems. As begging is essential for nestling growth and survival, it is likely that there is intense selection pressure on the development of this behaviour and each of the components that regulate it to allow individuals to compete successfully with their nestmates for limited parental food resources.

Hormones are regulators of physiological functions in both developing and adult organisms. The neuroendocrine regulatory systems for hormone secretion themselves differentiate and mature during the pre- and postnatal phases. The rates of neuroendocrine development and maturation vary with developmental mode, with precocial species hatching at more advanced stages of maturation than altricial species (McNabb et al. 1998). These differences in the development of hormonal control systems likely influence the role of hormones in the regulation of begging. We focus our discussion of the development of neuroendocrine control systems and hormonal regulation of begging upon altricial birds, but refer to semi-precocial and semi-altricial species when appropriate.

We have applied a functional, evolutionary perspective to describe the development of physiological systems that may influence begging performance. We first elaborate on the current state of knowledge regarding the neuromuscular and sensory substrates that may be involved in the begging response. We then describe the development of relevant endocrine regulatory systems that may influence the intensity, frequency and duration of begging behaviour of nestlings, including a description of the maturation of these systems during the nestling phase. We then discuss the effects of endogenous and maternally-derived hormones on the development, regulation and performance of begging and hence the ability of nestlings to compete with each other for food delivered by parents. Finally, we suggest several mechanistic, physiological approaches to the study of begging that may be useful for the empirical testing of evolutionary theories such as parent-offspring conflict, sibling rivalry, communication between parent and offspring and parental favouritism.

NEUROANATOMY OF THE BEGGING RESPONSE

Development of Begging Behaviour

The neuroanatomical substrates of the begging response include the brain, sensory pathways, motor pathways, skeletal muscles and the skeleton (Figure 1). Each of these components is poorly developed when altricial birds initially exhibit begging behaviour after hatching (Starck 1993). At this stage begging consists of assuming a rather stereotyped posture of raising and stretching the neck and head and opening the beak to receive food. Newly-hatched nestlings usually maintain this posture for only a brief period. In general, most newly-hatched nestlings appear to have some difficulty with the precise control of this posture, although there is substantial variation among nestmates in begging performance at this age. Within a few days, however, perhaps as neuromuscular systems mature, the begging posture is assumed more rapidly and more often, maintained for a longer time during each bout and appears to be under better control. Further maturation of the begging response appears to be associated with the functional maturation of the visual system. At this time nestlings clearly aim their response toward the food-delivering parent. At later stages nestlings will also jockey for specific positions in the nest to enhance their probability of being fed (Rydén & Bengtsson 1980; McRae et al. 1993). We describe below the neuroanatomical substrates of begging behaviour that we believe constitute an initially simple reflex arc that matures rapidly into a complex behaviour as input from functional sensory systems occurs. This input allows nestlings to perceive visual and auditory signals from nestmates, parents and the environment around the nest, and to modify their begging in response to these signals. The temporal relationships between these developmental events may be somewhat different in semi-precocial and semi-altricial species, due to the fact that those species typically hatch with more mature visual, auditory and motor capabilities than do altricial birds.

Skeletal Muscles

The *musculus complexus,* a large, dorsally located neck muscle that overlies the *m. spinalis* and *m. biventer cervicus*, serves two functions in avian development. During hatching it provides the force necessary to break the shell through dorsal and lateral head movements (Gross 1985). After hatching, the *m. complexus* allows for the dorsal flexion and extension of the neck during begging (Ashmore et al. 1973). The role of this muscle during hatching has been well studied in precocial birds, but relatively little is known about its function in the begging of altricial birds. The following descriptions of the *m. complexus* refer mainly to studies of precocial species.

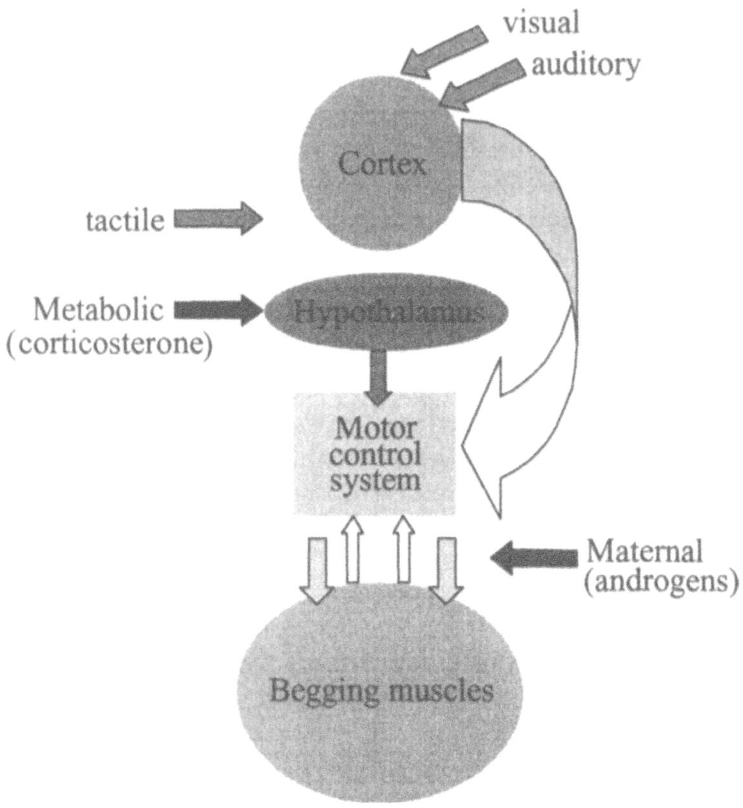

Figure 1. Diagram of the neuromuscular substrates (CNS motor control nuclei and muscles) of begging behaviour in a typical altricial bird and the putative sensory and metabolic inputs into the system. At the time of hatching the system is likely to represent a simple reflex arc composed of CNS motor control nuclei, motor neurons, the begging muscles and afferent fibres. Input into the system is likely to be limited to tactile sensation and metabolic state *via* the hypothalamus. Visual and auditory input into the system is likely to become important once the organs and processing areas for these modalities have matured during nestling development.

The *m. complexus* originates on the 3rd-5th cervical vertebrae and inserts along the posterior edge of the parietal bones. It is a striated, mainly phasic, muscle which consists of three segments separated by tendinous intersections. Two different fractions of myosin are present in the *m. complexus* fibres of the domestic goose (*Anser domesticus*, Fazekas et al. 1985). Bock and Hikida (1968) found that approximately 80% of the muscle fibres in the *m. complexus* of adult chickens (*Gallus domesticus*) are singly-innervated α, type II (fast-twitch) fibres, while the remaining 20% of the fibres are multiply innervated β, type I (slow-twitch) fibres. Wada et al. (1999) reported that in both adult pigeons (*Columba livia*) and adult chickens, the composition of muscle fibre types in the *m. complexus* is approximately 20% slow-twitch oxidative, 20% fast-twitch oxidative-glycolytic and 60% fast-twitch glycolytic. This composition of many fast-twitch and few slow-tonic fibres may allow a newly-hatched altricial bird to assume the begging posture quickly, but with little endurance. However, because the primary function of the *m. complexus* changes throughout the lifetime of a bird, it is possible that the composition of muscle fibres in the *m. complexus* differs between hatchlings and adults.

The *m. complexus* is noteworthy because of the morphological changes that it undergoes during development. Prior to hatching, the muscle enlarges as the result of an influx of lymph from adjacent lymph glands (Pohlman 1919; Fisher 1958). Indeed, large lymph glands are found along the lateral sides of the muscle at the time of hatching (Fisher 1958). In the chicken, the wet mass of the muscle increases as much as six-fold several days prior to hatching, declines to its pre-swelling mass after hatching, and then begins to increase again with normal muscular growth as the chick matures (Ashmore et al. 1973). Whether this change in mass is solely due to lymph infiltration or whether there are changes in fibre composition and orientation, associated with the changing function of the muscle, is not known. In domestic geese, the *m. complexus* attains its maximum size and weight, relative to body mass, at hatching (Fazekas et al. 1985). It is not known whether such hatching-related changes in the mass of the *m. complexus* also occur in altricial birds, although in the red-winged blackbird (*Agelaius phoeniceus*) the mass of the *m. complexus* decreases significantly within several hours of hatching (Lipar & Ketterson 2000), suggesting that this may indeed be the case. As reported below, testosterone influences the mass of this muscle at hatching, thus potentially influencing its function in hatching and begging.

Motor Pathways

The *m. complexus* is richly innervated by both the brain and the spinal cord in newly-hatched precocial birds. Innervation from the spinal cord arises in the cervical motor column (C1-C6). Each of the three muscle segments receives its own nerve branch (Kikuchi & Ashmore 1976; Gross 1985) with contributions by motor neurons from at least the first five cervical segments. In addition, axons of spinal accessory neurons in the upper cervical segments exit the spinal cord at the dorsal roots of C1 and C2 and innervate all three muscle segments. Innervation from the brain comes from the ipsilateral *nucleus supraspinalis* (*n. hypoglossus* in Karten & Hodos 1967) and also reaches each of the three muscle segments (Gross 1985; Watanabe & Ohmori 1988). The dorsal and facial motor nuclei innervate both the *m. complexus* and the muscles that control jaw movement (Gross 1985). The latter system may be important for the coordination of head raising and beak opening during the begging response.

The *n. supraspinalis* of both sexes of five-day-old nestling canaries (*Serinus canaria*) expresses androgen receptor mRNA (Gahr et al. 1996). This observation is particularly relevant to the hormonal regulation of begging behaviour because it suggests the presence of androgen receptors at this, and potentially earlier, stages of development. As outlined below, androgens in the eggs can influence begging performance. Other potential targets of androgens during development that might be implicated in producing variation in begging performance among hatchlings are cells of the spinal cord and the glycogen body, an organ for polysaccharide storage, both of which bind exogenous androgens in chicken embryos (Reid et al. 1981).

Afferent Pathways

Both the spinal cord and the brain receive sensory input from the *m. complexus*. All three of the muscle segments possess afferent fibres that terminate in the spinal cord (Gross 1985). In the brain, the tangential vestibular nucleus of the medulla appears to receive input from primary afferents. In addition, axons of the principal cells of the tangential nucleus project heavily to the upper cervical area of the spinal cord. This arrangement may comprise a potential feedback system between this brain nucleus and the *m. complexus* (Gross 1985).

Inputs from the Autonomic and Sensory Systems

Input from hypothalamic hunger and satiety networks to the begging control circuitry is likely to be present in newly-hatched nestlings and may influence the initiation, intensity and frequency of begging (see A.B. Clark this volume). This includes changes in hormone secretion in response to metabolic needs as described later in this chapter. Input from sensory systems other than the tactile system is unlikely to be important for eliciting the reflex-like begging response in newly-hatched nestlings, due to the fact that much of the maturation of the brain and sensory machinery occurs only after hatching (Starck 1993). The maturation of sensory perception and processing systems, however, allows auditory and visual signals to become important in the initiation and control of begging. As of yet, there have been no investigations of the neural pathways that may provide input from auditory, visual or tactile processing areas to the brain or spinal cord nuclei that control the *m. complexus*. Indeed, little is known about the development of these sensory systems in general. This would be a fruitful area of investigation for those who might like to study mechanisms of brain development in a functional, evolutionary context.

DEVELOPMENT OF HORMONAL SYSTEMS

Several classes of hormones, including sex steroids, glucocorticoids, thyroid hormone and growth hormone, have profound effects upon the physiological and behavioural development of birds (Adkins-Regan 1990; Kawata 1995; McNabb et al. 1998). As described below, these hormones may be of either endogenous or maternal origin and may influence the development of the anatomical substrates of begging, the probability of begging in response to external or internal stimuli and the performance of the behaviour, all of which may impact sibling competition. To understand the potential effects of these hormones on the development and control of begging behaviour, we first provide a brief overview of the development of hormonal systems, focusing on altricial birds. We then describe the current state of knowledge regarding the relationship between some of these hormones and the development of begging behaviour in the context of sibling competition and nestling survival. Our primary focus is on the effects of steroid hormones, with additional information being provided regarding thyroid hormone and growth hormone.

Sex Steroid Hormones

The sex steroids are produced primarily in the gonads. Included in this group of hormones are androgens (e.g. testosterone, androstenedione and 5α-dihydrotestosterone), oestrogens (e.g. 17β-oestradiol) and progestins (e.g. progesterone). The production of these hormones by the gonads is controlled by the actions of the hypothalamo-pituitary-gonadal (HPG) axis. In response to environmental or endogenous stimuli, the hypothalamus releases gonadotropin-releasing hormone (GnRH) which induces the release of the gonadotropins, luteinizing hormone (LH) and follicle-stimulating hormone (FSH), from the anterior pituitary. The gonadotropins, in turn, induce gonadal maturation, maintain gonadal function and stimulate gonadal sex steroid synthesis.

Information regarding the development of the HPG axis in altricial birds is limited. In the European starling (*Sturnus vulgaris*), the concentration of plasma LH is elevated in both sexes until four days after hatching (Williams et al. 1987), at which time it decreases and remains at a level comparable to that of non-breeding adults. The concentration of plasma testosterone is comparable to that of non-breeding adults throughout this period. In the great tit (*Parus major*), the plasma concentrations of LH, testosterone and 17β-oestradiol are all elevated at hatching (Silverin & Sharp 1996). Circulating LH decreases to basal levels by the age of nine days, while the concentrations of testosterone and 17β-oestradiol return to basal levels by the age of three and two days, respectively. In both sexes of the European robin (*Erithacus rubecula*) initially high plasma levels of 5α–dihydrotestosterone decrease during nestling development while the levels of testosterone increase during the latter portion of the nestling period, a time when begging and sibling competition are likely to be intense (Figure 2). The studies of these three species indicate that sex steroids, particularly androgens, are present in significant concentrations in the circulating plasma of newly-hatched and nestling altricial birds and that male and female nestlings exhibit similar levels and profiles.

Sex steroids exert a wide array of effects during vertebrate development. In the central nervous system, testosterone, 5α-dihydrotestosterone and 17β-oestradiol are all integral to the development of neuroanatomical sex differences that are the bases for differences in physiology and behaviour expressed later in the lifetime of the individual (Arnold & Gorski 1984; Adkins-Regan 1990; Arnold & Schlinger 1993; Kawata 1995). Steroids are also responsible for the control of muscular development. For example, testosterone has anabolic effects upon the development of the *levator ani* in rats (Joubert et al. 1994), the forelimb musculature in frogs (Sidor &

Blackburn 1998) and the sonic muscle in two species of teleost fish (Brantley et al. 1993; Connaughton & Taylor 1995). In addition, exogenous testosterone is metabolized *in vitro* by the cartilaginous tissue of chicken embryos (Murota & Tamaoki 1967). These observations indicate that testosterone influences the differentiation and development of the muscular component of motor control during embryonic development.

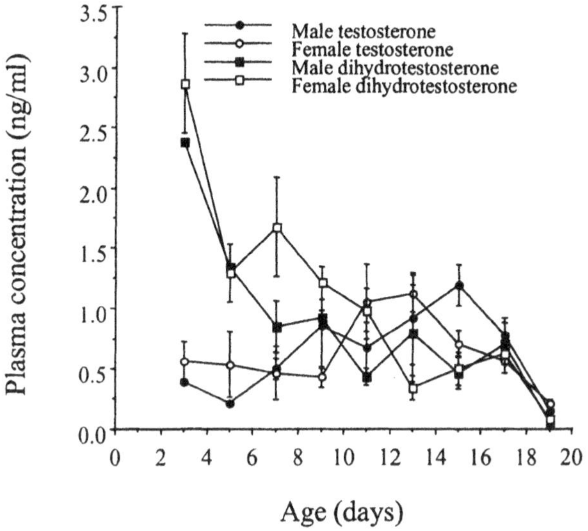

Figure 2. Plasma concentrations of testosterone and 5α-dihydrotestosterone in European robin nestlings (H. Schwabl & E.S. Hiatt unpublished data).

While many of these developmental effects are under the control of endogenously-produced hormones, steroids of maternal origin may also have early developmental influences. For example, the experimental elevation of 17β-oestradiol in the plasma of laying Japanese quail (*Coturnix japonica*) results in an increased incidence of right oviducts in their adult female offspring (Adkins-Regan et al. 1995). Similarly, short-term modification of maternal oestrogen levels in mice results in a dose-dependent increase in bone mass in adult offspring (Migliaccio et al. 1996), and the injection of testosterone proprionate into pregnant mice increases aggressive behaviour in their female offspring (Mann & Svare 1983). The recent discovery of sex steroid hormones in the yolk of avian eggs (e.g. Schwabl 1993, 1997; Schwabl et al. 1997; Gil et al. 1999; Lipar et al. 1999; Sockman & Schwabl 2000; Eising et al. 2001; French et al. 2001; Royle et al. 2001) suggests that avian development may be influenced by steroids of maternal origin as well as by those produced endogenously.

The concentrations of steroid hormones in both the plasma and follicles of females are known to undergo changes over the course of clutch formation (Shahabi et al. 1975; Wingfield & Farner 1978a, b; Donham 1979; Hammond et al. 1980; Johnson & van Tienhoven 1980; Etches & Cheng 1981; Bahr et al. 1983). In the canary, there is a correlation between maternal faecal testosterone concentration during yolk deposition of individual eggs and the concentration of yolk testosterone in those eggs after they have been laid (Schwabl 1996a). This may be indicative of a passive process of hormone deposition into the egg, with hormone levels in the eggs being a consequence of hormonal changes required for other reproductive functions in the female. Further studies are, however, required to determine the extent to which the female can control the accumulation of steroids in each egg and to what extent variation in the levels of yolk steroids is a by-product of normal female reproductive physiology (Birkhead et al. 2000). This is important and necessary for the conclusion that mothers manipulate offspring hormonally. Such studies should monitor female and yolk steroid levels under various environmental, social and maternal conditions.

Thyroid Hormone

Thyroxine (T_4) and triiodothyronine (T_3), collectively known as thyroid hormone, are produced in the thyroid gland. In birds, most thyroid hormone effects are mediated by T_3. This is due to the fact that, as in other vertebrates, avian thyroid receptors have a higher affinity (as much as ten-fold) for T_3 than for T_4 (Bellabarba et al. 1988). In addition, most T_4 is converted to T_3 after secretion from the thyroid gland (McNabb 1992). The regulation of the thyroid gland is analogous to the regulation of the gonads. Upon stimulation, the hypothalamus releases thyrotropin-releasing hormone (TRH), which induces the release of thyroid-stimulating hormone (TSH) from the anterior pituitary. TSH has a stimulatory effect upon the growth and secretory activity of the thyroid gland.

The development of the avian thyroid consists of three phases with respect to the concentration of plasma thyroid hormone (McNabb et al. 1998). The first phase is one in which plasma concentrations of thyroid hormone are extremely low. There is an increase in thyroid gland activity and plasma hormone levels during the second phase, while during the third phase thyroid hormone levels gradually adjust to adult levels. There are differences, with respect to timing, in the maturation of the HPT axis between altricial and precocial species (McNabb et al. 1998). In embryos of precocial species, the HPT axis matures about mid-incubation and appears to be responsible for increases in thyroid hormone during the latter half of incubation and at hatching (McNabb et al. 1998). In altricial birds, such as the ring dove (*Streptopelia risoria*) the plasma concentrations of thyroid hormone are low during the latter half of incubation and at hatching (McNabb et al. 1998). This is due to the fact that the maturation of the HPT axis occurs only after hatching (McNichols & McNabb 1988), coinciding with the development of endothermy in altricial species.

Thyroid hormone has been detected in the yolks of both chicken (Sechman & Bobek 1988) and quail (McNabb & Wilson 1997; Wilson & McNabb 1997) eggs. It is therefore possible that, in addition to any endogenously-produced hormone, there is embryonic exposure to maternally-derived thyroid hormone.

The effects of thyroid hormone on avian development are widespread and thyroid hormone is necessary for normal body growth (McNabb et al. 1998). Thyroid hormone does not generally have a direct effect upon cell proliferation, which would result in tissue growth (McNabb & King 1993). Rather, thyroid hormone acts in a permissive or interactive fashion with other hormones to promote body growth. One exception to this rule is that thyroid hormone induces cell proliferation during the development of the central nervous system (Nunez 1984). As mentioned previously, thyroid hormone is also important in the development of thermoregulation, often at the expense of body growth (McNabb et al. 1998). Thyroid hormone also has specific effects on tissue differentiation and maturation. For example, it is involved in the development of the skeletal system, particularly in the differentiation of cartilage and in calcification during bone formation (McNabb et al. 1998). In muscle, thyroid hormone induces the formation of specific actin and myosin isoforms that are associated with the maturation of contractile function (Stockdale & Miller 1987). Thyroid hormone may influence begging indirectly through these general developmental effects and through its function in regulating metabolism.

Glucocorticoids

The release of glucocorticoids (in birds primarily corticosterone) from the adrenal cortex is regulated by the actions of the hypothalamo-pituitary-adrenal (HPA) axis. This axis is analogous to the HPG and HPT axes mentioned previously. Higher brain centres, in response to endogenous or environmental cues, induce the hypothalamus to release corticotropin-releasing hormone (CRH). CRH stimulates the anterior pituitary to release adrenocorticotropic hormone (ACTH), which, in turn, stimulates the adrenal cortex to synthesize glucocorticoids.

The ontogeny of adrenal steroidogenic function in birds has been well investigated in both the domestic fowl and the mallard (*Anas platyrhynchos*). In these precocial species, the HPA axis is functional during embryonic development. The production of glucocorticoids increases in embryos midway through incubation and peaks at hatching (Wise & Frye 1973; Kalliecharan & Hall 1976; Marie 1981; Tanabe et al. 1986; Carsia et al. 1987). After hatching, plasma levels of corticosterone decline (Holmes & Kelley 1976; Wentworth & Hussein 1985; Carsia et al. 1987); however, mild stress due to restraint can precipitate increased corticosterone levels at that time (Holmes et al. 1992). The investigation of the development of the HPA axis of birds that are not precocial has been limited to a few species during the post-hatching period. In the altricial canary (Schwabl 1999), the plasma concentration of corticosterone is low in five-day-old nestlings but then increases in both sexes of older nestlings to levels that are comparable to the basal levels of adults. The plasma concentrations of these older nestlings appear to vary with hatching and laying order. The nestlings of several semi-precocial and semi-altricial species, including the black-legged kittiwake (*Rissa tridactyla*, Kitayski et al. 1999), the blue-footed booby (*Sula nebouxii*, de la Mora et al. 1996) and the American kestrel (*Falco sparverius*, Heath & Dufty 1998; Sockman & Schwabl 2001), show considerable plasma levels of corticosterone at early nestling stages. As a whole, these results indicate that the HPA axis is functional during early nestling stages in semi-precocial and semi-altricial birds and, at least at later stages, in altricial birds. As reported below, corticosterone may indeed be a factor in the hormonal control of begging in relation to nestling need. But differences among species depending on developmental mode are expected in the role of corticosterone during specific nestling stages.

Overall, glucocorticoids are important during development in the differentiation and maturation of organs, such as the lung and the gut, that occur at the time of hatching in anticipation of post-hatch respiration and feeding (Scanes et al. 1987). They are also involved in the shifts in osmoregulation that occur during hatching as the individual goes from an aquatic to a terrestrial environment (McNabb et al. 1998) and, as elaborated below, in nestling energy regulation.

Growth Hormone and Growth Factors

Avian growth hormone is secreted by the anterior pituitary in response to hormonal cues from the hypothalamus. In both precocial and altricial species, the production of growth hormone increases prior to hatching, peaks at some time after hatching, remains high during post-hatch growth, and then decreases gradually as the individual reaches maturity (McNabb et al. 1998). The timing of these events differs among species according to whether they are altricial or precocial.

Growth hormone is necessary for normal growth (King & Scanes 1986). During the post-hatching period, growth hormone is responsible for overall growth as well as for the proliferation and differentiation of various tissues, including muscle precursor cells and chondrocytes (McNabb et al. 1998). However, growth hormone does not act directly upon target tissues; rather, its effects are mediated by the actions of insulin-like growth factor-1, also known as IGF-1 (McNabb et al. 1998). As should be expected, the concentration of IGF-1 in the plasma parallels that of growth hormone throughout development (McNabb et al. 1998). Growth hormone stimulates IGF-1 production in the liver (O'Neill et al. 1990) and, to a lesser degree, in the pelvic cartilage (Burch et al. 1986). Some of the effects of growth hormone may be mediated by the actions of thyroid hormone, as growth hormone increases the production of T_3 early in the post-hatching period (McNabb et al. 1998). Insulin and IGF-1 have been detected in unfertilized chicken eggs (e.g. de Pablo et al. 1982; Scavo et al. 1989); an observation that might hint at maternal effects, *via* these growth regulators, upon the differentiation and growth of muscle and nervous tissue during early embryogenesis (de Pablo & de la Rosa 1995). These maternal effects could potentially influence a nestling's ability to beg. Again, if one is to argue for maternal manipulation, it will be important to distinguish how the hormone quantity in the egg is related to the quantity of hormone circulating in the mother, which may be influenced by her age, social, ecological and nutritional conditions, or whether the mother is able to control how much is delivered to each egg.

HORMONAL INFLUENCES ON BEGGING BEHAVIOUR AND PERFORMANCE

Two types of steroid hormones, androgens of maternal origin (Schwabl 1996b) and endogenously-produced corticosterone (Kitayski et al. 2001), have been shown to influence begging behaviour. The mechanisms and functions of these hormonal effects on begging are likely to be fundamentally distinct. The presence of maternal steroid hormones in the egg yolk may influence the maturation of the anatomical substrates of the begging response before normal steroidogenesis by embryonic or nestling tissue occurs. Maternal steroids could also have organizational effects on the activity of steroidogenic tissues, which in turn might modify the availability of hormones and thus begging at later nestling stages. Because nestling birds typically retain a portion of unabsorbed yolk for several days after hatching, yolk steroids may also have activational effects on the performance of begging behaviour even after hatching. In contrast, the effects of endogenously-produced corticosterone may be limited to activational influences on the expression of the behaviour in relation to metabolic state. From an evolutionary perspective, the influence of maternal steroids may reflect maternal interests while the influence of nestling corticosterone may reflect offspring needs.

Maternal Androgens and Begging Behaviour

Maternal steroid hormones, including testosterone, androstenedione, 17β-oestradiol and 5α-dihydrotestosterone, are present in the egg yolk of altricial as well as precocial species, including canaries (Schwabl 1993, 1996b), Japanese quail (Adkins-Regan et al. 1995), cattle egrets (*Bubulcus ibis*, Schwabl et al. 1997), house sparrows (*Passer domesticus*, Schwabl 1997), red-winged blackbirds (Lipar et al. 1999), zebra finches (*Taeniopygia guttata*, Gil et al. 1999), American kestrels (Sockman & Schwabl 2000), common terns (*Sterna hirundo*, French et al. 2001), lesser black-backed gulls (*Larus fuscus*, Royle et al. 2001) and black-headed gulls (*Larus ridibundus*, Eising et al. 2001). In many of these species there is a systematic, differential apportionment of these steroids to the eggs of a clutch according to their position in the laying order. For example, yolk testosterone concentrations increase with laying order in the canary (Schwabl 1993), the red-winged blackbird (Lipar et al. 1999), the American kestrel (Sockman & Schwabl 2000), the common tern (French et al. 2001), the lesser black-backed gull (Royle et al. 2001) and the black-headed gull (Eising et al. 2001), all species in which hatching asynchrony produces a size hierarchy among nestlings. This differential allocation of steroid hormones to the eggs of a clutch may represent a hormonal mechanism by which mothers influence the ability of nestmates to compete for parentally delivered food (Schwabl 1993; Schwabl et al. 1997).

While variation in egg mass and the onset of incubation – two other well-known maternal factors that affect nestling competition – impart their effects indirectly through the formation of a size or age hierarchy, the differential allocation of maternal steroids appears to directly affect the underlying neuromuscular substrates of begging behaviour. For example, in newly-hatched canaries, begging performance, as measured by the frequency of begging bouts, is enhanced by exposure to testosterone in the egg (Schwabl 1996b). This results in greater growth trajectories in both sexes of nestlings within 24 hours of hatching. These increased trajectories are maintained throughout most of the nestling phase. The enhancement of begging in last-hatched offspring through greater allocation of testosterone to the eggs that produce those offspring may serve to increase their probability of leaving the nest in good condition. The design of this study controlled for possible influences of increased parental feeding, but it was not possible to distinguish whether parents delivered more total food, or if food allocation to nestlings hatched from eggs exposed to greater amounts of testosterone was increased at the expense of nestlings hatched from control eggs with exposure to lower amounts of testosterone. In the black-headed gull, however, the growth of first-hatched chicks was reduced in the presence of two androgen-treated siblings, indicating that yolk androgens influence the distribution of limited parental food among the nestlings in a brood (Eising et al. 2001). Because third-hatched nestlings that hatched from eggs treated with androgens had a greater mass and tarsus length than third-hatched nestlings that hatched from control eggs, this field study, which included controls for egg quality, hatching asynchrony and parental feeding rates, also indicated that treatment of eggs with androgens results in an enhancement of nestling growth rates.

There are several conceivable mechanisms through which maternally-derived testosterone may enhance begging performance in altricial birds. First, embryonic exposure to testosterone may indirectly increase begging through an increase in metabolism. We have no empirical evidence for this type of an effect in embryos or nestlings. In fact, testosterone either decreased resting metabolic rates in adult birds (Wikelski et al. 1999) or had no effect (Deviche 1992). Second, testosterone may accelerate the differentiation and development of sensory systems for tactile or auditory cues that may elicit begging in hatchlings. At present there is no evidence in support of this mechanism. Third, androgens may influence brain functions underlying attention and persistence. Indeed, testosterone treatment of newly-hatched domestic fowl enhances their attention by increasing binocular fixation and persistence (Andrew 1975; Clifton et al. 1988). More evidence for general effects on the behavioural phenotype that might impact begging comes from the semi-precocial black-headed gull (Eising et al. in press). In this species nestlings compete for food that parents regurgitate near the nest. Nestlings from androgen-treated eggs scored higher in all measures of begging and competition for food, including the initial response to arrival of the parent, approach of the parent, bill pecking and pumping, which is the conspicuous begging display in this species (Eising et al. in press; C. Eising & T. Groothuis personal communication). Fourth, testosterone may impact the development of the muscles and motor control systems which underlie the initiation and maintenance of the begging posture. Several lines of evidence support this role of maternal testosterone in the establishment of differences among nestlings in the performance of begging behaviour. Yolk testosterone concentration, which increases with laying order independently of sex, is positively correlated with the mass of the *m. complexus,* relative to body mass, in altricial red-winged blackbird hatchlings (Lipar & Ketterson 2000). The injection of testosterone into eggs of this species results in an increase in the mass of the *m. complexus* at hatching, while the injection of flutamide, a testosterone antagonist, results in a reduction in the mass of the *m. complexus* (Lipar & Ketterson 2000). The increase in mass of the *m. complexus* that is found in those individuals exposed to higher levels of maternal testosterone may be due to a higher survival of myocytes and motor neurons. Moreover, as mentioned above, the *nucleus supraspinalis* (*n. hypoglossus*), which seems to be an integral part of the motor control pathway of begging, expresses androgen receptors in male and female canary nestlings (Gahr et al. 1996). As a whole, these results indicate that maternal androgens may enhance begging by influencing the development of the neuromuscular substrates of the begging reflex as well as those underlying the motivation to beg. These effects seem to be similar in both sexes, but we expect that species may differ depending on their developmental mode.

It is unknown whether endogenously-produced androgens in the embryo or nestling have similar effects on the anatomical substrates and the expression of begging behaviour mentioned above. But, as shown earlier, in some altricial species nestlings of both sexes show substantial circulating concentrations of androgens during times when sibling hierarchies may be established, or when begging and competition for food are intense. Experimental manipulations of androgen levels, carefully controlled for other factors such as laying order and maternal androgen levels, have to be performed to investigate the role of endogenous hormones. These methods could also be used to test competing hypotheses relating to the evolution and function of sibling rivalry, parent-offspring conflict, signal evolution and maternal favouritism.

Nestling Corticosterone, Begging Behaviour and Metabolic Requirements

Corticosterone is an important signal in the regulation of body condition and metabolic needs in adult and developing birds (see above). In altricial and semi-altricial birds the HPA axis, which is the regulatory system for corticosterone synthesis, appears to become fully functional only after hatching (Schwabl 1999 for the canary; Sockman & Schwabl 2001 for the American kestrel). It is, therefore, unlikely that corticosterone secretion plays an important role in the regulation of begging behaviour immediately after hatching. It may, however, play a major role in the regulation of begging during later stages of the nestling period. In the semi-precocial black-legged kittiwake, plasma corticosterone levels increase as the quality or quantity of food that is given to nestlings is decreased. Moreover, circulating corticosterone levels are negatively correlated with body condition in four-week-old nestlings of this species (Kitayski et al. 1999). Corticosterone levels also increase with experimental food restriction in the blue-footed booby (de la Mora et al. 1996) and a negative correlation between corticosterone levels and body condition already existed in five-day-old nestlings of the American kestrel (Sockman & Schwabl 2001). However, no correlation between these two variables was detected in the northern mockingbird (*Mimus polyglottos,* Sims & Holberton 2000), the canary (Schwabl 1999), or in another study of the American kestrel (Heath & Dufty 1998). A recent study showed that experimentally increased corticosterone facilitates begging in 15-day-old black-legged kittiwakes and that parents respond with changes in resource allocation to the young (Kitayski et al. 2001). Moreover, corticosterone elicited greater begging rates in young with a nestmate than in singletons. Thus, at least in this species, the secretion of corticosterone is likely to be one of the mediating mechanisms that translate metabolic need into begging. The regulation of begging through the secretion of corticosterone could indicate that begging is an honest signal that presents an evolutionarily stable resolution of parent-offspring conflict (Kitayski et al. 2001). Cheating may be prevented because elevated corticosterone levels, if they are required for continuous begging, may result in the manifold health costs of prolonged exposure to glucocorticoids (Sapolsky et al. 1996). These results are also consistent with predictions that begging intensity should be context dependent, i.e. change with the presence and condition of nestmates (Godfray 1995).

Potential hormonal mechanisms of the regulation of begging behaviour and sibling competition are emerging from these studies, but we are far from a well-founded understanding of the processes that regulate begging at the proximate level. Additional comparative studies of the hormonal control of begging, for example in altricial versus semi-precocial species, and during different post-hatching stages, are warranted before we can use this knowledge to manipulate begging behaviour in tests of functional and evolutionary hypotheses.

FUTURE DIRECTIONS

A mechanistic understanding of the neuroanatomical components and regulation of begging behaviour in hatchlings and nestlings will lay the ground for experimental tests of functional evolutionary hypotheses of begging. These include comparative studies of (1) the physiology of the begging muscle(s), neuromuscular interface, central nervous system nuclei and input from sensory processing areas of the brain; (2) the effects of maternally-derived and endogenously-produced hormones on the embryonic development and functional anatomy of these components; and (3) the neuroendocrine regulation of begging behaviour of nestlings. Equipped with this information we will then be in a position to design valid and rigorous experiments to test and discriminate between competing evolutionary hypotheses concerning begging as a signal to the parent, sibling competition, parent-offspring conflict and maternal effects. Presently, it is safe to say that maternal androgens that are deposited in the eggs of birds in variable amounts probably impact the development and function of the hardware of the begging response, such as muscles (Lipar & Ketterson 2000) and possibly neural control circuits. Small asymmetries in the development of hardware among newly-hatched siblings, caused by maternal effects, may be amplified *via* a positive feedback loop between begging performance, food intake, growth and the ability to compete for limited parental feedings. Such amplification and perpetuation of subtle differences at hatching have been observed (e.g. Schwabl 1996b; Sockman & Schwabl 2000). It therefore seems that differential hormone deposition into the eggs of a clutch is a means by which offspring performance in the nest may be regulated by the mother. It remains to be shown how ecological and evolutionary factors have shaped such maternal influences on sibling rivalry and whether selection acts on the female generation or on the offspring generation. Care has to be taken when using patterns of maternal androgen allocation among the eggs of a clutch to conclude which young may be favoured by the mother, because high androgen levels in eggs may not always be beneficial to the offspring (Sockman & Schwabl 2000). To date, maternal androgens and offspring corticosterone have been found to affect begging, but we would not be surprised to find that other hormones and growth factors of maternal as well as offspring origin also influence this behaviour. Because we currently have at least some limited understanding of the effects of maternal androgens on begging, manipulation of begging performance and nestling competitive abilities with physiological doses of androgens or their antagonists may be a method of choice for experimental tests of adaptive hypotheses. We also need studies that try to unravel whether selection acts on variation in hormone levels in the mother, with only indirect effects upon nestling fitness, or whether selection acts directly on the influences of maternal hormones upon the offspring.

We are also beginning to understand the role of endogenous hormonal secretions, such as corticosterone, in the regulation of begging behaviour in nestlings. Manipulations of this hormone may be useful to modify nestling begging behaviour and to monitor the responses of nestmates and parents (Kitayski et al. 2001), as well as the effects on growth and development. If manipulations of corticosterone as an endocrine signal of a nestling's metabolic state and need are to be performed, one should keep in mind that maternal androgens in the egg may modify the functional state of this hormonal axis in the nestling (Schwabl 1999; Sockman & Schwabl 2001). Such studies also need to consider differences among species in developmental mode, as this influences the function of endocrine control systems at hatching and thereafter. The integration of proximate physiological and ultimate evolutionary approaches in the study of begging behaviour will eventually enhance our understanding of broader biological questions such as conflict between the sexes or parents and offspring, sibling rivalry, maternal favouritism and reproductive decisions.

REFERENCES

Adkins-Regan, E. 1990. Hormonal bases of sexual differentiation in birds. In: *Hormones, Brain and Behaviour in Vertebrates. I. Sexual Differentiation, Neuroanatomical Aspects, Neurotransmitters, and Neuropeptides* (Ed. by J. Balthazart). Basel: S. Karger.

Adkins-Regan, E., Ottinger, M.A. & Park, J. 1995. Maternal transfer of estradiol to egg yolks alters sexual differentiation of avian offspring. *Journal of Experimental Zoology* **271**, 466-470.

Andrew, R.J. 1975. Effects of testosterone on the behaviour of the domestic chick. II. Effects present in both sexes. *Animal Behaviour* **23**, 156-168.

Arnold, A.P. & Gorski, R.A. 1984. Gonadal steroid induction of structural sex differences in the central nervous system. *Annual Review of Neuroscience* **7**, 413-442.

Arnold, A.P. & Schlinger, B.A. 1993. Sexual differentiation of brain and behavior: the zebra finch is not just a flying rat. *Brain, Behavior and Evolution* **42**, 231-241.

Ashmore, C.R., Addis, P.B., Doerr, L. & Stokes, H. 1973. Development of muscle fibers in the *complexus* muscle of normal and dystrophic chicks. *Journal of Histochemistry and Cytochemistry* **21**, 266-278.

Bahr, J.M., Wang, S.C., Huang, M.Y. & Calvo, F.O. 1983. Steroid concentrations in isolated theca and granulosa layers of preovulatory follicles during the ovulatory cycle of the domestic hen. *Biology of Reproduction* **29**, 326-334.

Bellabarba, D., Belisle, S., Gallo-Payet, N. & Lehoux, J.G. 1988. Mechanism of action of thyroid hormones during chick embryogenesis. *American Zoologist* **28**, 389-399.

Birkhead, T., Schwabl, H. & Burke, T. 2000. Testosterone and maternal effects: integrating mechanisms and function. *Trends in Ecology and Evolution* **15**, 86-87.

Bock, W.J. & Hikida, R.S. 1968. An analysis of twitch and tonus fibers in the hatching muscle. *Condor* **70**, 211-222.

Brantley, R.K., Marchaterre, M.A. & Bass, A.H. 1993. Androgen effects on vocal muscle structure in a teleost fish with inter- and intra-sexual dimorphism. *Journal of Morphology* **216**, 305-318.

Burch, W.M., Weir, S. & Van Wyk, J.J. 1986. Embryonic chick cartilage produces its own somatomedin-like peptide to stimulate cartilage growth *in vitro*. *Endocrinology* **119**, 1370-1376.

Carsia, R.V., Morin, M.E., Rosen, H.D. & Weber, H. 1987. Ontogenic corticosteroidogenesis of the domestic fowl: response of isolated adrenocortical cells. *Proceedings of the Society for Experimental Biology and Medicine* **184**, 436-445.

Clifton, P.G., Andrew, R.J. & Brighton, L. 1988. Gonadal steroids and attentional mechanisms in young domestic chicks. *Physiology and Behavior* **43**, 441-446.

Connaughton, M.A. & Taylor, M.H. 1995. Effects of exogenous testosterone on sonic muscle mass in the weakfish, *Cygnoscion regalis*. *General and Comparative Endocrinology* **100**, 238-245.

de la Mora, A.N., Drummond, H. & Wingfield, J.C. 1996. Hormonal correlates of dominance and starvation-induced aggression in chicks of the blue-footed booby. *Ethology* **102**, 748-761.

de Pablo, F. & de la Rosa, J. 1995. The developing CNS: a scenario for the action of proinsulin, insulin, and insulin-like growth factors. *Trends in Neuroscience* **18**, 143-150.

de Pablo, F., Roth, J., Hernandez, E. & Pruss, R.M. 1982. Insulin is present in chicken eggs and early chick embryos. *Endocrinology* **111**, 1909-1916.

Deviche, P. 1992. Testosterone and opioids interact to regulate feeding in a male migratory songbird. *Hormones and Behavior* **26**, 394-405.

Donham, R.S. 1979. Annual cycle of plasma luteinizing hormone and sex hormones in male and female mallards (*Anas platyrhynchos*). *Biology of Reproduction* **21**, 1273-1285.

Eising, C.M., Eikenaar, C., Schwabl, H. & Groothuis, T.G.G. 2001. Maternal androgens in black-headed gull (*Larus ridibundus*) eggs: consequences for chick development. *Proceedings of the Royal Society of London, Series B* **268**, 839-846.

Eising, C.M., Eikenaar, C. & Groothuis, T.G. in press. Maternal androgens and hatching asynchrony in black-headed gulls: an experimental field study. *Advances in Ethology*.

Etches, R.J. & Cheng, K.W. 1981. Changes in the plasma concentrations of luteinizing hormone, progesterone, oestradiol, and testosterone and in the binding of follicle-stimulating hormone to the theca of follicles during the ovulation cycle of the hen (*Gallus domesticus*). *Journal of Endocrinology* **91**, 11-22.

Fazekas, S., Fehér, G., Kondics, L., Óváry, I. & Székessy Hermann, V. 1985. Purification and properties of myosin from the "hatching muscle" (*m. complexus*) of geese. *Acta Physiologica Hungarica* **66**, 5-25.

Fisher, H.I. 1958. The "hatching muscle" in the chick. *The Auk* **75**, 391-399.

French, J.B. Jr., Nisbet, I.C.T. & Schwabl, H. 2001. Maternal steroids and contaminants in common tern eggs: a mechanism of endocrine disruption? *Comparative Biochemistry and Physiology C* **128**, 91-98.

Gahr, M., Metzdorf, R. & Aschenbrenner, S. 1996. The ontogeny of the canary HVC revealed by the expression of androgen and estrogen receptors. *Developmental Neuroscience* **8**, 311-315.

Gil, D., Graves, J., Hazon, N. & Wells, A. 1999. Male attractiveness and differential testosterone investment in zebra finch eggs. *Science* **286**, 126-128.

Godfray, H.C.J. 1995. Signaling of need between the parent and the young: parent-offspring conflict and sibling rivalry. *American Naturalist* **146**, 1-24.

Gross, G.H. 1985. Innervation of the *complexus* ("hatching") muscle of the chick. *Journal of Comparative Neurology* **232**, 180-189.

Hammond, R.W., Olson, D.M., Frenkel, R.B., Biellier, H.V. & Hertelendy, F. 1980. Prostaglandins and steroid hormones in plasma and ovarian follicles during the ovulation cycle of the domestic hen (*Gallus domesticus*). *General and Comparative Endocrinology* **42**, 195-202.

Heath, J.A. & Dufty, A.M. Jr. 1998. Body condition and the adrenal stress response in captive American kestrel juveniles. *Physiological Zoo*logy **71**, 67-73.

Holmes, W.N. & Kelley, M.E. 1976. The turnover and distribution of labelled corticosterone during post-natal development of the duckling, *Anas platyrhynchos*. *Pflügers Archive* **365**, 145-150.

Holmes, W.N., Cronshaw, J., Collie, M.A. & Rohde, K.D. 1992. Cellular aspects of the stress response in precocial neonates. *Ornis Scandinavica* **23**, 388-397.

Johnsen, A.L. & van Tienhoven, A. 1980. Plasma concentrations of six steroids and LH during the ovulatory cycle of the hen, *Gallus domesticus*. *Biology of Reproduction* **23**, 386-393.

Joubert, Y., Tobin, C. & Lebart, M.C. 1994. Testosterone-induced masculinization of the rat *levator ani* muscle during puberty. *Developmental Biology* **162**, 104-110.

Kalliecharan, R. & Hall, B.K. 1976. A developmental study of the levels of progesterone, corticosterone, cortisol, and cortisone in the adrenal glands of the embryonic chick. *General and Comparative Endocrinology* **30**, 404-409.

Karten, H. & Hodos, W. 1967. *A Stereotaxic Atlas of the Brain of the Pigeon (Columba livia)*. Baltimore: Johns Hopkins Press.

Kawata, M. 1995. Roles of steroid hormones and their receptors in structural organization in the nervous system. *Neuroscience Research* **24**, 1-46.

Kedar, H., Rodríguez-Gironés, M.A., Yedvab, S., Winkler, D.W. & Lotem, A. 2000. Experimental evidence for offspring learning in parent-offspring communication. *Proceedings of the Royal Society of London, Series B* **267**, 1723-1727.

Kikuchi, T. & Ashmore, C.R. 1976. Developmental aspects of the innervation of skeletal muscle fibers in the chick embryo. *Cell Tissue Research* **171**, 233-251.

King, D.B. & Scanes, C.G. 1986. Effect of mammalian growth hormone and prolactin on the growth of hypophysectomized chickens. *Proceedings of the Society for Experimental Biology and Medicine* **182**, 201-207.

Kitayski, A.S., Piatt, J.F., Wingfield, J.C. & Romano, M. 1999. The adrenocortical stress-response of black-legged kittiwake chicks in relation to dietary restrictions. *Journal of Comparative Physiology B* **169**, 303-310.

Kitayski, A.S., Wingfield, J.C. & Piatt, J.F. 2001. Corticosterone facilitates begging and affects resource allocation in the black-legged kittiwake. *Behavioral Ecology* **12**, 619-625.

Lipar, J.L. & Ketterson, E.D. 2000. Maternally derived yolk testosterone enhances the development of the hatching muscle in the red-winged blackbird *Agelaius phoeniceus*. *Proceedings of the Royal Society of London, Series B* **267**, 2005-2010.

Lipar, J.L., Ketterson, E.D. & Nolan, V. Jr. 1999. Intra-clutch variation in testosterone content of red-winged blackbird eggs. *The Auk* **116**, 231-235.

Mann, M.A. & Svare, B. 1983. Prenatal testosterone exposure elevates maternal aggression in mice. *Physiology and Behavior* **30**, 503-507.

Marie, C. 1981. Ontogenesis of the adrenal glucocorticoids and of the target function of the enzymatic tyrosine transaminase activity on the chick embryo. *Journal of Endocrinology* **90**, 193-200.

McNabb, F.M.A. 1992. *Thyroid Hormones*. Englewood Cliffs: Prentice Hall.

McNabb, F.M.A. & King, D.B. 1993. Thyroid hormone effects on growth, development and metabolism. In: *The Endocrinology of Growth, Development, and Metabolism of Vertebrates* (Ed. by M.P. Schreibman, C.G. Scanes & P.K.T. Pang). San Diego: Academic Press.

McNabb, F.M.A. & Wilson, C.M. 1997. Thyroid hormone deposition in avian eggs and effects on embryonic development. *American Zoologist* **37**, 553-560.

McNabb, F.M.A., Scanes, C.G. & Zeman, M. 1998. The endocrine system. In: *Avian Growth and Development: Evolution within the Altricial-Precocial Spectrum* (Ed. by J.M. Starck & R.E. Ricklefs). New York: Oxford University Press.

McNichols, M.J. & McNabb, F.M.A. 1988. Development of thyroid function and its pituitary control in embryonic and hatchling precocial Japanese quail and altricial ring doves. *General and Comparative Endocrinology* **69**, 109-118.

McRae, S.B., Weatherhead, P.J. & Montgomerie, R. 1993. American robin nestlings compete by jockeying for position. *Behavioral Ecology and Sociobiology* **33**, 101-106.

Migliaccio, S., Newbold, R.R., Bullock, B.C., Jefferson, W.J., Sutton, F.G. & McLachlan, J.A. 1996. Alterations of maternal estrogen levels during gestation affect the skeleton of female offspring. *Endocrinology* **137**, 2118-2125.

Murota, S.-I. & Tamaoki, B.I. 1967. Metabolism of progesterone and testosterone by chick cartilage *in vitro*. *Biochimica and Biophysica Acta* **137**, 347-355.

Nunez, J. 1984. Effects of thyroid hormones during brain differentiation. *Molecular and Cellular Endocrinology* **37**, 125-132.

O'Neill, I.E., Houston, B. & Goddard, C. 1990. Stimulation of insulin-like growth factor-I production in primary cultures of chicken hepatocytes by chicken growth hormone. *Molecular and Cellular Endocrinology* **70**, 41-47.

Pohlman, A.G. 1919. Concerning the causal factor in the hatching of the chick, with particular reference to the *musculus complexus*. *Anatomical Record* **17**, 89-104.

Reid, F.A., Gasc, J.M., Stumpf, W.E. & Sar, M. 1981. Androgen target cells in spinal cord, spinal ganglia, and glycogen body of chick embryos. *Experimental Brain Research* **44**, 243-248.

Royle, N.J., Surai, P.F. & Hartley, I.R. 2001. Maternally derived androgens and antioxidants in bird eggs: complementary but opposing effects? *Behavioral Ecology* **12**, 381-385.

Rydén, O. & Bengtsson, H. 1980. Differential begging and locomotory behaviour by early and late hatched nestlings affecting the distribution of food in asynchronously hatched broods of altricial birds. *Zeitschrift für Tierpsychologie* **53**, 209-224.

Sapolsky, R.M., Krey, L.C. & McEwen, B.S. 1996. The neuroendocrinology of stress and aging: the glucocorticoid cascade hypothesis. *Endocrine Reviews* **7**, 284-301.

Scanes, C.G., Hart, L.E., Decuypere, E. & Kuhn, E.R. 1987. Endocrinology of the avian embryo: an overview. *Journal of Experimental Zoology* **1**, 253-264.

Scavo, L., Alemany, J., Roth, J. & de Pablo, F. 1989. Insulin-like growth factor-I activity is stored in the yolk of the avian egg. *Biochemical and Biophysical Research Communications* **162**, 1167-1173.

Schwabl, H. 1993. Yolk is a source of maternal testosterone for developing birds. *Proceedings of the National Academy of Sciences USA* **90**, 11446-11450.

Schwabl, H. 1996a. Environment modifies the testosterone levels of the female bird and its eggs. *Journal of Experimental Zoology* **276**, 157-163.

Schwabl, H. 1996b. Maternal testosterone in the avian egg enhances postnatal growth. *Comparative Biochemistry and Physiology* **114A**, 271-276.

Schwabl, H. 1997. The contents of maternal testosterone in house sparrow *Passer domesticus* eggs vary with breeding conditions. *Naturwissenschaften* **84**, 406-408.

Schwabl, H. 1999. Developmental changes and among-sibling variation of corticosterone levels in an altricial avian species. *General and Comparative Endocrinology* **116**, 403-408.

Schwabl, H., Mock, D.W. & Gieg, J.A. 1997. A hormonal mechanism for parental favouritism. *Nature* **386**, 231.

Sechman, A. & Bobek, S. 1988. Presence of iodothyronines in the yolk of the hen's egg. *General and Comparative Endocrinology* **69**, 99-105.

Shahabi, N.A., Norton, H.W. & Nalbandov, A.V. 1975. Steroid levels in follicles and the plasma of hens during the ovulatory cycle. *Endocrinology* **96**, 962-968.

Sidor, C.A. & Blackburn, D.G. 1998. Effects of testosterone administration and castration on the forelimb musculature of male leopard frogs, *Rana pipiens*. *Journal of Experimental Zoology* **280**, 28-37.

Silverin, B. & Sharp, P. 1996. The development of the hypothalamic-pituitary-gonadal axis in juvenile great tits. *General and Comparative Endocrinology* **103**, 150-166.

Sims, C.G. & Holberton, R.L. 2000. Development of the corticosterone stress response in young northern mockingbirds (*Mimus polyglottos*). *General and Comparative Endocrinology* **119**, 193-201.

Sockman, K.W. & Schwabl, H. 2000. Yolk androgens reduce offspring survival. *Proceedings of the Royal Society of London, Series B* **267**, 1451-1456.

Sockman, K.W. & Schwabl, H. 2001. Plasma corticosterone in nestling American kestrels: effects of age, handling stress, yolk androgens, and body condition. *General and Comparative Endocrinology* **122**, 205-212.

Starck, J.M. 1993. Evolution of avian ontogenies. *Current Ornithology* **10**, 275-366.

Stockdale, F.E. & Miller, J.B. 1987. The cellular basis of myosin heavy chain isoform expression during development of avian skeletal muscles. *Developmental Biology* **123**, 1-9.

Tanabe, Y., Saito, N. & Nakamura, T. 1986. Ontogenetic steroidogenesis by testes, ovary, and adrenals of embryonic and postembryonic chickens (*Gallus domesticus*). *General and Comparative Endocrinology* **63**, 456-463.

Wada, N., Miyata, H., Tomita, R., Ozawa, S. & Tokuriki, M. 1999. Histochemical analysis of fiber composition of skeletal muscles in pigeons and chickens. *Archives Italiennes de Biologie* **137**, 75-82.

Watanabe, T. & Ohmori, Y. 1988. Location of motoneurons supplying upper neck muscles in the chicken studied by means of horseradish peroxidase. *Journal of Comparative Neurology* **270**, 271-278.

Wentworth, B.C. & Hussein, M.O. 1985. Serum corticosterone levels in embryos, newly hatched and young turkey poults. *Poultry Science* **65**, 2195-2201.

Wikelski, M., Lynn, S., Breuner, C., Wingfield, J.C. & Kenagy, G.J. 1999. Energy metabolism, testosterone and corticosterone in white-crowned sparrows. *Journal of Comparative Physiology A* **185**, 463-470.

Williams, T.D., Dawson, A., Nicholls, T.J. & Goldsmith, A.R. 1987. Reproductive endocrinology of free-living nestling and juvenile starlings, *Sturnus vulgaris* - an altricial species. *Journal of Zoology* **212**, 619-628.

Wilson, C.M. & McNabb, F.M.A. 1997. Maternal thyroid hormones in Japanese quail eggs and their influence on embryonic development. *General and Comparative Endocrinology* **107**, 153-165.

Wingfield, J.C. & Farner, D.S. 1978a. The annual cycle of plasma irLH and steroid hormones in feral populations of the white-crowned sparrow, *Zonotrichia leucophrys gambelli. Biology of Reproduction* **19**, 1046-1056.

Wingfield, J.C. & Farner, D.S. 1978b. The endocrinology of a natural breeding population of the white-crowned sparrow (*Zonotrichia leucophrys pugetensis*). *Physiological Zoology* **51**, 188-205.

Wise, P.M. & Frye, B.E. 1973. Functional development of the hypothalamo-hypophyseal-adrenal cortex axis in the chick embryo, *Gallus domesticus. Experimental Zoology* **185**, 277-292.

13.

IMMUNITY AND BEGGING

Nicola Saino[1] & Anders Pape Møller[2]

[1]*Dipartimento di Biologia, Università degli Studi di Milano, I-20133 Milano, Italy (n.saino@mailserver.unimi.it)*

[2]*Laboratoire d'Ecologie Evolutive Parasitaire, Université Pierre et Marie Curie, F-75252 Paris Cedex 5, France (amoller@hall.snv.jussieu.fr)*

ABSTRACT

Offspring begging signals may have evolved to provide reliable information about the health status of individual offspring to their parents, either because such signals are condition-dependent or because they are revealing handicaps. Since immune responses are directly related to probability of survival, we hypothesize that parents may maximize their reproductive success by allocating limited effort to offspring displaying signals that reliably reveal their immunocompetence. Offspring health status will be influenced by maternal, additive genetic and environmental effects, and these components are likely to contribute to phenotypic variation in offspring begging signals. We provide examples using gape colour, begging behaviour and begging calls to illustrate the point that begging signals may reliably reveal the health status and hence the reproductive value of an individual offspring.

INTRODUCTION

In this chapter we investigate potential relationships between offspring immunity and begging behaviour. We assume throughout that parasites both past and present constitute a strong selective force shaping the evolution of life history traits and signals involved in a diverse array of social interactions within and between species. In particular, we develop the idea that begging signals of offspring may reflect their ability to raise an efficient immune response towards an attack from a parasite. Since the immune system is strongly condition-dependent, and since many signals also are

J. Wright and M.L. Leonard (eds.), The Evolution of Begging, 245–267.

condition-dependent in their expression, this opens up the possibility that begging signals reliably reflect the condition of the sender. This has important implications for receivers of such signals.

This chapter starts with a brief introduction to the immune system, allowing us to put subsequent sections into perspective, and a description of the direct relationship between immunity and viability. In the following sections we expand on a range of specific themes: how begging displays, including gape colour and features of solicitation calls relate to offspring need and immune status, the relevance of maternally transmitted immunity to offspring performance and behaviour, and the relationship between parental response to offspring begging and how this relates to immunity. Finally, we provide a list of some gaps in our knowledge of the relationship between begging and immunity.

THE IMMUNE SYSTEM AND WHY IT IS IMPORTANT

The immune system provides the most diverse, specific and effective set of antiparasite defence mechanisms in vertebrates, and birds are no exception (Tizard 1991; Pastoret et al. 1998). The two fundamental components of immunity are 'innate' and 'acquired' (also called 'adaptive') immunity. They are partly integrated, with most molecules and cells having a role in both components (Roitt et al. 1996; Janeway & Travers 1997).

Innate immunity provides a first line of protection from, and clearance of, pathogens by components of the immune system that also play a role in the pathogen-specific acquired immune response. Hosts are isolated by mechanical barriers (skin and epithelia) from the external environment that is the source of pathogenic challenge. The innate immune system consists of the physical and chemical environment at the external surface of the host, immune processes mediated by cells that express non-specific phagocytic ability (e.g. the macrophages, monocytes, heterophils) and inflammatory processes. Innate mechanisms of defence are thus rapid and non-specific means of getting rid of pathogens.

Vertebrates are unique in having evolved acquired immunity, which generally takes longer to activate. The first distinctive feature of acquired immunity is that it develops the ability to recognize non-self elements during maturation, paving the way for adaptive immune response to non-self and 'immunological tolerance' of self antigens. Second, the acquired immune system is responsible for the generation of a tremendous diversity of recognition molecules, which ensure that almost any parasite can be detected *via* recognition of portions of the molecules that it expresses, i.e. its epitopes. Third, acquired immunity allows for a specific response to the pathogen upon recognition of its epitope(s). This is achieved by processing and presentation of the antigen by antigen-presenting cells, in association with molecules expressed by the major histocompatibility complex (MHC) of genes and activation of effector T- and B-cells. Fourth, immunity can be passively transferred to offspring, thus enhancing protection during maturation of their naïve immune system. Last, but not least, acquired immunity allows for long-lasting protection and response upon re-encounter with the same pathogen, based on memory of previously encountered antigens. During subsequent exposure to a given parasite, high affinity for antigenic determinants by immune cells and differentiation of 'memory' cells will allow a fast and strong secondary immune response.

There are two main branches of acquired immune response: cell-mediated and humoral immunity. When naïve T-cells are primed by interaction with an antigen presented in the context of self-MHC molecules at the surface of professional antigen-presenting cells (APCs) and in association with co-stimulatory factors, they proliferate and differentiate into 'armed' T-cells, ready to interact with their target cells presenting the specific antigen to which they are committed. However, T-cells can differentiate into effector cytotoxic or helper T-cells, respectively. In turn, differentiation of helper T-cells into either of the two main subsets of TH1 or TH2 depends on a set of factors including the release of chemical mediators (cytokines), and will determine the prevailing cell-mediated versus humoral nature of the ensuing immune response. Cytotoxic T-cells can kill infected target cells. TH1 cells activate macrophages to eliminate intracellular parasites, but also stimulate production of immunoglobulins of the G isotype (IgG), whereas TH2 have their major function in activating B-cells to produce humoral factors (antibodies) that are involved in immune response against extracellular pathogens. Binding of the antibodies to the pathogen mediates its destruction in a number of ways, including intervention of accessory phagocytic and natural killer cells, and molecules of the complementary system of plasma proteins, which coat the pathogen thus facilitating destruction by phagocytes.

Several experiments, mainly on poultry, have shown both genetic and environmental sources of variation in the ability to raise innate and acquired immune responses (i.e. for high immunocompetence) by young birds. This variation has therefore not been depleted by strong directional selection for increased immunocompetence. In addition, some elements of the immune system, such as the MHC of genes, may be under balancing selection. Hence, additive genetic variation in immunocompetence seems to exist, but parents of altricial species can influence the immunity of their progeny by means of parental strategies (e.g. Dietert et al. 1994; Saino et al. 1997b; Pinard-van der Laan et al. 1998).

Is there any particular reason why we should expect offspring immunity to be a matter of special concern for the parents and the offspring themselves? Immunological defence against parasites can be considered a crucial component of individual fitness, independent of host species and life stage (Wakelin 1996), given the ubiquitous nature of parasitism and its dramatic impact on host condition (Price 1980; Loye & Zuk 1991; Wikel 1996; Clayton & Moore 1997). Furthermore, there are special reasons why immunity should be crucial particularly for young individuals. First, upon hatching birds are faced with an extraordinarily diverse fauna of virulent parasites to which they have not previously been exposed. Hence, young birds have had no previous chance of developing specific autonomous protective immunity, and their immune response will inevitably be a weak and slow 'primary' one. Second, young birds may possess immature immune organs and functions in the early stages of ontogeny. Birds possess two primary lymphoid organs: the bursa of Fabricius, which is the site where B-lymphocytes mature, and the thymus, which is the site of maturation of T-lymphocytes. These organs are still developing by the time of hatch, as are secondary immune organs such as the spleen, and other lymphoid organs and tissues in various regions of the body. Acquired immune functions involve lengthy and energetically costly processes to develop, and also involve contact with parasite-borne antigens, implying that hatchlings do not possess the entire immunological repertoire of antiparasite defence that will be available later in life.

Third, parasites have evolved to match their reproductive cycle to that of their host to exploit host reproduction as a major opportunity of transmission, thus rendering their impact on hosts particularly strong soon after birth (Noble & Noble 1976; Cox 1982). Finally, nestlings do not possess the entire repertoire of behavioural defence mechanisms of adults (Hart 1997), and the nest is a very suitable environment for parasites. Thus, immunity could well be a main determinant of offspring performance, and rapid development of the immune system and maturation of efficient immune functions are major determinants of fitness (Tizard 1991; Apanius 1997; Pastoret et al. 1998).

Development, maintenance and functioning of the immune system may be energetically costly, although little quantified (Klasing & Leshchinsky 1999; Lochmiller & Deerenberg 2000), and resources available for immune processes may be limiting. The costs of immunity may thus result in phenotypic trade-offs between responses to different parasites, and between immune response and other activities (Møller 1997). Surprisingly few studies have investigated phenotypic trade-offs within the immune system. Barn swallow (*Hirundo rustica*) nestlings challenged with sheep red blood cells (SRBC, a multigenic antigen commonly used to assess the ability to raise a T-lymphocyte-dependent humoral response) had weaker cell-mediated immunity, assessed as the response to an intradermal injection of a lectin (phytohemagglutinin, PHA which is mitogenic to T-cells), compared to their control siblings (N. Saino unpublished data). This effect of SRBC suggests that mounting a humoral immune response negatively affected cell-mediated immunocompetence. Trade-offs between humoral and cell-mediated components of the immune response may be caused by resources limiting simultaneous functioning of different branches of immunity. At a mechanistic level, the negative association between humoral and cell-mediated response may be governed by the subset of helper T-cells involved in the immune response. Commitment to a type of helper T-cell (TH2) pathway resulting in a humoral response is associated with production of chemical messengers, particularly interleukin-4, that inhibits cell-mediated response. Conversely, commitment to a TH1 pathway resulting in a cell-mediated response involves the release of other messengers, such as interferon-γ, that inhibit humoral response.

In another experiment, adult female red jungle fowl (*Gallus gallus*) experimentally infected with an intestinal nematode had more eosinophils (granulocytes involved in immune response to metazoan endoparasites), but also a reduced response to PHA (Johnsen & Zuk 1999). In addition, circulating levels of immunoglobulins were negatively correlated with the intensity of the response to PHA in parasitized individuals, but not in uninfected birds. This implies a trade-off between humoral and cellular immunity when parasite infestation depressed host condition.

Trade-offs have also been shown to occur between immunity and somatic growth and developmental stability. Juvenile Japanese quail (*Coturnix japonica*) challenged with SRBC or either of two T-cell-independent antigens (*Mycoplasma synoviae* and Newcastle disease virus) had reduced body mass, feather growth and, in the case of challenge with *M. synoviae*, increased fluctuating asymmetry (Fair et al. 1999). In an experiment on the sand martin (*Riparia riparia*), infestation of nests with ticks reduced reproductive success and caused an immune response by nestlings, which had larger concentrations of immunoglobulins and circulating leukocytes than nestlings in sprayed nests (Szép & Møller 1999). Nestlings from infested nests had faster feather growth but smaller skeletal body size. Hence, parasitism caused a developmental response, translated into a trade-off between immune response and feather growth on the one hand, which allowed for defence against the parasite and earlier fledging, and reduced bone growth on the other. Since this phenotypic response allows nestlings to escape parasites, this excludes the possibility that the response was a parasite-mediated manipulation.

Parental immunity might also be traded off against reproductive effort, with an experimental increase in parental effort resulting in impaired immune response. That was the case for impaired antibody response to different antigens by female collared flycatchers (*Ficedula albicollis,* Nordling et al. 1998), and blue tits (*Parus caeruleus,* Råberg et al. 2000) and compromised T-cell-mediated immune response in the pied flycatcher (*Ficedula hypoleuca*, Moreno et al. 1999) and the barn swallow (Saino et al. in press). On the other hand, induction of a humoral immune response by injection of a non-pathogenic antigen and hence the production of a primary immune response had no effect on reproductive performance in the starling (*Sturnus vulgaris,* Williams et al. 1999), while it depressed reproductive success in the pied flycatcher (Ilmonen et al. 2000).

Development, maintenance and functioning of the immune system are contingent upon nutritional condition (Chandra & Newberne 1977; Glick et al. 1981; Gershwin et al. 1985; Klasing 1988; Lochmiller et al. 1993). For example, nestling barn swallows fed more frequently by their parents had larger cellular immune responses than control nestlings (Saino et al. 1997b). Quality rather than quantity of food is particularly important, for example, in terms of content of specific amino acids or proteins (e.g. Lochmiller et al. 1993; Saino et al. 1997b) or micronutrients such as carotenoids (Bendich 1989; Chew 1993; Møller et al. 2000).

Immunity can also be compromised by extrinsic factors acting as sources of stress since the adrenocortical response frequently leads to immune suppression (reviews in Sapolsky 1992; Apanius 1998; Wingfield et al. 1998). Low temperatures in conjunction with poor parental care, for example, may act as a stressor and suppress components of immunity (e.g. Dietert et al. 1994; Dabbert et al. 1997). Hence, the extensive dependence of immunity on environmental conditions (see Dietert et al. 1994) suggests that there is ample scope for parent birds to influence the immunity of their offspring through parental care. However, parental effort and thus offspring immunity may be traded off against parental condition and immunity.

Immunity may therefore mediate parent-offspring conflict, since parents and offspring may have to rely on the same nutritional and energy resources limiting their immunity. In addition, females may have to directly trade off the amount of protective immune factors delivered to their eggs against their own immunity.

EFFECTS OF IMMUNITY ON VIABILITY

Immunocompetence, i.e. the ability to raise an efficient immune response, is assumed to increase viability. There are, however, no field studies analysing the survival consequences of experimentally altered immunity of birds. A few correlational studies in wild birds show a positive relationship between immunocompetence and the survival of young (Christe et al. 1998, 2001) and adult birds (Saino et al. 1997a; Gonzalez et al. 1999; Soler et al. 1999).

Immunocompetence of young birds correlates positively with phenotypic traits that commonly predict survival, such as body mass or condition of nestlings or fledglings (Saino et al. 1997b; Christe et al. 1998). In addition, there is ample evidence from the veterinary literature that greater immunocompetence enhances viability (Tizard 1991; Pastoret et al. 1998). High immune responsiveness, however, does not necessarily imply high immunocompetence. An excessively intense immune response, like that observed in hypersensitivity reactions, can be harmful to the host, having a negative impact on its fitness. Hence, a distinction should be drawn between immunocompetence, which has by definition a positive effect on host fitness, and immune responsiveness, which if it is too large might entail immunopathological costs for the host.

Response to artificial selection demonstrates large amounts of additive genetic variation even in economically important strains of poultry, which have presumably been subject to intense artificial selection for resistance to pathogens (review in Pinard-van der Laan et al. 1998). Maintenance of genetic variation in immunity may rely on evolutionary trade-offs between different components of immunity, between immunity and other life history traits, coevolutionary interactions between hosts and parasites or be caused by balancing selection on MHC genes.

Negatively correlated responses to artificial selection for single or multiple immune traits have been shown in poultry, suggesting negative genetic correlations between different components of immunity (see Pinard-van der Laan et al. 1998). Artificial selection on immune traits can also produce a correlated response in life history traits. For example, selection for enhanced response to SRBC in poultry has reduced their growth rate (Pinard-van der Laan et al. 1998). A negatively correlated response of growth rate to selection on immune traits is relevant for the study of begging displays, sib-sib competition and the evolution of parental strategies. Indeed, selection for high immunocompetence may result in low growth rates and thus reduced ability to prevail in sib-sib competition for parental care. However, balancing selection on immunocompetence may vary in intensity among offspring in a brood owing to size hierarchies, for example, arising from hatching asynchrony.

BEGGING DISPLAYS, NEED AND IMMUNITY

Begging displays consist of a diverse array of visual and acoustic signals by young birds directed to their parents (or close relatives), or hosts (in the case of brood parasitism), predominantly during food provisioning (e.g. Evans 1992; Kilner & Johnstone 1997).

Begging displays may evolve as manipulative signals used by offspring to obtain more care than is optimal for their parents. However, evolutionary theory concerning parental investment predicts that natural selection has favoured the ability of parents to assess offspring quality, since fitness rewards may vary depending on allocation of the same amount of resources to offspring differing in quality. For example, body size may directly reveal offspring quality. Alternatively, parents may infer offspring quality indirectly through condition-dependent signals, such as begging. Nestling barn swallows in broods that had been enlarged produced begging calls at a higher pitch than nestlings from broods reduced in size, thus potentially signalling their relatively small body mass (Sacchi et al. in press). Pitch of begging call can reliably signal body mass, since there is a negative relationship between the frequency of the sound and the size of the source (e.g. Ryan & Brenowitz 1985). Parents are predicted to use only signals that reliably reflect condition, and reliability of begging signals could be enforced by their costs (e.g. energy consumption or increased risk of predation; Leech & Leonard 1996; Briskie et al. 1999; M.A. Chappell & G.C. Bachman this volume; D. Haskell this volume). This should lead to a differentially larger benefit, in terms of parental care, of a given level of begging relative to that of more needy offspring (Godfray 1995a,b; reviews in Kilner & Johnstone 1997; Mock & Parker 1997; R.A. Johnstone & H.C.J. Godfray this volume).

Variation in intensity of begging displays has been assumed to reflect variation in offspring 'need', with 'needier' offspring begging more. Need, however, has been assigned different meanings. Theoreticians use need as a measure of the marginal benefit that offspring accrue from obtaining additional resources (e.g. Godfray 1995b). On the other hand, empirical studies typically equate need to hunger or food deprivation (see Kilner & Johnstone 1997).

It can be argued, for example, that it is not only offspring hunger that should be a matter of concern. Parental fitness gains depend on an array of factors, and decisions that maximize fitness will be influenced by the effects of need and other aspects of the state of offspring. The effect on parental fitness of allocating more care to a food-deprived offspring may even change in sign depending upon components of offspring fitness and current ecological conditions (e.g. food abundance). Therefore, we use need to indicate the amount of food required to achieve satiation, thus distinguishing this aspect of the state of offspring from other aspects (broadly defined as condition) that will affect their fitness. Offspring in prime condition will have a physiological or immunological state that maximizes fitness under a given set of ecological conditions (Saino et al. 2000a). Need, however, may not strictly co-vary with condition. Two equally satiated nestlings may differ in condition if they differ in the level of infection by a parasite. Alternatively, previously established phenotypic variation among siblings, sexual dimorphism, variation in metabolic efficiency or age may produce different needs, but no difference in condition. Need and condition may vary on very different time scales, since satiation may be a matter of minutes, whereas factors affecting condition such as infection will vary over days or weeks (Saino et al. 2000a). Several other problems make the relationship between need and condition complicated. The amount of food requested for optimal performance may differ when different activities are considered. For example, maximal somatic growth in poultry is attained at lower levels of food intake compared to those that ensure maximal immune response (Dietert et al. 1994). Crucial aspects of condition may be affected by micronutrients resulting in well-fed nestlings suffering from defects in immune function and thus being in poor condition. Finally, the expression of signals of need can be constrained by condition, because a nestling in poor condition cannot afford to beg at high intensity, so that at the two extremes of the range of nestling condition, the expression of a costly begging signal may be similar.

We hypothesize that begging consists of different condition-dependent signals that reveal both current need and offspring condition, consistent with what has been proposed for multiple secondary sexual characters (Møller & Pomiankowski 1993). Since immunity is likely to be a major component of condition, we hypothesize that begging also has evolved under selection by parents for condition-dependent signals revealing this crucial component of offspring reproductive value.

MATERNAL EFFECTS ON EGG QUALITY AND THEIR CONSEQUENCES

Maternal effects may affect offspring immunity very early in life (Mousseau & Fox 1998). Some maternal effects have already been subjected to experimentation, whereas others remain a possibility in need of investigation.

Eggs contain maternal hormones, notably androgens (e.g. testosterone; Schwabl 1996, 1997; Gil et al. 1999; H. Schwabl & J. Lipar this volume).

Testosterone, for instance, can modify the expression of begging behaviour. Schwabl (1996) has shown that nestling canaries (*Serinus canaria*) hatched from eggs inoculated with testosterone produce hatchlings that beg more than control siblings. Interestingly, testosterone concentration naturally increases with the laying order of eggs in this species, suggesting that females deliver more testosterone to nestlings that are likely to hatch last and thus be disadvantaged in competition for food with their nestmates (Schwabl 1996). This has potential consequences for immunity of individual offspring since immunity is partly determined by success in acquisition of critical food resources, which in turn depends on the expression of begging displays. In addition, testosterone enhances nestling growth (Schwabl 1996). Nestlings often fight to occupy the position in the nest that ensures easy access to food (e.g. Kacelnik et al. 1995; Kilner 1995). Since the ability of young birds to successfully compete for parental care is also a matter of body size, larger offspring usually outcompete their smaller siblings (e.g. Price & Ydenberg 1995). Differential allocation of testosterone to eggs will ultimately result in variation in immunity within broods as a consequence of differential access to food resources critical to immune function.

Androgens have also been repeatedly assigned a negative effect on immunity (e.g. Grossman 1985), although complexity of the interrelationships between the endocrine and immune systems makes it difficult to reach firm and general conclusions (Hillgarth & Wingfield 1997). Under the largely untested assumption that androgens negatively affect the development and functioning of the immune system in very early life stages, offspring may experience immunological costs of maternal testosterone. Hence, maternal testosterone has the potential to affect offspring immunity, and these effects may be mediated by begging, competitive interactions among nestmates and physiological effects of androgens.

Micronutrients are also passed from the mother into the egg. The yellow to bright orange colour of the yolk is determined by carotenoids, which have major roles in immune system regulation and function (Blount et al. 2000; Møller et al. 2000). Thus, differential allocation of carotenoids among the eggs of a clutch may cause variation in immune function among offspring, although the role played by maternal carotenoids in the early life of the offspring has yet to be elucidated.

A potentially crucial type of maternal effect is passive transfer of immunity to eggs. Birds deliver an array of protective immune factors, including immunoglobulins (Ig) of the main isotypes (IgG, IgA, and IgM), to the egg (Rose & Orlans 1981; Naqui et al. 1983; Tizard 1991; Graczyk et al. 1994; Smith et al. 1994; Pastoret et al. 1998). For example, immunization of hens against the virus causing bursal disease elicits a humoral response, and part of the antibodies against the virus are transferred to the offspring which can therefore be protected against the viral infection (Ahmad & Siddique 1998; see also Fadly & Smith 1991; Homer et al. 1992). Similarly, nestling barn swallows received antibodies from their mother against the Newcastle disease virus in proportion to the concentration in the mothers' plasma after experimental immunization (N. Saino, P. Dell'Ara, M. Incagli, R. Ambrosini, R. Martinelli & A.P. Møller, unpublished data).

Transfer of immunity *via* the egg can be considered a long-lasting maternal effect that influences the general state of offspring during ontogeny. For example, plasma concentration of antibodies in female barn swallows peaks around the day that egg laying begins, whereas no such temporal pattern is observed in males (Saino et al. 2001a). Thus, variation in immunoglobulin concentration is functionally linked to egg laying, and by raising their levels of circulating antibodies just prior to the onset of laying females do not compromise their own immunity. One possible functional interpretation of this finding is that females raise their circulating levels of antibodies to provide an adequate amount of protective immune factors to their offspring *via* the egg without reducing their own level of defence. Alternatively, just prior to laying females may be more susceptible to infection, or exposed to parasites owing to more frequent contact with food items and the nest. In another study, concentration of circulating immunoglobulins, and T-cell-mediated immunocompetence of nestling barn swallows increased with hatch order (Figure 1). In the barn swallow, hatching is relatively synchronous, with most clutches hatching within 24 hours. A size hierarchy is established among siblings with late-hatched nestlings being consistently smaller throughout the nestling period, although not differing in skeletal body size (Saino et al. 2001b). However, late-hatched nestlings beg for food more frequently than early-hatched ones (Saino et al. 2001b), begging rate reflects hunger and parents feed begging nestlings more (Saino et al. 2000a).

Hence, a high begging rate is associated with relatively late hatching, small body mass and size, but also large cellular immunity and a high concentration of circulating antibodies. This suggests that late-hatched nestlings allocate a larger share of parental resources to immune function compared with early-hatched ones, thus sacrificing somatic growth. Other proximate mechanisms may generate the positive association between hatching order and immune variables. In the barn swallow, lay order and hatch order of eggs are strongly positively correlated (Saino et al. 2001b). Mothers may allocate more immunoglobulins and other factors crucial for the development of immune function (e.g. carotenoids) to the last eggs in order to enhance immunity of their late-hatching offspring that otherwise may be disadvantaged in the competition for food with their nestmates. This interpretation is not particularly likely because nestlings hatched from the first eggs laid in a clutch received more anti-NDV antibodies, suggesting that mothers allocate more antibodies to the first rather than the last eggs in their clutch (N. Saino, P. Dell'Ara, M. Incagli, R. Ambrosini, R. Martinelli & A.P. Møller unpublished data).

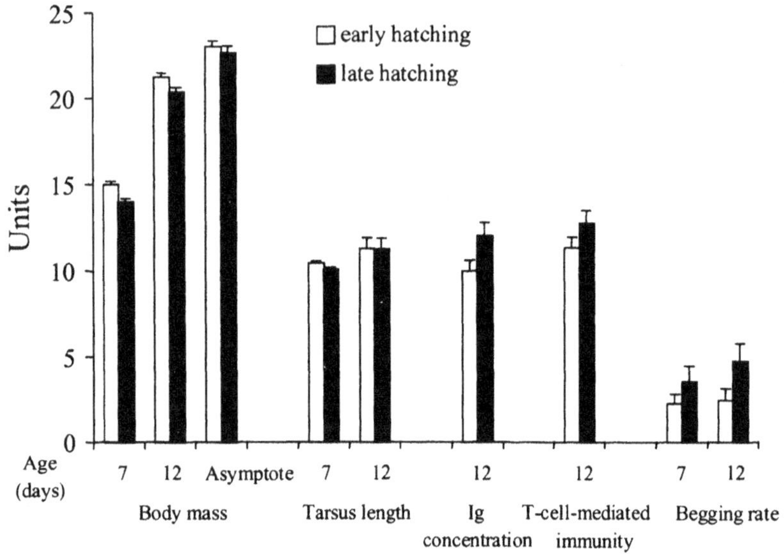

Figure 1. Mean (± SE) body mass (g), tarsus length (mm), index of T-cell-mediated immunity, immunoglobulin concentration (mg/ml) and begging rate (events/minute) in relation to hatch order of nestling barn swallows measured at different ages (days from hatching; from Saino et al. 2001b). Asymptotic body mass has been estimated for the two groups of nestlings separately by fitting a logistic curve to data recorded at age 1, 4, 7 and 12 days. Within each brood, late-hatched nestlings have smaller body mass and beg more frequently for food, but have larger concentrations of circulating immunoglobulins and T-cell-mediated immunocompetence, than their siblings which hatched early. However, hatch order may strictly reflect lay order of the eggs. Hence, part of the variation in nestling phenotype in relation to hatch order may be due to early maternal effects mediated *via* the egg.

GAPE COLOUR

Conspicuously bright gape colours, ranging from yellow to red are ubiquitous in nestlings of almost all passerines and other nidicolous birds (Ficken 1965; Kilner & Davies 1998; Kilner 1999). Mouth colour of otherwise dully coloured nestlings functions to solicit parental care as shown in experimental studies of great tits (*Parus major*), canaries and barn swallows (Götmark & Ahlström 1997; Kilner 1997; Saino et al. 2000b). Parents may also have selected for the evolution of other exaggerated characters such as orange ornamental plumes on the head of chicks of the American coot (*Fulica americana*, Lyon et al. 1994).

What is the adaptive function of such parental preference for feeding offspring with brightly coloured mouths? The answer could be none, if offspring were simply exploiting a bias in the sensory system of parents toward a particular colour (Lyon et al. 1994). A second possibility is that a brightly coloured mouth enhances detectability of the offspring, and particularly so in dark nest sites. This early suggestion by Ficken (1965) was only partly confirmed in a comparative study of 31 European species, mostly passerines, where darkness of the nest was found to be positively associated with relative size and brightness of the flange, rather than colour of the mouth *per se* (Kilner & Davies 1998).

Alternatively, condition-dependent expression of mouth colour could play a role in the resolution of parent-offspring conflict, if it signals inherent quality of offspring such as their need for food or other components of general state. For example, begging displays may have evolved as honest signals of need (Parker & Mock 1987; Parker et al. 1989; Godfray 1991, 1995a,b; Kilner & Johnstone 1997) on which parents base their adaptive parental decisions.

The aspect of offspring genetic or phenotypic quality being reflected by mouth coloration may, however, depend on the mechanism responsible for gape coloration. Contingent need for food can result in gapes suddenly becoming redder as the nestling starts begging for food, although this has only been observed in seed-regurgitating finches (Kilner 1997; Kilner & Davies 1998). If mouth colour is mainly determined by blood flow, a redder mouth may reliably indicate that relatively little blood is being diverted to the gut for digestion and, thus, that the nestling is needy (i.e. it is not satiated; Kilner 1997). Similarly, mouth colour could signal satiation if any anatomical or physiological mechanism exists that reduces blood influx to the mouth when the gut is full (Kilner 1997).

Mouth colour may also be determined by natural pigments such as carotenoids. These pigments are acquired by animals through the diet and circulate in the blood and/or are stored in other tissues such as the visible parts of the mouth. Carotenoids colour most ornamental feathers in birds (Møller et al. 2000), but also cause yellow to red coloration of skin, eyes, bill and the oral cavity (Ficken 1965; Fox 1979; Goodwin 1984; Hill 1992; Stradi 1998). For example, nestling barn swallows had more brightly coloured gapes when fed additional lutein (the main carotenoid in barn swallows; Saino et al. 1999a), compared with control nestlings (Saino et al. 2000b).

Which component of offspring state might be reflected by a carotenoid-based begging signal? Carotenoids are not just biochromes. They play a multitude of roles in stimulation and modulation of immune function, and in detoxification processes, including scavenging of free oxygen radicals, which are produced by the immune system as a powerful weapon to destroy parasites. Hence, carotenoids affect the immune system and protect host molecules and cells from oxidative stress, either generated by host metabolism or by extrinsic factors (see Ames 1983; Bendich 1989; Chew 1993; Kim et al. 2000a,b; Møller et al. 2000).

Parasitism can depress carotenoid-based coloration in birds and other vertebrates (Møller et al. 2000). There is ample evidence from correlational studies that infected birds have duller coloration due to carotenoids, although exceptions exist (Møller et al. 2000). However, the few manipulative studies available indicate that infection with pathogens (e.g. *Coccidia* in poultry; Bletner et al. 1966; Allen 1997), or challenge with antigens eliciting an immune response but not simulating the virulent consequences of real parasites, can depress carotenoid-based coloration. For example, nestling barn swallows injected with SRBC had duller gapes compared with their control siblings (Saino et al. 2000b; Figure 2). Apparently, removal of lutein from mouth tissues of SRBC-injected nestlings was partly responsible for depressed gape coloration, since artificial provisioning of lutein to SRBC-injected nestlings reversibly restored gape coloration (Saino et al. 2000b; Figure 2). Barn swallow parents preferred to feed nestlings with an experimentally reddened mouth colour, suggesting that they allocated more food to offspring showing no sign of infection (Saino et al. 2000b). This is the only study showing a direct link between immune status and expression of a begging signal.

What prevents nestlings injected with SRBC from displaying brighter gapes, thereby obtaining a larger share of parental care? Similarly, what mechanisms prevent infected nestlings from signalling at the same level as healthy siblings, thus concealing their poor state? Dietary carotenoids may be available in limiting amounts, and those allocated to mouth coloration may be unavailable to immune function and detoxification. Young birds may thus trade off the expression of begging signals against immunity. Signalling theory predicts that reliability can be enforced in a condition-dependent signal such as mouth colour when the cost is larger for infected than uninfected individuals. This prevents infected nestlings from signalling at the same level as their healthy nestmates (Heywood 1989; Grafen 1990). Alternatively, nestling signals may be revealing handicaps (*sensu* Hamilton & Zuk 1982) that directly reveal the condition of offspring. Limitation of dietary carotenoids is crucial if immune status is signalled by gape colour. Carotenoids may generally be limiting micronutrients (Saino et al. 2000b). A few experimental supplementation studies suggest that carotenoids are indeed limiting (Møller et al. 2000). The only evidence for dietary carotenoids limiting gape coloration comes from a study showing that supplementation of the diet with carotenoids produced a brighter gape in barn swallow nestlings compared with that of their unprovisioned siblings (Saino et al. 2000b). Thus, carotenoids available in the natural diet of barn swallows do not maximize expression of gape colour. Rapid turnover of carotenoids during severe infection (Allen 1997) may exacerbate the trade-off between expression of carotenoid-based coloration and immunity (Møller et al. 2000; but see Hill 1999).

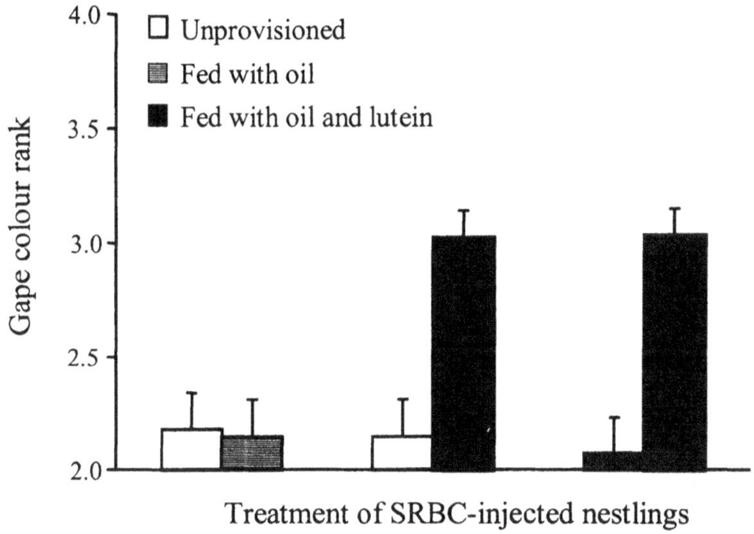

Figure 2. Mean (± SE) gape coloration of nestling barn swallows injected with SRBC and provisioned with a carotenoid (lutein), fed with the solvent, or unprovisioned (from Saino et al. 2000b). Nestlings provisioned with lutein ranked higher in gape coloration compared with their siblings in both control groups.

We conclude that brightly coloured gapes, and other colourful signals of young birds, may have evolved under selection for reliable signalling of a diverse array of components of offspring general state, such as food deprivation or infection and immune status. The mechanisms suggested to enforce reliability of these condition-dependent signals remain speculative.

PARENTAL RESPONSE TO OFFSPRING BEGGING: THE ROLE OF IMMUNITY

In the first section we showed how parents incur an immunocompetence cost through attending offspring. Whether this cost is mediated by parental responses to begging, and whether there are fitness consequences of such impaired parental immunity, remain to be elucidated. If parents experience an immunological cost, a conflict of interest may occur within families. From the perspective of the parent, the resolution of this conflict may consist of a trade-off between their own immunocompetence and that of their progeny.

This intergenerational trade-off has only been investigated once. Both male and female barn swallow parents that raised offspring capable of a strong cell-mediated immune response had smaller chances of surviving to the next breeding season than parents with less responsive offspring (Saino et al. 1999b). This suggests that high immunological performance of offspring imposes a viability cost upon parents (Saino et al. 1999b). In addition, males but not females were less likely to survive when their brood was enlarged. Offspring from enlarged broods had lower T-cell-mediated immunocompetence compared with siblings which had been transferred to reduced broods (Saino et al. 1997b). Furthermore, male parents with a reduced brood had greater T-cell-mediated immunocompetence than males with control or enlarged broods (Saino et al. in press). Since (1) nestlings in reduced broods beg less for food than nestlings in enlarged broods (Saino et al. 2000a); (2) parents feed enlarged broods more often than reduced broods (although feeding rate per nestling is smaller than that in reduced broods; Saino et al. 1997b); and (3) a larger *per capita* feeding rate results in enhanced immunocompetence of offspring (Saino et al. 1997b), enhanced begging increases immunocompetence of nestlings, but simultaneously compromises parental immunity. Hence, parental effort has negative effects upon immunocompetence of parents, while enhancing that of offspring.

Parental care should be allocated depending on offspring need, but also upon other components of offspring state reflecting their reproductive value. In addition, parental decisions may be affected by current ecological conditions, since these will affect the fitness rewards of a particular parental strategy. We recently manipulated brood size in barn swallows to create differences in food availability, and experimentally altered two components of nestling state: immune condition and hunger (Saino et al. 1997b; Saino et al. 2000a). The immune system of half the brood was challenged by injection with SRBC, while the other nestlings were sham-inoculated.

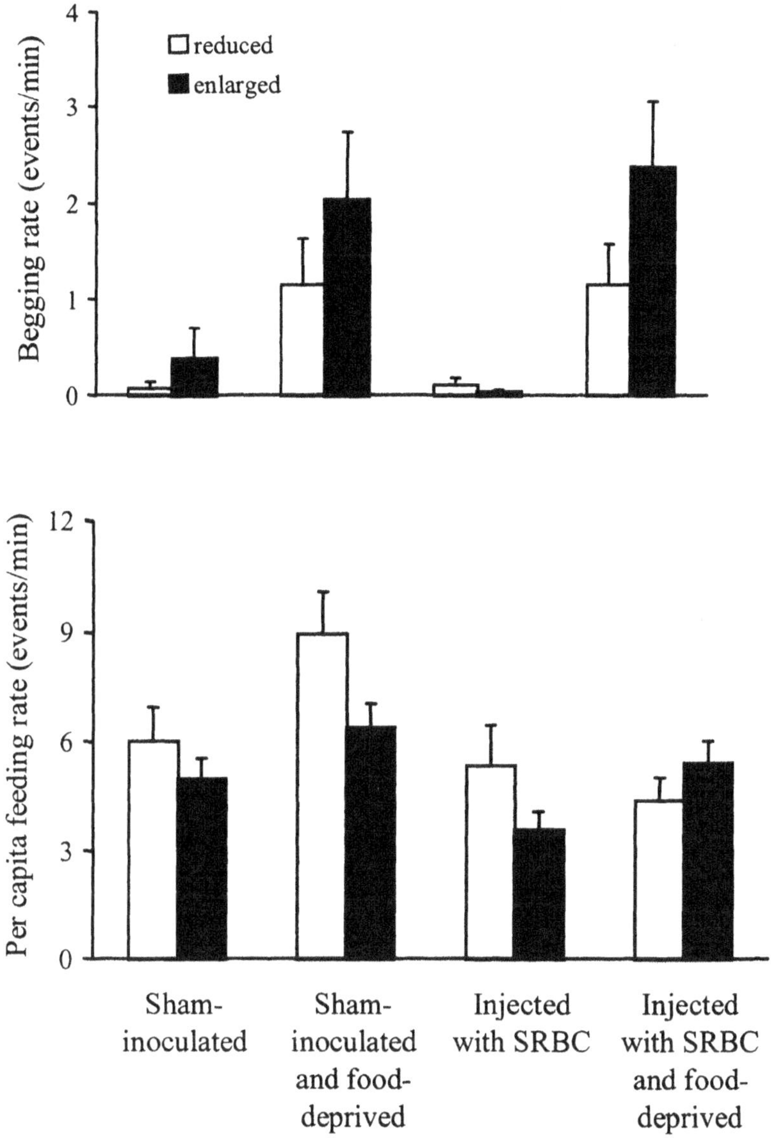

Figure 3. Mean (± SE) begging and feeding rates in relation to food deprivation and SRBC injection in enlarged and reduced broods of the barn swallow. Food deprivation, but not SRBC injection, had a significant effect on begging rate in both reduced and enlarged broods. Parents fed individual nestlings more frequently in reduced compared with enlarged broods. In reduced broods, parents fed relatively more food-deprived nestlings, but only if they had not been injected with SRBC, whereas no effect of food deprivation or SRBC injection existed in enlarged broods (from Saino et al. 2000a).

Approximately one week later, a subset of the SRBC-injected and control nestlings were randomly assigned to a food deprivation procedure, while their nestmates were allowed to continuously receive food, thus generating a full factorial design. Frequency of begging was not significantly affected by injection of the antigen, while it was enhanced by food deprivation and brood size enlargement. Adults responded to food deprivation by delivering more feedings to hungry nestlings, but only when these had not been injected with SRBC and belonged to a brood reduced in size (Figure 3). These results demonstrate that parent barn swallows base their decisions on the frequency of begging episodes, but also on other cues of offspring condition. Presumably these include gape colour, which is reduced in brightness by inoculation with SRBC (Saino et al. 2000a). Hence, parents apparently respond to different components of offspring state (e.g. hunger and immune condition), which in turn may be revealed by different condition-dependent signals.

FUTURE DIRECTIONS

In the preceding pages we have reviewed the literature potentially relating begging to immune function. This literature is still very much in its infancy, and we would like to conclude by emphasizing areas of research in need of further investigation. First, we lack information on whether begging is determined by offspring need or other components of general condition relevant to viability and thus parental decisions. What is the relative importance of the different aspects of offspring state? Second, why are begging signals so diverse? Do different components of begging reflect different aspects of condition (see R.M. Kilner this volume)? Third, from the perspective of the offspring, it would be interesting to know how begging translates into condition and ultimately immune function. Does begging result in acquisition of resources critical to offspring immunity? Are intergenerational conflicts of interest for access to resources limiting immune function mediated by begging displays? Fourth, how does level of begging translate into viability? Finally, what limits begging signals? Are the costs incurred in terms of predation, immunosuppression or another currency?

We have shown in a series of experiments that many of these questions indeed are tractable. What is most needed at the moment is information on the diversity and the generality of the relationships between begging and immune function. Future studies will allow us to determine to what extent offspring begging behaviour is affected by coevolutionary interactions between parasites and their hosts.

ACKNOWLEDGEMENTS

We thank Jon Wright for inviting us to contribute this chapter and Marty Leonard, Jon Wright and an anonymous referee for constructive comments. We are grateful to the many people who helped with our fieldwork. Our studies have been funded by the Italian Ministero per l'Università e la Ricerca Scientifica e Tecnologica and Consiglio Nazionale delle Ricerche.

REFERENCES

Ahmad, M. & Siddique, M. 1998. Kinetics of antibodies in serum, egg yolk and day-old chicks against infectious bursal disease in chicken broiler breeders. *Pakistan Veterinary Journal* **18**, 187-191.

Allen, P.C. 1997. Production of radical species during infections in chickens. *Poultry Science* **76**, 814-821.

Ames, B.N. 1983. Dietary carcinogens and anticarcinogens: oxygen radicals and degenerative diseases. *Science* **221**, 1256-1264.

Apanius, V. 1997. The immune system. In: *Avian Growth and Development: Evolution Within the Altricial-Precocial Spectrum* (Ed. by J.M. Starck & R.E. Ricklefs). Oxford: Oxford University Press.

Apanius, V. 1998. Stress and immune response. In: *Stress and Behavior* (Ed. by A.P. Møller, M. Milinski & P.J.B. Slater). San Diego: Academic Press.

Bendich, A. 1989. Carotenoids and the immune response. *Journal of Nutrition* **119**, 112-115.

Bletner, J.K., Mitchell, R.P. & Tugwell, R.L. 1966. The effect of *Eimeria maxima* on broiler pigmentation. *Poultry Science* **45**, 689-694.

Blount, J.D., Møller, A.P. & Houston, D.C. 2000. Why egg yolk is yellow. *Trends in Ecology and Evolution* **15**, 47-49.

Briskie, J.V., Martin, P.R. & Martin, T.E. 1999. Nest predation and the evolution of nestling begging calls. *Proceedings of the Royal Society of London, Series B* **266**, 2153-2159.

Chandra, R.K. & Newberne, P.M. 1977. *Nutrition, Immunity, and Infection*. New York: Plenum Press.

Chew, B.P. 1993. Role of carotenoids in the immune response. *Journal of Dairy Science* **76**, 2804-2811.

Christe, P., Møller, A.P. & de Lope, F. 1998. Immunocompetence and nestling survival in the house martin: the tasty chick hypothesis. *Oikos* **83**, 175-179.

Christe, P., de Lope, F., González, G., Saino, N. & Møller, A.P. 2001. The influence of weather conditions and ectoparasites on immune responses, morphology and recapture probability of nestling house martins (*Delichon urbica*). *Oecologia* **126**, 333-338.

Clayton, D.H. & Moore, J. 1997. *Host-Parasite Evolution: General Principles and Avian Models*. Oxford: Oxford University Press.

Cox, F.E.G. 1982. *Modern Parasitology*. Oxford: Blackwell Scientific Publications.

Dabbert, C.B., Lochmiller, R.L. & Teeter, R.G. 1997. Effects of acute thermal stress on the immune system of the Northern Bobwhite (*Colinus virginianus*). *The Auk* **114**, 103-109.

Dietert, R.R., Golemboski, K.A. & Austic, R.E. 1994. Environment-immune interactions. *Poultry Science* **73**, 1062-1076.

Evans, R.M. 1992. Embryonic and neonatal vocal elicitation of parental brooding and feeding responses in American white pelicans. *Animal Behaviour* **44**, 667-675.

Fadly, A.M. & Smith, E.J. 1991. Influence of maternal antibody on avian leukosis virus infection in white leghorn chickens harboring endogenous virus 21. *Avian Disease* **35**, 443-451.

Fair, J.M., Hansen, E.S. & Ricklefs, R.E. 1999. Growth, developmental stability and immune response in juvenile Japanese quails (*Coturnix coturnix japonica*). *Proceedings of the Royal Society of London, Series B* **266**, 1735-1742.

Ficken, M.S. 1965. Mouth colour of nestling passerines and its use in taxonomy. *Wilson Bulletin* **77**, 71-75.

Fox, D.L. 1979. *Biochromy*. Berkeley: University of California Press.

Gershwin, M.E., Beach, R.S. & Hurley, L.S. 1985. *Nutrition and Immunity*. Orlando: Academic Press.

Gil, D., Graves, J., Hazon, N. & Wells, A. 1999. Male attractiveness and differential testosterone investment in zebra finch eggs. *Science* **286**, 126-128.

Glick, B., Day, E.J. & Thompson, D. 1981. Calorie-protein deficiencies and the immune response of the chicken. I. Humoral immunity. *Poultry Science* **60**, 2494-2500.

Godfray, H.C.J. 1991. Signalling of need by offspring to their parents. *Nature* **352**, 328-330.

Godfray, H.C.J. 1995a. Evolutionary theory of parent-offspring conflict. *Nature* **376**, 133-138.

Godfray, H.C.J. 1995b. Signaling of need between parents and young: parent-offspring conflict and sibling rivalry. *American Naturalist* **146**, 1-24.

Gonzalez, G., Sorci, G., Møller, A.P., Ninni, P., Haussy, C. & de Lope, F. 1999. Immunocompetence and condition-dependent sexual advertisement in male house sparrows (*Passer domesticus*). *Journal of Animal Ecology* **68**, 1225-1234.

Goodwin, T.W. 1984. *The Biochemistry of the Carotenoids. Vol.II. Animals*. London: Chapman and Hall.

Götmark, F. & Ahlström, M. 1997. Parental preference for red mouth of chicks in a songbird. *Proceedings of the Royal Society of London, Series B* **264**, 959-962.

Graczyk, T.K., Cranfield, M.R., Shaw, M.L. & Craig, L.E. 1994. Maternal antibodies against *Plasmodium* spp. in African black-footed penguin chicks. *Journal of Wildlife Disease* **30**, 365-371.

Grafen, A. 1990. Biological signals as handicaps. *Journal of Theoretical Biology* **144**, 517-546.

Grossman, C.J. 1985. Interactions between the gonadal steroids and the immune system. *Science* **227**, 257-261.

Hamilton, W.D. & Zuk, M. 1982. Heritable true fitness and bright birds: a role for parasites? *Science* **218**, 384-387.

Hart, B.J. 1997. Behavioural defence. In: *Host-Parasite Evolution: General Principles and Avian Models* (Ed. by D.H. Clayton & J. Moore). Oxford: Oxford University Press.

Heywood, J.S. 1989. Sexual selection by the handicap mechanism. *Evolution* **43**, 1387-1397.

Hill, G.E. 1992. The proximate basis of variation in carotenoid pigmentation in male house finches. *The Auk* **109**, 1-12.

Hill, G.E. 1999. Is there an immunological cost to carotenoid-based ornamental coloration? *American Naturalist* **154**, 589-595.

Hillgarth, N. & Wingfield, J.C. 1997. Parasite-mediated sexual selection: endocrine aspects. In: *Host-Parasite Evolution: General Principles and Avian Models* (Ed. by D.H. Clayton & J. Moore). Oxford: Oxford University Press.

Homer, B.L., Butcher, G.D., Miles, R.D. & Rossi, A.F. 1992. Subclinical infectious bursal disease in an integrated broiler production operation. *Journal of Veterinary Diagnosis & Investigation* **4**, 406-411.

Ilmonen, P., Täärna, T. & Hasselquist, D. 2000. Experimentally activated immune defence in female pied flycatchers results in reduced breeding success. *Proceedings of the Royal Society of London, Series B* **267**, 665-670.

Janeway, C.A. & Travers, P. 1997. *Immunobiology*. London: Current Biology Ltd.

Johnsen, T. & Zuk, M. 1999. Parasites and tradeoffs in the immune response of female red jungle fowl. *Oikos* **86**, 487-492.

Kacelnik, A., Cotton, P.A., Stirling, L. & Wright, J. 1995. Food allocation among nestling starlings: sibling competition and the scope of parental choice. *Philosophical Transactions of the Royal Society of London, Series B* **259**, 259-263.

Kilner, R. 1995. When do canary parents respond to nestling signals of need? *Proceedings of the Royal Society of London, Series B* **260**, 343-348.

Kilner, R. 1997. Mouth colour is a reliable signal of need in begging canary nestlings. *Proceedings of the Royal Society of London, Series B* **264**, 963-968.

Kilner, R. 1999. Family conflicts and the evolution of nestling mouth colour. *Behaviour* **136**, 779-804.

Kilner, R. & Davies, N.B. 1998. Nestling mouth colour: ecological correlates of a begging signal. *Animal Behaviour* **56**, 705-712.

Kilner, R. & Johnstone, R.A. 1997. Begging the question: are offspring solicitation behaviours signals of need? *Trends in Ecology and Evolution* **12**, 11-15.

Kim, H.W., Chew, B.P., Wong, T.S., Park, J.S., Weng, B.B., Byrne, K.M., Hayek, M.G. & Reinhart, G.A. 2000a. Dietary lutein stimulates immune response in the canine. *Veterinary Immunology and Immunopathology* **74**, 315-327.

Kim, H.W., Chew, B.P., Wong, T.S., Park, J.S., Weng, B.B., Byrne, K.M., Hayek, M.G. & Reinhart, G.A. 2000b. Modulation of humoral and cell-mediated immune responses by dietary lutein in cats. *Veterinary Immunology and Immunopathology* **73**, 331-341.

Klasing, K.C. 1988. Influence of acute feed deprivation or excess feed intake on immunocompetence of broiler chicks. *Poultry Science* **67**, 626-634.

Klasing, K.C. & Leshchinsky, T.V. 1999. Functions, costs, and benefits of the immune system during development. In: *Proceedings of the 22nd International Ornithological Congress, Durban* (Ed. by N.J. Adams & R.H. Slotow). Johannesburg: BirdLife South Africa.

Leech, S.M. & Leonard, M.L. 1996 . Is there an energetic cost to begging in nestling tree swallows (*Tachycineta bicolor*)? *Proceedings of the Royal Society of London, Series B* **263**, 983-987.

Lochmiller, R.L. & Deerenberg, C. 2000. Trade-offs in evolutionary immunology: just what is the cost of immunity? *Oikos* **88**, 87-99.

Lochmiller, R.L., Vestey, M.R. & Boren, J.C. 1993. Relationship between protein nutritional status and immunocompetence in Northern Bobwhite chicks. *The Auk* **110**, 503-510.

Loye, J.E. & Zuk, M. 1991. *Bird-Parasite Interactions*. Oxford: Oxford University Press.

Lyon, B.E., Eadie, J.M. & Hamilton, L.D. 1994. Parental choice selects for ornamental plumage in American coot chicks. *Nature* **371**, 240-243.

Mock, D.W. & Parker, G.A. 1997. *The Evolution of Sibling Rivalry*. Oxford: Oxford University Press.

Møller, A.P. 1997. Parasitism and the evolution of host life history. In: *Host-Parasite Evolution: General Principles and Avian Models* (Ed. by D.H. Clayton & J. Moore). Oxford: Oxford University Press.

Møller, A.P. & Pomiankowski, A. 1993. Why have birds got multiple sexual ornaments? *Behavioral Ecology and Sociobiology* **32**, 167-176.

Møller, A.P., Biard, C., Blount, J.D., Houston, D.C., Ninni, P., Saino, N. & Surai, P.F. 2000. Carotenoid-dependent signals: indicators of foraging efficiency, immunocompetence or detoxification ability? *Poultry and Avian Biology Reviews* **11**, 137-159.

Moreno, J., Sanz, J.J. & Arriero, E. 1999. Reproductive effort and T-lymphocyte cell-mediated immunocompetence in female pied flycatchers *Ficedula hypoleuca*. *Proceedings of the Royal Society of London, Series B* **266**, 1105-1109.

Mousseau, T.A. & Fox, C.W. 1998. *Maternal Effects*. New York: Oxford University Press.

Naqui, S.A., Marquez, B. & Sahin, N. 1983. Maternal antibody and its effect on infectious bursal disease virus. *Avian Diseases* **25**, 831-838.

Noble, E.R. & Noble, G.A. 1976. *Parasitology*. 4th Edtn. Philadelphia: Lea and Fabiger.

Nordling, D., Andersson, M., Zohari, S. & Gustafsson, L. 1998. Reproductive effort reduces specific immune response and parasite resistance. *Proceedings of the Royal Society of London, Series B* **265**, 1291-1298.

Parker, G.A. & Mock, D.W. 1987. Parent-offspring conflict over clutch size. *Evolutionary Ecology* **1**, 161-174.

Parker, G.A., Mock, D.W. & Lamey, T.C. 1989. How selfish should stronger sibs be? *American Naturalist* **133**, 846-868.

Pastoret, P., Gabriel, P., Bazin, H. & Govaerts, A. 1998. *Handbook of Vertebrate Immunology*. San Diego: Academic Press.

Pinard-van der Laan, M.H., Siegel, P.B. & Lamont, S.J. 1998. Lessons from selection experiments on immune response in the chicken. *Poultry and Avian Biology Review* **9**, 125-141.

Price, K. & Ydenberg, R. 1995. Begging and provisioning in broods of asynchronously-hatched yellow-headed blackbird nestlings. *Behavioral Ecology and Sociobiology* **37**, 201-208.

Price, P.V. 1980. *Evolutionary Biology of Parasites*. Princeton: Princeton University Press.

Råberg, L., Nilsson, J.Å., Ilmonen, P., Stjernman, M. & Hasselquist, D. 2000. The cost of an immune response: vaccination reduces parental effort. *Ecology Letters* **3**, 382-386.

Roitt, I., Brostoff, J. & Male, D. 1996. *Immunology*. London: Mosby.

Rose, M.E. & Orlans, E. 1981. Immunoglobulins in the egg, embryo and young chick. *Developmental Comparative Immunology* **5**, 315-319.

Ryan, M.J. & Brenowitz, E.A. 1985. The role of body size, phylogeny, and ambient noise in the evolution of bird song. *American Naturalist* **126**, 87-100.

Sacchi, R., Saino, N. & Galeotti, P. in press. Spectrographic features of begging calls reveal general condition and need for food of barn swallow (*Hirundo rustica*) nestlings. *Behavioral Ecology*.

Saino, N., Bolzern, A.M. & Møller, A.P. 1997a. Immunocompetence, ornamentation and viability of male barn swallows (*Hirundo rustica*). *Proceedings of the National Academy of Sciences USA* **97**, 579-585.

Saino, N., Calza, S. & Møller, A.P. 1997b. Immunocompetence of nestling barn swallows in relation to brood size and parental effort. *Journal of Animal Ecology* **66**, 827-836.

Saino, N., Stradi, R., Ninni, P., Pini, E. & Møller, A.P. 1999a. Carotenoid plasma concentration, immune profile and plumage ornamentation of male barn swallows (*Hirundo rustica*). *American Naturalist* **154**, 441-448.

Saino, N., Calza, S., Ninni, P. & Møller, A.P. 1999b. Barn swallows trade survival against offspring condition and immunocompetence. *Journal of Animal Ecology* **68**, 999-1009.

Saino, N., Ninni, P., Incagli, M., Calza, S., Sacchi, R. & Møller, A.P. 2000a. Begging and parental care in relation to offspring need and condition in the barn swallow (*Hirundo rustica*). *American Naturalist* **156**, 637-649.

Saino, N., Ninni, P., Calza, S., Martinelli, R., De Bernardi, F. & Møller, A.P. 2000b. Better red than dead: carotenoid based mouth coloration reveals infection in barn swallow nestlings. *Proceedings of the Royal Society of London, Series B* **267**, 57-61.

Saino, N., Martinelli, R. & Møller, A.P. 2001a. Immunoglobulin plasma concentration in relation to egg laying and mate ornamentation of female barn swallows (*Hirundo rustica*). *Journal of Evolutionary Biology* **14**, 110-119.

Saino, N., Incagli, M., Martinelli, R. & Møller, A.P. 2001b. Immunity, growth and begging behaviour of nestling barn swallows (*Hirundo rustica*) in relation to hatching order. *Journal of Avian Biology 32*, 263-270.

Saino, N., Incagli, M., Martinelli, R. & Møller, A.P. in press. Immune response of male barn swallows in relation to parental effort, corticosterone plasma levels and sexual ornamentation. *Behavioral Ecology.*

Sapolsky, R.M. 1992. Neuroendocrinology of the stress-response. In: *Behavioural Endocrinology* (Ed. by J.B. Becker, S.M. Breedlove & D. Crews). Cambridge: MIT Press.

Schwabl, H. 1996. Maternal testosterone in the avian egg enhances postnatal growth. *Comparative Biochemistry and Physiology* **114**, 271-276.

Schwabl, H. 1997. The contents of maternal testosterone in house sparrow *Passer domesticus* eggs vary with breeding conditions. *Naturwissenschaften* **84**, 406-408.

Smith, N.C., Wallach, M., Miller, C.M.D., Morgenstern, R., Braun, R. & Eckert, J. 1994. Maternal transmission of immunity to *Eimeria maxima*: enzyme-linked immunosorbent assay analysis of protective antibodies induced by infection. *Infection and Immunity* **62**, 1348-1357.

Soler, M., Martín-Vivaldi, M., Marín, J.M. & Møller, A.P. 1999. Weight lifting and health status in the black wheatear. *Behavioral Ecology* **10**, 281-286.

Stradi, R. 1998. *The Colour of Flight.* Milan: Solei.

Szép, T. & Møller, A.P. 1999. Cost of parasitism and host immune defence in the sand martin *Riparia riparia*: a role for parent-offspring conflict? *Oecologia* **119**, 9-15.

Tizard, I. 1991. *Veterinary Immunology.* Philadelphia: W.B. Saunders.

Wakelin, D. 1996. *Immunology to Parasites.* Cambridge: Cambridge University Press.

Wikel, S.K. 1996. *The Immunology of Host-Ectoparasitic Arthropod Relationships.* Wallington: CAB International.

Williams, T.D., Christians, J.K., Aiken, J.J. & Evanson, M. 1999. Enhanced immune function does not depress reproductive output. *Proceedings of the Royal Society of London, Series B* **266**, 753-757.

Wingfield, J.C., Hunt, K., Breuner, C., Dunlap, K., Fowler, G.S., Freed, L. & Lepson, J. 1998. Environmental stress, field endocrinology, and conservation biology. In: *Behavioral Approaches to Conservation in the Wild* (Ed. by J.R. Clemmons & R. Buchholz). Cambridge: Cambridge University Press.

SIBLING COMPETITION

14.

BEGGING AND ASYMMETRIC NESTLING COMPETITION

Barb Glassey & Scott Forbes
Department of Biology, University of Winnipeg, Winnipeg MB R3B 2E9, Canada (bglassey@escape.ca; s.forbes@uwinnipeg.ca)

ABSTRACT

Genotypically, all offspring are created equal, but maternal manipulations of phenotype can render some offspring more equal than others (e.g. differences in egg size or composition, hormonal titre or hatching interval). A phenotypic handicap results when such variation impairs an individual's competitive status. Here we examine both the causes and consequences of manipulations of such phenotypic handicaps. Hatching asynchrony is the primary handicap; differences in egg size and hormonal manipulations play secondary roles, unless offspring hatch synchronously. Begging strategies are role-dependent: last-hatched marginal offspring generally beg harder, but receive less food than earlier-hatched core offspring, consistent with phenotype-limited models of begging behaviour. There are alternative, though not mutually exclusive, explanations for such behaviour. Smaller nestlings may simply be hungrier, or influenced by different hormonal titres. Future work should focus on role-dependent begging strategies, such as whether marginal nestlings modulate their begging effort according to their prospects of winning.

INTRODUCTION

Siblings compete for parental resources in many birds and mammals, and success in intrabrood competition is a powerful determinant of offspring growth and survival (Mock & Parker 1997). Nestlings compete by begging; a behaviour that involves both visual (e.g. wing flapping, gaping, neck stretching and jostling) and vocal elements (Kilner & Johnstone 1997). In

J. Wright and M.L. Leonard (eds.), The Evolution of Begging, 269–281.

most passerine birds, sibling rivalry is described as scramble competition, so that an individual's success in begging competition is determined by its performance relative to its nestmates (Parker 1985; Harper 1986; Parker et al. 1989; Mock & Parker 1997).

Although from their parent's perspective all offspring begin life as genetic equals, parents often create within-brood variation in phenotype. If such variation impairs an individual's competitive behaviour (e.g. as measured by food reception, growth and/or fitness), then it qualifies as a handicap (Parker 1982). Altricial birds routinely confer phenotypic handicaps upon some of their progeny and advantages upon others. These take three main forms: (1) differences in egg size; (2) differences in hormonal titre; and (3) variation in the timing of hatching (nestling size and development).

Our objectives in this chapter are threefold: First, we shall examine the origins of phenotypic variation. Second, we shall explore which type(s) of phenotypic variation meet the definition of a phenotypic handicap. Finally, we shall identify phenotypic limited begging strategies and examine the consequences, if any, of phenotypic variation for the outcome of begging competitions.

Causes of Phenotypic Handicaps in Altricial Birds

Hatching Asynchrony

Age differences lead directly to asymmetries in both size and development that may result in disparities in growth, mortality and ultimately fitness. Altricial birds routinely hatch their broods asynchronously, creating initial inequalities in nestling size and development (Mock 1984; Magrath 1990). The phenotypic handicap of hatching asynchrony exerts a profound influence upon the outcome of sibling competition, although the direct effects of size and development may differ.

Whether by design or not, variation in hatching interval (nestling size and development), egg size or hormonal titre serves to modulate the within-brood competitive balance among asynchronously hatched broods. Of these, size differences associated with hatching interval seem to trump all others, and represent the only true phenotypic handicap (Parker 1982). We are aware of few studies showing that younger progeny enjoy superior prospects for growth, survival or fitness compared to earlier-hatched nestmates, even though they may hatch from larger eggs or be fortified by maternal steroids. That is not to say that these secondary components are unimportant. Alterations to egg size or steroid dose may exaggerate or diminish (but not erase) the competitive disadvantage of smaller offspring. Evidence bearing upon this conjecture is, however, scant. Whenever the joint effects of egg size and hatching asynchrony (nestling size) have been examined simultaneously, hatching asynchrony is almost always the more important of the two, and by a wide margin (Williams 1994). Similarly, although hormones may serve to mitigate the effects of hatching asynchrony early in the nestling period, the effect is temporary and cannot overcome the size differential between nestlings created by hatching asynchrony (Schwabl 1996; H. Schwabl & J. Lipar this volume).

Egg Size

Egg size variation within clutches of birds is both widespread and puzzling (Williams 1994). Both adaptive and non-adaptive explanations have been invoked for this phenomenon. Adaptive explanations include modulation of within-brood competitive asymmetries (Howe 1976; Slagsvold et al. 1984), role-dependent phenotypes of core versus marginal eggs (Forbes & Lamey 1996; Forbes 1999), adaptive coin-flipping (Kaplan & Cooper 1984), adjustment of incubation duration and hatching asynchrony and hence the probability of brood reduction (Parsons 1972) and preferential investment in eggs with higher survival prospects (Ankney 1980; Quinn & Morris 1986; Williams et al. 1993; Viñuela 1997). Non-adaptive explanations emphasize the role of energy/nutrient availability (Rydén 1978; Magrath 1992; Sydeman & Emslie 1992), hormonal constraints (Leblanc 1987; Williams et al. 1993) and varying environments during egg-laying (Ojanen et al. 1981; Järvinen & Ylimaunu 1986; Magrath 1992). As Stearns (1992) cautions, some variability is expected due to chance processes and requires no adaptive explanation. From the perspective of this paper, the consequences in egg size variation (see below) are of greater import than the causes.

Hormonal Titres

Size is important in mixed-age broods. First-hatched nestlings enjoy preferred access to resources in nearly every case examined (as measured by higher growth rates). There is, however, reason to believe that size is not the only determinant. Here the recent work on hormonal bias is of direct interest (see H. Schwabl & J. Lipar this volume). The more frequent begging of later-hatched canary (*Serinus canaria*) nestlings associated with higher hormonal titres in last-laid eggs resulted in access to more food early in the nestling period and initially faster growth of nestlings (Schwabl 1996). More frequent begging did not, however, make a lasting difference; these initial advantages were lost, as later-hatched nestlings could not overcome size differences associated with hatching asynchrony.

Female birds deposit hormones in their eggs in varied patterns. Cattle egrets (*Bubulcus ibis*) and zebra finches (*Taeniopygia guttata*) deposit more testosterone in first-laid eggs, and the amount decreases successively in later-laid eggs (Schwabl et al. 1997; Gil et al. 1999). The opposite pattern occurs in canaries, American kestrels (*Falco sparverius*) and red-winged blackbirds (*Agelaius phoeniceus*), where females add successively more testosterone to later-laid eggs (Schwabl 1993, 1996; Lipar et al. 1999; Sockman & Schwabl 2000). Zebra finch females also add extra steroids to all eggs when they mate with more attractive males, while still retaining the intra-clutch pattern (Gil et al. 1999).

Recently, costs associated with high doses of hormones have been identified, suggesting that parents need not always play favourites by supplying extra androgens. Experimentally increasing the levels of androgens in first-laid kestrel eggs to those of the better supplied last-laid egg had an adverse effect on the growth, mass and survival of older nestlings (Sockman & Schwabl 2000), and high testosterone may suppress the immune system (reviewed in Gil et al. 1999).

Consequences of Phenotypic Variation for Begging

Hatching Asynchrony and Size

Older, first-hatched nestlings consistently make better competitors by virtue of their larger size (Rydén & Bengtsson 1980; Göttlander 1987; McRae et al. 1993; Kacelnik et al. 1995; Price & Ydenberg 1995; but see Stamps et al. 1985; Leonard & Horn 1996). Consequently, the mass hierarchy within a brood, first established by asynchronous hatching (review in Magrath 1990) is maintained through competition. The smallest nestlings in a brood are usually the poorest competitors and typically grow more slowly and die sooner and more often (Mock 1984; Magrath 1990).

Larger nestlings are able to access a greater share of food primarily by reaching higher than their younger siblings. If size (neck height) surpasses all other components of begging as the chief determinant of who gets fed, then the issue is largely resolved from the outset. Large nestlings will win unless an accident befalls an older sibling or parents preferentially feed smaller nestlings (Stamps et al. 1985; Göttlander 1987; Leonard & Horn 1996). How high a nestling stretches its neck remains a chief determinant of begging success in a variety of altricial birds (Smith & Montgomerie 1991; Teather 1992; Leonard & Horn 1996; Dearborn 1998) and is determined primarily by size. If larger, older siblings choose to exert their full advantage, they will win begging competitions that are decided on this basis. Parents can, and sometimes do, resort to the obvious counterstrategy of playing favourites. The best evidence for this comes from parrots (Stamps et al. 1987; Krebs et al. 1999; E.A. Krebs this volume).

In begging competitions among mixed-age broods of red-winged blackbirds, neck height was the strongest determinant of feeding success (Teather 1992; Glassey 2000). First-hatched nestlings that ranked first in neck height received two thirds of first food offers; later-hatched nestlings did nearly as well, receiving three fifths of first food offers when they ranked first in neck height (Glassey 2000). What then was the difference between these offspring? First-hatched nestlings were twice as likely to rank first in neck height as later-hatched nestlings. Thus, if, and when, the oldest nestlings chose to exert their full advantage, they won the size-related begging competition. This outcome is a direct result of the prenatal parental manipulation of phenotype.

Hatching Asynchrony and Development

Little attention has focused upon how developmental disparities influence the outcome of begging competition. Unlike the size differential imposed by hatching asynchrony, physiological changes such as endothermy associated with development are initiated midway through the nestling period, potentially exaggerating the competitive advantage(s) of larger, first-hatched core offspring over smaller and later-hatched marginal offspring. For example, first-hatched nestlings acquire and improve motor skills ahead of their later-hatched siblings, and reach key developmental landmarks, such as the acquisition of sight and the initiation of thermoregulation, sooner (Marsh & Wichler 1982; Choi & Bakken 1990; Olson 1992). Sensory maturation governs which cues will evoke begging behaviour, and motor activation determines the speed of response (Khayutin 1985). The acquisition of sight, in combination with enhanced auditory sensitivity, shortens the interval between signal reception (e.g. a parent returning with food) and begging response during the latter half of the nestling period (Khayutin 1985). Similarly, enhanced motor unit recruitment begins midway through the nestling period, imparting greater strength to the neck and gastrocnemius muscles that are used to initiate and maintain begging (Marsh & Wichler 1982; Olson 1994). Consequently, the begging response of older, endothermic nestlings, is faster than the response of younger, ectothermic nestlings (Choi & Bakken 1990). Thus, differential begging behaviour (intrabrood variation in feeding response due to divergent sensory and physiological abilities of older and younger nestlings) is assumed to occur in mixed-age broods (Bengtsson & Rydén 1981; Pijanowski 1992).

Hill and Beaver (1982) identify a window during the development of red-winged blackbird broods when differential development is theoretically most likely to occur. At this stage, which we call the transitional phase (Glassey 2000), first-hatched nestlings have initiated endothermy, while later-hatched offspring are still ectothermic (Hill & Beaver 1982). It is at the transitional phase that developmental asymmetries are potentially most important to last-hatched progeny as the needs of first- and later-hatched nestlings begin to diverge, and older nestlings require more food to meet their increasing metabolic requirements.

Secondary Handicaps: Egg Size and Hormones

Most studies find that although initial differences in egg size may influence growth early in the nestling period, the effects are not lasting and are overridden by nestling size asymmetries caused by hatching asynchrony (Schifferli 1973; Bryant 1978; Bancroft 1984; Stokland & Amundsen 1988; Magrath 1992; Ohlsson & Smith 1994). Thus, by the end of the nestling period, differences in egg size do not appear to translate into differences in offspring quality. Differences in egg size may, however, make a substantial difference early in the nestling period (review in Williams 1994).

Similarly, changes in hormone titre may mitigate the size differential, but they do not appear to overcome the effect of hatching asynchrony. Unlike androstenedione, which does not directly affect nestling begging behaviour, testosterone temporarily creates better competitors by increasing the rate of begging (Schwabl 1996) and by enhancing the *musculus complexus*, a group of muscles used to bring the bird to an upright position (Lipar & Ketterson 2000; H. Schwabl & J. Lipar this volume).

The Core versus Marginal Divide

Hatching asynchrony creates a structured sibship, effectively dividing the brood into separate castes of core and marginal offspring (Mock & Forbes 1995; Mock & Parker 1997). The chief significance of this is the functional and behavioural separation of core and marginal progeny. The core brood represents the minimum subset that parents can or will rear across breeding attempts. These are progeny that show low variance in growth and survival in the face of varying ecological or developmental conditions (Forbes et al. 1997, 2001; Forbes & Glassey 2000). Here the size differential resulting from hatching asynchrony is key, buffering the core brood from extrinsic variation (e.g. in resource levels, weather conditions) and intrinsic variation in developmental fate (e.g. failure to hatch, congenital defects, pathogens or parasites). What happens to the core brood is largely, though not completely, divorced from what happens to the marginal brood. The reverse is not true.

Why is the distinction between core and marginal progeny important? Because games among offspring occur within a structured sibship. When one looks, for example, at the relationship between size and begging success (e.g. by assessing size rank within the brood) one misses the key biological distinction between core and marginal offspring. The size disparities among the core brood are much less important than size disparities between the core and marginal brood. In effect, there are two games at play within the brood; that among core offspring, and that between core and marginal offspring (and a third, among the marginal brood, if there is more than one marginal offspring).

This conceptual shift has important practical implications. It is less useful, for example, to talk about overall brood size than it is about the size of the core and marginal brood. In many ways, a brood of three core nestlings has more in common with a brood of three core nestlings and one marginal nestling than it has with a brood of two marginal nestlings and one core nestling (Figure 1). Since the presence of the marginal offspring has little effect on core offspring, the growth and survival of core nestlings in the first two broods will be more similar than between broods of the same nominal size but differing numbers of core nestlings.

The core-marginal dichotomy is most useful when considering altricial passerines where a single marginal nestling is most common (Clark & Wilson 1981). In the much rarer cases, where complete brood hierarchies exist, the distinction between core and marginal blurs; the senior-most members of the marginal brood will most closely resemble the core brood in their developmental fate. Here the functional differences between nestlings become more continuous than the core-marginal dichotomy implies. The traditional view, however, of treating all offspring as more or less equal in their parent's eyes is even less suitable in such cases. Offspring are not all created equal and the core-marginal taxonomy is conceptually important to make this clear.

This core-marginal dichotomy has direct application to the study of nestling begging. Unless incubation commences with the first egg, which is rare among altricial birds, there will be two or more roughly coequal nestlings within the brood. That is, a core brood. What happens to the core brood is very different from what happens to the marginal brood, and is something that is not captured if one simply uses size rank to measure the effect of hatching asynchrony. That is, the trivial differences in competitive ability among core brood members will tower over differences between core and marginal nestlings, providing growth and survival are in any way related to behaviour (Forbes et al. 1997, 2001; Forbes & Glassey 2000).

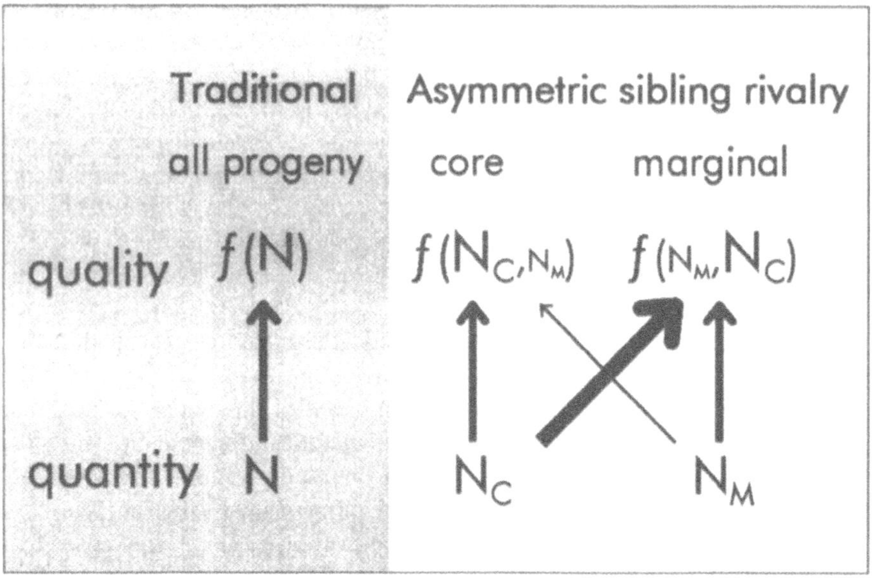

Figure 1. Schematic view of asymmetric sibling rivalry. Under the traditional view of avian family size, all offspring are created equal, and a simple number versus quality trade-off achieves the optimization of brood size. Creating more offspring, N, reduces quality, f (N), and the optimum is that maximizing N x f (N). Under asymmetric sibling rivalry, not all offspring are created equal. The brood consists of N_C core offspring and N_M marginal offspring that are not competitively equal. Rather, core offspring are superior competitors to marginal offspring. This renders what was a one-dimensional optimization of brood size now four-dimensional. The life history optimization facing parents is twofold: how many core offspring to make, and how many marginal offspring to make. There are now four potential interactions between offspring number and quality (arrows). The effects of asymmetric sibling rivalry on begging behaviour are only now being explored.

Phenotype-Limited Begging Strategies

In asynchronously hatched broods, smaller marginal nestlings generally beg more, but receive less for their efforts than their older and larger nestmates (Budden & Wright 2001). Four, non-mutually exclusive explanations for this pattern immediately suggest themselves.

First, theoretical models of scramble competition under the influence of a competitive handicap predict that handicapped progeny should beg more for a smaller reward (Parker et al. 1989; Godfray 1995; Mock & Parker 1997). Smaller yellow-headed blackbirds (*Xanthocephalus xanthocephalus*) beg more intensely than their larger broodmates (Price & Ydenberg 1995) and experiments showed that individuals of this species can assess their relative size and alter begging tactics accordingly. Pairing a smaller focal nestling with an older, larger nestmate triggered more prolonged and intense begging (Price et al. 1996). Similarly, experimental studies of barn swallows (*Hirundo rustica*, Lotem 1998) and European starlings (*Sturnus vulgaris*, Cotton et al. 1999) have shown that younger nestlings beg harder but receive less food for their effort, which again is broadly consistent with theory.

Second, differences in begging behaviour may arise directly from the phenotype manipulation by hormonal titre. Last-hatched canaries, for example, experimentally fortified with maternal steroids, open their eyes one day earlier and beg more frequently (Schwabl 1996).

Third, smaller later-hatched nestlings may beg harder simply because they are hungrier. When experimentally deprived of food, nestlings beg more vigorously, both physically (Kacelnik et al. 1995; Kilner 1995; Leonard & Horn 1998) and vocally (Henderson 1975; Price & Ydenberg 1995). Last-hatched progeny routinely grow more slowly and die sooner than older broodmates, often due to starvation. To suggest that hunger levels differ among offspring, and contribute to differences in begging behaviour, is surely uncontroversial.

Finally, smaller offspring may beg harder in order to induce parents to increase overall levels of provisioning (Kacelnik et al. 1995). Studies involving experimental food deprivation (von Haartman 1953; Bengtsson & Rydén 1983; Whittingham & Robertson 1993) and playback of recorded begging calls (Henderson 1975; Burford et al. 1998; Price 1998; Wright 1998; but see Clark & Lee 1998) indicate that parents increase foraging rates in response to elevated vocal solicitations by the brood. Discussions of nestling begging versus phenotypic handicaps inevitably converge on the issue of need. We shall defer comment on this and simply refer the reader to A.B. Clark's chapter in this volume.

FUTURE DIRECTIONS

The role of phenotypic handicaps in nestling begging competition is largely unexplored, as is the role of developmental asynchrony in the onset of endothermy. This last issue is potentially important because more rapid motor responses may play a key part in determining the outcome of begging competition. Also not resolved is whether phenotypic handicaps alter begging competition from scrambles to systems of hierarchical resource allocation (see Parker et al. 1989). When and why do parents play favourites within their brood *via* preferential feeding, and why do they not do it more often? Do last-hatched marginal offspring modulate their behaviour to reflect short-term variation in the prospects of winning begging contests? If core nestlings beg at or near maximum levels, there is often little chance for a marginal nestling to win. A prudent beggar might choose to conserve its resources both within and across begging bouts, for example, by choosing to commence begging later in a begging bout, or to beg harder when core sibs are satiated. These types of games among asymmetric siblings pose significant challenges for both empiricists and theoreticians.

ACKNOWLEDGEMENTS

We thank Jon Wright, Marty Leonard and an anonymous reviewer for providing valuable comments and thoughtful criticisms of earlier versions of this chapter.

REFERENCES

Ankney, C.D. 1980. Egg weight, survival and growth of lesser snow goose goslings. *Journal of Wildlife Management* **44**, 174-182.

Bancroft, G.T. 1984. Patterns of variation in size of Boat-tailed Grackle *Quiscalus major* eggs. *Ibis* **126**, 496-509.

Bengtsson, H. & Rydén, O. 1981. Development of parent-young interaction in asynchronously hatched broods of altricial birds. *Zeitschrift für Tierpsychologie* **56**, 255-272.

Bengtsson, H. & Rydén, O. 1983. Parental feeding rate in relation to begging behavior in asynchronously hatched broods of the great tit *Parus major. Behavioral Ecology and Sociobiology* **12**, 243-251.

Bryant, D.M. 1978. Establishment of weight hierarchies in the broods of house martins *Delichon urbica. Ibis* **120**, 16-26.

Budden, A.E. & Wright, J. 2001. Begging in nestling birds. *Current Ornithology* **16**, 83-118.

Burford, J.E., Friedrich, T.J. & Yasukawa, K. 1998. Response to playback of nestling begging in the red-winged blackbird, *Agelaius phoeniceus. Animal Behaviour* **56**, 555-561.

Choi, I.-H. & Bakken, G.S. 1990. Begging response in nestling red-winged blackbirds (*Agelaius phoeniceus*): effect of body temperature. *Physiological Zoology* **63**, 965-986.

Clark, A.B. & Lee, W.-H. 1998. Red-winged blackbird females fail to increase feeding in response to begging call playbacks. *Animal Behaviour* **56**, 563-570.

Clark, A.B. & Wilson, D.S. 1981. Avian breeding adaptations: hatching asynchrony, brood reduction, and nest failure. *Quarterly Review of Biology* **56**, 253-277.

Cotton P.A., Wright, J. & Kacelnik, A. 1999. Chick begging strategies in relation to brood hierarchies and hatching asynchrony. *American Naturalist* **153**, 412-420.

Dearborn, D.C. 1998. Begging behavior and food acquisition by brown-headed cowbird nestlings. *Behavioral Ecology and Sociobiology* **43**, 259-270.
Forbes, L.S. 1999. Within-clutch variation in propagule size: the double-fault model. *Oikos* **85**, 146-150.
Forbes, L.S. & Glassey, B. 2000. Asymmetric sibling rivalry and nestling growth in red-winged blackbirds. *Behavioral Ecology and Sociobiology* **48**, 413-417.
Forbes, L.S. & Lamey, T.C. 1996. Insurance, developmental accidents, and the risks of putting all your eggs in one basket. *Journal of Theoretical Biology* **180**, 247-256.
Forbes, L.S., Thornton, S., Glassey, B., Forbes, M.J. & Buckley, N. 1997. Why parent birds play favourites. *Nature* **390**, 351-352.
Forbes, S., Glassey, B., Thornton, L. & Earle, L. 2001. The secondary adjustment of clutch size in red-winged blackbirds (*Agelaius pheoniceus*). *Behavioral Ecology and Sociobiology* **50**, 37-44.
Gil, D., Graves, J., Hazen, N. & Wells, A. 1999. Male attractiveness and differential testosterone investment in zebra finch eggs. *Science* **286**, 126-128.
Glassey, B. 2000. Resource competition among nestling red-winged blackbirds. PhD thesis, University of Manitoba.
Godfray, H.J.C. 1995. Signaling of need between parents and young: parent-offspring conflict and sibling rivalry. *American Naturalist* **146**, 1-24.
Göttlander, K. 1987. Parental feeding behaviour and sibling competition in the pied flycather *Ficedula hypoleuca*. *Ornis Scandinavica* **18**, 269-276.
Harper, A.B. 1986. The evolution of begging: sibling competition and parent-offspring conflict. *American Naturalist* **128**, 99-141.
Henderson, B.A. 1975. Role of the chick's begging behavior in the regulation of parental feeding behavior of *Larus glaucescens*. *Condor* **77**, 488-492.
Hill, R.W. & Beaver, D.L. 1982. Inertial thermostability and thermoregulation in broods of redwing blackbirds. *Physiological Zoology* **55**, 250-260.
Howe, H.F. 1976. Egg size, hatching asynchrony, sex and brood reduction in the common grackle. *Ecology* **57**, 1195-1207.
Järvinen, A & Ylimaunu, J. 1986. Intraclutch egg-size variation in birds: physiological responses of individuals to fluctuations in environmental conditions. *The Auk* **103**, 235-237.
Kacelnik, A., Cotton, P.A., Stirling, L. & Wright, J. 1995. Food allocation among nestling starlings: sibling competition and the scope of parental choice. *Proceedings of the Royal Society of London, Series B* **259**, 259-263.
Kaplan, R.H. & Cooper, W.S. 1984. The evolution of developmental plasticity in reproductive characteristics: An application of the 'adaptive coin-flipping' principle. *American Naturalist* **123**, 393-410.
Khayutin, S.N. 1985. Sensory factors in the behavioral ontogeny of altricial birds. *Advances in the Study of Behavior* **15**, 105-152.
Kilner, R. 1995. When do canary parents respond to signals of need? *Proceedings of the Royal Society of London, Series B* **260**, 343-348.
Kilner, R. & Johnstone, R.A. 1997. Begging the question: are offspring solicitation behaviours signals of need? *Trends in Ecology and Evolution* **12**, 11-15.
Krebs, E.A., Cunningham, R.B. & Donnelly, C.F. 1999. Complex patterns of food allocation in asynchronously hatching broods of crimson rosellas. *Animal Behaviour* **57**, 753-763.
Leblanc, Y. 1987. Intraclutch variation in egg size of Canada Geese. *Canadian Journal of Zoology* **65**, 3044-3047.
Leonard, M. & Horn, A. 1996. Provisioning rules in tree swallows. *Behavioral Ecology and Sociobiology* **38**, 341-347.
Leonard, M.L. & Horn, A.G. 1998. Need and nestmates affect begging in tree swallow. *Behavioral Ecology and Sociobiology* **42**, 431-436.

Lipar, J.L. & Ketterson, E.D. 2000. Maternally derived yolk testosterone enhances the development of the hatching muscle in the red-winged blackbird *Agelaius phoeniceus*. *Proceedings of the Royal Society of London, Series B* **267**, 2005-2010.

Lipar, J.L., Ketterson, E.D. & Nolan, V.Jr. 1999. Intraclutch variation in testosterone content of Red-winged Blackbird eggs. *The Auk* **116**, 231-235.

Lotem, A. 1998. Higher levels of begging behavior by small nestlings: a case of a negatively correlated handicap. *Israel Journal of Zoology* **44**, 29-45.

Magrath, R.D. 1990. Hatching asynchrony in altricial birds. *Biological Reviews* **65**, 587-622.

Magrath, R.D. 1992. The effect of egg mass on the growth and survival of blackbirds: a field experiment. *Journal of Zoology* **227**, 639-653.

Marsh, R.L. & Wichler, S.J. 1982. The role of muscle development in the transition to endothermy in nestling bank swallows, *Riparia riparia*. *Journal of Comparative Biology B* **149**, 99-105.

McRae, S.B., Weatherhead P.J. & Montgomerie, R. 1993. American robin nestlings compete by jockeying for position. *Behavioral Ecology and Sociobiology* **33**, 101-106.

Mock, D.W. 1984. Infanticide, siblicide, and avian nestling mortality. In: *Infanticide: Comparative and Evolutionary Perspectives* (Ed. by G. Hausfater & S.B. Hrdy). New York: Aldine Publishing Co.

Mock, D.W. & Forbes, L.S. 1995. The evolution of parental optimism. *Trends in Ecology and Evolution* **10**, 130-134.

Mock, D.W. & Parker, G.A. 1997. *The Evolution of Sibling Rivalry*. Oxford: Oxford University Press.

Ohlsson, T. & Smith, H.G. 1994. Development and maintenance of nestling size hierarchies in the European starling. *Wilson Bulletin* **106**, 448-455.

Ojanen, M., Orell, M. & Väisänen, R.A. 1981. Egg size variation within passerine clutches: effects of ambient temperature and laying sequence. *Ornis Fennica* **58**, 93-108.

Olson, J.M. 1992. Growth, the development of endothermy, and the allocation of energy in red-winged blackbirds (*Agelaius phoeniceus*) during the nestling period. *Physiological Zoology* **65**, 124-152.

Olson, J.M. 1994. The ontogeny of shivering thermogenesis in the red-winged blackbird (*Agelaius phoeniceus*). *Journal of Experimental Biology* **191**, 59-88.

Parker, G.A. 1982. Phenotype-limited evolutionarily stable strategies. In: *Current Problems In Sociobiology* (Ed. by B.R. Bertram, T.H. Clutton-Brock, R.I.M. Dunbar, D.I. Rubenstein & R. Wrangham). Cambridge: Cambridge University Press.

Parker, G.A. 1985. Models of parent-offspring conflict. V. Effects of the behaviour of the two parents. *Animal Behaviour* **33**, 519-533.

Parker, G.A., Mock, D.W. & Lamey, T.C. 1989. How selfish should stronger sibs be? *American Naturalist* **133**, 846-868.

Parsons, J. 1972. Egg size, laying date, and incubation period in the herring gull. *Ibis* **114**, 536-541.

Pijanowski, B.C. 1992. A revision of Lack's brood reduction hypothesis. *American Naturalist* **139**, 1270–1292.

Price, K. 1998. Benefits of begging for yellow-headed blackbird nestlings. *Animal Behaviour* **56**, 571-577.

Price, K. & Ydenberg, R. 1995. Begging and provisioning in broods of asynchronously-hatched yellow-headed blackbird nestlings. *Behavioral Ecology and Sociobiology* **37**, 201-208.

Price, K., Harvey, H. & Ydenberg, R. 1996. Begging tactics of nestling yellow-headed blackbirds *Xanthocephalus xanthocephalus* in relation to need. *Animal Behaviour* **51**, 421-435.

Quinn, J.S. & Morris, R.D. 1986. Intraclutch egg weight apportionment and chick survival in Caspian Terns. *Canadian Journal of Zoology* **64**, 2116-2122.

Rydén, O. 1978. Egg weight in relation to laying sequence in a south Swedish urban population of the blackbird *Turdus merula*. *Ornis Scandinavica* **9**, 80-86.
Rydén, O. & Bengtsson, H. 1980. Differential begging and locomotory behaviour by early and late hatched nestlings affecting the distribution of food in asynchronously hatched broods of altricial birds. *Zeitschrift für Tierpsychologie* **53**, 209-224.
Schifferli, L. 1973. The effect of egg weight on the subsequent growth of nestling Great Tits *Parus major*. *Ibis* **115**, 549-558.
Schwabl, H. 1993. Yolk is a source of maternal testosterone for developing birds. *Proceedings of the National Academy of Sciences USA* **90**, 11446-11450.
Schwabl, H. 1996. Maternal testosterone in the avian egg enhances postnatal growth. *Comparative Biochemistry and Physiology* **114A**, 271-276.
Schwabl, H., Mock, D.W. & Gieg, J.A. 1997. A hormonal mechanism of parental favouritism. *Nature* **386**, 231.
Slagsvold, T., Sandvik, J., Rofstad, G., Lorentsen, O. & Husby, M. 1984. On the adaptive value of intra-clutch egg-size variation in birds. *The Auk* **101**, 685-697.
Smith, H.G. & Montgomerie, R. 1991. Nestling American robins compete with siblings by begging. *Behavioral Ecology and Sociobiology* **29**, 307-312.
Sockman, K. & Schwabl, H. 2000. Yolk androgens reduce offspring survival. *Proceedings of the Royal Society of London, Series B* **267**, 1451-1456.
Stamps, J.A., Clark, A.B., Arrowood, P. & Kus, B. 1985. Parent-offspring conflict in budgerigars. *Behaviour* **94**, 1-39.
Stamps, J., Clark, A., Kus, B. & Arrowood, P. 1987. The effects of parent and offspring gender on food allocation in budgerigars. *Behaviour* **101**, 177-199.
Stearns, S. 1992. *The Evolution of Life Histories*. Oxford: Oxford University Press.
Stokland, J.N. & Amundsen, T. 1988. Initial size hierarchy in broods of the shag: relative significance of egg size and hatching asynchrony. *The Auk* **105**, 308-315.
Sydeman, W.J. & Emslie, S.D. 1992. Effects of parental age on hatching asynchrony, egg size and third chick disadvantage in western gulls. *The Auk* **109**, 242-248.
Teather, K.L. 1992. An experimental study of competition between male and female nestlings of the red-winged blackbird. *Behavioral Ecology and Sociobiology* **31**, 81-87.
Viñuela, J. 1997. Adaptation vs. constraint: intraclutch egg-mass variation in birds. *Journal of Animal Ecology* **66**, 781-792.
von Haartman, L. 1953. Was reizt den Trauerfliegenschnäpper zu füttern? *Die Vogelwarte* **16**, 157-164.
Whittingham, L.A. & Robertson, R.J. 1993. Nestling hunger and parental care in red-winged blackbirds. *The Auk* **110**, 240-246.
Williams, T.D. 1994. Intraspecific variation in egg composition in birds: effects on offspring fitness. *Biological Reviews* **68**, 35-59.
Williams, T.D., Lank, D.B. & Cooke, F. 1993. Is intraclutch egg-size variation adaptive in the lesser snow goose? *Oikos* **67**, 250-256.
Wright, J. 1998. Helpers-at-the-nest have the same provisioning rule as parents: experimental evidence from play-backs of chick begging. *Behavioral Ecology and Sociobiology* **42**, 423-429.

15.

SIBLING COMPETITION AND THE EVOLUTION OF BROOD SIZE AND DEVELOPMENT RATE IN BIRDS

Robert E. Ricklefs
Department of Biology, University of Missouri-St. Louis, St. Louis, MO 63121-4499, USA (ricklefs@umsl.edu)

ABSTRACT

Parents and their offspring may come into conflict over both the level of food provisioning to the brood and the distribution of food among nestlings within a brood. Sibling competition within a brood favours nestlings whose rapid development increases their competitive status. This may in turn reduce the overall fitness of parents by increasing the food requirements of individual nestlings. Parents can reduce conflict with their offspring by reducing the number of young in each brood and by hatching their eggs asynchronously, so that competitive status is determined by age and not by growth rate. Under these circumstances, benefits of slow development, possibly including a more responsive immune system or delayed senescence, may also result in slower growth and reduced food requirements of nestlings. Future studies of parent-offspring relationships should include evolutionary responses to competition between siblings and parental mechanisms for controlling these responses.

INTRODUCTION

Nestlings of altricial species of birds beg to obtain food from their parents, which respond to solicitation from their nestlings (Henderson 1975; Bengtsson & Rydén 1983; Whittingham & Robertson 1993; Ottosson et al. 1997). Experimentally increasing the intensity of nestling solicitation, for example by adding young to nests, depriving nestlings of food or playing

J. Wright and M.L. Leonard (eds.), The Evolution of Begging, 283–301.

begging vocalizations, generally stimulates parents to increase provisioning of food to the nest (Hussell 1988). This solicitation-response system should ensure that nestlings are optimally nourished with respect to maximizing their inclusive fitness and that parents distribute food among their offspring optimally with respect to their personal fitness. Parents and their offspring, however, may come into conflict over the allocation of food because of the basic asymmetry of genetic relationship between a nestling and its sibling (Trivers 1974). Godfray (1991, 1995) demonstrated that parents and their offspring could achieve a joint evolutionarily stable strategy with respect to solicitation and delivery of food. He did not, however, investigate situations in which nestlings compete directly within a brood, which has the potential to reduce evolutionary fitness of both parent and offspring by selecting traits, such as growth rate, that are useful in direct competition, but which reduce fitness overall.

In this chapter, I consider the effects of sibling competition on the growth rate and food requirements of nestlings and the consequences of variation in growth rate on the fitness of parents and their offspring. I also examine the more general relationship between food requirements of nestlings and parental provisioning and fitness, and the ways in which adults can adapt their parental strategies in response. Moreover, if parents were programmed behaviourally to satisfy nestling demands, this responsiveness could under stressful circumstances reduce lifetime reproductive success, particularly when nestlings acquire additional food at a low marginal return in personal fitness. Even when nestlings do not solicit or accept resources beyond the intake that is optimal for their parents (Cotton et al. 1996), reduced parental fitness could result from a mismatch between the demands of the brood and the capacity of the parents to supply the needs of their nestlings. Although such circumstances could arise in several ways, the most important may be unpredictable variation in the food supply. If females adjusted their brood size to average feeding conditions, periods of below average food availability could lead to undernourishment of the nestlings and excessive stress or risk to the parents as they try to satisfy nestling demands.

FOOD REQUIREMENT AND PARENTAL PROVISIONING

The food requirements of a brood are the product of the requirements of individual nestlings and the number of nestlings per brood. Individual requirements depend on many factors, including growth rate, heat production to regulate body temperature and activity at the nest (Dunn 1980; Weathers et al. 1997). Furthermore, requirements are not absolute. Rather, above a minimum threshold required for survival, additional food consumption by nestlings is thought to contribute incrementally to fitness at a diminishing rate.

Presumably, the female parent controls the number of nestlings per brood, while growth rate and activity are primarily strategies of nestlings. Due to the basic asymmetry of genetic relationships among siblings (Trivers 1974), nestlings and their parents may come into conflict with respect to such nestling adaptations as growth rate and such parental adaptations as brood size and responsiveness to solicitation. Rapid growth benefits nestlings directly by reducing time-dependent mortality, which may be high during the period of nestling development (Ricklefs 1969, 1984; Trivers 1974; Bosque & Bosque 1995; Martin 1995). Moreover, rapid growth may improve the competitive position of a nestling within the brood when it achieves large size at an early age (Magrath 1990; Stouffer & Power 1990). Rapid growth and consequent high food requirements may indirectly influence the inclusive fitness of a particular nestling by reducing the number of its siblings, either within its own brood or over the lifetime of its parents (Konarzewski et al. 1998). Selection on nestling traits always tends to maximize nestling inclusive fitness, regardless of costs to parental lifetime reproductive success.

Figure 1 shows how a mismatch between the requirements of the brood and the ability of parents to provide food could reduce parental fitness. This mismatch would most often arise owing to uncertainty regarding the state of the environment and of hatching success. Food shortage relative to demand influences fitness along two paths. The one to the right in Figure 1 is based on the direct link between nestling solicitation and parental response, which, in times of food shortage, can stress the parent and reduce lifetime reproductive success (Ottosson et al. 1997). This pathway presumes that parents do not withhold care from their offspring even when failing to do so reduces their own survival and future reproduction. That is, a parent's drive to satisfy its offspring's needs overrides its physiological assessment of its own personal condition. The evidence for this is strong, but not unanimous. Food deprivation usually leads to increased food solicitation by nestlings (Hussell 1988; Price & Ydenberg 1995; Christe et al. 1996; Leonard & Horn 1998), although the competitive environment may also influence the intensity of solicitation by a particular nestling (Leonard et al. 2000; Roulin et al. 2000; A. Roulin this volume) and nestlings low in the size hierarchy of a brood may passively endure starvation (M. Konarzewski personal communication). Brood enlargement experiments often show enhanced offspring production (Linden & Møller 1989; Dijkstra et al. 1990) and reduced survival or future fecundity in parents (Dijkstra et al. 1990; Cucco & Malacarne 1995; Jacobsen et al. 1995). Sometimes, however, adults appear to rear larger broods without adverse effects (Ydenberg & Bertram 1989; Congdon 1990).

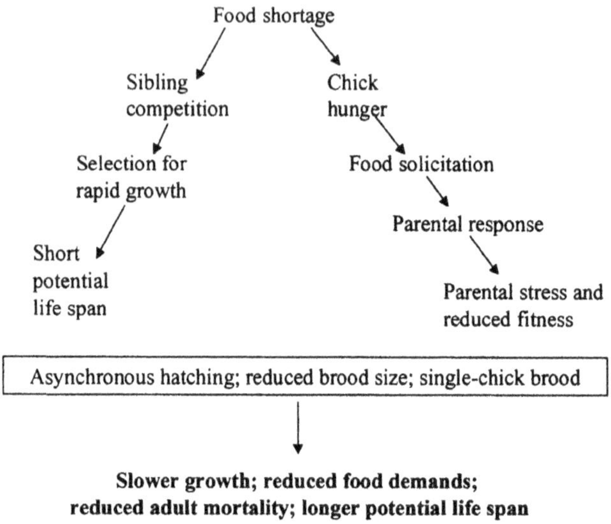

Figure 1. Food shortages influence fitness by imposing stress on adults mediated by nestling solicitation and parental response (right-hand path), and by the evolution of rapid growth through sibling competition (left-hand path). Reduced brood size and asynchronous hatching can eliminate parent-offspring conflict over the distribution of food and increase both parental fitness and the inclusive fitness of the nestlings.

The path to the left in Figure 1 is based on the advantage of size in sibling competition, which selects strongly for rapid growth. This can be shown by the observation that species with single-nestling broods have slower growth rates than those with multi-nestling broods (Werschkul & Jackson 1979; Ricklefs 1982). Rapid growth rate increases the rate of optimal resource provisioning per nestling (Ricklefs 1984; Klaassen & Bech 1992) and presumably forces parents to reduce brood size when food is limited. Rapid growth is also correlated in comparative analyses with reduced maximum potential life span (Ricklefs & Scheuerlein 2001), perhaps through connections between early development and ageing (Ricklefs 1993). Models of the relationship between fitness and growth rate suggest that there is little conflict between parent and offspring when the primary cause of selection for rapid growth is time-dependent mortality (Konarzewski et al. 1998). Direct competition between broodmates, in contrast, can apply strong selection to increase the rate of growth, even in the absence of strong time-dependent mortality. In this case, small changes in growth rate can have large effects on fitness that are independent of surviving time-dependent mortality factors. Because fitness depends on growth rate relative to broodmates, selection will tend to increase growth rate continuously until it is limited by physiological or other intrinsic factors or until it is constrained by inclusive fitness considerations (Briskie et al. 1994; Anderson & Ricklefs 1995). This high growth rate increases food demands on the parents without providing compensating increases in brood survival.

PARENTAL CONTROL OVER PARENT-OFFSPRING CONFLICT

Is the parent bird helpless in its conflict with its own nestlings? Several responses would seem to be possible. Parents can reduce the effect of nestling hunger on parental fitness by (1) moderating their responsiveness to nestling solicitation; (2) reducing brood size to reduce sibling competition and parent-offspring conflict, as well as parental stress; and (3) manipulating the competitive hierarchy among nestlings so that no advantage accrues from intense begging or rapid growth.

Can parents distribute food among nestlings in a manner that maximizes the total recruitment from the brood as a whole rather than the probability of recruitment of any particular individual nestling? To do so, the parent would have to discriminate between the stronger begging signals of better fed larger nestlings and the weaker signals of smaller but hungrier nestlings, in order to prevent first-hatched nestlings from controlling the distribution of food within the brood. Quantitative models have shown that parents can increase their personal fitness by providing better nourished nestlings with less food than needed to maximize a nestling's inclusive fitness, and using the balance to increase the survival of more poorly nourished siblings (Forbes 1993; Anderson & Ricklefs 1995). Several studies indicate that parents of some species preferentially feed smaller, less dominant nestlings (Stamps et al. 1985; Leonard & Horn 1996; Krebs et al. 1999).

By reducing the size of their brood, parents can bring the food requirements of the nestlings into line with the parents' abilities to provide food. Adjusting brood size can maximize the personal fitness of the parent in the context of parent-offspring conflict, but it does not directly diminish conflict produced by nestling adaptations. That is, growth rate and activity are still selected to maximize the inclusive fitness of the individual nestling at the expense of parental fitness. When brood size is reduced so much that nestlings no longer compete for resources, and thus do not conflict with parents over allocation of food, parental fitness may be reduced below what it would be when food is allocated optimally among nestlings in a larger brood.

When parents create conditions that minimize the selective advantage of sibling competition, nestlings may then reduce growth rate and other competitive traits to obtain other personal or inclusive fitness benefits. Slower growth also may increase the personal fitness of the parent in parallel with that of their nestlings, through larger potential brood size or reduced stress. In general, when there is no sibling competition, there is no parent-offspring conflict over the distribution of resources within a brood. Parents of many species appear to reduce the fitness consequences of sibling competition within broods by reducing brood size, so that there are adequate resources for all nestlings, and by asynchronous hatching of the nestlings. The latter mechanism determines the nestling dominance hierarchy by age (i.e. size) beyond the ability of nestling adaptations to change an individual's status within the brood (Ricklefs 1993). Both mechanisms are under parental control and provide the adult with potentially effective means of reducing the adverse consequences of parent-offspring conflict when parents are generally responsive to nestling solicitation. These mechanisms are consistent with variation in growth rates among species (Werschkul & Jackson 1979; Ricklefs 1982, 1993) and with the observation that many species can rear more offspring than they normally do (Ydenberg & Bertram 1989).

Responsiveness to Nestling Solicitation

Becoming less responsive to nestling solicitation has its own dangers in terms of inadequately providing for nestlings through mismatched solicitation-response behaviour. I presume here that nestling solicitation honestly signals the degree to which its optimal nutritional level is satisfied at any particular time (Godfray 1991; Cotton et al. 1996). The optimal level should be adjusted to maximize the inclusive fitness of the nestling, although this may not correspond to the maximum personal fitness of the parent. I also presume that parents are selected to use the information concerning the nutritional state of their offspring to allocate food in a manner that maximizes their own personal fitness. In doing so, a parent may favour an undernourished nestling more than is desirable for the dominant, better nourished nestling(s) in a brood. This preference establishes selection on marginally nourished nestlings to overemphasize their poor condition, in a classic case of parent-offspring conflict (Johnstone 1996).

I can see no way for families to resolve this conflict other than through nestling begging hyperbole. A well-nourished nestling has no selection upon it to acquire additional resources and therefore begging signals should honestly indicate lack of hunger (Godfray 1995). Yet, when two hungry nestlings coexist in a brood and signal with the same relationship between begging intensity and undernourishment, discerning parents will allocate food between the nestlings to maximize parental fitness at the expense of maximizing the inclusive fitness of any one individual nestling. Thus, when nestlings compete strongly and relatively evenly within a brood, selection should drive the intensity of begging to progressively higher levels, restrained only when costs of extremely intense begging increase rapidly (Johnstone 1996). When nestling solicitation is inflated by conflict among offspring, it should be difficult for parents to discern the true nutritional level of their nestlings and to allocate food so as to maximize their own personal fitness. Note that inflation is not the same as dishonesty when there is no within-brood competition (Godfray 1995). What signal would be used, if relative body size (or age) cannot accurately predict the position of a nestling in the competitive hierarchy of a brood and thus its likely level of undernourishment (Harper 1986)?

A Model of Sibling Competition

A simple model relating nestling personal fitness to food provisioning illustrates the fitness consequences for parents and offspring of different levels of food allocation within a brood (Parker et al. 1989; Forbes 1993; Anderson & Ricklefs 1995). Let X be the food acquired by a nestling, and define the nestling's fitness by $W = 1 - X^c$, where c [>0] determines how rapidly fitness increases with an increase in food acquisition above a threshold level of $X = 1$ ($W = 0$). This function increases at a progressively slowing rate from $W = 0$ at a threshold level of food provisioning ($X = 1$) to an asymptote of $W = 1$ at high values of X. Given this relationship between fitness and food consumption, when R is the total food delivered to a two-nestling brood and q is the proportion of R consumed by the dominant A nestling, the inclusive fitness of the A nestling (I_A) is maximized when:

$$\hat{q} = \frac{1}{1+2^{-\frac{1}{c+1}}} \tag{1}$$

This result pertains so long as the fitness of the A nestling when it dominates its sibling ($q = 1$) is less than its inclusive fitness when it shares food, that is, $W_{A(q=1)} < I_A$, and the food share of the B nestling is greater than 1, that is, $(1 - q)\ R > 1$. Otherwise, the optimum proportion of food consumed by the A nestling is $\hat{q} = 1$ (Anderson & Ricklefs 1995). The value of $\hat{q}$ varies between 0.50 at c approaching ∞ (fitness rises rapidly with increasing food consumption) and 0.67 as c approaches 0. When $c = 3$, the A nestling should monopolize food ($q = 1$) at levels of delivery (R) less than 2.76 and otherwise share food at level $\hat{q} = 0.554$. These relationships are shown in Figure 2. This figure also shows the relative fitnesses of parents and offspring when the fitness of the parent is maximized ($q = 0.5$ when $R >$ 2.47) and when the inclusive fitness of the A nestling is maximized ($q = \hat{q}$ when $R > 2.76$).

One can see that the consequences of parent-offspring conflict for the parents are substantial only for $R > 2.47$ and approaching 2.76, at which point the parent loses about 20% of its potential fitness. Above $R = 2.76$, the A nestling shares food and the maximum fitness difference is about 3%. The differential between the inclusive fitnesses of the A and B nestlings at this point is about 11%, which constitutes strong selection on the B nestling to improve its competitive position, especially compared to the relatively weak selection on parents to overcome parent-offspring conflict.

Brood Size

A second possibility for parents is to reduce the size of their broods, which increases the probability that nestlings are adequately nourished and reduces the probability that parents stress themselves in providing for their offspring. When conditions are relatively constant, parents could rear the young in a brood to a high level of fitness without reducing brood size substantially below the maximum they could nourish. Slightly undershooting the optimum reproductive output may be a good strategy when the costs of overshooting are high. When nestlings are provisioned at a high level, rapid growth is no longer favoured by sibling competition and personal and inclusive fitness factors favouring slow growth can predominate. Slower growth further reduces food requirements and may allow parents to increase the size of their broods. As discussed below, slow growth in multi-nestling broods is relatively common, especially among tropical birds.

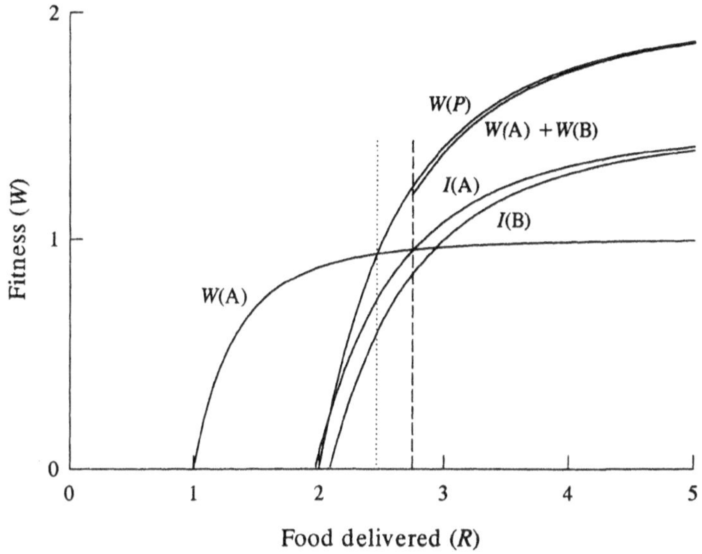

Figure 2. The relationship of various measures of fitness to food delivery to the nest, where $R = 1$ is the minimum for reproductive success > 0 (Anderson & Ricklefs 1995). In this model, two nestlings (A and B) compete for food. *W*(A) is the fitness of the A nestling when it monopolizes all the food. *I*(A) is the inclusive fitness of the A nestling when it controls food allocation and maximizes its inclusive fitness. *I*(B) is the inclusive fitness of the B nestling when the A nestling maximizes its inclusive fitness. *W*(P) is the combined fitness of the parents when food is distributed evenly between nestlings. *W*(A) + *W*(B) is the combined fitness of the A and B nestlings when A shares food with B but maximizes its own inclusive fitness. The vertical dotted line is the lower limit of parent-offspring conflict over the allocation of food. To the left of this line only a single nestling is reared. The vertical dashed line indicates the point where the A nestling switches between food monopoly and food sharing with its younger B nestling, hence the upper bound of the region of maximum parent-offspring conflict.

A difficulty inherent in adjusting brood size is that the number of nestlings changes by integers, thus providing only a coarse adjustment (Ricklefs 1968). This is especially difficult for the transition between two nestlings and one nestling, where the productivity of the brood is nominally halved. However, this transition has occurred most commonly in pelagic seabirds and it has particularly interesting implications for parent-offspring relationships. Nestlings in broods of two compete for limiting resources and presumably they grow rapidly to maintain a size advantage in this competition. This results in a high food requirement during the growth period. When the food available to parents decreases below the level required by two rapidly growing nestlings, brood size must be reduced to one and reproductive output drops by one half. Following the two-nestling/one-nestling transition, however, sibling competition no longer exists within the brood and selection favours reduced growth rate, particularly where time-dependent mortality is low. This in turn allows the parents to exploit less rich or less rewarding food supplies, such as the open ocean at great distance from the nesting colony. The result is the situation typical of pelagic seabirds, with a single, slowly growing nestling and, in some cases, excess parental feeding capacity.

Once the single-nestling brood has been established, it may be difficult to return to a multi-nestling brood. As the food supply increases, either because of improving conditions or a change in the foraging ecology of the parents, individuals may pass the point at which they can rear two slowly growing nestlings. If a two-nestling brood is then produced, however, sibling competition between the nestlings can select for rapid nestling growth rate and food requirements, in which case the parents may no longer be able to nourish their young adequately. Thus, phylogenetic comparisons should reveal more frequent cases of transition from two nestlings to one.

The distribution of clutch size on the phylogeny of Sibley and Ahlquist (1990) suggests that single-egg clutches were derived independently in many groups, including the tropicbirds (Phaethontidae), boobies (Sulidae), frigatebirds (Fregatidae), petrels (Procellariiformes), New World vultures and the hawks and eagles. However, more detailed phylogenies based on DNA sequencing indicate that single-nestling broods may also be ancestral in some lineages, as in the case of some cotingas (Prum et al. 2000) and the blue-footed (*Sula nebouxii*) and Peruvian boobies (*Sula variegata*) (Friesen & Anderson 1997). Such a reversal might occur when a newly exploited food supply is much more abundant than the former food supply.

Optimum Brood Size under Variable Food Conditions

When conditions vary unpredictably from one reproductive attempt to another, the point at which a parent sets its maximum reproductive potential (i.e. brood size) depends on the penalty imposed by trying to raise a large brood under poor feeding conditions. When future reproductive success falls off rapidly as adults are pushed to satisfy demands in excess of their capacity, parents should reduce brood size to correspond more closely to the poorest conditions they are likely to experience (Boyce & Perrins 1987; Lalonde 1991). One consequence of such an adaptive response would be that adults could raise enlarged broods during most years and rarely show signs of reproductive stress.

We can see this effect in a simple model in which the capacity of the parents to deliver food to their brood (R) varies from season to season and adult survival, nestling survival, or both depend on the ratio of nestling food demand (D), which is proportional to brood size, to R. In this model, R is log-normally distributed with the mean of $\log R$ equal to 0 (that is, R = 1). Survival decreases with the demand/resource ratio according to $S = \exp(-a[D/R]^b)$. We define fitness as adult survival, plus recruitment. Parent survival is the product of S_{NR}, the component of survival unrelated to reproduction, and $S_P = f(D/R)$ as above. Recruitment is the product of food demand D and offspring survival $S_O = f(D/R)$. When $a = 0.5$, $b = 3$, and both parents and nestlings suffer equally under a high demand/resource ratio, the optimum brood size is less than the mean resource availability and it decreases as the reproduction-independent survival increases (Figure 3). Using a standard deviation of $\log R$ of 0.3 (a factor of ± 2.0), the optimum brood size is about 1 SD unit below the mean when S_{NR} = 0.50. Thus, resources would exceed demand in 87% of seasons and would be twice the demand in 9% of seasons. This effect increases as the non-reproductive survival increases. This model could be made more complete by including a threshold of resource abundance below which adults abandon their broods (Erikstad et al. 1998). The simpler model, however, makes the point that brood size may often be smaller than parents can rear in a particular season.

Brood Reduction

An alternative to setting a low brood size is facultative brood reduction (Lack 1954; Ricklefs 1965; O'Connor 1978; Mock & Forbes 1994). In this case, brood size is set close to the maximum number of offspring that parents can rear during good years. In bad years, young are selectively starved, beginning with the youngest in the brood, until brood demand is brought into line with the ability of the parents to provide food. The problem with this strategy is that when parents cannot control their response to food solicitation, they are subjected to the constant physiological stress and risk of predation caused by having to provide for an overly large brood, even though the size of the brood may decrease over time. If this is the case, then examples of brood reduction become difficult to explain except during unusually poor feeding conditions.

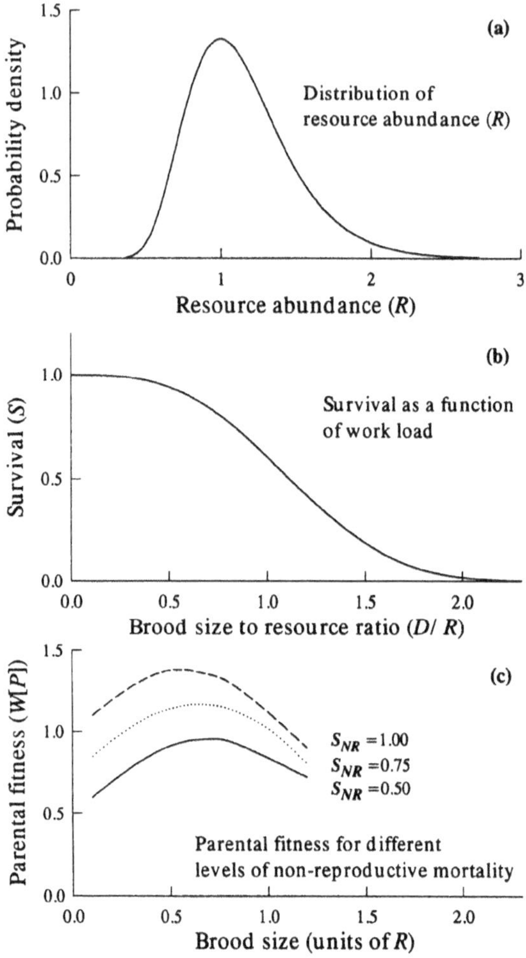

Figure 3. (a) Probability density distribution of resource abundance when resources are distributed log-normally with mean 0 (R = 1) and SD = 0.3. (b) Survival of adult or nestling according to the expression $S = \exp(-a[D/R]^b)$ when a = 0.5 and b = 3. (c) Parental fitness is the sum of adult survival plus number of offspring reared with the survival curves in (b) applied to both nestlings and adults and brood size scaled to units of R under fluctuating environmental conditions (a). Curves in (c) are shown for different levels of survival due to factors unrelated to reproduction (S_{NR}).

Manipulation of the Competitive Hierarchy

The third general mechanism available to parents is to manipulate the competitive hierarchy among nestlings so the nestling cannot gain a fitness advantage through increased competitive advantage. Parents can accomplish this by establishing a rigid competitive hierarchy through asynchronous hatching (Anderson 1989; Lessells & Avery 1989). Competitive success within a brood is determined primarily by body size (Hahn 1981; Bryant & Tatner 1990; Stouffer & Power 1990; Mock & Parker 1997). By hatching the brood asynchronously, parents can establish size differences between nestlings that cannot be overcome by rapid embryonic or postnatal growth (Ricklefs 1993). This mechanism applies most clearly to complete asynchrony within the brood, but may also be effective when only a part of the brood hatches asynchronously (Stouffer & Power 1990). By beginning full incubation after the first egg is laid, parents cause the nestlings to hatch in the order of laying at intervals equal to the laying interval. The spread between hatching can thus be increased by increasing the time between successive eggs. Although a single day is the most common laying interval in birds, intervals of two or more days are common, particularly among species in which the nestlings hatch asynchronously.

Ricklefs (1993) noted a strong association between the length of the incubation period and hatching asynchrony. He suggested that if size differences between nestlings could not be made up by variation in growth rate among nestlings, there would be no selective advantage to rapid growth resulting from increased competitive ability within a brood. Consequently, whatever advantages resulted from slow embryonic development would predominate and one would observe the evolution of long incubation periods. Because prolonged incubation provides no immediate benefit to either the nestling or the parent, and confers many disadvantages, Ricklefs (1993) suggested that a substantial benefit must be obtained through longer potential life span or higher reproductive rate. Potential mechanisms by which this could be accomplished would be to use the longer developmental period to build a better immune system (Ricklefs 1992) or nervous system. Direct evidence bearing on these hypotheses is not available.

The same argument pertains to nestling growth rate as to embryonic growth rate. Asynchronous hatching can eliminate the fitness consequences of variation in postnatal growth rate in a competitive environment and allow advantages of slow growth to predominate. In addition to the possibility of building a higher quality individual, slow growth also reduces food requirements. This may benefit the parent by allowing an increase in brood size, reduced stress of reproduction, or the exploitation of marginal food resources. It may also benefit the nestling directly by reducing the probability of undernourishment during poor feeding conditions, and indirectly by increasing inclusive fitness, that is, the number of siblings reared. When nestlings cannot increase their personal fitness through greater competitive ability, the inclusive fitness of the nestling approaches the personal fitness of the parent and parent-offspring conflict diminishes or perhaps even disappears.

The foregoing arguments suggest that slow postnatal development and reduced parent-offspring conflict should occur when single nestlings are reared and when nestlings hatch asynchronously. There are few exceptions to the strong association between single-nestling broods and slow embryonic and postnatal growth (Werschkul & Jackson 1979; Ricklefs 1982, 1993). Asynchronous hatching and slow embryonic and postnatal growth are also strongly associated (Ricklefs 1993; R.E. Ricklefs unpublished data). Indeed, exceptions to the general pattern in fact support the hypothesis or highlight alternative mechanisms for controlling sibling competition and parent-offspring conflict. Several groups of birds have asynchronous hatching and rapid postnatal growth. Without exception, these are either brood parasites (Cuculiformes, *Buphagus*, *Molothrus*, Viduinae), the hosts of brood parasites, or communal nesters (Coliidae). Where unrelated nestlings occur together in the same nest there is no kin-selected restraint upon competition through inclusive fitness effects. In this case, the most effective strategy for the parent and its nestling is to hatch as soon as possible by beginning incubation with the first egg laid and to grow as rapidly as possible.

Exceptions to the general pattern also occur in the opposite direction, that is, slow development with synchronous hatching. Ricklefs (1993) found that several groups of birds had synchronous hatching and long incubation periods: Alcedinidae (kingfishers), Criniferinae (go-away birds), Acanthisittidae (riflemen), Pipridae (manakins) and Acanthizini (thornbirds). None of these groups is well known, but detailed studies of the common kingfisher (*Alcedo atthis*) and rifleman (*Acanthisitta chloris*) found little or no starvation within broods (Gray 1969; Boag 1982). Thus, it is possible that food supplies for these species are predictable and adults have reduced brood size enough to eliminate sibling competition for food.

The general picture that emerges from comparative studies is that parents manipulate sibling competition to reduce development rate and increase personal fitness under two circumstances. The first, employed by pelagic seabirds, large raptorial birds, and a few tropical species (e.g. cotingas and some doves), enables parents to exploit marginal food resources. Reducing brood size to one nestling eliminates sibling competition and allows for the evolution of slow development. This in turn reduces demand for food delivery and allows parents to utilize even more marginal food resources. Somewhat surprising is the absence of transitional stages having two-nestling broods, asynchronous hatching, and slow development. Reducing clutch size produces a decreased food requirement quickly and reliably. Asynchronous hatching followed by selection for slow development may require a long period of evolution, at the beginning of which there is no reduction in food requirement, unless strong sibling hierarchies reduce the intensity of food solicitation and direct competition between nestlings.

The second circumstance occurs among generally long-lived species of primarily tropical birds, including most of the Coraciiformes, all Psittaciformes and Apodiformes, and some suboscine South American Passeriformes. Ricklefs (1993) argued that the primary advantage of long incubation periods in these groups is based on a link between slow embryonic development and long potential life span. Slow postnatal development would convey a further advantage to parents by reducing the food requirements of their nestlings. One of the chief difficulties of this hypothesis is that the benefits resulting from asynchronous hatching do not appear immediately, but follow many generations of evolution of slow growth rate. Incubation period is phylogenetically conservative, most of the variance in incubation period residing at the ordinal level (Ricklefs 1993). This difficulty would largely disappear if asynchronous hatching and slow development were ancestral, but there is no clear indication that this is so. The basal, non-raptorial altricial land birds, the Piciformes, uniformly have synchronous hatching, rapid embryonic development, and moderate to rapid post-hatching growth.

THE TRANSITION BETWEEN LIFE HISTORIES WITH AND WITHOUT PARENT-OFFSPRING CONFLICT

Parent birds can reduce or eliminate sibling competition and parent-offspring conflict by reducing the size of the brood, hatching the eggs asynchronously or both. Because changes in development rate evolve slowly, the primary advantages to smaller brood size and asynchronous hatching must be realized through reduced reproductive stress on the adults resulting from reduced demand at the nest. By itself, complete asynchronous hatching may reduce food requirement by reducing competitive behaviour within the brood as a direct consequence of a nestling's position within a size hierarchy. A size hierarchy within the brood may also help adults to distribute food optimally among nestlings under the range of food delivery rates in which parent-offspring conflict would be strong (see Figure 2). Subsequent advantages of slow development to the parent or nestling could be realized over longer evolutionary periods.

These considerations raise the question of why many species lack mechanisms to reduce sibling competition and parent-offspring conflict. Where potential adult life span is short owing to high mortality unrelated to reproduction, benefits of higher reproductive success from large broods may outweigh costs of additional adult mortality, and brood size may increase. Even with larger brood sizes, however, complete asynchronous hatching would appear to convey advantages when it facilitates brood reduction and reduces the costs of competition between nestlings. Balancing these benefits are several potential costs, including management of a brood having a spread of ages, especially at the time of fledging. In addition, sibling competition selects for rapid development, which may increase the fitness of parents under high time-dependent nestling mortality or where breeding seasons are short (Ricklefs 1984). Thus, under some circumstances, parents may use sibling competition to promote adaptations that increase their personal fitness at the expense of inclusive fitness of individual nestlings.

FUTURE DIRECTIONS

The study of parent-offspring interactions would benefit from more detailed modelling of the fitness consequences of asynchronous hatching and reduced brood size when evolutionary responses of nestling growth rate are included as factors that influence nestling food requirements and feed back on parental fitness through food provisioning rates. Such studies should be combined with additional empirical comparisons and experimental analyses of the relationship between parent and offspring fitness under varying food requirements and satisfaction of those requirements. Analysis of parent-offspring relationships in a phylogenetic context might provide insights into evolutionary transitions in brood size and hatching asynchrony. Nestling begging and parental responses should also parallel the control of sibling competition by parents, which could be addressed by comparative studies of begging in species with synchronous and asynchronous hatching. Where sibling competition and active begging attract predators to the nest, parents might be expected to control sibling competition even at the expense of brood size.

A major related problem is to understand the significance of slow development rates in some tropical species in spite of high time-dependent mortality of eggs and nestlings. Little is known of the dependence of food requirements on growth rate or the potential fitness benefits of slower growth. Although the evolution of slow growth with strongly asynchronous hatching is relatively straightforward, its occurrence in a few cases with synchronous hatching is puzzling. One might predict that such species would exhibit a high mean and a low within-brood variance in offspring fitness, indicating that young cannot improve their personal fitness through enhanced within-brood competition. Finally, empirical studies of brood-size-dependent nesting success and adult survival under varying conditions might shed light on reduced brood size as an adult strategy to improve lifetime reproductive success.

ACKNOWLEDGEMENTS

This chapter benefited from the insightful comments and suggestions of Marty Leonard, Kate Lessells and Jon Wright. The author's research is supported in part by NIH grant AG16895-01.

REFERENCES

Anderson, D.J. 1989. The role of hatching asynchrony in siblicidal brood reduction of two booby species. *Behavioral Ecology and Sociobiology* **25**, 363-368.

Anderson, D.J. & Ricklefs, R.E. 1995. Evidence of kin-selected tolerance by nestlings in a siblicidal bird. *Behavioral Ecology and Sociobiology* **37**, 163-168.

Bengtsson, H. & Rydén, O. 1983. Parental feeding rate in relation to begging behavior in asynchronously hatched broods of the great tit *Parus major*. An experimental study. *Behavioral Ecology and Sociobiology* **12**, 243-251.

Boag, D. 1982. *The Kingfisher*. Dorset: Blandford Press.

Bosque, C. & Bosque, M.T. 1995. Nest predation as a selective factor in the evolution of developmental rates in altricial birds. *American Naturalist* **145**, 234-260.

Boyce, M. & Perrins, C.M. 1987. Optimizing great tit clutch size in a fluctuating environment. *Ecology* **68**, 142-153.

Briskie, J.V., Naugler, C.T. & Leech, S.M. 1994. Begging intensity of nestling birds varies with sibling relatedness. *Proceedings of the Royal Society of London, Series B* **258**, 73-78.

Bryant, D.M. & Tatner, P. 1990. Hatching asynchrony, sibling competition and siblicide in nestling birds: studies of swiftlets and bee-eaters. *Animal Behaviour* **39**, 657-671.

Christe, P., Richner, H. & Oppliger, A. 1996. Begging, food provisioning, and nestling competition in great tit broods infested with ectoparasites. *Behavioral Ecology* **7**, 127-131.

Congdon, B.C. 1990. Brood enlargement and post-natal development in the black noddy *Anous minutus*. *Emu* **90**, 241-247.

Cotton, P.A., Kacelnik, A. & Wright, J. 1996. Chick begging as a signal: are nestlings honest? *Behavioral Ecology* **7**, 178-182.

Cucco, M. & Malacarne, G. 1995. Increase of parental effort in experimentally enlarged broods of pallid swifts. *Canadian Journal of Zoology* **73**, 1387-1395.

Dijkstra, C., Bult, A., Bijlsma, S., Daan, S., Meijer, T. & Zijlstra, M. 1990. Brood size manipulations in the kestrel (*Falco tinnunculus*): effects on offspring and parent survival. *Journal of Animal Ecology* **59**, 269-285.

Dunn, E.H. 1980. On the variability of energy allocation of nestling birds. *The Auk* **97**, 19-27.

Erikstad, K.E., Fauchald, P., Tveraa, T. & Steen, H. 1998. On the cost of reproduction in long-lived birds: the influence of environmental variability. *Ecology* **79**, 1781-1788.

Forbes, L.S. 1993. Avian brood reduction and parent-offspring "conflict". *American Naturalist* **142**, 82-117.

Friesen, V.L. & Anderson, D.J. 1997. Phylogeny and evolution of the Sulidae (Aves: Pelecaniformes): a test of alternative modes of speciation. *Molecular Phylogenetics and Evolution* **7**, 252-260.

Godfray, H.C.J. 1991. Signalling of need by offspring to their parents. *Nature* **352**, 328-330.

Godfray, H.C.J. 1995. Signaling of need between parents and young: parent-offspring conflict and sibling rivalry. *American Naturalist* **146**, 1-24.

Gray, R.S. 1969. Breeding biology of rifleman at Dunedin. *Notornis* **16**, 5-22.

Hahn, D.C. 1981. Asynchronous hatching in the laughing gull: cutting losses and reducing rivalry. *Animal Behavior* **29**, 421-427.

Harper, A.B. 1986. The evolution of begging: sibling competition and parent-offspring conflict. *American Naturalist* **128**, 99-114.

Henderson, B.A. 1975. Role of the chick's begging behavior in the regulation of parental feeding behavior of *Larus glaucescens*. *Condor* **77**, 488-492.

Hussell, D.J.T. 1988. Supply and demand in tree swallow broods: a model of parent-offspring food-provisioning interactions in birds. *American Naturalist* **131**, 175-202.

Jacobsen, K.O., Erikstad, K.E. & Sæther, B.E. 1995. An experimental study of the costs of reproduction in the kittiwake *Rissa tridactyla*. *Ecology* **76**, 1636-1642.

Johnstone, R.A. 1996. Begging signals and parent-offspring conflict: do parents always win? *Proceedings of the Royal Society of London, Series B* **263**, 1677-1681.

Klaassen, M. & Bech, C. 1992. Resting and peak metabolic rates of arctic tern nestlings and their relations to growth rate. *Physiological Zoology* **65**, 803-814.

Konarzewski, M., Kooijman, S.A.L.M. & Ricklefs, R.E. 1998. Models for avian growth and development. In: *Avian Growth and Development. Evolution within the Altricial-Precocial Spectrum* (Ed. by J.M. Starck & R.E. Ricklefs). Oxford: Oxford University Press.

Krebs, E.A., Cunningham, R.B. & Donnelly, C.F. 1999. Complex patterns of food allocation in asynchronously hatching broods of crimson rosellas. *Animal Behaviour* **57**, 753-763.

Lack, D. 1954. *The Natural Regulation of Animal Numbers*. Oxford: Clarendon Press.

Lalonde, R.G. 1991. Optimal offspring provisioning when resources are not predictable. *American Naturalist* **138**, 680-686.

Leonard, M. & Horn, A. 1996. Provisioning rules in tree swallows. *Behavioral Ecology and Sociobiology* **38**, 341-347.

Leonard, M.L. & Horn, A.G. 1998. Need and nestmates affect begging in tree swallows. *Behavioral Ecology and Sociobiology* **42**, 431-436.

Leonard, M.L., Horn, A.G., Gozna, A. & Ramen, S. 2000. Brood size and begging intensity in nestling tree swallows. *Behavioral Ecology* **11**, 196-201.

Lessells, C.M. & Avery, M.I. 1989. Hatching asynchrony in European bee-eaters *Merops apiaster*. *Journal of Animal Ecology* **58**, 815-835.

Linden, M. & Møller, A.P. 1989. Cost of reproduction and covariation of life history traits in birds. *Trends in Ecology and Evolution* **4**, 367-371.

Magrath, R.D. 1990. Hatching asynchrony in altricial birds. *Biological Reviews* **65**, 587-622.

Martin, T.E. 1995. Avian life history evolution in relation to nest sites, nest predation and food. *Ecological Monographs* **65**, 101-127.

Mock, D.W. & Forbes, L.S. 1994. Life-history consequences of avian brood reduction. *The Auk* **111**, 115-123.

Mock, D.W. & Parker, G.A. 1997. *The Evolution of Sibling Rivalry*. Oxford: Oxford University Press.

O'Connor, R.J. 1978. Brood reduction in birds: selection for fratricide, infanticide and suicide? *Animal Behavior* **26**, 79-96.

Ottosson, U., Backman, J. & Smith, H.G. 1997. Begging affects parental effort in the pied flycatcher, *Ficedula hypoleuca*. *Behavioral Ecology and Sociobiology* **41**, 381-384.

Parker, G.A., Mock, D.W. & Lamey, T.C. 1989. How selfish should stronger sibs be? *American Naturalist* **133**, 846-868.

Price, K. & Ydenberg, R. 1995. Begging and provisioning in broods of asynchronously-hatched yellow-headed blackbird nestlings. *Behavioral Ecology and Sociobiology* **37**, 201-208.

Prum, R.O., Rice, N.H., Mobley, J.A. & Dimmick, W.W. 2000. A preliminary phylogenetic hypothesis for the cotingas (Cotingidae) based on mitochondrial DNA. *The Auk* **117**, 236-241.

Ricklefs, R.E. 1965. Brood reduction in the curve-billed thrasher. *Condor* **67**, 505-510.

Ricklefs, R.E. 1968. On the limitation of brood size in passerine birds by the ability of adults to nourish their young. *Proceedings of the National Academy of Sciences USA* **61**, 847-851.

Ricklefs, R.E. 1969. An analysis of nesting mortality in birds. *Smithsonian Contributions to Zoology* **9**, 1-48.

Ricklefs, R.E. 1982. Some considerations on sibling competition and avian growth rates. *The Auk* **99**, 141-147.

Ricklefs, R.E. 1984. The optimization of growth rate in altricial birds. *Ecology* **65**, 1602-1616.

Ricklefs, R.E. 1992. Embryonic development period and the prevalence of avian blood parasites. *Proceedings of the National Academy of Sciences USA* **89**, 4722-4725.

Ricklefs, R.E. 1993. Sibling competition, hatching asynchrony, incubation period, and lifespan in altricial birds. *Current Ornithology* **11**, 199-276.

Ricklefs, R.E. & Scheuerlein, A. 2001. Comparison of aging-related mortality among birds and mammals. *Experimental Gerontology* **36**, 845-857.

Roulin, A., Kölliker, M. & Richner, H. 2000. Barn owl (*Tyto alba*) siblings vocally negotiate resources. *Proceedings of the Royal Society of London, Series B* **267**, 459-463.

Sibley, C.G. & Ahlquist, J.E. 1990. *Phylogeny and Classification of Birds*. New Haven: Yale University Press.

Stamps, J., Clark, A.B., Arrowood, P. & Kus, B. 1985. Parent-offspring conflict in budgerigars. *Behaviour* **94**, 1-40.

Stouffer, P.C. & Power, H.W. 1990. Density effects of asynchronous hatching and brood reduction in European starlings. *The Auk* **107**, 359-366.

Trivers, R.L. 1974. Parent-offspring conflict. *American Zoologist* **14**, 249-264.

Weathers, W.W., Hodum, P.J. & Anderson, D.J. 1997. Is the energy cost of begging by nestling passerines surprisingly low? *The Auk* **114**, 133.

Werschkul, D.F. & Jackson, J.A. 1979. Sibling competition and avian growth rates. *Ibis* **121**, 97-102.

Whittingham, L.A. & Robertson, R.J. 1993. Nestling hunger and parental care in red-winged blackbirds. *The Auk* **110**, 240-246.

Ydenberg, R.C. & Bertram, D.F. 1989. Lack's clutch size hypothesis and brood enlargement studies on colonial seabirds. *Colonial Waterbirds* **12**, 134-137.

16.

FEEDING CHASES IN PENGUINS: BEGGING COMPETITION ON THE RUN?

Javier Bustamante[1], P. Dee Boersma[2] & Lloyd S. Davis[3]
[1]*Department of Applied Biology, Estación Biológica de Doñana, CSIC, Sevilla, Spain (busta@ebd.csic.es)*
[2]*Department of Zoology, University of Washington, Seattle, Washington 98195-1800, USA (boersma@u.washington.edu)*
[3]*Department of Zoology, University of Otago, Dunedin, New Zealand (adelie@stonebow.otago.ac.nz)*

ABSTRACT

Feeding interactions in Pygoscelid penguins normally involve chases where a parent runs and is closely followed by its chicks. Several hypotheses have been advanced to explain this behaviour. Studies with chinstrap and Adélie penguins indicate that feeding chases are mostly associated with a one-parent/two-chick situation. Chases are almost absent in one-chick broods, or situations in which one sibling is removed from the crèche and only the remaining sibling begs for food. Data are consistent with the hypothesis that chases serve to separate siblings in two-chick broods in order to feed them more efficiently.

INTRODUCTION

Penguins are colonial seabirds nesting in burrows or in open nests and raising one or two chicks. The nestling period is divided into an initial guarding stage, in which the chicks remain at the nest and are guarded by one of the parents, and a crèche stage, in which both parents forage at sea simultaneously. At the latter stage, chicks move from the nest site and join other chicks of the colony in groups or crèches, with parents regularly coming ashore to feed. During the crèche stage these feeding interactions are frequently associated with chases in which a parent runs and is closely

J. Wright and M.L. Leonard (eds.), The Evolution of Begging, 303–318.

followed by begging chicks. When an adult comes ashore it approaches its nest site and gives a series of display calls. If not already at the nest site, its chicks will emerge from the crèche, approach the vocalizing adult and beg for food. The adult may feed its offspring at this point or run off pursued by the begging chicks, which it stops intermittently to feed. It takes 10 to 15 minutes for an adult to feed one or two chicks about a kilogram of food, divided into 15 to 20 feeds. Chases can take place before the first feed, between consecutive feeds, or after the last feed as the parent leaves the colony or returns to the sea (Table 1).

Table 1. Comparison of feeding chases in Adélie, chinstrap and gentoo penguins. When two figures are given they correspond to one-chick and two-chick broods respectively.

	Adélie[1]	chinstrap[2]	gentoo[3]
Chick age at earliest chases (d)		32.8[5]	
Meal* duration (s)	584-673	779[6]	840
No. chases/meal	4.8-7.8	2.1-9.0	1.7
No. chases before first feed†	0.3-0.6	0.1-2.3	0.5
No. chases during feeds		0.5-4.6	0.3
No. chases after last feed		1.5-2.1	1.0
No. of feeds per meal	14.8-16.1	17.0-17.7	19
Mean maximum distance to the nest during a meal (m)	43.9-84.5[4]		

[1]Boersma & Davis 1997; [2]Bustamante et al. 1992; [3]Lundberg & Bannasch 1983; [4]Thompson 1981; [5]Moreno et al. 1998; [6]Moreno et al. 1996; *meal = complete feeding interaction from parent arrival to parent departure; †feed = each single ball of food transferred to a chick during a meal.

Limited forms of chases (or chicks following the adult before or after being fed) have been described occasionally in other species of penguins (Table 2), but frequent and well-developed feeding chases seem to take place only in the Adélie, chinstrap and gentoo penguins. These Pygoscelids raise one or two chicks, are open-nesters and have chicks that join crèches. Feeding chases are similar in the three species and occur once chicks join crèches at about two to five weeks of age.

Table 2. Published references and studies of feeding chases in penguins.

Species	References to feeding chases	Descriptive studies	Experimental studies
Emperor penguin *Aptenodytes forsteri*	1*		
King penguin *Aptenodytes patagonica*	2*		
Adélie penguin *Pygoscelis adeliae*	3,4,5	6,7,8	
Gentoo penguin *Pygoscelis papua*	9,10	7	
Chinstrap penguin *Pygoscelis antarctica*	11	7,12,13	14
Crested penguins *Eudyptes spp.*	15		
Erect-crested penguin *Eudyptes sclateri*	16		
Fiordland crested penguin *Eudyptes pachyrhynchus*	17*		
Jackass penguin *Spheniscus demersus*	18		
Yellow-eyed penguin *Megadyptes antipodes*	19		

1 Prévost 1961; 2 Stonehouse 1960; 3 Sladen 1958; 4 Taylor 1962; 5 Penney 1968; 6 Thompson 1981; 7 Lundberg & Bannasch 1983; 8 Boersma & Davis 1997; 9 Pettingill 1964; 10 Van Zinderen Bakker 1971; 11 W.J.L. Sladen personal communication in Thompson 1981; 12 Bustamante et al. 1992; 13 Moreno et al. 1998; 14 Moreno et al. 1996; 15 J. Warham personal communication 1971 in Thompson 1981; 16 Warham 1972; 17 Warham 1974; 18 Kearton 1931 in Thompson 1981; 19 Richdale 1957 in Thompson 1981; * the behaviour described could be a feeding chase, but is not defined as such by the author.

In other penguin species that raise a single chick and have open nests and crèching behaviour, (genus *Aptenodytes* and *Eudyptes*), the chick may follow its parent out of the crèche for a short distance (5-10 m) before being fed. This is probably to avoid disturbance from other chicks or adults, but this behaviour has not been studied in detail. In species of penguins raising one or two chicks, nesting in burrows and with less developed crèching behaviour (e.g. genus *Spheniscus*), parents seem to control the transfer of food by using their flippers to counteract sibling competition (Boersma 1991) and chases are rare (Seddon 1990; Seddon & Van Heezik 1991).

HYPOTHESES TO EXPLAIN THE FUNCTION OF FEEDING CHASES

Chases Prevent Competition or Interference from Foreign Chicks or Adults

First proposed by Sladen (1958) and later by others (Penney 1968; Müller-Schwarze & Müller-Schwarze 1977), this hypothesis suggests that chases are initiated by parents to separate their chicks from the rest of the crèche and thus avoid competition from foreign chicks or interference by territorial adults. The hypothesis predicts that (1) feeding chases should be more common in species with crèching behaviour than those without; (2) within species, feeding chases should be more common when breeding at higher densities; and (3) feeding chases should lead chicks out of the crèche where feeding can take place in the absence of competition or interference.

In relation to the first prediction, feeding chases are common in Pygoscelids, but are not so common or well developed in *Aptenodytes* or *Eudyptes* species that also form crèches. Species that normally do not form crèches like *Spheniscus spp.*, yellow-eyed penguins and little penguins (*Eudyptula minor*) do not have feeding chases. Regarding the second prediction, no studies have tested the prevalence of feeding chases at different breeding densities or in crèches of different sizes within a species. In relation to the third prediction, in chinstrap penguins chases do not always lead out of the crèche. Almost all feeds occurred inside the colony in one-chick families, as do at least half of those in two-chick families (Bustamante et al. 1992). Similarly, in Adélie penguins feeding away from the colony is more common in two-chick than one-chick situations (Boersma & Davis 1997). As interference from foreign chicks should not be dependent on brood size, this hypothesis cannot explain why in one-chick broods feeding outside the colony is less frequent and feeding chases are less frequent (Bustamante et al. 1992; Boersma & Davis 1997), and why similar patterns were found in two-chick broods following the temporary experimental removal of one chick (Moreno et al. 1996).

Summing up, observational data do not provide much support for the hypothesis that feeding chases occur because they reduce competition from foreign chicks or interference from territorial adults.

Chases Facilitate Parent-Offspring Recognition

Parents could potentially recognise their chicks by the latter's propensity to run after them (N.J. Volkman & S.G. Trivelpiece personal communication, cited in Lundberg & Bannasch 1983). This hypothesis predicts (1) that chases should be more common in species with crèching behaviour because chicks that abandon the nest site cannot be recognized by their location. It also predicts (2) that chases should be independent of brood size, as the need to recognize offspring once they are in a crèche is independent of whether there are one or two chicks; and (3) chases should take place before the first feed. In accordance with the first prediction chases are present in species with crèching behaviour, but are not common in all of them (see above). Contrary to the second prediction chases are also clearly associated with two-chick broods. In addition, in contradiction to the third prediction many feeding chases in the chinstrap (Lundberg & Bannasch 1983; Bustamante et al. 1992), Adélie (Lundberg & Bannasch 1983; Boersma & Davis 1997) and gentoo penguins (Lundberg & Bannasch 1983) take place after the first feed (i.e. after any offspring recognition would already have taken place).

If chases are a means whereby parents recognize their own chicks, we have to assume that chicks cannot be recognized in a less costly way. Vocal cues, however, allow offspring to recognize their parents in the Adélie penguin (Penney 1968; Davis & McCaffrey 1989). Although chicks may beg for food from adults other than their parents, they are often aggressively repelled both in the chinstrap penguin (Bustamante et al. 1992) and in the Adélie penguin (Boersma & Davis 1997). This demonstrates that adults recognize foreign chicks by means other than the chase. Even when foreign chicks participate in chases, they very rarely obtain any food, indicating that they are not able to deceive the adults. In chinstrap penguins, Bustamante et al. (1992) observed nine individually marked chicks participating in feeding chases with adults that were not their parents and none obtained a feed. In conclusion, predictions from this hypothesis are not supported by observational evidence and it is unlikely that feeding chases are necessary for offspring recognition.

Chases Reduce Sibling Competition and Improve the Efficiency of Food Transfer

Müller-Schwarze and Müller-Schwarze (1977) and later Thompson (1981) suggested that chases temporarily separate siblings in two-chick broods, avoiding unnecessary and wasteful sibling competition and ensuring the efficient feeding of both chicks. This hypothesis predicts that (1) chases should be uncommon in species raising a single chick. (2) In those raising two chicks chases should be more common in two-chick broods than in one-chick broods, in which no sibling separation is necessary. (3) Parents should also be more inclined to run when their offspring are closer together than when they are further apart (as they can feed one chick without interference from its sibling). (4) Feeding chases should not take place if only one of the chicks from a two-chick brood is present at the arrival of the parent. (5) Chases should tend to separate siblings and parents should wait until siblings are separated before they start feeding one chick. (6) Chases should create a symmetric distribution of food within the brood.

Feeding chases are rare or less developed in *Aptenodytes* and *Eudyptes* species that raise only one chick, in accordance with our first prediction. In agreement with the second prediction, both in the Adélie (Thompson 1981; Boersma & Davis 1997) and in the chinstrap penguin (Bustamante et al. 1992) chases are more frequent in two-chick than in one-chick broods. Supporting the third prediction feeding chases in the Adélie penguin are more likely to occur when siblings are less than 0.5 m apart than when they are farther apart (Boersma & Davis 1997). The fourth prediction is supported by an experimental study by Moreno et al. (1996) which showed that chases did not take place when only one sibling of a two-chick brood was present at the arrival of the adult. Bustamante et al. (1992) found that chases in which the two siblings participated ended with the separation of the chicks before being fed in only half of the instances, which to some extent in disagrees with prediction number 5. However, better data by Moreno et al. (1996) demonstrated that chasing intensity drops dramatically and food transfer rate increases after siblings are separated, thereby supporting prediction number 5 (Figure 1). In support of the last prediction, Boersma & Davis (1997) showed that in Adélies the chick that is fed tends to alternate after each feeding chase, and one sibling monopolized all feeds in only 12% of meals. They suggested that chases create a more symmetric distribution of food within the brood.

Observational and experimental data are in strong agreement with the predictions from this hypothesis. The hypothesis does not explain why, apart from Pygoscelids, other species raising two chicks do not rely on feeding chases to reduce sibling competition and transfer food efficiently. A possible reason why temperate species of penguins do not exhibit feeding chases is the difference in food availability. At the high latitudes where Pygoscelid penguins breed, food is more abundant than it is at lower latitudes. There is a considerable amount of evidence that food is often limited for temperate penguins (Boersma 1978; Boersma et al. 1990). When food is more limited feeding chases might become too energetically costly, so it may not be surprising that temperate penguins do not have well-developed feeding chases.

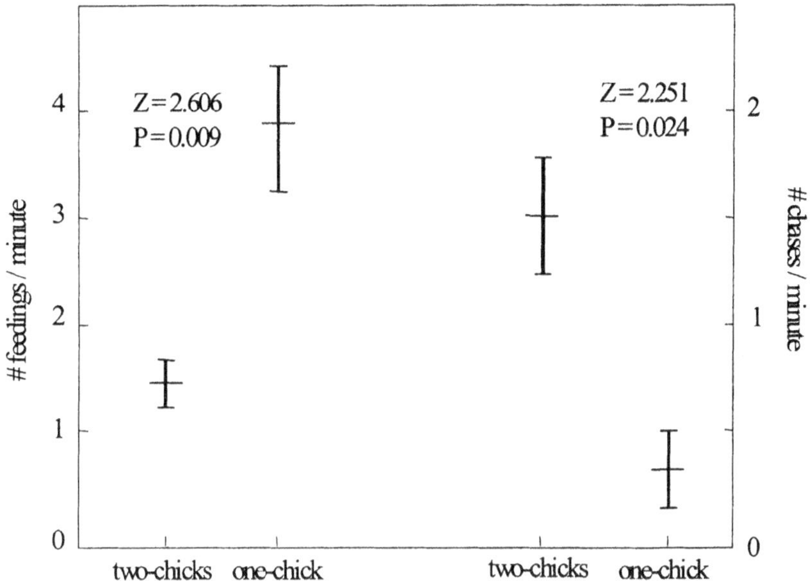

Figure 1. Effects of sibling separation on feeding and chasing rates in the chinstrap penguin. Modified from Moreno et al. (1996). Two-chicks refers to the first part of a feeding chase, when both chicks were chasing, whereas one-chick refers to the final part of a chase when only one chick still followed the parent. Results of Wilcoxon matched-pairs tests; n = 9.

Chases Facilitate Brood Reduction

In direct opposition to the previous hypothesis, feeding chases could be a mechanism to create asymmetries and facilitate brood reduction if they lead to preferential feeding of the strongest chick when food is scarce (Lundberg & Bannasch 1983). In any case, feeding chases cannot be considered as the main brood reduction mechanism in Pygoscelid penguins because they do not occur during the guarding stage, when brood reduction is common, and can only operate at the crèche stage, when chick mortality is low.

Most of the general predictions of this hypothesis are the same as those of the 'Efficient Food Transfer Hypothesis' (see above). They only differ in that this hypothesis predicts an increase in feeding chase intensity and more asymmetric food distribution within broods when food is limited.

As with the 'Efficient Food Transfer Hypothesis', the observational data are in agreement with the general predictions of this hypothesis. There are, however, no adequate data to test if chases are more frequent, or food distribution more asymmetric, when food is limited. In the Adélie penguin, Boersma and Davis (1997) found that the largest chick was fed first more often in broods where size differences were obvious, but the larger chick did not obtain a greater number of feeds. Bustamante et al. (1992) describe a case of brood reduction during the crèche stage in a situation in which a single adult was feeding two chicks. On two occasions chick A chased the most and obtained all the feedings. Chick B was found dead a few days later with a very low weight indicating starvation.

Mortality during the crèche stage was rare in the Adélie at Cape Bird and chicks over 30 days had very little likelihood of dying (Davis & McCaffrey 1986). In this situation, feeding chases played no role in brood reduction. On the other hand, chicks in two-chick broods of chinstrap penguins at Deception Island had a 0.6% daily mortality rate during the crèche stage (Moreno et al. 1994), and most mortality was due to starvation. Under this scenario feeding chases could facilitate brood reduction by providing an asymmetric distribution of food.

Observational evidence does not allow us to reject this hypothesis completely, but brood reduction being so uncommon during the crèche stage suggests that it cannot be the main cause of feeding chases.

Adults Initiate Chases to Avoid Harassment by their Chicks

Lundberg and Bannasch (1983) suggest that chases could be the result of parents being unable to cope with the intense begging activity of their offspring. Chicks are always fed in front of the adult while begging at the tip of the parent's bill. Chicks push against each other to gain a feeding position and frequently interrupt feeds by pushing in front of the chick being fed.

Although this hypothesis is usually considered as a proximate cause of feeding chases, harassment avoidance could be considered a function in itself, if harassment had a negative effect on the well-being of the parents (Moreno et al. 1996). Also, factors other than harassment by chicks (acoustic stimulation, proximity between siblings and to the parent) could be the proximate factors that initiate feeding chases.

Specific predictions from the 'Harassment Avoidance Hypothesis' are that (1) feeding chases should increase with chick age, because heavier competing chicks will be more difficult to separate; (2) the initiation of chases should be associated with active competition between siblings (siblings pushing against each other to gain a feeding position) and not merely with the distance between them; (3) duration of chases should be a consequence of the intensity of parental reaction, with no relation to the time it takes to separate the two siblings; (4) the propensity to feed after a chase should not depend on whether siblings have become separated, but merely on whether or not they compete.

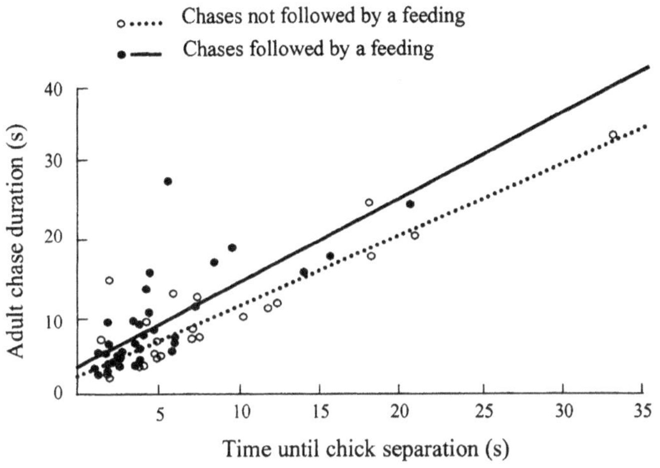

Figure 2. Time parents spent running as a function of the time it takes for the siblings to be separated during a feeding chase. Data are from chinstrap penguins in Deception Island (modified from Moreno et al. 1998). The two regression lines represent chases followed by a feeding ($y = 3.87 + 1.09x$, $F = 42.64$, $P < 0.001$, $R = 0.75$, $n = 35$), and chases not followed by a feeding ($y = 2.63 + 0.91x$, $F = 95.47$, $P < 0.001$, $R = 0.89$, $n = 27$).

In chinstrap penguins chases do not increase with chick age (Bustamante et al. 1992) and chases are frequently initiated without clear signs of sibling competition; when siblings are less than 1 m apart and away from their parent (Moreno et al. 1998). The time the parent spends running is determined by the time it takes until siblings are separated, which would not be the case if the parent ran just as a reaction to harassment by chicks. If the parent stops briefly (i.e. less than four seconds) after chicks become separated, the second chick will frequently catch up, but the chase will not be followed by a feeding (Figure 2). Parents avoid delivering food in the presence of two begging chicks, independently of their competitive disposition (Moreno et al. 1998). In the Adélie penguin, feeding chases were more likely to occur when siblings were less than 0.5 m apart than when they were further apart (Boersma & Davis 1997).

The specific predictions of this hypothesis are not supported by the observational data from the chinstrap penguin; the only species for which data to test the predictions are adequate.

Chases Allow Parents to Assess the Nutritional Status of their Offspring

By provoking a feeding chase the parent could evaluate how hungry the chicks are and thereby determine how much food to give them (Thompson 1981). The parent could decide how much food to give its offspring and how much to keep and, in the two-chick situation, how to divide the food between the two chicks. Information regarding nutritional status is required because adults feed chicks independently of their mates and do not know when each chick has received its last feeding. In addition, feeding frequency by penguins is low (usually less than once per day), making it difficult to track changes in offspring need.

The hypothesis predicts that (1) chasing effort by chicks is a good index of hunger. (2) Feeding chases should take place both in species raising one chick and those raising or two chicks; and (3) in those typically raising two chicks, in both one-chick and two-chick broods.

Thompson (1981) showed experimentally in Adélie penguins that chick persistence in a chase was a function of the time since the last feed, in agreement with prediction number 1. This relationship between hunger and persistence has also been observed in chinstraps (J. Moreno personal communication). Bustamante et al. (1992) indicate that this hypothesis predicts that chases should be most obvious prior to the first feed of a meal, which was not the case in their study of chinstrap penguins. Once chick hunger state has been ascertained, further chasing seems wasteful, but parents could still use chases between successive feedings as a way of testing if the amount of food transferred to that chick has reduced its hunger level. The apparent lack of feeding chases in single-brooded species contradicts the second prediction. In partial agreement with the third prediction there are some chases in one-chick broods, as observed both in the Adélie (Boersma & Davis 1997) and in the chinstrap penguin (Bustamante et al. 1992), which take place mainly after the first feeding. This hypothesis predicts (4) that a chick that has been recently fed should not be as persistent in a chase as a sibling that has not been fed. There are currently no data to test this prediction, but Bustamante et al. (1992)

observed that in some instances (6 out of 26) the arrival of a parent of a two-chick brood elicited no response from one of the siblings. This chick could have been satiated and therefore not interested in chasing.

This hypothesis could explain why some feeding chases take place in one-chick broods of Pygoscelid penguins. If this was the main reason for feeding chases to evolve, however, one would expect them to be as frequent as in the two-chick broods, and to appear in species raising only one chick.

Chases are an Expression of Parent-Offspring Conflict

Chases could be an expression of parent-offspring conflict prior to chick independence. Parents will become increasingly reluctant to invest in (i.e. feed) chicks, which are themselves interested in maintaining the level of parental contribution (Trivers 1974). This hypothesis predicts (1) an increase in chasing intensity with chick age. It also predicts (2) that other indicators of parent-offspring conflict, such as frequency of begging or parental aggression toward chicks, should increase with chick age and the frequency of feedings should decrease. It does not predict that (3) there should be differences in feeding chases between species that raise one or two chicks, nor (4) between one- or two-chick broods. In the chinstrap penguin, Bustamante et al. (1992) found no significant increase in feeding chases or in any other indicator of parent-offspring conflict with chick age and no significant decrease in feeding frequency with chick age, in disagreement with predictions 1 and 2. No other studies have investigated in detail other expressions of parent-offspring conflict in penguins. The lack of chases in species raising a single chick and the low incidence in one-chick broods, although not inconsistent with this hypothesis, do not provide strong support.

Chases are a Secondary Expression of Other Evolutionary Pressures upon Chicks

Chases could be a secondary result of muscle training, exploring new ground, finding the location of departure routes to the sea or acquiring social experiences during the period before independence (Sladen 1958; Müller-Schwarze & Müller-Schwarze 1977; Thompson 1981). Although all of these could be secondary consequences of feeding chases, none of them are clearly the main cause. Thompson (1981) found that chases in the Adélie penguin were directed preferentially to the beach that would be used as a departure route at fledging, but Bustamante et al. (1992) found no clear association between feeding chase direction in the chinstrap penguin and the location of the departure beach. In addition, chicks did not remain away from the crèche exploring new ground, but instead returned to the colony soon after the feeding interaction. Single chicks or groups of chicks also moved between crèches when no feeding chases were taking place, so chases are not a requirement for exploratory behaviour. Finally, locomotor training by chicks can be observed throughout the day, with chicks running and beating their flippers during intense bouts of activity in the absence of parents or any association with feeding chases.

DISCUSSION

Of the different hypotheses advanced to explain the occurrence of feeding chases in Pygoscelid penguins, several are clearly not supported by observational or experimental data. Feeding chases do not reduce competition from foreign chicks or territorial adults nor do they facilitate parent-offspring recognition. They are also not a clear expression of parent-offspring conflict. Harassment of parents by offspring does not seem to be the main proximate cause of feeding chases and the data provide little support for the hypothesis that feeding chases are a secondary result of other adaptive processes.

Feeding chases do appear to separate offspring in two-chick broods, so parents can feed them more efficiently. Observational and experimental data are in agreement with this hypothesis (Bustamante et al. 1992; Moreno et al. 1996, 1998; Boersma & Davis 1997). Efficient feeding is necessary, because food is sometimes dropped and never retrieved when parents try to feed two siblings that are begging and pushing against each other. Feeding chases are more common in two-chick compared to one-chick broods. They are also more common when both siblings of a two-chick brood are present compared to when only one sibling is present. Feeding visits by parents take longer, and parents need more time to transfer the same amount of food, when there are competitive interactions between chicks. Parents start to run as a direct consequence of the distance between siblings and stop as soon as siblings become separated. Chasing rate declines and food transfer rate increases when siblings become separated compared to when they are together. This hypothesis also explains why feeding chases are rare in species rearing only one chick (e.g. *Aptenodytes, Eudyptes*), but it does not explain why other species of penguin raising two chicks (i.e. yellow-eyed penguins, little penguins and *Spheniscus* penguins) do not rely on feeding chases for efficient food transfer. It could be that they use other means to reduce the costs of food transfer (e.g. using their flippers to counteract sibling competition; Boersma 1991). It is also possible that chases cannot develop in breeding colonies typically sited within areas of dense vegetation (e.g. yellow-eyed penguins, little penguins and *Speniscus* penguins), rugged topography (e.g. crested penguins) or inside caves (e.g. little penguins). Even in Pygoscelid penguins feeding chases tend to be uncommon when colonies are situated in rugged topography (J. Moreno personal communication).

Feeding chases might amplify existing chick asymmetries and additionally facilitate brood reduction in cases of food scarcity, but only during the second half of the nestling period. Although brood reduction could be an adaptive advantage, it is unlikely to be the primary reason for feeding chases. Brood reduction is only reasonable as an explanation of feeding chases in situations in which laying parents cannot easily predict food availability during the crèche period. Food supply could be difficult to predict because of fluctuations in food levels within the season or because of the death of one of the parents. In both situations, parents might not know during the initial breeding stages (incubation to guarding) whether they could raise one or two chicks to fledging. Most brood reduction in chinstrap and Adélie penguins takes place during the guarding stage, before feeding chases start (Davis & McCaffrey 1986; Moreno et al. 1994). This reduces the possible role of brood reduction as a general explanation for feeding chases in Pygoscelid penguins. Feeding chases could, however, still have a secondary brood-reducing function in some instances. For example, seal predation causes relatively high adult mortality in chinstraps at Deception Island and brood reduction during the crèche stage is common. Therefore, feeding chases could facilitate brood reduction in this case.

Feeding chases could possibly be used by parents to assess the nutritional status of their offspring. For instance, Thompson (1981) and J. Moreno (unpublished data) found that persistence in a chase by a chick is proportional to its hunger level. In this way, feeding chases could be one part of a costly begging display in Pygoscelid penguins. It has not been proved, however, that parents use this information either to regulate food distribution between siblings, decide how much food to keep for themselves, or to adjust provisioning effort. This hypothesis can explain why there are feeding chases in some one-chick broods (something not explained by the 'Efficient Food Transfer Hypothesis'). If feeding chases are part of a begging signal, they are probably one of the few cases of offspring solicitation in which the signal is forcibly extracted from the offspring by the parents. This hypothesis also does not explain why feeding chases have only evolved in Pygoscelid penguins, and why other penguin species rely on alternative begging signals to assess the nutritional status of their offspring.

FUTURE DIRECTIONS

Current research on feeding chases in penguins has shown that they facilitate the efficient transfer of food from parent to offspring in two-chick broods of Pygoscelid penguins. It is still unclear, however, whether feeding chases could occasionally facilitate brood reduction, and if they are also a component of an honest begging signal. Future research should try to address why feeding chases do not occur in other species raising two chicks, for example in yellow-eyed penguins, little penguins or *Spheniscus* penguins, in which there should also be a need for efficient food transfer. It would also be interesting to know more about the feeding chases described in *Aptenodytes* and *Eudyptes* penguins.

Feeding chases are likely to be a costly behaviour. Penguins may spend up to 14% of the feeding visit in chases, and running necessarily entails some energetic cost for both parents and offspring. Chases take chicks away from the protection of the crèche, which might increase predation risk (Boersma & Davis 1997). Future studies should try to measure the cost of feeding chases, both in terms of the energetic cost of running and the increase in predation risk. The costs associated with the extra time spent during the feeding interaction and food dropped to the ground should also be determined. Unfortunately, this is not an easy task because parents will not feed their chicks when they are not allowed to run (e.g. if they are kept inside an enclosure; J. Moreno personal communication).

To know if feeding chases play a role in brood reduction, it would be necessary to compare situations in which food is and is not limited. Feeding chases should be more frequent in years when food is scarce, but one would have to experimentally control for chick hunger level (which would increase with decreasing food provision). It would be possible to study feeding chases and food distribution in two-chick broods in which one parent has accidentally died, or to perform experiments by temporarily removing one parent (although it may also be necessary to know if parents are aware whether their mate is still provisioning the chicks, other than through chick hunger level). Chicks could also be directly provided with extra food to manipulate hunger level.

Feeding chases could be used by parents to gather information on long-term offspring need. The important point is whether parents use such information in their provisioning behaviour. Future studies should address whether parents retain some food in their stomach when chasing persistence is low. This could be estimated with ultrasonography. In addition, it is of interest to determine whether parents modify the amount of time devoted to resting and foraging depending on the persistence of chick feeding chases. Studying in detail the feeding chases that take place in one-chick broods, which cannot be related to efficient food transfer, could give some indication of whether parents use feeding chases as an honest begging signal.

ACKNOWLEDGEMENTS

The studies of feeding chases at Deception Island were funded by the Spanish Plan Nacional Antártico (project # ANT91-1264 from CICYT - Comision Interministerial de Ciencia y Tecnología, Spain). Funding to JB during the writing of this chapter was provided by projects PB97-1154 and 1DF97-0648 from the CICYT and the European Commission. PDB was supported by the University of Washington. LSD was supported by an Otago Research Grant.

REFERENCES

Boersma, P.D. 1978. Galapagos penguins as indicators of oceanographic conditions. *Science* **200**, 1481-1483.

Boersma, P.D. 1991. Asynchronous hatching and food allocation in the Magellanic penguin *Spheniscus magellanicus*. *Acta XX Congressus Internationalis Ornithologici* **2**, 961-973.

Boersma, P.D. & Davis, L.S. 1997. Feeding chases and food allocation in Adélie penguins, *Pygoscelis adeliae*. *Animal Behaviour* **54**, 1047-1052.

Boersma, P.D., Stokes, D.L. & Yorio, P. 1990. Reproductive variability and historical change of Magellanic Penguins (*Spheniscus magellanicus*) at Punta Tombo, Argentina. In: *Biology of Penguins* (Ed. by L. Davis & J. Darby). New York: MacMillan.

Bustamante, J., Cuervo, J. & Moreno, J. 1992. The function of feeding chases in the chinstrap penguin (*Pygoscelis antarctica*). *Animal Behaviour* **44**, 753-759.

Davis, L.S. & McCaffrey, F.T. 1986. Survival analysis of eggs and chicks of Adélie penguins *Pygoscelis adeliae*. *The Auk* **103**, 379-388.

Davis, L.S. & McCaffrey, F.T. 1989. Recognition and parental investment in Adélie penguins. *Emu* **89**, 155-158.

Lundberg, V.U. & Bannasch, R. 1983. Beobachtungen und analysen zum futterwettlauf bei pinguinen. *Zoologische Jahrbücher Abteilung für Allgemaine Zoologie und Physiologie der Tiere* **87**, 391-404.

Moreno, J., Carrascal, L.M., Sanz, J.J., Amat, J.A. & Cuervo, J.J. 1994. Hatching asynchrony, sibling hierarchies and brood reduction in the chinstrap penguin *Pygoscelis antarctica*. *Polar Biology* **14**, 21-30.

Moreno, J., Carrascal, L.M. & Sanz, J.J. 1996. Parent-offspring interaction and feeding chases in the chinstrap penguin *Pygoscelis antarctica*. *Bird Behaviour* **11**, 31-34.

Moreno, J., Amat, J.A., Sanz, J.J. & Carrascal, L.M. 1998. Determinants of feeding chases in the chinstrap penguin *Pygoscelis antarctica*. *Emu* **98**, 192-196.

Müller-Schwarze, D. & Müller-Schwarze, C. 1977. *Pinguine*. Wittenberg: Neue Brehm Bücherei.

Penney, R.L. 1968. Territorial and social behavior in the Adélie penguin. *Antarctic Research Series* **12**, 83-131.

Pettingill, O.S.J. 1964. Penguins ashore at the Falkland Islands. *Living Bird* **3**, 45-64.

Prévost, J. 1961. *Écologie du Manchot Empereur* Aptenodytes forsteri *Gray*. Paris: Hermann.

Seddon, P.J. 1990. Behaviour of the yellow-eyed penguin chick. *Journal of Zoology* **220**, 333-343.

Seddon, P.J. & Van Heezik, Y.M. 1991. Hatching asynchrony and brood reduction in the jackass penguin - an experimental study. *Animal Behaviour* **42**, 347-356.

Sladen, W.J.L. 1958. The Pygoscelid penguins. Parts 1 and 2. *Falkland Islands Dependency Survey Science Report* **17**, 1-97.

Stonehouse, B. 1960. The king penguin of South Georgia. *Falkland Islands Dependency Survey Science Report* **23**, 1-81.

Taylor, R.H. 1962. The Adélie penguin *Pygoscelis adelia* at Cape Royds. *Ibis* **104**, 176-204.

Thompson, D.H. 1981. Feeding chases in the Adélie penguin. *Antarctic Research Series* **30**, 105-122.

Trivers, R.L. 1974. Parent-offspring conflict. *American Zoologist* **14**, 249-264.

Van Zinderen Bakker, E.M. Jr. 1971. A behaviour analysis of the gentoo penguin. In: *Marion and Prince Edwards Islands: Report on the South African Biological and Geological Expedition 1965-1966* (Ed. by E.M. Van Zinderen Bakker Jr., J.M. Winterbottom & R.A. Dyer). Cape Town: A.A. Balkema.

Warham, J. 1972. Aspects of the breeding biology of the erect crested penguin *Eudyptes sclateri*. *Ardea* **60**, 145-184.

Warham, J. 1974. The fiordland crested penguin *Eudyptes pachyrhynchus*. *Ibis* **116**, 1-27.

17.

SIBLING COMPETITION AND PARENTAL CONTROL: PATTERNS OF BEGGING IN PARROTS

Elizabeth A. Krebs
Department of Zoology and Entomology, University of Queensland, Brisbane 4072, Australia (ekrebs@zen.uq.edu.au)

ABSTRACT

Begging and food allocation patterns are the outcome of complex and repeated interactions between parents and young. In most systems studied, food allocation is regulated by begging and scramble competition. In contrast, little is understood about how nestling solicitation behaviours will evolve in systems where parents engage in complex patterns of food allocation. Parrots appear to be an excellent group in which to examine the shifting balance between sibling competition and parental control. Studies to date have shown that levels of sibling competition within parrot broods are low, possibly in response to parental control over food distribution. I assess what is known about the function of nestling begging in parrots and evaluate why begging signals appear to function differently in this group.

INTRODUCTION

In altricial birds, nestling begging strategies and parental feeding strategies are the result of complex and repeated interactions between nestling behaviour and parental responses. Nestlings attempt to acquire food by the intensity of their vocal and postural signalling, through scramble competition with their siblings or by enforcing a dominance hierarchy (see Budden & Wright 2001). Parents, in contrast, can distribute food based on the outcome of interactions between young or can allocate food directly to specific nestlings. The optimal strategy for a nestling will be shaped both by the

J. Wright and M.L. Leonard (eds.), The Evolution of Begging, 319–336.

behaviour of its siblings and the responses of parents. Three broad patterns of interactions and food allocation in birds emerge. At one extreme, nestlings may form a strictly enforced dominance hierarchy that determines access to food (e.g. raptors, herons and boobies; Mock & Parker 1997). In this situation food distribution favours dominant young, and parents show little selectivity. Alternatively, nestlings may engage in begging and scramble competition and parents distribute food mainly based on the intensity of begging signals and nestling position (e.g. passerines; Kilner & Johnstone 1997). At the other extreme, parents may engage in active food allocation by directly allocating food to specific young and respond little to nestling behaviour (e.g. Stamps et al. 1985). Active food allocation is poorly understood and studies on nestling behaviour have focused on species that fall mainly into the first two categories.

So, in general, parents either allow nestling behaviour to dictate the distribution of food or they control it directly. Parents can allocate food directly by selectively feeding nestlings based on physical cues such as size (Cotton et al. 1999), age (Braun & Hunt 1983), sex or condition (Stamps et al. 1985) or by feeding all young equally (Krebs et al. 1999). There are several potential benefits to direct control of food distribution. Selective feeding allows parents to rapidly alter the distribution of food within the brood in response to changes in food availability. It can also reduce the costs of sibling competition by preventing large nestlings from monopolizing food. Selective feeding will, however, increase the costs of food distribution because parents must identify and choose which nestling to feed, reducing feeding efficiency (Stamps et al. 1985; Göttlander 1987). Allocating food in a manner so as counter to nestling interests may also be costly, because it increases the level of conflict between parents and offspring (Forbes 1993).

The circumstances under which direct parental control of food distribution should evolve are currently unknown. It is clear, however, that the extent to which nestling solicitation behaviours accurately reflect hunger is influenced by the competitive asymmetries within the brood (Kilner 1995). If there are large size differences, larger nestlings will dominate in scramble competition. In addition, the costs and benefits of begging displays are also likely to vary with nestling size (Parker et al. 1989; B. Glassey & S. Forbes this volume), suggesting that the intensity of displays will be a less reliable signal of nestling hunger, as size hierarchies increase within the brood. Even within a species, begging rates vary considerably between broods, consistent with the idea that there is plasticity in nestling behaviour. A recent experimental study, for example, has shown that nestlings learn to alter their begging postures when the rewards obtained from differing postures were varied (Kedar et al. 2000).

The species that best exemplifies the nuances and complexities of parent-offspring interactions over food distribution is none other than our ubiquitous pet, the budgie. Fifteen years ago, Judy Stamps and colleagues set out to test the basic assumptions of the function of begging, using captive budgerigars (*Melopsittacus undulatus*) as the model system. Far from confirming simple relationships between nestling solicitation and parental food distribution, Stamps et al. discovered surprising complexity in both parental and offspring behaviour (Stamps et al. 1985, 1987, 1989). Male and female parents engaged in differing patterns of food allocation, and nestlings responded in a complex way to parental cues. I believe the life history characteristics of parrots make them an ideal system to unravel this tension between parental control and sibling competition. In this chapter, I examine what is known about food allocation and begging in parrots, explore some of the reasons for the evolution of parental control in this group and examine the implications of strong parental control of food allocation for the evolution of nestling solicitation behaviours.

Life History of Parrots

Among birds, parrots are a morphologically distinct and ancient group with no close living relatives (Smith 1975; Sibley & Ahlquist 1990). Despite their superficial similarities, the 333 known species of parrot have adapted to life in a variety of habitats ranging from tropical to alpine. Populations can also be sedentary, nomadic or migratory (Forshaw 1981, 1989; Collar & Juniper 1992). Parrots are typically socially monogamous, and appear to have long-term pair bonds (Forshaw 1989). Most species are secondary cavity-nesters and consequently are dependent on the availability of hollows to breed.

The breeding biology of parrots is highly variable. For example, clutch sizes range from a single egg in macaws (*Ara* spp., Munn 1992) and cockatoos (subfamily: Cacatuinae, Forshaw 1981), to up to ten eggs in green-rumped parrotlets (*Forpus passerinus*, Waltman & Beissinger 1992) and budgerigars (Wyndham 1981). In Australian parrots the median clutch size is four (Saunders et al. 1984). Parrots typically lay their eggs at two-day intervals (Forshaw 1981, 1989; Sindel & Gill 1992). Large clutch sizes combined with two-day laying intervals result in a highly protracted laying period in some species. Thus, by varying the onset of incubation, parrots have the potential to greatly stagger or compress hatching intervals.

Hatching intervals in parrots are typically large and also variable. Detailed studies on hatching patterns in the wild have only been carried out on two species (green-rumped parrotlets, Stoleson & Beissinger 1997; and crimson rosellas, *Platycercus elegans*, Krebs 1999). Hatching patterns in parrots range from complete asynchrony, where incubation begins with the first egg laid and the hatching period is equal to the laying period (Grenier & Beissinger 1999), to a more typical pattern, where full incubation commences mid-laying and nestlings hatch over about one day intervals (Krebs 1999; see Low 1978 and Sindel & Gill 1992 for avicultural examples). Thus, hatching in parrots is considerably more protracted than in most passerines (>85% of the species listed in Clark & Wilson 1981), and staggered hatching produces large size asymmetries between nestlings in a brood.

Clutch sizes and incubation periods of parrots are similar to that of other non-passerines of the same size, however, parrots appear to have relatively long nestling periods (Saunders et al. 1984). For example, the nestling period of a 400g pigeon is about 25 days, and for a raptor or owl it is about 36 days, whereas the nestling period of a parrot is estimated at 58 days (Saunders et al. 1984). Lower growth rates and long nestling periods may have evolved in parrots because predation at the nest is uncommon, as in other cavity-nesting birds (Lack 1968). Alternatively, growth rates may evolve in response to the levels of sibling competition within the brood. Higher growth rates may be selected for in species with high levels of sibling competition because being larger increases a nestling's competitive ability and access to food (Ricklefs 1993; Royle et al. 1999; R.E. Ricklefs this volume). Thus, the lower growth rates observed in parrots may also reflect relatively low levels of sibling competition within the brood.

Patterns of Begging and Food Allocation in Parrots

The only detailed studies on begging and food allocation in parrots to date are on budgerigars and crimson rosellas (Stamps et al. 1985, 1989; Krebs 2001). Both species are endemic to Australia and feed primarily on seeds (Forshaw 1981). Budgies and rosellas lay a median of five or six eggs, which hatch highly asynchronously, and young remain in the nest for approximately 35 days. Nestlings attempt to obtain food in two ways, by begging (vocalizing, posturing and quivering their wings) and by scramble competition with their siblings to obtain a position close to a feeding parent. Parents feed nestlings by direct bill contact, and regurgitate a gruel of seeds directly into their crops. Feeding visits are relatively long (mean $\pm$ SE visit length in seconds for rosellas: males = 140 ± 8, n = 105; females = 166 ± 13, n = 95; E.A. Krebs unpublished data) and in rosellas parents can deliver up to 35g of food (25% of their body mass) in a single visit (Krebs et al. 1999). All nestlings are typically fed in a feeding visit, and parents can regurgitate up to 50 times per visit (Stamps et al. 1985; Krebs et al. 1999). Parents move around the nest during feeding visits, and seek out particular nestlings, ignore actively begging nestlings and refuse to feed nestlings grabbing at their bill (Stamps et al. 1985; Krebs et al. 1999). Thus, feeding visits in parrots are very dynamic events with both parents and young moving around the nest cavity.

The Problem of Measuring Begging in Parrots

In order to test hypotheses about the function of solicitation behaviours by nestlings, it is essential to accurately measure the intensity of these behaviours and their relationship to food allocation. Quantifying begging and scramble competition in birds where parents feed primarily a single nestling per visit is a relatively straightforward affair. Prior to food allocation, a nestling's position and aspects of its begging (e.g. gape order, height) can be measured, and comparisons made between the behaviour of nestlings who were fed and those who were not. Feeding visits by parrots are, however, protracted and consist of many transfers of food, making it difficult to assign a precise measure to begging or the intensity of scramble competition. For example, both parents and nestlings move continuously during feeding visits; consequently any measure of a nestling's proximity to a feeding parent is simultaneously a measure of parental and nestling movements. So, although one can easily measure changes in nestling position over a feeding visit, it is difficult to assess whether changes are due to changes in parental behaviour, nestling behaviour, or both!

Stamps et al. (1985, 1987, 1989) quantified begging in budgerigar broods by examining the number of begging bouts, or continuous periods of begging, which occurred over a feeding visit. They did not directly consider the duration or intensity of begging that occurred, but only the number of episodes per feeding visit. They used the number of begging bouts to calculate a rate of begging for different nestlings. In rosellas, Krebs (2001) assessed the relative intensity of begging and position of target nestlings relative to the parent in the intervals prior to each food transfer, and used these measures to assign a mean begging score and position over a feeding visit. This approach allowed the relative intensity of nestling begging to be compared with the amount of food allocated to each nestling over the entire feeding visit. Neither of these approaches, however, accurately reflected the changes in nestling behaviour over the duration of a feeding visit, nor attempted to disentangle nestling versus parental movements. Future studies of parrots should carefully consider how solicitation behaviours should be assessed. One potential approach is to remove experimentally the effects of scramble competition by restricting the movement of nestlings (e.g. Kilner 1995).

Caveats aside, four major patterns appear to characterize food allocation and begging in parrots.

Begging Rates Are Affected by Hatching Rank

In both budgerigars and rosellas, small nestlings within the brood beg more than large nestlings. In budgies, last-hatched nestlings beg at twice the rate of first-hatched nestlings, after controlling for differences in nestling age (Stamps et al. 1989). Last-hatched rosella nestlings begged 25% more than first-hatched nestlings, but only when the brood was less than two weeks old. When begging was measured one week later, large nestlings had increased their begging, and begging intensity did not differ between nestlings (Krebs 2001). It is difficult to infer from Stamps et al.'s data whether last-hatched nestlings are begging more, or first-hatched nestlings are begging less. They report a positive correlation between begging rate and hatch order, suggesting that first-hatched budgie nestlings may also reduce their begging. However, because Stamps et al. controlled for differences in nestling age by comparing early- and late-hatched nestlings at the same age (i.e. when each was 20-days-old) and broods hatched over an average of eight days, measurements for last-hatched nestlings were taken when early-hatched nestlings were starting to fledge, making it difficult to compare patterns between nestlings.

Large nestlings reduce their begging in several other asynchronously hatching species (e.g. starlings, *Sturnus vulgaris*, Cotton et al. 1999; yellow-headed blackbirds, *Xanthocephalus xanthocephalus,* Price & Ydenberg 1995). If competitive interactions determine access to food, then models suggest that the high competitive ability of large nestlings allows them to reduce their begging costs and maintain their share of food (Parker et al. 1989). This is consistent with the pattern observed in some species; reduced begging by large nestlings did not result in a reduced share of food in yellow-headed blackbirds (Price & Ydenberg 1995) or starlings (Cotton et al. 1999). By contrast, in budgies, last-hatched nestlings beg more, but are also fed more, leading Stamps et al. (1989) to conclude that there was no evidence suggesting that beg/feed ratios differed based on hatching order. In crimson rosellas, parental responses to begging were complex (see below), although increased begging by last-hatched nestlings led to increased food allocation by male parents (Krebs 2001).

The Relationship between Begging and Food Allocation Is Complex and Not Strongly Linked to Nestling Hunger

Relative increases in nestling begging should correspond to decreased food allocation, if begging functions to honestly signal offspring hunger. This pattern is observed in many species; nestlings beg harder when they are deprived of food and reduce their begging when fed (Bengtsson & Rydén 1983; Redondo & Castro 1992; Price et al. 1996; Leonard & Horn 1998). In contrast, rosella nestlings did not beg more when they were hungry relative to the rest of the brood (Krebs 2001). Begging rates in parrots are highly variable between broods and do not relate to individual hunger in a simple way. Counter to expectation, begging rates within budgie broods were positively correlated to feeding rates but the strength of this relationship varied considerably between broods (r_s = 0.2-1.0, Stamps et al. 1989). Overall, feeding rates in rosellas did not increase when begging within the brood was high (Krebs & Magrath 2000). This relationship, however, was also variable in rosellas, since five of nine pairs increased their feeding rate when the brood was hungry. Thus, there appears to be underlying variation in begging rates between broods even in a captive population fed *ad lib.* This suggests that begging in parrot broods reflects variation in parental responses to nestlings more than variation in nestling responses to hunger.

Parental responses to begging may differ because parents vary in their quality. Parental quality may affect the costs of increasing parental effort, for example, older or more experienced parents may be efficient at collecting food, and easily able to increase their feeding rate, whereas younger or less experienced birds may be unable to do so (Wright & Cuthill 1989). Parental age and experience is likely to be highly variable in species such as parrots since they are long-lived (Forshaw 1981). In addition, long-lived birds like parrots are expected to be conservative in their reproductive effort (Stearns 1992). Budgerigars, for example, have a nomadic lifestyle, and breed only when and where conditions are suitable (Wyndham 1981). Rosellas are relatively sedentary and defer breeding if conditions are poor (Krebs 1998). Thus, parrots should be sensitive to the costs of feeding.

Variability in begging rates between parrot broods occurs partly because male and female nestlings beg at different rates and may have different begging strategies. For example, female nestlings in budgerigar broods begged 50% more than males (Stamps et al. 1989). This pattern is also observed in rosellas, and seems to be due to the differing strategies of male and female nestlings, since female nestlings beg much more intensely in response to increased brood hunger than males (Figure 1; E.A. Krebs unpublished data). Since Stamps et. al.'s work on budgerigars was conducted in captivity with *ad lib.* food, similar patterns might occur in budgies if food availability was altered. In addition to individual sex effects, sex ratio appears to influence brood begging rates. Stamps et al. (1989) observed a 300% increase in begging in female-biased broods compared to male-biased broods. The same qualitative pattern occurs in rosellas; nestlings in female-biased broods begged intensely 60% of the time during feeding visits, whereas nestlings in male-biased broods begged intensely for only 41% of the time (E.A. Krebs unpublished data). Although cause and effect are difficult to tease apart, differences in begging by male- and female-biased broods may arise because parental food allocation patterns are influenced by the sex ratio of the brood and nestlings learn to respond to the differing patterns of food allocation (Stamps et al. 1985; Kedar et al. 2000).

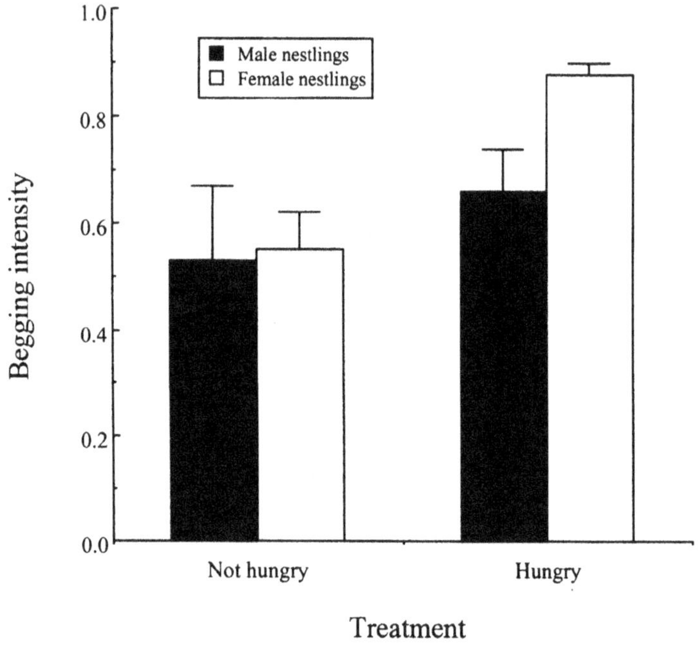

Figure 1. Changes in begging in response to increased brood hunger for male and female crimson rosella nestlings. Begging intensity was measured as the proportion of time that a nestling begged at a maximum intensity (i.e. vocalizing, posturing and flapping wings; for details see Krebs 2001). The differences in begging intensity for males and females were not attributable to differences in hatching ranks.

Sex differences in begging have been observed in two other species (red-winged blackbirds, *Agelaius phoeniceus*, Teather 1992; yellow-headed blackbirds; Price & Ydenberg 1995). In both species males are considerably larger than females, and male nestlings beg more intensely, especially when hungry, suggesting their begging reflects higher nutritional requirements. Female parrots are not, however, larger than males and are unlikely to have higher nutritional requirements. It is not clear why female nestlings might beg more in parrots since begging intensity does not correlate with food allocation in a straightforward way. There is some evidence that female parrots benefit from improved condition or early fledging. In budgies, early-fledging females were more likely to breed as one-year-olds than late-fledging females (Stamps et al. 1987). This pattern may also occur in rosellas, since female rosellas are frequently sighted breeding in juvenile plumage, suggesting they can breed as first year birds, whereas males do not breed until two years old (Krebs 1998).

Position Within the Nest Does Not Strongly Reflect Nestling Hunger

In some species the primary determinant of feeding success is a nestling's position within the nest (e.g. Bengtsson & Rydén 1983; Greig Smith 1985; Lessells & Avery 1989). If food allocation is regulated by scramble competition, larger and older nestlings will generally be favoured, because they are able to obtain a position close to the feeding parent. In parrots, since food is distributed directly *via* bill to bill contact, nestlings that are fed by necessity are closest to the feeding parent. Because both parents and nestlings are moving around the nest, however, the efficacy of scramble competition as a strategy for obtaining food is limited. For example, even if a nestling obtains a position close to a parent at the beginning of a feeding visit, the relative advantage of that position will change as the parent moves. Nevertheless, large nestlings are still somewhat advantaged in scramble competition because they can reposition themselves more quickly within a brood than can small nestlings.

The dynamics of nestling position and food allocation are not known for budgies, but in rosellas neither large nor small hungry nestlings were more likely to be positioned close to a feeding parent, even when the whole brood was hungry and scramble competition was presumably high (Krebs & Magrath 2000). Rosella nestlings did position themselves closer to the parent that was more likely to feed them under normal circumstances; large nestlings were observed to be closer to males and small nestlings were closer to females. However, since a nestling did not get closer to a parent when it was hungrier than the rest of the brood, it appears that changes in parental behaviour largely determined the relative position of a nestling during feeds (Krebs 2001). As expected if parental behaviour regulates the relative distance to young during feeding visits, the distance a parent moved between consecutive food transfers increased when the brood was hungry (Krebs & Magrath 2000).

If the pay-off to a nestling from engaging in scramble competition is low and parents distribute food based on non-behavioural cues (see below), this should lead to a reduction in sibling competition. Thus, parental control of food allocation in budgies and rosellas may affect nestling behaviour by reducing the benefits of scramble competition. In many other altricial birds the ability of parents to control the distribution of food appears to decrease as the brood ages (e.g. Lessells & Avery 1989; Michaud & Leonard 2000). Thus, parental control of food allocation in parrots is likely to be strongest when nestlings are young, have limited mobility and the parent can compensate for any scramble competition by moving or refusing to feed nestlings. When nestlings are large and mobile it is more difficult for parents to control food distribution because high levels of scramble competition lead to interference by non-feeding young. In rosellas, high levels of scramble competition lead to food transfers being frequently interrupted and shortened feeding visits, potentially reducing the food for all nestlings. Only under these circumstances did parents feed nestlings that were closest to them.

Parents Have Differing Food Allocation Strategies

Last-hatched nestlings have growth equal to earlier-hatched nestlings in some species of parrots (white-tailed black cockatoos, *Calyptorhynchus funereus,* Saunders 1982; crimson rosellas, Krebs 1999). Equal growth of nestlings cannot be explained by food allocation based on competitive interactions and suggests that parents selectively feed later-hatched nestlings. Selective feeding of last-hatched nestlings has been shown in budgies (Stamps et al. 1985), crimson rosellas (Krebs et al. 1999) and anecdotally reported in golden-shouldered parrots (*Psophotus chrysopterygius*, S. Garnett personal communication). Selective feeding of last-hatched nestlings has also been observed in several other asynchronously hatching species including pied flycatchers (*Ficedula hypoleuca*, Göttlander 1987), tree swallows (*Tachycineta bicolor*, Leonard & Horn 1996) and red-winged blackbirds (Westneat et al. 1995). Given the large number of studies on passerines, however, and the small number of studies on parrots, it appears to be relatively common within the latter.

In both passerines and parrots, females are primarily responsible for the selective feeding of small nestlings (see Westneat et al. 1995 for an exception). It is unclear why females should be more likely to selectively feed last-hatched young. Potentially, the costs of feeding last-hatched nestlings, such as discriminating between young, are relatively low for females. Alternatively, the benefits from improved growth of last-hatched nestlings may be relatively greater for females. Increased female mortality may cause females to invest more in current reproduction, maximizing the number of current young fledged (see Slagsvold 1997 for a review of hypotheses). If food is abundant, equal growth rates within the brood should maximize the number of young produced. Female budgies fed *ad lib.* appear to overcompensate for reduced growth rates of any nestling by feeding later-hatched and undersized nestlings more than other nestlings within the brood (Stamps et al. 1985). Female rosellas in the wild appear to be sensitive to food availability since they normally feed small nestlings, but divert food away from them when brood hunger is high (Krebs & Magrath 2000).

Male parrots feed large nestlings in the brood more than small nestlings, and in rosellas, especially favour first-hatched male nestlings (Stamps et al. 1985; Krebs et al. 1999). In part, the complex food allocation patterns observed in parrots are mediated by differential responses to begging. In budgerigars, fathers provisioned the brood in response to brood begging intensity, whereas mothers provisioned at a relatively fixed rate (Stamps et al. 1985, 1989). Rosella parents were also not equally responsive to begging. Mothers did not redistribute food to an intensely begging nestling. Fathers, however, reallocated food to last-hatched nestlings in proportion to their begging intensity (Krebs 2001).

Why Are Parrots Different?

The distinguishing feature of parent-offspring interaction in parrots is that parents actively control the distribution of food within the brood. This strategy is likely to have evolved because the costs of direct food allocation by parents are low, and the benefits high. Four aspects of parrot breeding biology tip the balance in favour of parental control of food allocation.

Last-Hatched Nestlings Have High Reproductive Value

In most asynchronously hatching birds, last-hatched young are doubly disadvantaged. Their smaller size puts them at a competitive disadvantage in gaining access to food, resulting in lower growth rates. In addition, despite being younger, last-hatched young in most species still fledge at close to the same time as their siblings. This means that they need to grow at a faster rate than their older siblings in order to be ready to fledge. In species where fledging is synchronous, this may reduce parental returns on investment in later-hatched young compared with early-hatched young, and lead to preferential investment in larger nestlings in the brood (e.g. Bengtsson & Rydén 1983). In these species, last-hatched nestlings may primarily represent insurance value for their parents (Forbes et al. 1997).

In parrots, however, four factors are likely to increase the reproductive value of later-hatched young. First, fledging occurs asynchronously in parrots. This allows last-hatched nestlings to remain in the nest and complete their development at the same rate as earlier-hatched nestlings. Second, growth rates of nestlings are flexible, so nestlings can catch up after periods of lower growth. For example, several studies on parrots have observed that smaller nestlings within the brood initially had lower growth rates but by the end of the nestling period reached equal (or greater) asymptotic masses as compared to their older siblings (Stamps et al. 1985; Rowley & Chapman 1991; Stoleson & Beissinger 1997). Third, the cost of selectively feeding smaller nestlings is low. A long nestling period minimizes the costs of selective feeding, by allowing time for parents to compensate for any deviations in brood growth rates. Fourth, small and large nestlings are equally likely to survive to independence. Studies on wild populations have consistently found no relationship between a nestling's mass at fledging and its subsequent survival (long-billed corellas, *Cacatua pastinator*, Smith 1991; Major Mitchell's cockatoos, *Cacatua leadbeateri*, Rowley & Chapman 1991; green-rumped parrotlets, Stoleson & Beissinger 1997; crimson rosellas, Krebs 1999). Thus, small variations in fledging mass do not appear to reduce offspring fitness, in contrast to the situation in many other avian species (Magrath 1991).

The Benefits of Efficient Food Distribution Are Low

Selective feeding of young is costly to parents because it reduces the efficiency of food distribution. Rather than feeding the loudest or most prominent nestling, parents must discriminate between nestlings and locate the desired one to feed; thus increasing the time required to distribute food (Göttlander 1987). This pattern is observed in both budgies and rosellas where females, who feed selectively, are slower than males in transferring food. Extra time spent distributing food can be costly to parents because it reduces their foraging rate or increases the risk of predation at the nest. Both these costs will be relatively low for parrots because parents feed infrequently, and deliver a large amount of food during each visit. Thus, any additional time at the nest is a very low proportion of the total available for foraging. Similarly, the increased risk of predation at the nest is likely to be extremely small, because parrots are cavity-nesters with few nest predators.

Parents Can Assess Nestling Hunger Using Non-Behavioural Cues

One reason for parents to distribute food based upon begging signals is that begging allows parents to assess aspects of offspring condition that would otherwise not be apparent ('cryptic condition'; Godfray 1991). The value of begging signals, however, will be lowered if parents are able to assess the short-term hunger and/or condition of young directly. Since there are large size asymmetries between nestlings in parrot broods, the outcome of scramble competition will be determined by the relative competitive ability of nestlings more than by their hunger levels. Changes in the begging intensity of individual nestlings may provide information on nestling hunger, but are likely to be overwhelmed by the large differences in competitive ability that exist within the brood. In this case, nestling hunger could be more accurately determined by a physical cue. One physical cue in parrot nestlings is the size of their large and prominent crop. Parents distributing food according to relative crop size should be able to reliably assess nestling hunger, as well as the previous distribution of food. Allocating food based upon nestling crop size would not only overcome any influence of sibling competition on solicitation behaviours, but would also allow a parent to coordinate complex food allocation strategies with its mate.

The Feeding Ecology of Parrots Reduces the Benefits of Sibling Competition

Parrots distribute food by transferring a large number of small parcels of food to nestlings. The degree to which food is divided appears to be under parental control since the average size of transfers increases with brood age (Stamps et al. 1985; E.A. Krebs unpublished data). Dividing food into many small parcels may increase the costs of monopolizing food, and reduce the benefits of sibling competition for parrot nestlings. This is an inverted variant of the 'Prey Size Hypothesis' proposed by Mock (1985) to explain differences in sibling aggression between different species of Ardeids.

Mock (1985) argued that the intensity of sibling competition was linked to the monopolizability of food by individual nestlings. In Ardeids, food was most easily monopolized when prey items were small because they could be obtained directly from the parent's bill, pre-empting competitors. In contrast, when large prey items were deposited onto the nest floor, food was difficult for individuals to monopolize (see Mock & Parker 1997 for a summary and Drummond 2001 for a critique).

The 'Prey Size Hypothesis' predicts that food should be easily monopolizable in parrots. The nutritional benefits, however, of a single transfer are low, whereas the costs required to obtain it through scramble competition with siblings are high. Thus, distributing food in small transfers actually reduces the benefits of monopolizing food. Why does the same apparent pattern of food distribution produce opposite results? One important difference is that for Ardeid young, even small prey items are relatively large, and only a few boluses are produced at each feeding visit (maximum boluses/feed for great egrets, *Casmerodius albus* = 8, Mock 1985; maximum transfers/feed for crimson rosellas = 50, Krebs et al. 1999). Thus, the costs and benefits of monopolizing food may be linked to the levels of sibling competition in a species (see H. Drummond this volume).

FUTURE DIRECTIONS

If nothing else I hope this chapter convinces the reader that parent-offspring interactions in parrots are intriguing, but desperately in need of more behavioural studies. Parrots in many habitats are not difficult to study in the wild. The use of nestboxes combined with recent developments in remote video technology and computers can allow detailed monitoring of behaviour in the wild. Parrots are relatively long-lived birds, consequently it will be difficult to understand the consequences of patterns of parental care or of food allocation in the nest without long-term monitoring of a marked population. Aside from tackling the basics of parent-offspring interaction in other species of parrots, several questions remain unanswered.

Why Do Parrot Broods Not Form Dominance Hierarchies?

In most species, large size asymmetries between young lead to strong dominance hierarchies within the brood. Parrots, despite large size asymmetries between nestlings, show little sign of an enforced (or subtle, for that matter) dominance hierarchy. In fact, in both budgies and rosellas, older nestlings sometimes feed younger nestlings (Stamps et al. 1985; E.A. Krebs unpublished data). This lack of a dominance hierarchy is perplexing when contrasting parrots with other highly asynchronous non-passerines. For example, the broods of many Ardeids and raptors are characterized by clear dominance hierarchies that sometimes have fatal consequences for last-hatched young. Perhaps in parrots there are longer-term benefits of having siblings, such as a correlation between post-fledging associations and subsequent recruitment. Examining nestling interactions in nests with naturally or experimentally varying hatching spreads may help to identify the factors that suppress the formation of dominance hierarchies in parrots.

What Constrains Recruitment in Parrot Societies?

The majority of studies on altricial birds concentrate on the easily studied nestling period. Although this typically is a period of intense parental investment, interactions in the post-fledging period may be more critical in determining survival to independence or recruitment. In addition, parrots fledge their young highly asynchronously, in contrast to most passerines. Although age at fledging and post-fledging associations appear to influence the future reproductive success of at least females in captive budgerigars (Stamps et al. 1990), we do not understand either the costs and benefits of asynchronous fledging, or the factors which predict recruitment into the breeding population. Understanding the factors that regulate recruitment in parrot societies may clarify why such differing patterns of parent-offspring interaction have evolved in this group.

How do Differences in Parental Ability Affect Food Allocation Decisions?

Parrots are relatively long-lived birds, breeding in long-term pair bonds. These attributes may have facilitated the evolution of complex role specialization in parental care. Male and female parrots both invest highly in parental care but appear to have entirely different strategies. Whether this reflects sexual conflict due to the differing costs and benefits of parental care for males and females, or a complex form of parental cooperation, is unclear. Long-term studies on marked populations of abundant, yet sedentary, parrots would allow better understanding of the effects of age and pair bond duration on parental allocation decisions.

ACKNOWLEDGEMENTS

I thank Rob Magrath for the initial inspiration to work on crimson rosellas and for his input over the study. Rosella fieldwork could not have been completed without the help of Kate Trumper, Benj Whitworth and David

Green. Thanks also to Hugh Drummond, David Green, Marty Leonard and Jon Wright for their comments on previous versions of the manuscript.

REFERENCES

Bengtsson, H. & Rydén, O. 1983. Parental feeding rate in relation to begging behaviour in asynchronously hatched broods of the Great tit *Parus major*. *Behavioral Ecology and Sociobiology* **12**, 243-251.

Braun, B.M. & Hunt, G.L. Jr. 1983. Brood reduction in black-legged kittiwakes. *The Auk* **100**, 469-476.

Budden, A.E. & Wright, J. 2001. Begging in nestling birds. *Current Ornithology* **16**, 83-118.

Clark, A.B. & Wilson, D.S. 1981. Avian breeding adaptations: hatching asynchrony, brood reduction, and nest failure. *Quarterly Review of Biology* **56**, 253-277.

Collar, N.J. & Juniper, A.T. 1992. Dimensions and causes of the parrot conservation crisis. In: *New World Parrots in Crisis: Solutions from Conservation Biology* (Ed. by S.R Beissinger & N.F.R. Snyder). Washington: Smithsonian Institution Press.

Cotton, P.A., Wright, J. & Kacelnik, A. 1999. Chick begging strategies in relation to brood hierarchies and hatching asynchrony. *American Naturalist* **153**, 412-420.

Drummond, H. 2001. A revaluation of the role of food in broodmate aggression. *Animal Behaviour* **61**, 517-526.

Forbes, L.S. 1993. Avian brood reduction and parent-offspring conflict. *American Naturalist* **142**, 82-117.

Forbes, S., Thornton, S., Glassey, B., Forbes, M. & Buckley, N.J. 1997. Why parent birds play favourites. *Nature* **390**, 351-352.

Forshaw, J. 1981. *Australian Parrots*. Melbourne: Lansdowne.

Forshaw, J. 1989. *Parrots of the World*. Melbourne: Lansdowne.

Godfray, H.C. J. 1991. Signalling of need by offspring to their parents. *Nature* **352**, 328-330.

Göttlander, K. 1987. Parental feeding behaviour and sibling competition in the pied flycatcher *Ficedula hypoleuca*. *Ornis Scandinavica* **18**, 269-276.

Greig-Smith, P. 1985. Weight differences, brood reduction and sibling competition among nestling stonechats *Saxicola torquata* (Aves: Turdidae). *Journal of Zoology* **205**, 453-465.

Grenier, J.L. & Beissinger, S.R. 1999. Variation in the onset of incubation in a neotropical parrot. *Condor* **101**, 752-761.

Kedar, H., Rodríguez-Gironés, M.A., Yedvab, S., Winkler, D.W. & Lotem, A. 2000. Experimental evidence for offspring learning in parent-offspring communication. *Proceedings of the Royal Society of London, Series B* **267**, 1723-1727.

Kilner, R. 1995. When do canary parents respond to nestling signals of need? *Proceedings of the Royal Society of London, Series B* **260**, 343-348.

Kilner, R. & Johnstone, R.A. 1997. Begging the question: are offspring solicitation behaviours signals of need? *Trends in Ecology and Evolution* **12**, 11-15.

Krebs, E.A. 1998. Breeding biology of crimson rosellas *Platycercus elegans* on Black Mountain, Australian Capital Territory. *Australian Journal of Zoology* **46**, 119-136.

Krebs, E.A. 1999. Last but not least, nestling growth and survival in asynchronously hatching crimson rosellas. *Journal of Animal Ecology* **68**, 266-281.

Krebs, E.A. 2001. Begging and food distribution in crimson rosella (*Platycercus elegans*) broods: why don't hungry chicks beg more? *Behavioral Ecology and Sociobiology* **50**, 20-30.

Krebs, E.A. & Magrath, R.D. 2000. Food allocation in crimson rosella broods: parents differ in their responses to chick hunger. *Animal Behaviour* **59**, 739-751.

Krebs, E.A., Cunningham, R.B. & Donnelly, C.F. 1999. Complex patterns of food allocation in asynchronously hatching broods of crimson rosellas. *Animal Behaviour* **57**, 753-763.

Lack, D. 1968. *Ecological Adaptations for Breeding in Birds*. Bristol: Methuen.

Leonard, M. & Horn, A. 1996. Provisioning rules in tree swallows. *Behavioral Ecology and Sociobiology* **38**, 341-347.

Leonard, M.L. & Horn, A.G. 1998. Need and nestmates affect begging in tree swallows. *Behavioral Ecology and Sociobiology* **42**, 431-436.

Lessells, C.M. & Avery, M.I. 1989. Hatching asynchrony in European bee-eaters *Merops apiaster*. *Journal of Animal Ecology* **58**, 815-835.

Low, R. 1978. *Lories and Lorikeets: The Brush-Tongued Parrots*. Melbourne: Inkata Press.

Magrath, R.D. 1991. Nestling weight and juvenile survival in the blackbird, *Turdus merula*. *Journal of Animal Ecology* **60**, 335-351.

Michaud, T. & Leonard, M. 2000. The role of development, parental behavior and sibling competition in fledging by nestling tree swallows. *The Auk* **117**, 1000-1006.

Mock, D.W. 1985. Siblicidal brood reduction: the prey-size hypothesis. *American Naturalist* **125**, 327-343.

Mock, D.W. & Parker, G. 1997. *The Evolution of Sibling Rivalry*. Oxford: Oxford University Press.

Munn, C.A. 1992. Macaw biology and ecotourism, or "when a bird in the bush is worth two in the hand". In: *New World Parrots in Crisis: Solutions from Conservation Biology* (Ed. by S.R. Beissinger & N.F.R. Snyder). Washington: Smithsonian Institution Press.

Parker, G.A., Mock, D.W. & Lamey, T.C. 1989. How selfish should stronger sibs be? *American Naturalist* **133**, 846-868.

Price, K. & Ydenberg, R. 1995. Begging and provisioning in broods of asynchronously-hatched yellow-headed blackbird nestlings. *Behavioral Ecology and Sociobiology* **37**, 201-208.

Price, K., Harvey, H. & Ydenberg, R. 1996. Begging tactics of nestling yellow-headed blackbirds, *Xanthocephalus xanthocephalus*, in relation to need. *Animal Behaviour* **51**, 421-435.

Redondo, T. & Castro, F. 1992. Signalling of nutritional need by magpie nestlings. *Ethology* **92**, 193-204.

Ricklefs, R.E. 1993. Sibling competition, hatching asynchrony, incubation period, and lifespan in altricial birds. *Current Ornithology* **11**, 199-276.

Rowley, I. & Chapman, G. 1991. The breeding biology, food, social organisation, demography and conservation of the Major Mitchell or pink cockatoo, *Cacatua leadbeateri*, on the margin of the Western Australian wheatbelt. *Australian Journal of Zoology* **39**, 211-261.

Royle, N.J., Hartley, I.R., Owens, I.P.F. & Parker, G.A. 1999. Sibling competition and the evolution of growth rates in birds. *Proceedings of the Royal Society of London, Series B* **266**, 923-932.

Saunders, D.A. 1982. The breeding behaviour and biology of the short-billed form of the white-tailed black cockatoo *Calyptorhynchus funereus*. *Ibis* **124**, 422-455.

Saunders, D.A., Smith, G.T. & Campbell, N.A. 1984. The relationship between body weight, egg weight, incubation period, nestling period and nest site in the Psittaciformes, Falconiformes, Strigiformes and Columbiformes. *Australian Journal of Zoology* **32**, 57-65.

Sibley, C.G. & Ahlquist, J.E. 1990. *Phylogeny and Classification of Birds. A Study in Molecular Evolution*. London: Yale University Press.

Sindel, S. & Gill, J. 1992. *Australian Grass Parakeets*. Australia: Singil Press.

Slagsvold, T. 1997. Brood division in birds in relation to offspring size, sibling rivalry and parental control. *Animal Behaviour* **54**, 1357-1368.

Smith, G.A. 1975. Systematics of parrots. *Ibis* **117**, 18-68.

Smith, G.T. 1991. Breeding ecology of the western long-billed corella, *Cacatua pastinator pastinator*. *Wildlife Research* **18**, 91-110.

Stamps, J., Clark, A.B., Arrowood, P. & Kus, B. 1985. Parent-offspring conflict in budgerigars. *Behaviour* **94**, 1-40.

Stamps, J., Clark, A., Kus, B. & Arrowood, P. 1987. The effects of parent and offspring gender on food allocation in budgerigars. *Behaviour* **101**, 177-199.

Stamps, J., Clark, A., Arrowood, P. & Kus, B. 1989. Begging behavior in budgerigars. *Ethology* **81**, 177-192.

Stamps, J., Kus, B., Clark, A. & Arrowood, P. 1990. Social relationships of fledgling budgerigars, *Melopsittacus undulatus*. *Animal Behaviour* **40**, 688-700.

Stearns, S.C. 1992. *The Evolution of Life Histories*. New York: Oxford University Press.

Stoleson, S.H. & Beissinger, S.R. 1997. Hatching asynchrony, brood reduction, and food limitation in a neotropical parrot. *Ecological Monographs* **67**, 131-154.

Teather, K.L. 1992. An experimental study of competition for food between male and female nestlings of the red-winged blackbird. *Behavioral Ecology and Sociobiology* **31**, 81-87.

Waltman, J.R. & Beissinger, S.R. 1992. Breeding biology of the green-rumped parrotlet. *Wilson Bulletin* **104**, 65-84.

Westneat, D.F., Clark, A.B. & Rambo, K.C. 1995. Within-brood patterns of paternity and paternal behaviour in red-winged blackbirds. *Behavioral Ecology and Sociobiology* **37**, 349-356.

Wright, J. & Cuthill, I. 1989. Manipulation of sex differences in parental care. *Behavioral Ecology and Sociobiology* **25**, 171-181.

Wyndham, E. 1981. Breeding and mortality of budgerigars (*Melopsittacus undulatus*). *Emu* **81**, 240-243.

18.

BEGGING VERSUS AGGRESSION IN AVIAN BROODMATE COMPETITION

Hugh Drummond
Instituto de Ecología, Universidad Nacional Autónoma de México, A.P. 70-275, 04510 D.F., Mexico (hugh@servidor.unam.mx)

ABSTRACT

The begging of broodmates that use aggression to compete for food has seldom been studied. Generally, subordinate broodmates beg as frequently as dominants, but receive less food. Overall relative begging frequencies of broodmates may influence parental food distributions, but these frequencies are not the main factor governing distributions. Rather, observations of boobies and diverse species suggest that aggression limits the effectiveness of begging by subordinate young by confining its timing, location or form. Although aggression is a powerful means of confining a broodmate's begging and outcompeting it for food, offspring of few avian species use aggression. Aggression needs to be both effective and profitable. Effectiveness may depend on slow food transfer and aggressive potential, and profitability may depend on spatial and temporal concentration of food and small brood size. Conceivably, potential to influence subsequent competitiveness is also important. A tentative framework combining these factors is proposed.

INTRODUCTION

There is a dramatic difference between the behaviour of most avian young, which compete with their nestmates by jostling and begging, and that of a minority of species whose young use violent aggression (hereafter, aggressive species). Aggressive young do not renounce begging, implying that it can be profitable for an offspring to divide its energies between petitioning providers and subduing competitors. Parents rarely interfere

J. Wright and M.L. Leonard (eds.), The Evolution of Begging, 337–360.

directly in aggression, but the tripartite relationship between petitioner, competitor and provider potentially includes complications. For example, begging could draw an aggressive or begging response from competitors and broodmate aggression could influence how much food parents provide or how they allocate it among offspring. In the first half of this chapter, I review begging in aggressive species and assess the relationship between food ingestion, aggression, begging and food distribution among broodmates. Begging has seldom been explicitly studied in aggressive species, so I draw heavily from those few studies (mostly of boobies, *Sula* spp.) that report quantifications, and lightly from numerous studies that provide qualitative information. This approach will document some (hopefully representative) patterns and put a premium on the question that occupies the second half of the chapter: since aggression can be highly effective for overriding begging, why is it not more widespread among avian species?

I will focus mainly on two categories of aggressive species with parental feeding. Obligate brood reducers, which are species where the single junior (younger) chick almost always dies within a few days of hatching. Facultative brood reducers are species where the junior chicks (the one or two youngest brood members) are more likely to die than senior chicks (the one or two oldest brood members), depending on the level of parental food provisioning (Mock 1984). Typically, junior chicks of species in both brood reducing categories are subordinate to seniors, and juniors feed less frequently and grow more slowly than seniors (e.g. Safriel 1981; Braun & Hunt 1983; Cash & Evans 1986; Mock & Parker 1997; Drummond 2001a). The nature of the dominance relationship between broodmates varies from the one-sided, all-out violence seen in some obligate brood reducers to learned dominance and subordinance, mediated partly by displays, in some facultative brood reducers (Drummond 1999). In facultative brood reducers, at least, an important function of much aggression may be to secure dominant status (Drummond 2001a). Lastly, there are species whose precocial young feed themselves, but nonetheless develop an agonistic hierarchy (Drummond 2001a). Although these self-feeding chicks do not beg, their behaviour will be discussed when it helps to explain the behaviour of parentally-fed young.

Equitable comparison of broodmates hatched a few days apart sometimes requires comparing them at the same age, using scores from different dates. These will be referred to as 'same-aged' comparisons.

BEGGING IN AGGRESSIVE SPECIES

The Incidence of Begging and Intimidation

Begging during the period of broodmate agonism has been reported in at least 14 species of facultative brood reducers and at least four species of obligate brood reducers (Table 1). I know of no cases in which aggressive, parentally-fed young do not beg. Begging has also been reported in goslings of the magpie goose (*Anseranas semipalmata*). Young of this species mainly feed themselves, but also feed from parents and show aggression (Johnsgard 1961).

The general feeding advantage of dominant booby chicks is not because they beg more frequently than subordinates. Begging in blue-footed boobies, involving oscillatory head movements and rhythmic vocalizations (Nelson 1978), was quantified by observing six two-chick broods for 12 hours/day and recording in each successive two-minute interval whether each chick begged (one-zero record). Between hatching and age 60 days, during which period only a few feeding bouts occurred each day, dominants begged during an average of 2.29 intervals/hour, and same-aged subordinates during an average of 2.66 intervals/hour (Wilcoxon test: $T = 10$, $n = 4$, $P = 0.12$). The begging rates of both chicks were nearly twice as high during the first three weeks after hatching, when feeding was most frequent, than they were subsequently (Drummond et al. 1986; Drummond & García Chavelas 1989; Anderson & Ricklefs 1995). Begging in brown booby hatchlings showed a similar pattern: same-aged dominant and subordinate young begged at similar rates of 3.27 and 3.97 intervals/hour, respectively, when observed 24 hours/day during the first three days of each hatchling's life (Wilcoxon test: $T = 0$, $n = 5$, $P = 0.06$; Cohen Fernández 1988).

In other species, the feeding advantage of dominant young may arise partly because they beg more frequently overall. Begging in cattle egrets, in which chicks scissor their parent's beak with their own, peaked just before the start of independent foraging (Fujioka 1985a), and the most senior chick scissored the parental beak more frequently than the most junior chick (Ploger & Mock 1986; Mock & Ploger 1987). Scissoring does, however, combine begging and access. Similarly, black-legged kittiwake dominants begged more frequently than subordinates in two-chick broods (1.96 versus 1.30, 30-second intervals/hour; one-tailed Sign test: $x = 1$, $n = 9$, $P = 0.02$), but dominants were on average more than one day older than subordinates when measured (Braun & Hunt 1983) which may explain the difference in rates.

Table 1. Species reported to show both broodmate aggression and begging.

Species	Inhibition of Begging[1]	Reference
Facultative Brood Reducers		
Great egret *Casmerodius albus*	Yes	Mock 1985
Cattle egret *Bubulcus ibis*	Yes	Fujioka 1985a, b; Ploger & Mock 1986
Great blue heron *Ardea herodias*	Yes	David & Berril 1987; Mock et al. 1987a
Grey heron *Ardea cineria*	?	Milstein et al. 1970
Bald ibis *Geronticus eremita*	Yes	Hirsch 1979; Oliver et al 1979
Osprey *Pandion haliaetus*	Yes	Poole 1979, 1982; Forbes 1991
Black-legged kittiwake *Rissa tridactyla*	Yes	Cullen 1957; Braun & Hunt 1983
South Polar skua *Catharacta maccormicki*	Yes	Young 1963; Spellerberg 1971
Blue-footed booby *Sula nebouxii*	Yes	Nelson 1978; Drummond & García Chavelas 1989
Double-crested cormorant *Phalacrocorax auritus*	?	Lewis 1929
Darter *Anhinga melanogaster*	?	Vestjens 1975
Brown pelican *Pelecanus occidentalis*	?	Pinson & Drummond 1993
Western grebe *Aechmophorus occidentalis*	Yes	Nuechterlein 1981
Oystercatcher *Haematopus ostralegus*	Yes	Safriel 1981
Obligate Brood Reducers		
Black eagle *Aquila verreauxii*	Yes	Rowe 1947; Gargett 1978
Lesser spotted eagle *Aquila pomarina*	Yes	Meyburg 1977
Brown booby *Sula leucogaster*	?	Cohen Fernández 1988
White pelican *Pelecanus erythroryhnchos*	Yes	Cash & Evans 1986

[1]Aggression by dominants causes subordinates to stop begging, beg less frequently or intensely or beg in a less favourable location.

According to qualitative observations on numerous species, the dominant chick's current or earlier aggression can inhibit subordinate chicks' begging and access to the parent. Subordinates reportedly declined to beg, begged less frequently or intensely, or begged in a less favourable location, in at least ten species of facultative brood reducers and at least three species of obligate brood reducers (Table 1). Great egret juniors hesitated to adopt a full posture or scissor promptly or declined to solicit altogether. A lesser spotted eagle subordinate chick often fled from the nest or stood "with its back to the middle of the nest, begging loudly but not even daring to look at the adult" (Meyburg 1977). Whereas western grebe dominants emerged from the parent's dorsal feathers and begged at every playback of a parental food call, their subordinate broodmates were only half as likely to emerge and often simply begged from under the feathers. When subordinate grebes stuck out their heads they were usually pecked, so they learned to postpone intense begging until after their antagonists were satiated (Nuechterlein 1981). In the precocial and highly mobile broods of the oystercatcher, subordinates conceded feeding priority to dominants by declining to approach parents, running away from parents when approached by dominants and hiding. Physical aggression was rare (in broods not confined by fences), so this inhibition of subordinate oystercatchers probably testifies to the enduring effects of earlier aggression (Safriel 1981).

How Does Begging Affect the Distribution of Food?

Brown booby parents do not allocate food in proportion to the relative begging frequencies of their two hatchlings. During the few days before junior hatchlings were fatally expelled from the nest, they begged, if anything, more frequently than their (larger and more mature) nestmates, but they received roughly four to ten times fewer mouth-to-mouth feeds (Figure 1). This does not reflect the lesser needs of the younger chick because juniors actually received considerably less food than seniors had received at the same age, despite begging just as frequently as those same-aged seniors (Figure 2). Similarly, over the first 60 days of each broodmate's life, blue-footed booby dominants received a privileged share of parental food, although they did not beg more frequently than subordinates (Drummond et al. 1986; Guerra & Drummond 1995). However, in an experiment where both broodmates' necks were encircled with tapes for two to three days to prevent ingestion, both at baseline and during food deprivation, the more frequently the dominant begged relative to its sibling, the higher its proportion of parental food offers to the brood (Drummond & García

Chavelas 1989). This relationship held for both young broods (seniors, 8-42 days old) and old broods (seniors, 43-72 days old).

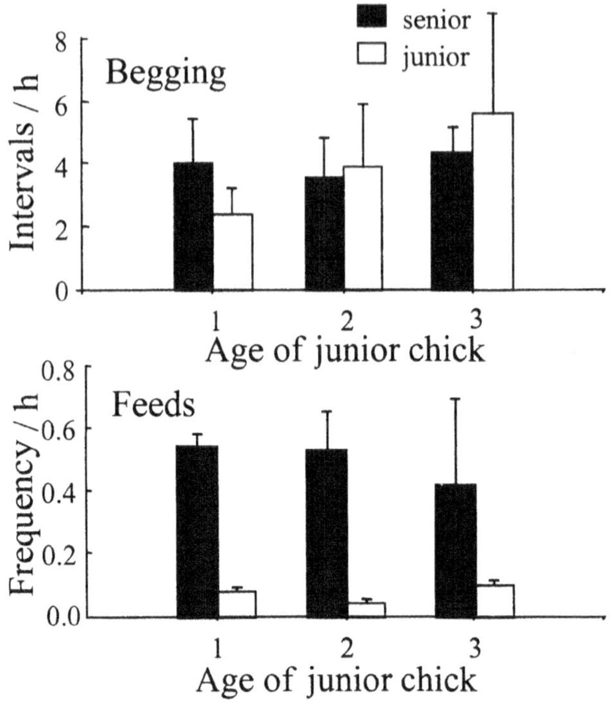

Figure 1. Mean (+SE) begging rates (two-minute intervals with begging) and feeding frequencies by brown booby broodmates, at the age (days post-hatch) of the junior chick. Each pair of bars represents the two broodmates on the same date, but with the senior an average of five days older than the junior (modified from Cohen Fernández 1988).

Hence, although dominant blue-footed boobies commonly receive more food offers than subordinates without begging more than subordinates overall, the proportion that dominants receive rises with their begging effort relative to that of their broodmate. Taken together, the findings for the two species of booby indicate that while the overall relative begging frequencies of broodmates may well have some influence on parental food distributions, other (unidentified) variables can be much more important.

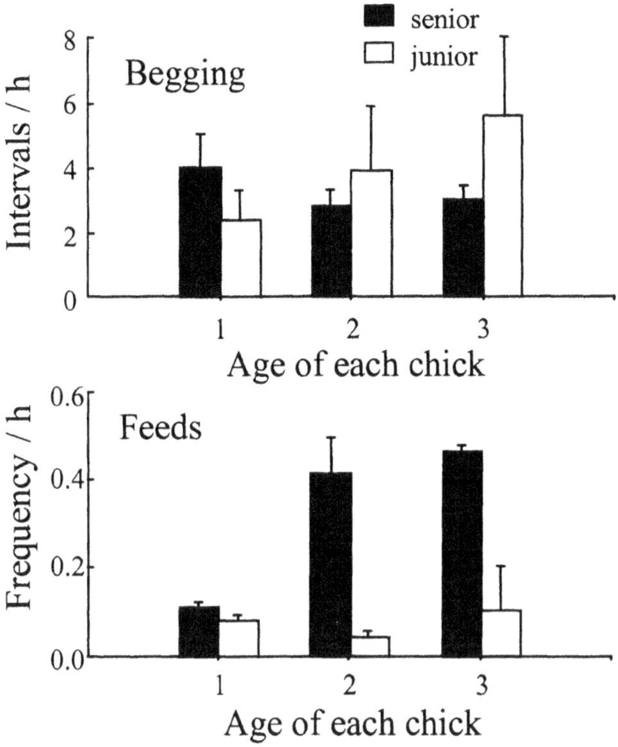

Figure 2. Mean (+SE) begging rates (two-minute intervals with begging) and feeding frequencies by brown booby broodmates at the same age (days post-hatch). Each pair of bars represents sampling on dates an average of five days apart (modified from Cohen Fernández 1988).

I suggest that parents in aggressive species respond mostly to the magnitude and location of each chick's begging relative to that of its broodmates, as well as to the timing of that begging. Dominants are usually larger and better coordinated than their rivals, enabling them to perform more salient begging displays, and their dominance can also give them considerable control over the timing, location and form of their rivals' begging. For example, senior white pelican chicks habitually begged right under the parent's breast after violently consigning their single rival to the back of the nest (Cash & Evans 1986). Parents seemed to feed only the chick in the conspicuous anterior location, even when the other chick was begging intensely. When the dominant chick allowed its subordinate broodmate to move into that location, parents would then feed the subordinate. Likewise, when a bald ibis parent arrived at the nest with food, the senior chick pecked its begging siblings into submissive head-down postures before it was fed. Only afterwards did each sibling in turn get its chance to beg and feed (Hirsch 1979). More extreme examples of control include great egret (Mock 1985) and great blue heron broods (David & Berrill 1987) where subordinates that were violently beaten into prostration neither solicited nor were offered food in the ongoing or succeeding feeding bout.

The blue-footed booby provides an example of less selfish control. When both chicks begged alongside each other (an unexceptional event), parental transfer of food to the subordinate chick was sometimes prevented by the dominant chick, but was frequently allowed to proceed (personal observation). Studies in Mexico and Galapagos have concluded that dominant blue-footed booby chicks effectively tolerate limited food-sharing with their subordinate broodmates, despite having the wherewithal to aggressively obtain a greater share for themselves (Drummond et al. 1986; Anderson & Ricklefs 1995).

The underfeeding of subordinate chicks can be further exacerbated by dominant chicks intercepting food offered to subordinates, and even by subordinates becoming too intimidated to accept offered food. One in every six of the (scarce) parental food offers to brown booby junior hatchlings was intercepted by the dominant (and much larger) chick (Cohen Fernández 1988). Similarly, the subordinate lesser spotted eagle chick observed by Meyburg (1977) was cowed by beatings to the point where "the adult female could not feed it, even though she bent over the chick, exerting herself and displaying astonishing patience in order to offer it pieces of meat".

In many respects the patterns of relative begging and relative feeding of unequal broodmates are similar in non-aggressive species (e.g. Rydén & Bengtsson 1980), but the extreme feeding bias imposed by senior brown booby chicks is testimony to the special efficacy of aggression. Aggressors deny parents the possibility of allocating food in response to simple begging competition by distorting the pattern of relative begging and by limiting parental choice over which broodmate will receive food.

How Does Ingestion Affect Begging?

There is experimental evidence that booby chicks, like passerines, beg more when they are deprived of food. Over the two or three days when blue-footed booby chicks wore tapes around their necks to prevent swallowing, and steadily lost weight, both broodmates increased the frequency of two-minute intervals during which they begged; reaching two to five times above the baseline rate (Drummond & García Chavelas 1989). Control broods, whose tapes were removed during feeding bouts to permit swallowing, showed no increase in begging.

However, there is a potential complication in these results. Reduced ingestion could also provoke greater aggression by the dominant chick, which might stifle attempts by the subordinate chick to increase its begging. According to the 'Food Amount Hypothesis' (Mock et al. 1987b), reduced ingestion should elicit an increase in aggressiveness. This hypothesis has now been confirmed by experiments on four species, including blue-footed boobies, and may apply to facultative brood reducers generally (review in Drummond 2001b, but see Mock et al. 1987b; Mock & Parker 1997). The precise causal pathway from food restriction to increased aggression has not yet been explored. Both deprivation *per se* and frustration (thwarting of feeding responses) are plausible stimuli. It is also conceivable that dominants increase their hostility partly in response to increased begging by their broodmate (Drummond 2001b).

Deprivation-induced aggression by the dominant booby chicks did not seem to suppress overall begging by their rivals, although it seems to have somehow guaranteed them greater feeding priority. In young broods, the increase in frequency of two-minute intervals with begging (mean $\pm$ SE from baseline to the first and second days of deprivation) was 142 $\pm$ 43% in dominants and 129 $\pm$ 37% in subordinates (Wilcoxon test: $T = 82$, $n = 18$, $P = 0.88$). In old broods the increase in begging (from baseline to the second and third days of deprivation) was 314 $\pm$ 110% in dominants and 474 $\pm$ 157% in subordinates ($T = 49$, $n = 15$, $P = 0.83$). Despite similar increases in the begging of food-deprived dominants and subordinates, the feeding advantage of dominants rose from 37% to 57% more attempted parental feeds than went to subordinates (Drummond & García Chavelas 1989). Presumably, intensified aggression secured enhanced feeding priority for the dominant chick, not by suppressing the subordinate chick's overall rate of begging, but by preventing it from begging at critical moments (perhaps when regurgitation was imminent) and from receiving food.

Why do parents apparently collude with their dominant offspring's attempts to skew allocations in their own favour? The aggressively enforced food distributions may, in fact, be in the parents' own interest (although this has not been demonstrated). Another possibility is that when the intensity of the dominant chick's aggression is food-dependent, parents may appease it by boosting its share of food, in order to protect their subordinate offspring (Drummond & García Chavelas 1989).

WHY IS AGGRESSION NOT MORE WIDESPREAD?

If by using aggression chicks can deny their rivals opportunity to beg intensely or in profitable moments and places, and thereby gain more food, why do not young of all species supplement begging with aggression? Presumably, for most birds such aggressive control of broodmates is not possible or, taking associated costs into account, would not be sufficiently profitable (Lamey & Mock 1991). Which factors, then, characterize the species that use aggression? To answer this question, Mock (1985) offered a general 'Prey Size Hypothesis', and Mock and Parker (1997) listed five 'evolutionary precursors' of resource-based broodmate aggression: resource limitation, resource monopolizability, weaponry, site topography and intrabrood asymmetries. The first four precursors were considered essential preconditions for the evolution of sibling aggression (Mock et al. 1990). I shall consider which of these, and other plausible factors, may predict use of aggression.

Resource Limitation

Aggression in most species with parental feeding is almost certainly an evolutionary response to food limitation, even when fighting starts before parental provisioning ability is exceeded by brood demand (e.g. Evans & McMahon 1987; Mock et al. 1990). Their aggression is essentially food competition by other means. However, food limitation is probably equally widespread among non-aggressive species with parental feeding. This factor predicts that broodmates will compete, but not whether they will do so by begging or aggression.

Feeding Method

The 'Feeding Method Hypothesis' (originally, Prey Size Hypothesis) holds that aggressive competition is favoured when food is transferred directly from the parent's mouth into the offspring's mouth (Mock 1985; anticipated by Werschkul 1979). Thus, when food parcels are small enough for direct transfer, a chick can monopolize access to them by aggressively displacing rivals; when parcels are too big for direct transfer, parents dump them on the substrate and chicks compete by scrambling for the food. The proximate role of feeding method in determining aggression was doubted by Drummond

(2001b). Here, I argue that there is little support for feeding method explaining the distribution of broodmate aggression among species.

The empirical support for a correlation across species between indirect feeding and non-aggression is negligible. The great blue heron is the only species cited in support, yet both reports of populations feeding large fish indirectly to chicks actually documented aggression (Mock 1985; David & Berrill 1987). Mock (1985) and Mock and Parker (1997) concluded that those populations illustrate the non-aggressive condition associated with indirect feeding because aggression was milder and less common than in great egrets and siblicide (evidenced by lesions) was uncommon. Use of this criterion, however, seems to reflect ambivalence over whether the 'Feeding Method Hypothesis' predicts mere aggression (see Mock & Parker 1997) or severe aggression amounting to siblicide (e.g. Mock 1985). The siblicide criterion was not applied when Mock and Parker (1997) found support for the 'Feeding Method Hypothesis' in the broodmate aggression of such direct-feeding species as the American black oystercatcher (*Haematopus bachmani*), the oystercatcher, the blue-footed booby, the magpie goose and the musk duck (*Biziura lobata*). In those species, aggression is similar to or milder than in great blue herons, and lesions and deaths caused directly by violence seem to be rare or absent (Johnsgard 1961; Davies 1963; Kear 1970; Safriel 1981; Groves 1984; Drummond et al. 1986). Certainly indirect feeding and non-aggression co-occur in some species (such as storks, Ciconiidae), but no association between those variables has been established.

The interspecific association between direct feeding and aggression is also dubious. Parental feeding is direct in most birds and yet only a small minority of them display broodmate aggression. Accordingly, Mock and Parker (1997) suggested that direct feeding is a necessary, but not sufficient condition for the use of aggression. There are, however, several counter-examples: broodmate aggression co-occurred with indirect feeding during most of the nestling period in grey herons (Milstein et al. 1970) and apparently also in great blue herons (Mock 1985). Aggression also co-occurred with indirect feeding early in the nestling period of great blue herons (Mock et al. 1987a), brown pelicans (Pinson & Drummond 1993) and great egrets (B. Ploger & M. Medeiros unpublished data), before the chicks were old enough to switch to direct feeding. For example, young great egret broods feeding indirectly on boluses of small prey, scored a daily average of 85 fights involving an average of 629 blows (B. Ploger & M. Medeiros unpublished data). Attacks were sometimes sufficient to exclude subordinate brood members from meals, showing that aggression can serve to monopolize indirect feeds and implying that struggle for dominance status was not the only function of aggression.

Apparently, both directly- and indirectly-fed chicks can often gain more food by attacking (and begging) than they can get by simply begging.

Spatial and Temporal Concentration of Food

It is obviously necessary for chicks to be close to each other for physical aggression to occur. If food is concentrated in space, as it usually is when provided by a parent (whether directly or indirectly), then at least the opportunity for aggression exists. Hence, the somewhat surprising occurrence of aggression among the precocial offspring of the magpie goose, the oystercatcher, the American black oystercatcher, the great northern diver (*Gavia immer*, Beebe 1907) and the sandhill crane (*Grus canadensis*) may be due to the existence of parental feeding in these species rather than to its directness (Drummond 2001b). Is parental feeding the only situation in which food becomes concentrated in space, and does parental feeding always imply spatial concentration? It would be interesting to know whether self-feeding young of any species use aggression to exclude broodmates from small and rich patches of food, and whether parents of any species make broodmate aggression unprofitable by spreading their feeds thinly in space. Johnsgard (1961) noted that magpie goslings often bit each other when begging simultaneously, and that "plant material and grain ... (were) scattered for them and ... would be picked up by the adults and then gradually allowed to dribble out of their bills as the goslings gathered around the adults' bills".

The size of food parcels is also likely to affect the economics of aggressive competition. The cost of aggression may be too high to justify its use when the food parcel or group of parcels obtained is very small. Whereas in most non-aggressive birds parents make numerous deliveries of small food parcels at short intervals, in species with broodmate aggression parental food parcels tend to be large and infrequent, and clustered in meals or bouts (temporal groups of parcels). Of course, when parcels or meals are very frequent, they tend to be small. Compare, for example, the starling's (*Sturnus vulgaris*) 25 parental nest visits/hour (Cotton et al. 1996) with the available daylight rates for aggressive species: young broods of great egrets, five feeding bouts/day (B. Ploger & M. Medeiros unpublished data); blue-footed booby dominants, 3.4 bouts (and 7.3 parcels)/day (Drummond et al. 1986); young broods of bald ibis, 16.7 bouts/day (Oliver et al. 1979); young brown pelicans, four parcels/hour, concentrated in bouts (Pinson & Drummond 1993); dominant kittiwake chicks, 2.3 parcels/hour, concentrated in bouts (Braun & Hunt 1983); young South Polar skua chicks, 0.4 parcels/hour (Young 1963); young dominant black eagle chicks, 0.5

bouts/hour (Rowe 1947); young black guillemot (*Cepphus grylle*) broods, 1.4 parcels/hour (Cook et al. 2000) and dominant chicks in old broods of American black oystercatchers, 7.4 parcels/hour (Groves 1984).

The food parcel is probably the relevant unit, except when an episode of aggression can secure for an individual all the parcels grouped in a bout, in which case the question is whether the cost of the episode of aggression is justified by the meal obtained. On this basis, the crimson rosella (*Platycercus elegans*) nestlings described by E.A. Krebs (this volume) are unlikely to compete aggressively for the large meals, provided at roughly hourly intervals, because each meal is transferred directly in about 15 small parcels during a two-minute visit (Krebs et al. 1999). A pre-existing dominance relationship probably reduces the cost of aggression over a particular food parcel but, in blue-footed boobies at least, the mere relationship often seems insufficient to secure priority, obliging dominant chicks to repeatedly attack and threaten during feeding bouts for each successive parcel (personal observation).

This hypothesis may partly account for the observation that broodmate aggression is largely confined to predatory species: those that feed their offspring relatively large animals rather than, say, small insects or seeds.

Slow Food Transfer

Aggression may not be an option if parents can transfer food parcels so rapidly that young have insufficient time to intervene aggressively. Perhaps the most startling difference between aggressive and non-aggressive species that I have observed is the speed with which a parcel passes from parent to offspring. This variable has seldom been quantified, but a booby or pelican parent seems to take several seconds to make a direct food transfer. The chick has time to strike a few blows and cow or displace its rival before completion and even before onset of transfer, which is anticipated by preparatory movements. By contrast, a passerine usually transfers a parcel in a fraction of a second, just after arriving at the nest. For example, video recordings at ten nests of each of three species showed the following median transfer times (from starting to dip down toward the nestlings until removal of the empty parental bill) of females and males, respectively: pied flycatcher (*Ficedula hypoleuca*), 520, 540 ms; great tit (*Parus major*), 1100, 900 ms; blue tit, (*P. caeruleus*), 720, 680 ms (R. Cordts, T. Limbourg & K. Lessells unpublished data). By the time a passerine parent starts to transfer food, it may be too late to frustrate the transfer by attacking the intended recipient, and a nestling that attacked before transfer started might miss the transfer because it was not begging. Relative speed is what counts: the quickness of the passerine parent relative to its nestling may give the parent control over which offspring receives each parcel (provided the target mouth is open).

Control through relative speed may prevail in species where (1) small and agile parents deliver (2) small food parcels (3) carried in the parental beak (4) directly into their offspring's mouth and (5) young are highly altricial. For example, young red-winged blackbird nestlings (*Agelaius phoeniceus*) open their mouths and leave transfer to the parents; they cannot see or grab a parcel from a parent until about six days post-hatch, almost two thirds of the way to fledging (A.B. Clark unpublished data). Limited parental control is expected in species where parents regurgitate, because transfer is slow and anticipated by postural and other movements. Regurgitating egrets lower their bills slowly to their chicks (Mock & Parker 1997), and in boobies and pelicans, transfer typically involves ponderous contortions of the parent and difficult (sometimes unsuccessful) coordination with the target chick to link oral cavities.

Aggressive Potential

The ability to physically cow, displace or injure rivals is necessary if aggression is to be used. Predatory weapons evidently are not required, since even ducks and geese attack their broodmates effectively within days of hatching (Drummond 2001a). Aggressive potential is a better term for this ability than weaponry, since morphology, sensory abilities, motor coordination and body size may determine a chick's ability to deliver a forceful blow. Aggressive potential also depends on the age disadvantage and physical vulnerability of the victim.

Probably in most birds the aggressive potential of adults and fledglings is sufficient for aggression to be effective, but the age at which sufficient potential emerges in the nestling bird must vary greatly. I suspect that in most passerines the aggressive potential of nestlings is seriously limited during at least the first half of the nestling period, by low head mass, soft blunt bills, blindness and motor immaturity. In contrast, blue-footed boobies can see and direct effective pecks at nestmates at <10% of fledging age (roughly eight days post-hatch), and broodmate attacks in parentally-fed and self-feeding species usually begin shortly after hatching of the victim (Drummond 2001a).

In some passerines specialized weaponry for wounding and even killing young broodmates has evolved. For example, blue-throated bee-eaters (*Merops viridis*) have lethal temporary hooks on the upper mandibles (Bryant & Tatner 1990), although their agonistic behaviour has not been directly observed. The presence of such weapons can be interpreted as evidence that the baseline aggressive potential of passerines is limited, but it also shows that this limitation can be overcome by evolution.

Site Topography

It has been proposed that the evolution of siblicidal aggression requires that the topography of the nest or nest site constrains the victim's avenues for escape from aggression (Mock 1985; Mock et al. 1990; Mock & Parker 1997). However, aggressive competition for parentally provided food occurs among avian broodmates that are highly mobile and by no means confined to the nest or its vicinity (e.g. sandhill crane, Miller 1973; magpie goose, Kear 1970; South Polar skua, Young 1963; oystercatcher, Safriel 1981; American black oystercatcher, Groves 1984; blue-footed boobies old enough to leave the nest territory, personal observation). In the last four of these species, aggressive exclusion results in differential mortality of subordinate broodmates due to starvation, predation, thermal stress or attacks by adult neighbours. Nonetheless, deaths caused directly by broodmate violence (narrowly defined siblicide) are presumably more common under topographical constraint.

Nest site topography has diverse effects on the efficacy and consequences of aggression. Anderson (1995) suggested that the function of the blue-footed booby's deep nest scrape may be to constrain and delay the siblicidal expulsion of subordinate hatchlings. Mock and Parker (1997) suggested that the constriction of nest burrows and holes can help dominant chicks to aggressively control access to parents, and also that nooks in the vegetation enclosing egret nests can provide subordinates with safe hiding places.

In sum, although site topography can affect the amount of violence absorbed by a chick and the manner of its death, confinement is not required for aggression to function as a supplement to begging, nor for aggression to lead to death. Thus, the role of site topography in the evolution of aggression may vary.

Small Brood Size

Drummond (2001a) speculated that in species with parental feeding, small brood size might be a predictor of broodmate aggression. This was based on the observation that among accipitrids and (less clearly) among ardeids and species with precocial young, aggression appears to be more likely or intense when broods are small. For example, accipitrids tend to have (1) large bodies, small broods and intense aggression; (2) medium bodies, medium-sized broods and variable aggression; or (3) small bodies, large broods and no aggression (Newton 1979). Aggression could be less effective for gaining feeding priority in larger broods if aggressors find it more difficult or costly to intimidate or displace several competitors simultaneously, or to concentrate punishment on particular individuals in a throng. Inability to target individuals could arise from (accidental) obstruction by other broodmates or from a limited capacity to discriminate individuals. In a brood of two, an attack can instantly prevent the single rival from feeding, whereas in a brood of four, attacking one rival risks creating an instant opening for the other two to feed.

At the proximate level, this hypothesis predicts that, within species, young should respond facultatively to greater brood size by reducing the proportion of energy applied to agonism rather than begging. Such an effect could be masked if offspring in larger broods also tend to receive less food *per capita* and consequently increase their aggression (and begging) in response to food deprivation. A quite different and no less plausible hypothesis holds that since greater brood size probably predicts less food *per capita*, aggression should intensify facultatively in response to greater brood size itself (Mock & Lamey 1991). Some descriptive observations and an experiment on cattle egrets support this idea (Mock & Lamey 1991), but neither completely excluded the alternative interpretation that aggression increased with brood size in response to decreased *per capita* food intake (Drummond 2001a). More experiments are clearly needed to disentangle the proximate effects of brood size and food deprivation on the intensity and proportions of agonism versus begging.

Intrabrood Asymmetries

Mock and Parker (1997) proposed that asymmetries among brood members in age and size facilitate the evolution of aggression, because asymmetry reduces the cost of aggression (Mock et al. 1990). However, agonistic hierarchies occur naturally in broods where hatching intervals and initial asymmetries are minimal. For example, in grouse (Tetraonidae) and in ducks and geese (Anatidae), although clutches usually hatch completely in a matter of hours (Carboneras 1992; de Juana 1994) several species have evolved aggression and exhibit dominance hierarchies among broodmates (Drummond 2001a). Also, when asymmetries in facultative brood reducers are minimized by swapping hatchlings between nests, agonistic hierarchies nonetheless emerge, albeit with intensified fighting (Fujioka 1985b; Mock & Ploger 1987; Osorno & Drummond 1995).

Hence, although ancestral symmetry among broodmates might slow down the evolution of agonism, it seems unlikely to prevent it. Rather, traits such as egg size, laying interval and incubation regime are likely to evolve to enhance or reduce asymmetry (R.E. Ricklefs this volume) and thereby influence the outcomes of broodmate agonism. Such evolution is the likely explanation for large asymmetries in aggressive species, which do indeed reduce the cost of competition (Fujioka 1985b; Mock & Ploger 1987; Osorno & Drummond 1995).

Potential to Influence Subsequent Competitiveness

In birds with self-feeding young, aggression among broodmates has been reported, and competitive interactions over food, brooding or space have occasionally been inferred (e.g. Rajala 1962). However, Kalas (1977) de-emphasized competition for current resources and suggested that the main function of broodmate agonism in the greylag goose (*Anser anser*) is to obtain status and the benefits of status that could accrue over the long-term in the wider flock. In many species with self-feeding chicks, social life after parental care is complex and includes rich opportunities for established dominance relations or enhanced agonistic prowess to influence competitive and cooperative interactions with broodmates and others (e.g. Davies 1963; Watts & Stokes 1971; Lamprecht 1985; Black & Owen 1989a, b; Lorenz 1991).

There is evidence for durable effects of early aggression on status, and there are hints that status can become more important after independence. Agonistic interactions among broodmates of at least one parentally-fed species train individuals into the habitual agonistic role of 'winner' or 'loser' during the nestling period (Drummond & Osorno 1992; Drummond & Canales 1998). In some self-feeding species, broodmate dominance relationships tend to endure for at least several weeks (Baumer 1955 in Nice 1962; Collias & Collias 1956; Davies 1963; Rushen 1982) and possibly influence later interactions with unfamiliar peers (Boag & Alway 1980, but see Black & Owen 1987). In western capercaillie (*Tetrao urogallus*) and black grouse (*Tetrao tetrix*) broods, agonistic hierarchies are established shortly after hatching, but fighting does not intensify until after independence. Both species show polygyny, group courtship and presumably intense sexual selection on adult males, in contrast to the socially monogamous willow grouse (*Lagopus lagopus*) whose chicks apparently do not fight until much later, at age three weeks (Rajala 1962).

Hence, another factor that could favour the evolution of broodmate aggression is the potential to influence subsequent competitiveness, itself a function of the social structure of the population and ontogenetic sensitivity to agonistic experience during the nestling period. In principle, influence on subsequent competitiveness could partially drive the evolution of agonism not only in self-feeding chicks, but also in those with parental feeding (cf. Simmons 1988). For example, there is scope for this in the bald ibis where nestmates form a linear agonistic hierarchy during the period of parental feeding and subsequently fledglings from several broods associate in hierarchical groups (Pegoraro & Thaler 1993), and in passerine species where juvenile birds display dominance when contesting food sources (e.g. Arcese & Smith 1985). Examples of sibling rivalry after the normal age for dispersal are discussed by Mock and Parker (1997).

A FRAMEWORK FOR INTEGRATION

In species whose young compete for food provided by parents, aggression should evolve when enabling factors such as slow food transfer and aggressive potential make it effective, provided that economic factors such as spatio-temporal concentration of food and small brood size also make it profitable (Table 2). Offspring of numerous species may be denied the option of using aggression because quick food transfer gives parents incontestable control over allocations or because young lack aggressive potential. Greater spatio-temporal concentration of food should increase the benefit of aggression, while larger brood sizes should reduce its effectiveness and increase its cost. Factors should interact to determine profitability. For example, even a well-armed nestling may find that violence does not pay when there are several rivals in the nest and parents can place food in their mouths in a fraction of a second. Conceivably, aggression may also evolve, even in the absence of food competition, when there is potential to influence subsequent competitiveness, provided aggressive potential is adequate.

Obligate brood reduction systems may be different. If the only function of aggression was to promptly kill the broodmate then only aggressive potential would seem necessary. It takes a few days, however, for a broodmate to succumb to all-out aggression, and if the broodmate had a feeding advantage during this time, superior nutrition might enable it to invert the dominance relationship. Hence, the evolution of aggression in obligate brood reducers may also require both enabling and economic factors (Table 2).

Table 2. Possible factors favouring evolution of broodmate aggression.

Type of Factor	Factor	Facultative Brood Reducers	Obligate Brood Reducers
Enabling[1]	Aggressive potential	YES	YES
	Slow food transfer	YES	YES[2]
Economic	Spatial/temporal concentration of food	YES	YES[2]
	Brood size	YES	YES[2]
Social	Potential to influence subsequent competitiveness	?	?

[1]Restricted parental access to the brood (e.g. in nest burrows and holes) may be an additional factor. [2]Favours aggression if the senior chick requires a feeding advantage to maintain dominance.

These factors imply multiple barriers to most passerines using aggression early in development. When passerine nestlings are young, parents control allocations by making instantaneous food transfers, and highly altricial nestlings may be too immature to use aggression effectively. Moreover, throughout the nestling period aggression is likely to be unprofitable because food parcels are usually small and broods tend to be large. But the enabling and economic factors are liable to change during the development of a brood if, for example, parental quickness relative to that of offspring diminishes, aggressive potential of young grows, food parcels increase in size and brood size diminishes. At the age (if any) when aggression becomes both effective and profitable, aggressiveness should emerge.

There is indeed some evidence that aggressiveness can emerge late in the development of nestlings of some passerines and other taxa. Reynolds (1996) inferred from wounds on dead black-billed magpie (*Pica pica*) nestlings that broodmate aggression emerged in roughly the last third of the nestling period, although behaviour was not observed and some nestlings that starved during the first third bore suspicious bruises on head and nape (and cannibalism of older victims complicates interpretation). M.L. Leonard (personal communication) observed repeated agonistic exchanges between two fledglings of the (taxonomically distinct) Australian magpie (*Gymnorhina tibicen*) competing for parental feeds, the larger one frequently lunging and pecking at the smaller to drive it back when it begged for food. Intriguingly, juvenile hummingbirds (Trochilidae) often attack their single broodmate with their feet when competing for food provided by parents (Schuchmann 1999). Theoretically, the advantage of being the first broodmate to start attacking (e.g. Rushen 1982) could sometimes result in an evolutionary 'onset race'. The benefits of dominant status could drive the evolution of very early aggression even when the aggression is initially less profitable, in energetic terms, than simple begging.

Monopolizability of food parcels, a key concept introduced by Mock (1985), should probably be understood in relative terms. In principle, a begging chick should use aggression if that increases its share of the food on offer. When a chick competes for food deposited on the nest floor, aggression may sometimes win it the whole parcel, but if the parcel is divisible an increased percentage may be the best that can be achieved, especially when there are several competitors. What matters then, is whether using aggression will yield a greater net gain than mere begging, jostling and scrambling. In practice, young of several species do fight when fed indirectly, and I suspect that this aggression often garners a merely increased share rather than the whole parcel, although effects of fighting on the division of indirect food parcels have not been analysed.

FUTURE DIRECTIONS

To understand begging in aggressive species, we need analyses of the relationship between begging by the brood and overall parental food provisioning. We also need to understand how aggression, begging and food division are related. A focus on individual feeding bouts and on the roles of brood members of different dominance rank is desirable.

To test whether the factors proposed in this chapter have indeed favoured the evolution of aggression in parentally-fed species, and to explore the interactions of those factors, we need formal comparative studies that control for phylogeny (e.g. Harvey & Pagel 1991). We can also test for proximate effects of those factors on the tendency to use aggression and on its effectiveness and profitability, using both descriptive analyses and experimental tests. In addition, comparison of the energetic costs of aggression and begging could help us understand the economics of using violence.

ACKNOWLEDGEMENTS

For helpful comments, ideas and discussion, I thank Marty Leonard, Jon Wright, Anne Clark, Barb Glassey, Amber Budden, Kate Lessells, Trond Amundsen and Gregynog workshop participants generally, as well as José Luis Osorno, Ian Newton, Francisco Sánchez Tortosa, Roxana Torres, Jaime Zaldívar and Sylvia Rojas. Anne Clark, Kate Lessells, Ruedigern Cordts and Tobias Limbourg kindly provided unpublished observations, and Cristina Rodríguez and Isabel Fuentes helped with literature searches and manuscript preparation. The field research in which this review is grounded was financed by grants from the Universidad Nacional Autónoma de México (DGAPA-IN211491), CONACYT (D112-903581, PCCNCNA-031528, 071PÑ-1297, 31973-H), the National Geographic Society (3065-85, 4535-91) and the Conservation and Research Foundation.

REFERENCES

Anderson, D.J. 1995. The role of parents in siblicidal brood reduction of two booby species. *The Auk* **112**, 860-869.

Anderson, D.J. & Ricklefs, R.E. 1995. Evidence of kin-selected tolerance by nestlings in a siblicidal bird. *Behavioral Ecology and Sociobiology* **37**, 163-168.

Arcese, P. & Smith, J.N.M. 1985. Phenotypic correlates and ecological consequences of dominance in song sparrows. *Journal of Animal Ecology* **54**, 817-830.

Beebe, C.W. 1907. Notes on the early life of loon chicks. *The Auk* **24**, 34-41.

Black, J.M. & Owen, M. 1987. Determinants of social rank in goose flocks: acquisition of social rank in young geese. *Behaviour* **102**, 129-146.

Black, J.M. & Owen, M. 1989a. Parent-offspring relationships in wintering barnacle geese. *Animal Behaviour* **37**, 187-198.

Black, J.M. & Owen, M. 1989b. Agonistic behaviour in goose flocks: assessment, investment, and reproductive success. *Animal Behaviour* **37**, 199-209.

Boag, D.A. & Alway, J.H. 1980. Effects of social environment within the brood and dominance rank in gallinaceous birds. *Canadian Journal of Zoology* **58**, 44-49.

Braun, B.M. & Hunt, G.L. Jr. 1983. Brood reduction in black-legged kittiwakes. *The Auk* **100**, 469-476.

Bryant, D.M. & Tatner, P. 1990. Hatching asynchrony, sibling competition and bee-eaters. *Animal Behaviour* **39**, 657-671.

Carboneras, C. 1992. Family Anatidae (ducks, geese and swans). In: *Handbook of the Birds of the World. Vol. 1* (Ed. by J. del Hoyo, A. Elliott & J. Sargatal). Barcelona: Lynx Editions.

Cash, K.J. & Evans, R.M. 1986. Brood reduction in the American white pelican (*Pelecanus erythrorhynchos*). *Behavioral Ecology and Sociobiology* **18**, 413-418.

Cohen Fernández, E.J. 1988. La reduccion de la nidada en el bobo café (*Sula leucogaster nesiotes* Heller and Snodgrass 1901). Undergraduate thesis, Universidad Nacional Autónoma de México.

Collias, N.E. & Collias, E. 1956. Some mechanisms of family integration in ducks. *The Auk* **73**, 378-400.

Cook, M.I., Monaghan, P. & Burns, M.D. 2000. Effects of short-term hunger and competitive asymmetry on facultative aggression in nestling black guillemots *Cepphus grylle*. *Behavioral Ecology* **11**, 282-287.

Cotton, P.A., Kacelnik, A. & Wright, J. 1996. Chick begging as a signal: are nestlings honest? *Behavioral Ecology* **7**, 178-182.

Cullen, E. 1957. Adaptations in the kittiwake to cliff-nesting. *Ibis* **99**, 275-302.

David, S. & Berrill, M. 1987. Siblicidal attacks by great blue heron, *Ardea herodias*, chicks in a southern Ontario heronry. *Canadian Field Naturalist* **101**, 105-107.

Davies, S.J.J.F. 1963. Aspects of the behaviour of the Magpie Goose, *Anseranas semipalmata*. *Ibis* **105**, 76-98.

de Juana, E. 1994. Family Tetraonidae (grouse). In: *Handbook of the Birds of the World. Vol. 2* (Ed. by J. del Hoyo, A. Elliott & J. Sargatal). Barcelona: Lynx Editions.

Drummond, H. 1999. Agonism and dominance in nestling birds. In: *Proceedings of the 22nd International Ornithological Congress* (Ed. by N. Adams & R. Slotow). Johannesburg: Birdlife South Africa.

Drummond, H. 2001a. The control and function of agonism in avian broodmates. *Advances in the Study of Behavior* **30**, 261-301.

Drummond, H. 2001b. A revaluation of the role of food in nestling aggression. *Animal Behaviour* **61**, 1-10.

Drummond, H. & García Chavelas, C. 1989. Food shortage influences sibling aggression in the blue-footed booby. *Animal Behaviour* **37**, 806-819.

Drummond, H. & Osorno, J.L. 1992. Training siblings to be submissive losers: dominance between booby nestlings. *Animal Behaviour* **44**, 881-893.

Drummond, H. & Canales, C. 1998. Dominance between booby nestlings involves winner and loser effects. *Animal Behaviour* **55**, 1669-1676.

Drummond, H., Gonzalez, E. & Osorno, J.L. 1986. Parent-offspring cooperation in the blue-footed booby (*Sula nebouxii*): social roles in infanticidal brood reduction. *Behavioral Ecology and Sociobiology* **19**, 365-372.

Evans, R.M. & McMahon, B.F. 1987. Within-brood variation in growth and condition in relation to brood reduction in the American white pelican. *Wilson Bulletin* **99**, 190-201.

Forbes, L.S. 1991. Hunger and food allocation among nestlings of facultatively siblicidal ospreys. *Behavioral Ecology and Sociobiology* **29**, 189-195.

Fujioka, M. 1985a. Sibling competition and siblicide in asynchronously hatching broods of the cattle egret *Bubulcus ibis*. *Animal Behaviour* **33**, 1228-1242.

Fujioka, M. 1985b. Food delivery and sibling competition in experimentally even-aged broods of the cattle egret. *Behavioral Ecology and Sociobiology* **17**, 67-74.

Gargett, V. 1978. Sibling aggression in the black eagle in the Matopos, Rhodesia. *Ostrich* **49**, 57-63.

Groves, S. 1984. Chick growth, sibling rivalry, and chick production in American black oystercatchers. *The Auk* **101**, 525-531.

Guerra, M. del C. & Drummond, H. 1995. Reversed sexual size dimorphism and parental care: minimal division of labour in the blue-footed booby. *Behaviour* **132**, 479-496.

Harvey, P.H. & Pagel, M.D. 1991. *The Comparative Method in Evolutionary Biology*. Oxford: Oxford University Press.

Hirsch, U. 1979. Studies of West Palearctic birds: Bald Ibis. *British Birds* **72**, 313-324.

Johnsgard, P.A. 1961. The breeding biology of the Magpie Goose. *Wildfowl Trust Annual Reports* **12**, 92-103.

Kalas, V.S. 1977. Ontogenie und funktion der rangordnung innerhalb einer geschwisterschar von graugänsen (*Anser anser* L.). *Zeitschrift für Tierpsychologie* **45**, 174-198.

Kear, J. 1970. The adaptive radiation of parental care in waterfowl. In: *Social Behaviour of Birds and Mammals* (Ed. by J.H. Crook). London: Academic Press.

Krebs, E.A., Cunningham, R.B. & Donnelly, C.F. 1999. Complex patterns of food allocation in asynchronously hatching broods of crimson rosellas. *Animal Behaviour* **57**, 753-763.

Lamey, T.C. & Mock, D.W. 1991. Nonaggressive brood reduction in birds. *Acta XX Congressus Internationalis Ornithologici* **III**, 1741-1751.

Lamprecht, J. 1985. Structure and causation of the dominance hierarchy in a flock of bar-headed geese (*Anser indicus*). *Behaviour* **96**, 28-48.

Lewis, H.F. 1929. *The Natural History of the Double-crested Cormorant.* Ottawa: Ru-Mi-Lou Books.

Lorenz, K. 1991. *Here Am I – Where Are You?* New York: Harcourt Brace Jovanovich.

Meyburg, B.U. 1977. Sibling aggression and cross-fostering of eagles. In: *Endangered Birds* (Ed. by S.A. Temple). Madison: University of Wisconsin Press.

Miller, R.S. 1973. The brood size of cranes. *Wilson Bulletin* **85**, 436-441.

Milstein, P. LeS., Prestt, I. & Bell, A.A. 1970. The breeding cycle of the grey heron. *Ardea* **58**, 171-257.

Mock, D.W. 1984. Infanticide, siblicide and avian nestling mortality. In: *Infanticide: Comparative and Evolutionary Perspectives* (Ed. by G. Hausfater & S.B. Hrdy). New York: Aldine Publishing Company.

Mock, D.W. 1985. Siblicidal brood reduction: the prey-size hypothesis. *American Naturalist* **125**, 327-343.

Mock, D.W. & Ploger, B.J. 1987. Parental manipulation of optimal hatch asynchrony in cattle egrets: an experimental study. *Animal Behaviour* **35**, 150-160.

Mock, D.W. & Lamey, T.C. 1991. The role of brood size in regulating egret sibling aggression. *American Naturalist* **138**, 1015-1026.

Mock, D.W. & Parker, G.A. 1997. *The Evolution of Sibling Rivalry.* Oxford: Oxford University Press.

Mock, D.W., Lamey, T.C., Williams, C.F. & Pelletier, A. 1987a. Flexibility in the development of heron sibling aggression: an intraspecific test of the prey-size hypothesis. *Animal Behaviour* **35**, 1368-1393.

Mock, D.W., Lamey, T.C., Williams, C.F. & Ploger, B.J. 1987b. Proximate and ultimate roles of food amount in regulating egret sibling aggression. *Ecology* **68**, 1760-1772.

Mock, D.W., Drummond, H. & Stinson, C.H. 1990. Avian siblicide. *American Scientist* **78**, 438-439.

Nice, M.M. 1962. Development of behavior in precocial birds. *Transactions of the Linnean Society* **VIII**, 1-211.

Nelson, J.B. 1978. *The Sulidae: Gannets and Boobies.* London: Oxford University Press.

Newton, I. 1979. *Population Ecology of Raptors.* Berkhamsted: A.D. Poyser.

Nuechterlein, G.L. 1981. Asynchronous hatching and sibling competition in western grebe broods. *Canadian Journal of Zoology* **59**, 994-998.

Oliver, W.L.R., Mallet, M.M., Singleton, B.A. & Ellet, J.S. 1979. Observations on the reproductive behaviour of a captive colony of bare-faced ibis *Geronticus eremita. Dodo* **16**, 11-35.

Osorno, J.L. & Drummond, H. 1995. The function of hatching asynchrony in the blue-footed booby. *Behavioral Ecology and Sociobiology* **37**, 265-273.

Pegoraro, von K. & Thaler, E. 1993. Postembryonic development and juvenile phase of the Northern Bald Ibis or Waldrapp Ibis *Geronticus eremita* (L.). *Ecology of Birds* **15**, 155-192.

Pinson, D. & Drummond, H. 1993. Brown pelican siblicide and the prey-size hypothesis. *Behavioral Ecology and Sociobiology* **32**, 111-118.

Ploger, B.J. & Mock, D.W. 1986. The role of sibling aggression in distribution of food to nestling cattle egrets (*Bubulcus ibis*). *The Auk* **103**, 768-776.

Poole, A.L. 1979. Sibling aggression among nestling ospreys in Florida Bay. *The Auk* **96**, 415-417.

Poole, A.L. 1982. Brood reduction in temperate and sub-tropical ospreys. *Oecologia* **53**, 111-119.

Rajala, P. 1962. On the ecology of the broods of Capercaillie (*Tetrao urogallus*), Black Grouse (*Lyrurus tetrix*), and Willow Grouse (*Lagopus lagopus*). *Suom Riista* **15**, 28-52.

Reynolds, P.S. 1996. Brood reduction and siblicide in black-billed magpies (*Pica pica*). *The Auk* **113**, 189-199.

Rowe, E.G. 1947. The breeding biology of *Aquila verreauxi* Lesson. *Ibis* **89**, 576-607.

Rushen, J. 1982. The peck orders of chickens: how do they develop and why are they linear? *Animal Behaviour* **30**, 1129-1137.

Rydén, O. & Bengtsson, H. 1980. Differential begging and locomotory behaviour by early and late hatched nestlings affecting the distribution of food in asynchronously hatched broods of altricial birds. *Zeitschrift für Tierpsychologie* **53**, 209-224.

Safriel, U.N. 1981. Social hierarchy among siblings in broods of the oystercatcher *Haematopus ostralegus*. *Behavioral Ecology and Sociobiology* **9**, 59-63.

Schuchmann, K.L. 1999. FamilyTrochilidae (hummingbirds). In: *Handbook of the Birds of the World. Vol. 5* (Ed. by J. del Hoyo, A. Elliott & J. Sargatal). Barcelona: Lynx Editions.

Simmons, R. 1988. Offspring quality and the evolution of cainism. *Ibis* **130**, 339-357.

Spellerberg, I.F. 1971. Breeding behaviour of the McCormick skua *Catharacta maccormicki* in Antarctica. *Ardea* **59**, 189-230.

Vestjens, W.J.M. 1975. Breeding behaviour of the darter at Lake Cowal, NSW. *Emu* **75**, 121-131.

Watts, C.R. & Stokes, A.W. 1971. The social order of turkeys. *Scientific American* **224**, 112-118.

Werschkul, D.F. 1979. Nestling mortality and the adaptive significance of early locomotion in the little blue heron. *The Auk* **96**, 116-130.

Young, E.C. 1963. The breeding behaviour of the South Polar skua, *Catharacta maccormicki*. *Ibis* **105**, 203-233.

BEGGING AND BROOD PARASITISM

19.

BEGGING BEHAVIOUR AND HOST EXPLOITATION IN PARASITIC COWBIRDS

Donald C. Dearborn[1] & Gabriela Lichtenstein[2]

[1]*Department of Biology and Program in Animal Behavior, Bucknell University, Lewisburg, PA 17837, USA (ddearbor@bucknell.edu)*

[2]*International Institute for Environment and Development (IIED-AL), Gral Paz 1180 1429 Buenos Aires, Argentina (gaby@interlink.com.ar)*

ABSTRACT

In this chapter we explore the begging behaviour of cowbirds, obligate brood parasites that are typically raised in mixed broods with host young. As 'strangers in the nest', cowbird nestlings present both challenges and opportunities to evolutionary biologists. After a brief overview, we delve into four main topics: (1) the means by which cowbird young achieve success in host nests; (2) the differences between cowbirds and non-parasitic nestlings in the way that natural selection limits the exaggeration of begging behaviour; (3) the extent to which cowbird nestlings are perceived and treated as host young by the provisioning adults; and (4) the extent to which the begging signals of cowbird young are honest indicators of need. The most striking conclusions from our synthesis are that cowbird nestlings do not always successfully manipulate host adults and that begging by cowbirds can be informative of their level of need.

INTRODUCTION

The theoretical focus of studies of begging behaviour has centred on parent-offspring conflict, sibling competition, the honesty of signals and the associated need for signals to be costly (R.A. Johnstone & H.C.J. Godfray this volume). One of the most fascinating systems in which to study begging behaviour is brood parasitism, an uncommon but successful life history strategy present in one percent of the world's bird species (Davies 2000). Interspecific brood parasites lay their eggs in nests of other species (the

J. Wright and M.L. Leonard (eds.), The Evolution of Begging, 361–387.

hosts), which then incubate the eggs and care for the young. Brood parasites provide an interesting context in which to examine theoretical issues in begging behaviour because brood-parasitic nestlings are not related to the adults that provision them or to their nestmates (except in the case of multiple parasitism). This lack of relatedness has the potential to change the costs, and hence the honesty, of signals produced by brood-parasitic young.

Obligate brood parasitism occurs in cuckoos (Neomorphinae and Cuculinae, T. Redondo & J.M. Zuñiga and M. Soler this volume), cowbirds (Icterinae), the cuckoo-finch (*Anomalospiza imberbis*), the whydahs (Viduinae, R.B. Payne & L.L. Payne this volume), the honeyguides (Indicatorinae) and the black-headed duck (*Heteronetta atricapilla*, Davies 2000). In this chapter, we examine the begging behaviour of three species of cowbirds. We show how cowbird begging may have been shaped by selective forces that are unique to brood parasites, and we also discuss how cowbird behaviour can provide insights into the evolution of begging.

Cowbirds provide an ideal opportunity for the study of begging behaviour. Unlike the common cuckoo (*Cuculus canorus*, Wyllie 1981), these brood parasites are raised with the host young. Thus, cowbirds allow the study of nestling competition and adult allocation decisions involving nestlings that are not related to each other and that often exhibit behavioural or morphological phenotypes different from the host's own young. Cowbirds also parasitize host species of different sizes, which permits a comparison of the begging behaviour of parasitic nestlings when receiving different amounts of food and competing with different sized host nestlings. Lastly, one cowbird species mimics the host young, which allows comparative tests of the importance of exaggerated begging as a means of offsetting the disadvantage of looking dissimilar to host young.

Cowbird Natural History and Host Specificity

Recent evidence indicates that the five brood-parasitic cowbird species comprise a monophyletic group (Lanyon 1992; Johnson & Lanyon 1999), and in our discussion we will follow the nomenclature that reflects the best-supported phylogeny (Johnson & Lanyon 1999).

The brood-parasitic cowbirds vary tremendously in the extent to which they specialize on particular host species (Table 1). The screaming cowbird is a specialist brood parasite. Although there are some records of screaming cowbirds parasitizing other hosts (brown-and-yellow marshbird, *Pseudoleistes virescens*, Mermoz & Reboreda 1996; Chopi blackbirds *Gnorimopsar chopi*, Fraga 1996), their main host is the bay-winged cowbird. A recent phylogeny based on mitochondrial DNA suggests that the bay-winged cowbird is not in a monophyletic group with the parasitic cowbirds (Johnson & Lanyon 1999). Screaming cowbird young are almost identical to the host young in appearance and vocalizations, even though adult plumage and calls of both species are very different (Hudson 1920; Fraga 1979) and adult screaming cowbirds are larger (Lichtenstein 1997). The similarity remains through the fledgling period until the parasitic nestlings attain nutritional independence, and therefore, may be considered a true case of mimicry.

Table 1. General characteristics of the cowbirds (modified from Davies 2000).

Species	Distribution	No. hosts	Main hosts	Mean adult mass (g)[a,b,c]
Screaming cowbird *Molothrus rufoaxillaris*	Bolivia, Paraguay, Uruguay, southern Brazil, Argentina	1	bay-winged cowbird	male: 58.0 female: 48.4
Giant cowbird *M. oryzivora*[1]	Northern Argentina to Mexico	7	icterids (caciques & oropendola)	male: 219 female: 162
Bronzed cowbird *M. aeneus*	Colombia through Mexico; and across southern USA	87[d]	passerines (20% are icterids)	male: 66.7 female: 57.4
Shiny cowbird *M. bonaeriensis*	South America, the Caribbean, Florida	214	passerines	male: 55.5 female: 44.9
Brown-headed cowbird *M. ater*	North America	226	passerines	male: 49.0 female: 38.8
Bay-winged cowbird *Oreopsa badius*[2]	Bolivia, Paraguay, Uruguay, SE & NE Brazil, Argentina[c]	0	non-parasitic	male: 46.6 female: 43.7

[1]formerly *Scaphidura oryzivora*; [2] formerly *Molothrus badius*; [a] Dunning 1993; [b] Mason 1987; [c] Fraga 1986; [d] Sealy et al. 1997.

The giant cowbird is known to parasitize seven species of oropendolas and caciques (Icterinae, Ortega 1998). Giant cowbird young have a beak and face-iris coloration similar to that of oropendola nestlings, that disappears after the young have attained nutritional independence (Redondo 1993).

The bronzed cowbird, shiny cowbird and brown-headed cowbird are all host-generalists. These cowbirds parasitize between 87 and 226 host species (although not all of these hosts are able to rear cowbirds). Suitable hosts of all three cowbird species comprise an enormously diverse group of passerines that vary in body size, nestling behaviour, adult behaviour, nestling diet and decision rules used by adults when allocating food. Generally, brown-headed cowbirds tend to parasitize smaller hosts whereas shiny cowbirds apparently prefer to parasitize hosts larger than themselves (Mason 1980; Wiley 1988; but see Lichtenstein 1998).

The distinction between host-specialists and host-generalists is an important one, in part because it affects the predictability of the competitive environment in which the cowbird nestlings must solicit food. Generalist cowbird young may require behavioural flexibility to acquire food from a host that could exhibit a wide range of behavioural and morphological phenotypes. High fecundity and cheap egg production (Kattan 1995) might enable female generalist cowbirds to pursue an opportunistic strategy, laying eggs in nests of poor hosts in the absence of higher quality hosts. Specialist cowbird nestlings do not experience such variation in host phenotype, but they face a host that may have evolved defences against the parasite (Rothstein 1990).

This chapter reviews our current understanding of nestling behaviour by both generalist and specialist cowbirds. First, we examine the success of cowbirds in mixed broods of parasite and host young to test for benefits of exaggerated begging. Second, we examine the predation costs associated with exaggerated begging by cowbirds. Third, we ask whether cowbirds are treated differently from host young. Fourth, we address whether the apparently exaggerated level of begging by cowbirds is an honest reflection of their hunger. We conclude with suggestions for future work.

SUCCESS OF COWBIRDS IN MIXED BROODS

It is especially fascinating to consider the mechanisms underlying the success of cowbird nestlings because: (1) cowbirds do not face the inclusive fitness costs that potentially shape the begging behaviour of non-parasitic young and (2) cowbirds must acquire food from, and usually compete against, individuals of a different species. These unique circumstances raise many exciting questions. Under what conditions are cowbird young successful and what mechanisms are involved in this success? How does cowbird behaviour affect host young? In this section, we will explore these issues, using data collected mainly from brown-headed cowbirds.

Ejection of Host Nestmates

In most brood-parasitic species of cuckoos and honeyguides, the parasitic nestling ejects or kills the host eggs or young, leaving the parasite as the sole occupant of the nest (Payne 1977; Rothstein 1990). There is one documented case and several suspected cases of a brown-headed cowbird nestling ejecting a host nestling from the nest (Dearborn 1996), but this behaviour is the exception rather than the rule among cowbirds. Thus, cowbird young typically spend part or all of the nestling period in the company of host young. Consequently, a main determinant of the success of cowbird nestlings is their ability to acquire food in the presence of host nestmates.

Early Hatching by Cowbirds

Early hatching relative to host hatching is one factor that appears to facilitate the success of cowbird nestlings, and this trait is widespread among the brood-parasitic cowbirds. The specialist screaming cowbird has a shorter incubation time than its host, the bay-winged cowbird (Fraga 1986; Briskie & Sealy 1990), and for the three generalist species (brown-headed, shiny and bronzed) examined by Briskie and Sealy (1990), the incubation period is shorter than expected based on body size analysis of parasitic and non-parasitic members of the Icterinae. Thus, for both generalist and specialist cowbirds, the nestling is predisposed to early hatching. The likelihood, however, of hatching before the host nestlings depends also upon the timing of laying by the female cowbird and upon the incubation period of the host species.

When a cowbird nestling hatches before its host nestmates, the head start afforded the parasite can create a large asymmetry in size (e.g. Marvil & Cruz 1989). Hosts accept cowbird nestlings even when they hatch many days earlier than host young (McMaster & Sealy 1998), and early hatching means an early start at food acquisition. Brown-headed cowbird nestlings receive food almost immediately after hatching, and they can double in mass during the first 24-hour period (Hatch 1983). In addition, early hatching by cowbird young may cause a delay in hatch date of the last-laid host egg (McMaster & Sealy 1999), which could create a further size disadvantage for last-hatched host nestlings.

In general, if a cowbird hatches early in the nest of hosts that are similar or smaller in body size, the cowbird nestling is all but assured of remaining larger than its nestmates. Several mechanisms may maintain or exaggerate this size asymmetry. Adults of some host species may preferentially provision larger nestlings (Price & Ydenberg 1995), or allocate food in response to a begging cue that is partially size-dependent (Teather 1992; Kilner 1995; Dearborn 1998) or more readily produced by nestlings that are developmentally advanced (Lee 1995). The persistence of size and/or developmental asymmetries can lead to the death of host nestlings due to either trampling by the cowbird (Friedmann 1929; Nolan 1978) or starvation of host young that cannot compete with the much larger cowbird (Marvil & Cruz 1989). In contrast, when parasitizing larger hosts, the cowbird's advantage of earlier hatching may not be enough to overcome the size advantage of the host young (Lichtenstein 1998).

Effects of Nestmates on Cowbird Growth

The death of host nestmates is potentially advantageous for cowbird nestlings. Studies of an array of non-parasitic species have shown that individual nestlings exhibit lower growth rates when reared in unmanipulated broods with many nestmates (Henderson 1975; Emms & Verbeek 1991; Filliater & Breitwisch 1997) or when broods were experimentally enlarged (Nur 1984; Smith et al. 1989; Sanz 1997). Few researchers, however, have looked at the effect of nestmates on cowbird growth. D.C. Dearborn and D.E. Burhans (unpublished data) studied the growth of brown-headed cowbirds in nests of the indigo bunting (*Passerina cyanea*), a small host (adult mass of 14.5g versus 43.9g for cowbirds; Dunning 1993). Both the mass gain and structural growth of cowbird nestlings was markedly reduced by the presence of another cowbird in the nest, but was unaffected by the presence or number of host nestmates. In general, we might expect a greater impact on cowbird growth from large host nestmates, but this dynamic is complicated by the tendency for large hosts to experience multiple parasitism more frequently than small hosts (Lorenzana & Sealy 1999).

The implication of these findings, and those of Kattan (1996), is that the success of cowbirds at soliciting food from small-bodied hosts is unrelated to the presence or number of competitors in the nest, unless those competitors are also cowbirds. Because cowbirds in the nests of small hosts appear to pay a cost for sharing a nest with another cowbird, adult female cowbirds should avoid laying eggs in small hosts' nests that are already parasitized. Although multiple parasitism is common in some study systems

(Elliott 1978 and Trine 1998 for brown-headed cowbird; Lichtenstein 1998, 2001a for shiny cowbird and screaming cowbird, respectively), there is some evidence that brown-headed cowbirds may avoid laying in already-parasitized nests (Ortega et al. 1994). When multiple parasitism does occur, the fitness consequences of competition between cowbird nestmates depends on the relatedness between the cowbirds; multiple parasitism may be due to egg laying by multiple females (Hahn et al. 1999), in which case cowbirds would be unrelated, or repeated laying by one female (Alderson et al. 1999), in which case cowbirds would be full or half siblings.

Relative Success of Cowbird Nestlings with Different Host Species

Some species are inappropriate cowbird hosts because they reject cowbird eggs (Rothstein 1975). Among those hosts that do accept and hatch cowbird eggs, there is still substantial variation in host suitability. The most dramatic examples of this are those host species in which nestling diet is not adequate for the survival and growth of nestling cowbirds. Like most passerines, cowbird nestlings require a high protein diet of arthropods (Lowther 1993). Thus, cowbird nestlings do not survive in the nests of cedar waxwings (*Bombycilla cedrorum*, Rothstein 1976), which feed their young primarily fruit, or American goldfinches (*Carduelis tristis*, Middleton 1977) and house finches (*Carpodacus mexicanus*, Kozlovic et al. 1996), which feed their young primarily seeds. Surprisingly, parasitism rates can still be relatively high for these species in some study areas (e.g. 24.4% for house finches in Ontario; Kozlovic et al. 1996).

Among hosts that can rear cowbirds, incubation period and body size appear to be primary determinants of the success of cowbird nestlings. Cowbirds are quite successful with hosts that are smaller or that hatch later (Robinson et al. 1995), primarily due to their success at food acquisition (Dearborn 1998; Lichtenstein 1998). Cowbirds also do well with similar-sized host species (e.g. shiny cowbirds with yellow-hooded blackbirds, *Agelaius icterocephalus*, Cruz et al. 1990; brown-headed cowbirds with song sparrows, *Melodia melospiza,* Smith 1981). In contrast, cowbirds fare poorly with large and/or early-hatching hosts (e.g. shiny cowbirds with rufous-bellied thrushes, *Turdus rufiventris,* Lichtenstein 1998). Among hosts that are well matched with cowbirds in adult body size, there may still be differences in the size of host and parasitic nestlings. For example, adult brown-headed cowbirds are similar in size to both northern cardinals (*Cardinalis cardinalis*) and red-winged blackbirds (*A. phoeniceus*), but these species differ in egg size and also in hatchling size (cowbird: 3.0g egg, 2.3g hatchling; cardinal: 4.9g, 3.5g; redwing: 4.0g, 2.3g; Wetherbee & Wetherbee 1961; Lowther 1993; Yasukawa & Searcy 1995; Scott & Lemon 1996). Cowbirds thus start out on a more equal footing with red-wing nestmates than they do with cardinal nestmates. Perhaps as a consequence, cowbirds seem to fare better in red-wing nests than in cardinal nests (Weatherhead 1989; Ortega & Cruz 1991; Scott & Lemon 1996). Later in the chapter, we will explore the details of cowbird food acquisition with hosts of different sizes.

Two alternative hypotheses could explain the differential success of cowbirds with cardinals and red-winged blackbirds (or similar pairs of host species). The first is the likely relationship between phylogenetic relatedness and similarity of begging behaviour. Redwings are close relatives of cowbirds whereas cardinals are not; consequently, there may be a difference in the similarity of begging behaviour between host and parasite and, hence, in the success of cowbirds at soliciting food from host adults. For example, cowbirds in indigo bunting nests sometimes flutter their wings while begging (Dearborn 1998). Indigo bunting adults appeared to ignore this behaviour (Dearborn 1998), but icterid hosts might preferentially feed wing-flapping young, as such behaviour is performed by non-parasitic icterid nestlings when they are very hungry (Lee 1995).

A second alternative explanation for the differential success of cowbird nestlings raised by cardinals and red-wings is similarity of appearance between parasitic and host young (Lichtenstein 1998). Brown-headed cowbird nestlings, which have gapes with red lining and white rictal flanges in most parts of their range (Rothstein 1978), are outcompeted by the slightly larger northern cardinals, which have orange-red lining and white rictal flanges (Scott & Lemon 1996). In contrast, brown-headed cowbirds do well at nests of red-wings and even larger hosts with a similar gape colour (e.g. yellow-headed blackbird, *Xanthocephalus xanthocephalus*, Ortega & Cruz 1992). The idea that parents might discriminate on the basis of gape colour is not new. In 1978, Rothstein hypothesized that the geographical variation in the colour of the rictal flanges of the cowbirds was influenced by the preferential feeding of those nestlings whose gape is most similar to that of the host's own young. This might explain why, when parasitizing larger hosts, shiny cowbirds are more successful at parasitizing species with similar looking young. Icterids, with their red gapes (Ficken 1965), are among the most common hosts of the shiny cowbird (Friedmann et al. 1977; Manolis 1982; Wiley 1988). The more specialist species of parasitic cowbirds also parasitize icterids (reviewed in Lichtenstein 1997).

Effects of Brown-Headed Cowbirds on Host Nestlings and Adults

Cowbird nestlings have a direct negative impact on small-bodied hosts, above and beyond the well-documented effects of host egg removal by adult cowbirds (Sealy 1992) and reduced hatching success of host eggs (Hofslund 1957; Petit 1991). Some effects are directly due to the cowbird's larger size, whereas others are due to the cowbird's success at food acquisition. Cowbird nestlings may reduce the growth or survival of host young (Mayfield 1960; Root 1969; Marvil & Cruz 1989; Weatherhead 1989), and host young that do survive the nestling period may still exhibit a lower recruitment probability than hosts from unparasitized nests (Payne & Payne 1998). Cowbirds can impact host nestlings by trampling them to death (Mayfield 1960; Nolan 1978), out-competing them for food (Dearborn 1998; Lichtenstein & Sealy 1998), stimulating host young to spend more time begging (Dearborn et al. 1998) or causing nestlings to be brooded less as parents spend more time foraging (Dearborn et al. 1998). In addition to these effects on nestlings, host adults may pay a cost for raising a cowbird. For small-bodied hosts that are capable of raising mixed broods (e.g. Wolf 1987), host adults provision parasitized broods at a much higher rate than unparasitized broods (Dearborn et al. 1998). Such an increase in reproductive effort would be expected to impose a cost upon residual reproductive value. Testing for evidence of such costs is often difficult. One study, however, that has examined this issue thoroughly (Payne & Payne 1998) found that indigo buntings raising cowbirds incurred only minimal costs in terms of same season renesting, adult survival to the subsequent year and reproductive success in the subsequent year. Costs to large-bodied hosts are much less pronounced, and sometimes there is no impact on host fledging success (e.g. Cruz et al. 1990).

EVOLUTIONARY LIMITS ON BEGGING BY HOSTS AND COWBIRDS

Cowbirds are quite successful at acquiring food in nests of indigo buntings, as outlined earlier, and also in nests of other small hosts (e.g. yellow warblers, *Dendroica petechia*, Lichtenstein & Sealy 1998). Their success may occur in part because cowbirds can afford to beg more intensely than hosts. Although the energetic costs of begging appear minimal (M.A. Chappell & G.C. Bachman this volume) and are not expected to differ between host nestlings and cowbird nestlings, there are likely to be pronounced differences in the inclusive fitness costs (Hamilton 1964) and the predation costs (Harper 1986; Dearborn 1999) of begging by cowbirds versus non-parasitic nestlings. These differences may allow cowbirds to beg more loudly than non-parasitic nestlings.

Evidence for Loud Brood Parasites

For many years, researchers have noted that cowbird nestlings seem unusually loud (Friedmann 1929; Nice 1939; Hofslund 1957; Carter 1986; Payne 1991; also see Redondo 1993). To more firmly test these anecdotal reports, four studies have now quantified the rate or loudness of begging by brood parasites, and one study has experimentally tested whether the vocalizations at parasitized nests increase the risk of nest predation.

Davies et al. (1998) measured the call rates of common cuckoos raised by reed warblers (*Acrocephalus scirpaceus*) and found that a single cuckoo nestling called as frequently (and was fed an equivalent amount) as the entire brood of reed warblers that it replaced. In an early study of the shiny cowbird, Gochfeld (1979a) found that cowbird begging calls were louder and longer than those of their host nestmates, Loyca meadowlarks (*Sturnella loyca*). In addition, cowbirds were more likely to vocalize spontaneously than were their meadowlark nestmates. Briskie et al. (1994) used a comparative approach to examine the loudness of begging calls in relation to the average level of nestmate relatedness, when controlling for phylogeny. They found that species with low levels of nestmate relatedness (including brown-headed cowbirds and also non-parasitic species with high rates of extra-pair paternity) exhibited louder begging calls than species in which nestmates were usually full siblings. Lastly, Dearborn (1999) found that the rate and loudness of begging calls at indigo bunting nests was higher in nests with a cowbird than in nests without a cowbird. In Dearborn's (1999) study, measurements were made of entire broods. Based upon measures of total time begging by cowbirds and by hosts in parasitized and unparasitized nests (Dearborn 1998; Dearborn et al. 1998), it seemed likely that the louder, more frequent calls at parasitized nests were due to a combination of direct effects of cowbird vocalizations and indirect effects of cowbird behaviour on the calling of bunting nestlings.

For three different species of brood parasites, there is now evidence for louder or more rapid calling by the parasite. For cowbirds in particular, this may carry consequences for nest predation, especially for those species that parasitize hosts which build cup nests.

Predation Cost of Loud Vocalizations at Parasitized Nests

In conjunction with noticing the apparent loudness of cowbird nestlings, many authors have suggested that such noisiness may increase the risk of nest predation (Friedmann 1929; Hofslund 1957; Nolan 1978; Gochfeld 1979a; Payne 1991; Robinson et al. 1995; D.G. Haskell this volume). Comparisons of predation rates at parasitized and unparasitized nests within a given host population, however, have yielded equivocal results. Some studies have found higher predation rates at parasitized nests (Nice 1937; Burhans & Thompson 1999; Dearborn 1999), but others have found no difference (e.g. Mayfield 1960) or a difference in the opposite direction (i.e. unparasitized nests suffering higher predation rates; Arcese et al. 1996). Furthermore, the increase in predation rates at parasitized nests relative to unparasitized nests is not always greater during the nestling period than during incubation (e.g. Dearborn 1999), as would have been expected if cowbird begging calls were responsible for the overall difference in predation risk at parasitized versus unparasitized nests. Few comparisons of predation rates include power analyses, so it is difficult to judge our ability to detect differences that may exist.

Although the evidence from nest-monitoring studies is equivocal, experimental work by Haskell (1994, 1999, this volume) has shown that nestling vocalizations can indeed affect predation risk. In the one experimental test conducted thus far that deals explicitly with brood parasites, Dearborn (1999) compared predation rates at artificial nests broadcasting cowbird begging calls (loud and frequent), indigo bunting begging calls (less loud, less frequent) and silent nests. There was an overall trend towards more frequent predation at nests playing cowbird calls, followed in turn by nests playing indigo bunting calls and then silent nests. Despite several methodological shortcomings, this study demonstrates the possibility for nest predation to set an upper limit to begging behaviour in a manner that differs between cowbirds and non-parasitic species. Because predator identity can vary among nest sites and even among microhabitats within nest sites (Thompson et al. 1999; Dijak & Thompson 2000), the possible effects of predation on the evolution of begging calls are likely to exhibit much variation (e.g. Haskell 1999), and such variability likely extends to the potential for cowbird and host begging behaviour to be differentially shaped by predation risk.

ARE COWBIRDS TREATED AS HOST YOUNG?

The sight of a parasitic nestling being fed by its foster parents appears to confound our understanding of adaptive animal behaviour. Why are cowbird nestlings accepted by their hosts and in some cases raised even to the exclusion of the host young? Unlike the hosts of evicting parasitic species, (1) cowbird hosts have their own young for comparison; (2) the benefits of discrimination are high, as the parents could save the present brood (as well as future ones; Davies & Brooke 1988); and (3) parents could learn to recognize their offspring, paying a relatively low cost of misimprinting (Lotem 1993). In the following sections, we examine whether host adults treat cowbird young as though they were host young. As mentioned before, body size appears to be a primary determinant of the success of cowbird nestlings, thus our discussion will be organized according to the relative size of host and parasitic nestlings.

Interaction Between Cowbird Nestlings and Host Young in Nests of Smaller Host Species

As we have pointed out, brown-headed cowbird nestlings fare especially well in the nests of smaller hosts. What mechanism underlies their success? Is it mainly a matter of their larger size or their ability to manipulate the parents by producing exaggerated begging signals? Experiments comparing the feeding success of brown-headed cowbirds and host nestlings in yellow warbler nests showed that the nestling that reached the highest was more likely to be fed, regardless of its species (Lichtenstein & Sealy 1998). Thus, parents seemed to follow a 'laissez-faire' strategy and fed the closest nestling (Teather 1992; McRae et al. 1993; Leonard et al. 1994). The relative height reached by the nestlings was influenced by their size and begging intensity. When both nestlings were placed at the same distance to the parent they had the same chance of getting fed, which indicates that neither body size nor begging intensity alone could explain who got fed. Cowbirds did not get fed preferentially unless they were the closest to the parent, and the same was true for the yellow warbler nestlings. The laissez-faire strategy followed by the yellow warblers was exploited by the parasitic nestlings, which, because of their larger size and intense begging, were generally closer to the parents. In the same experiment, the feeding success of brown-headed cowbird nestlings that hatched one or two days before yellow warbler nestlings was significantly greater than that of cowbird nestlings that hatched the same day as the host young. The greater success of older cowbirds was not related to more intense begging, but to the size advantage gained from earlier hatching.

The importance of relative size in determining the feeding success of cowbirds was also found in a study of brown-headed cowbirds parasitizing indigo buntings (Dearborn 1998). In this study, cowbirds received more than twice as much food as bunting nestlings in parasitized nests. Cowbirds outperformed their bunting nestmates in the production of several size-dependent begging cues (e.g. height reached) that were correlated with food acquisition by buntings in unparasitized nests. In addition, a nestling transfer experiment showed that six-day-old indigo buntings (which are the same size as two-day-old cowbirds) received more food than their two-day-old bunting nestmates and did not receive a different amount of food than did twp-day-old cowbirds. Dearborn's (1998) study, however, also suggested that signal exaggeration might contribute to cowbird success in bunting nests. Cowbirds begged during more parental visits and for more total time than did their bunting nestmates, and in a given feeding visit cowbirds usually begged sooner and stopped begging later than their bunting nestmates. These results suggest that the duration or intensity of begging is also a component of cowbird success with small hosts.

Interaction Between Cowbird Nestlings and Host Young in Nests of Similar-Sized Host Species

Empirical data show that nestlings of similar-sized host species are not outcompeted by cowbirds. The same is true for larger hosts with longer incubation periods (e.g. brown-and-yellow marshbird; Mermoz & Reboreda 1994), in which case host young and cowbird nestlings hatch at a similar size. Observations of cowbirds parasitizing a similar-sized host species, such as cardinals, showed that brown-headed cowbird nestlings did not differ from host nestlings in food acquisition (Dearborn 1998). Experiments looking at the interaction between cowbirds and similar-sized species would enable us to examine how food acquisition is affected by appearance (e.g. gape colour) or behaviour when size differences are eliminated.

Interaction Between Cowbird Nestlings and Host Young in Nests of Larger Host Species

Shiny cowbirds and rufous-bellied thrushes provide an opportunity to examine the interactions between a cowbird nestling and a larger host (Lichtenstein 2001b). Shiny cowbird nestlings do not mimic these host young either in appearance or in vocalizations. In addition to weighing half as much as the host young, the cowbirds are proportioned very differently, plus the bright red of the cowbird gape contrasts with the bright yellow of the thrush gape.

Videotapes were taken of parasitized nests containing one host nestling and one parasite, and of non-parasitized nests containing a big and a small thrush that mimicked the size difference between the host and the parasitic nestlings. Shiny cowbirds did badly in comparison with both large and small host nestlings. Far from being fed preferentially, cowbird nestlings were in fact disfavoured and shiny cowbird nestlings starved in 68.7% of the parasitized broods. Considering that in a natural situation shiny cowbirds must compete with between one and three thrush nestlings, this result probably overestimates cowbird success.

Cowbird nestlings were offered food significantly fewer times than the host young, being fed only when the thrush nestlings did not beg or begged with a low intensity. Only 53.6% of the food items offered to the cowbird resulted in successful feeds. This was a result of parents removing the food from the cowbird's gape to feed their own nestlings if their own young started begging. The greater handling time for large food items might have contributed to the cowbird's low feeding success, as it gave more time to the thrush nestlings to start begging. Most of the food items that parents brought to the nest were, however, small, and the fact that cowbird nestlings were significantly less successful than the thrush nestlings even in acquiring small food provides evidence that the inefficiency of food transfer was not simply related to prey size. An alternative hypothesis is that parents were more reluctant to give the prey items to the cowbirds than to their own young.

The most important variable in determining who got fed was the relative height the nestlings reached, but parents used different rules for feeding nestlings in the parasitized and non-parasitized nests. In non-parasitized nests parents fed the nestling that reached the highest, and if both nestlings reached the same height, they had the same probability of getting fed. In the parasitized nests, however, the host nestlings had a significantly greater chance of getting fed if they were higher or at the same height as the cowbird. When the cowbird was positioned higher than the thrush nestlings, there was no difference in the probability that either was fed. Shiny cowbird nestlings were usually fed only if the host nestlings were not begging. These results suggest that the poor success of shiny cowbird nestlings in nests of rufous-bellied thrushes is not simply due to competition with their larger nestmates, but the result of parental discrimination.

The mechanism for discrimination could be similar to that which operates in cases of brood reduction. If there is enough food available, the larger host young become satiated, and the smaller parasite then has an opportunity to be fed. Alternatively, if there is not enough food available, the host nestlings always beg, so the parasite does not get fed. No experimental study on brood reduction has shown whether the parents actively discriminate against runt young, or whether runts are not fed because of sibling competition alone.

However, the variation in nestling size or appearance has also not been as extreme. As cowbirds were not only smaller than the host young but also looked very different, this study could not determine whether the cowbirds were discriminated against because of their appearance, or a combination of appearance and small size. Experimental studies changing the appearance of different-sized thrush nestlings will be needed, as well as the creation of experimental nests in which cowbirds are larger than their thrush companions.

Cowbird Size and Host Discrimination

The previous results suggest that, if all other costs are equal, discrimination against generalist brood parasites may be more likely to evolve where the cost of a recognition error is low (Lichtenstein 2001a). According to this hypothesis, discrimination would be more likely to evolve in hosts larger than the brood parasite. In these cases, the cowbird is the smallest nestling in the parasitized nest. If a host mistakenly discriminated against its own smallest nestling, the host would lose the nestling with the lowest fitness return. By contrast, discrimination against a generalist brood parasite is less likely to evolve when the cost of a parental mistake is high. Thus, hosts smaller than the parasites are not expected to evolve discrimination. Bigger nestlings promise a larger marginal return (Lack 1954; O'Connor 1978; Parker et al. 1989) than smaller nestlings. The cost of making a mistake in unparasitized nests would be to lose the best nestling. This hypothesis led to the following predictions: (1) generalist parasitic nestlings will be discriminated against at nests of large hosts and (2) generalist parasitic nestlings will do well at nests of smaller hosts. This experiment should be replicated with other large host species (e.g. meadowlarks) to determine if the reluctance to feed shiny cowbird nestlings is widespread.

If we do not assume equal costs, the situation is more complicated. The cost of accepting a parasitic nestling is much higher for a small host than for a large host, because the nestlings of the small host might be outcompeted by the parasitic nestling (Carter 1986; Briskie & Sealy 1987; Marvil & Cruz 1989). So why do small hosts not discriminate against a larger parasitic nestling? One possibility is that large nestlings are a supernormal stimulus (Redondo 1993). Offspring size is important in determining food allocation within a brood (Redondo & Castro 1992a; Price & Ydenberg 1995), and adult birds might be constrained from evolving a rule for refusing to feed the largest nestling. An alternative hypothesis states that for small hosts whose nestlings fail to compete with the parasite, the cost of misimprinting may outweigh the benefits of correct learning (Lotem 1993). If, however, host and parasite young fledge in nests of larger hosts, learning to recognize nestlings should be adaptive. If the parents were able to discriminate, why was the cowbird fed at all?

The fact that parents still fed the cowbirds was probably the result of conservative decision-making, rather than a perceptual inability. The absence of discrimination does not necessarily mean that an individual is unable to perceive the difference between two items. In a normal situation, it might be adaptive to follow the rule of feeding any nestling in their nest. An example comes from reed warblers that increased the rejection rate of parasitic eggs when they were first presented with a dummy cuckoo (Davies & Brooke 1988). Without evidence of parasitism, the best choice was to accept all the eggs. The additional information provided by the presence of the cuckoo, however, meant that the best response was to reject (Davies et al. 1996). The results of this study fit with the prediction that discrimination is more likely to evolve when foster parents can save most of their young after rejecting the parasite (Davies & Brooke 1988). These higher benefits, along with the smaller cost of misimprinting (Lotem 1993), make discrimination adaptive.

THE HONESTY OF BEGGING SIGNALS

As we have seen in previous chapters, young altricial birds communicate their desire to be fed by begging using postures and calling. For young in any condition, the benefits of gaining more resources are balanced by the direct and indirect costs of the begging signal (Godfray 1995a). As mentioned earlier, the potential direct costs include an increase in energetic expenditure (but see M.A. Chappell & G.C. Bachman this volume) and in the risk of predation (Dearborn 1999; D.G. Haskell this volume). The escalation of begging is also limited by an indirect cost arising from the fact that nestlings are related to their parents and siblings. Depriving present or future siblings of food decreases a nestling's inclusive fitness (Hamilton 1964). In a comparative study, Briskie et al. (1994) demonstrated that the loudness of begging signals increased as the relatedness of brood members decreased.

A growing number of empirical studies support the argument that begging is an honest indication of need (Redondo & Castro 1992b; Kilner 1995; Cotton et al. 1996). What happens when a parasitic nestling disrupts this stable signalling system? Parasitic nestlings do not share any genes with the foster parents nor with their nestmates, so their begging intensity is constrained only by the direct costs of begging. In theory then, this enables parasitic young to exhibit a higher intensity of solicitation than the host young for the same degree of need. As parents are accustomed to a stable signalling system, they should interpret the intense begging of the parasites as a signal of intense hunger. Theoretical studies suggest that cheating by parasites can be part of a stable signalling system provided that its incidence is very low (Grafen 1990; Dawkins & Guilford 1991; Johnstone & Grafen 1993). In accordance with this theory, the begging calls of parasitic nestlings belonging to 12 genera have been described as intense, exaggerated or conspicuous (Redondo 1993). Furthermore, an experimental study controlling the degree of food deprivation showed that great spotted cuckoos (*Clamator glandarius*) begged significantly more than their magpie hosts (*Pica pica*) for the same degree of need (Redondo 1993). In the common cuckoo, nestlings begged as frequently as an entire brood of host young (Davies et al. 1998) and demanded a higher provisioning rate than the host young in relation to their daily energy budget (Kilner & Davies 1999).

Signal Exaggeration in Brown-Headed and Shiny Cowbirds

Anecdotal observations (Friedmann 1929; Nice 1939) have suggested that cowbird nestlings beg more intensively than host young, and this result has been substantiated by recent experimental studies (Gochfeld 1979b; Briskie et al. 1994; Dearborn 1998; Lichtenstein 1998). As the hunger level of the nestlings was not controlled in any of these studies, it is impossible to draw conclusions about the influence of hunger on the begging signal. There are three possible scenarios: (1) intense begging behaviour of cowbirds is not related to hunger; (2) intense begging behaviour is an honest indicator of hunger; (3) begging of parasitic nestlings is dishonest, but still informative.

To test these hypotheses, Lichtenstein (1997) studied cowbird begging behaviour when the parasitic nestling was significantly smaller than the host young (shiny cowbirds parasitizing rufous-bellied thrushes) and when the parasitic nestling was significantly larger than the host young (brown-headed cowbirds parasitizing yellow warblers). Hand-feeding and food deprivation experiments were conducted with the parasite and host young. If parasitic nestlings beg as predicted by the signalling models, both species of cowbirds are expected to beg more intensively than the host young, but their begging will vary with hunger level. Alternatively, if the parasites do not beg informatively, the prediction is that both species of parasites would beg more intensively than their hosts, but there would be no variation in their begging intensity following the experimental manipulations of their hunger. If begging was only an expression of hunger, both species of parasite should beg with intensities similar to host young for the same degree of food deprivation.

The results of the study showed that both species of cowbirds varied their begging intensity relative to their hunger. Parasitic nestlings decreased their begging intensity after being hand-fed and increased it after being food deprived. Shiny cowbird nestlings begged more intensively than the thrush nestlings during all the experimental treatments, which suggests that their begging behaviour was not an honest indication of their short-term need. Their begging behaviour fits with the prediction of signalling models as it is not honest, but is still informative. On the other hand, brown-headed cowbirds begged as intensively as the host young, which might suggest that their begging honestly reflected their short-term need. These results suggest that the intensity of begging by parasitic nestlings is influenced both by their lack of kinship with the hosts and by their hunger, which is influenced by the relative size of the host young with which they compete for food.

Differences in the cost of begging may also partially explain the results of this experiment. That is, assuming that the energetic costs of begging are the same for the two species of parasite, the high risk of predation (54.6 %; Hebert & Sealy 1993) for brown-headed cowbird nestlings may have limited the escalation of loud begging. A reduced begging intensity is predicted for populations that suffer high risks of predation (Godfray 1995b). Shiny cowbirds, on the other hand, were in relatively safer nests, in an environment with lower predation pressure (34.3%; Lichtenstein 1998). In addition, shiny cowbirds had to beg very intensively in order to get as close to the feeding adults as their larger nestmates. The begging intensity was probably also influenced by hunger. As they were getting such small amounts of food, the benefit of getting more food may have outweighed the risk of predation.

Signal Exaggeration in Screaming Cowbirds

The ideal situation for studying the relationship between begging intensity and hunger level is the interaction between the specialist screaming cowbird and its almost exclusive host, the bay-winged cowbird. Screaming cowbird nestlings coexist with the host young, ruling out the possibility that their intense begging occurs because they are the only occupants of the parasitized nest. Second, they look the same as the host young, so they do not need to compensate for their odd appearance. Third, they are similar in size and weigh as much as the host young during the nestling period, so presumably they have the same metabolic rate and would not be hungrier than the host nestlings after receiving the same amount of food.

Hand-feeding experiments done in the laboratory showed that the begging behaviour of bay-winged cowbirds was influenced by their hunger level. As the period of food deprivation increased, both cowbird species decreased their latency to start begging and increased their begging intensity, the period they spent begging and the number of pellets consumed (Lichtenstein 1997). A positive correlation between begging intensity and hunger is consistent with other studies (Redondo & Castro 1992b; Price & Ydenberg 1995; Leonard & Horn 1998). The novel result here was that the begging intensity of the parasitic nestlings was also influenced by their hunger. Both species of cowbird exhibited long latencies to start begging following brief periods of food deprivation and the latencies got shorter as the nestlings became hungrier. These results contrast with those reported by Redondo (1993) for great spotted cuckoos, which did not vary their begging behaviour in relation to their hunger, and support the prediction of signalling models for informative begging signals in both parasitic and host nestlings (Johnstone & Grafen 1992; Godfray 1995b).

Although the begging behaviour of screaming cowbird nestlings was a reliable signal, they still exaggerated their need. Hand-feeding and food deprivation experiments indicated that screaming cowbirds begged more intensively and for a longer period than the host young for the same degree of food deprivation (Lichtenstein 2001a). When offered food *ad libitum*, they ate the same number of pellets as host young, indicating that they were not hungrier. Given the similarity in size and appearance between the parasitic nestlings and the host young, and thus similarities in the direct costs of begging (as suggested for the common cuckoo and its hosts, Lotem 1998), the exaggerated begging of the screaming cowbird can be explained solely by its lack of relatedness to its host.

Further evidence in favour of the reliability of the begging signal comes from a result suggesting that the begging of a nestmate did not influence the begging intensity of a focal nestling (Lichtenstein 2001a). Thus, both host and parasitic nestlings begged in accordance with their own need. A similar result was found for yellow warblers, rufous-bellied thrushes, shiny and brown-headed cowbirds (Lichtenstein 1997) and starlings (*Sturnus vulgaris*, Cotton et al. 1996). In contrast, studies on American robins (*Turdus migratorius*, Smith & Montgomerie 1991), yellow-headed blackbirds (Price et al. 1996) and tree swallows (*Tachycineta bicolor*, Leonard & Horn 1998) found that nestlings increased their begging in response to the begging of their nestmates. As this was not tested in a standard way across species, the disparity in results could be due to differences between the species that were tested or to the different experimental designs.

In summary, begging intensity results from an equilibrium between the direct and indirect costs of begging and the benefit of being fed. Whereas parasites are generally not related to their nestmates or to the host parents, and so do not share the indirect costs of begging with the host young, they do have to pay the direct costs of begging. The different species studied may have reached different equilibria, depending on factors such as the probability of being related to the rest of the brood, the probability of being fed when begging at certain intensities, hunger and risk of predation. The begging intensity exhibited by screaming cowbird nestlings suggests that in situations in which the costs of begging are relatively low (the nests of bay-winged cowbirds have a low risk of predation), parasitic nestlings engage in signal overplay.

These results provide a better understanding of the interaction between host parents and parasitic young. Parasitic nestlings exploit a pre-existing honest signalling system. Hosts are trapped because this system is adaptive as long as signals are, on average, a reliable indication of their own offspring's need. Not feeding young that are begging intensively is a cost they cannot afford. Coevolution between the specialist screaming cowbird and its major host seems to have arrived at a stable equilibrium, in which the parasitic nestlings are winning the arms race. The interaction between the generalist shiny and brown-headed cowbirds and their hosts, however, might still leave room for discrimination by the hosts and for further adaptation (probably in terms of better host choice) by the parasites.

FUTURE DIRECTIONS

Two widely held views are challenged in this chapter. First, it was generally accepted that parasitic nestlings are always successful at manipulating the host parents to feed them: the experiments with shiny cowbirds and rufous-bellied thrushes show that this view is incorrect, as cowbirds were actually fed less than host young by adult thrushes. Second, observations and empirical data previously suggested that parasitic nestlings always beg intensively, independently of their hunger level. This was shown not to be the case for cowbirds, as parasitic nestlings were found to beg informatively to the extent that they increased their begging intensity with their hunger level, as predicted by signalling models. Below we provide some ideas for further research in this area.

Nestling Discrimination

Work with shiny cowbirds and rufous-bellied thrushes supported the notion that discrimination against generalist parasitic nestlings is more likely to occur in nests of large hosts, where costs of recognition errors are lower. Future work could replicate these experiments with other large host species to determine whether discrimination against cowbirds by large hosts is a widespread phenomenon.

Additional work is also needed to further clarify the role of phylogenetic relatedness in the success of cowbirds with particular hosts. To what extent have cowbirds evolved mimicry *per se*, as opposed to exhibiting similarity to certain hosts more as a matter of phylogeny? This is an important issue in understanding relative success with different hosts and the apparent preference of some cowbird species for parasitizing other icterids.

Solicitation and Provisioning

Higher levels of solicitation result in higher nest visit rates, and this is particularly true at some parasitized nests. After arriving at the nest, however, some host parents prefer to feed their own young. If the host nestlings were able to assess their competitive ability with regard to that of their nestmate, or their feeding success in relation to the effort put into begging, host young could 'free-ride' upon the begging effort of the parasitic nestlings. The parasite makes all the effort to encourage parents to visit the nest frequently, but once the parent is in the nest, the host young are preferred. It would be interesting to compare the begging intensity and fledging mass of host nestlings when accompanied by a similar-sized and smaller nestmate (of the size of a cowbird), or a cowbird under controlled levels of hunger. Bearing in mind, however, that parasitic nestlings are usually accompanied by more than one host young, competition among host young might dilute the benefit of having a noisy cowbird in the nest.

Additional research is also needed on the relationship between begging intensity of brown-headed cowbirds and the nestlings of small hosts. In particular, we need to better assess whether competition with cowbirds might increase the need of host young such that they increase their own begging in a manner that honestly reflects their need.

Brood Parasitism and Nest Predation

Finally, more work is needed on the link between cowbird parasitism and nest predation. This issue is important from the standpoint of signal evolution and also with regard to avian conservation, as the impact of cowbird nestlings on host reproductive success may depend on synergistic effects of food acquisition and nest predation.

Several key questions remain unanswered. To what extent are the louder, more frequent vocalizations at parasitized nests due to direct versus indirect effects of brown-headed cowbird nestlings? How strong is the link between these vocalizations and predation risk across an array of hosts, nest types and predator communities? And lastly, what is the relative importance of vocal cues in determining food allocation within a nest? This question stems from a general methodological discrepancy between studies of food allocation and studies of nest predation. When studying the allocation of food among nestmates, most researchers examine non-vocal cues (but see A.G. Horn & M.L. Leonard this volume), primarily due to the difficulty of measuring the properties of individual calls and assigning these calls to the correct offspring in the nest; the effect of begging behaviour on nest predation, in contrast, is due to vocal cues. These research areas need to be merged to allow a comprehensive view of the factors shaping the evolution of begging.

ACKNOWLEDGEMENTS

We thank Marty Leonard, Spencer Sealy and Jon Wright for useful feedback on this chapter. Thanks also to John Faaborg and Nick Davies for support, ideas and encouragement.

REFERENCES

Alderson, G.W., Gibbs, H.L. & Sealy, S.G. 1999. Determining the reproductive behaviour of individual brown-headed cowbirds using microsatellite DNA markers. *Animal Behaviour* **58**, 895-905.

Arcese, P., Smith, J.N.M. & Hatch, M. 1996. Nest predation by cowbirds and its consequences for passerine demography. *Proceedings of the National Academy of Sciences USA* **93**, 4608-4611.

Briskie, J.V. & Sealy, S.G. 1987. Responses of least flycatchers to experimental inter- and intraspecific brood parasitism. *Condor* **89**, 899-901.

Briskie, J.V. & Sealy, S.G. 1990. Evolution of short incubation periods in the parasitic cowbirds, *Molothrus* spp. *The Auk* **107**, 789-794.

Briskie, J.V., Naugler, C.T. & Leech, S.M. 1994. Begging intensity of nestling birds varies with sibling relatedness. *Proceedings of the Royal Society of London, Series B* **258**, 73-78.

Burhans, D.E. & Thompson, F.R.III. 1999. Habitat patch size and nesting success of yellow-breasted chats. *Wilson Bulletin* **111**, 210-215.

Carter, M.D. 1986. The parasitic behaviour of the bronzed cowbird in South Texas. *Condor* **88**, 11-25.

Cotton, P.A., Kacelnik, A. & Wright, J. 1996. Chick begging as a signal: are nestlings honest? *Behavioral Ecology* **7**, 178-182.

Cruz, A., Manolis, T.D. & Andrews, R.W. 1990. Reproductive interactions of the shiny cowbird *Molothrus bonariensis* and the yellow-hooded blackbird *Agelaius icterocephalus* in Trinidad. *Ibis* **132**, 436-444.

Davies, N.B. 2000. *Cuckoos, Cowbirds and Other Cheats*. London: T. & A.D. Poyser.

Davies, N.B. & Brooke, M. de L. 1988. Cuckoos versus reed warblers: adaptations and counteradaptations. *Animal Behaviour* **36**, 262-284.

Davies, N.B., Brooke, M. de L. & Kacelnik, A. 1996. Recognition errors and probability of parasitism determine whether reed warblers should accept or reject mimetic cuckoo eggs. *Proceedings of the Royal Society of London, Series B* **263**, 925-931.

Davies, N.B., Kilner, R.M. & Noble, D.G. 1998. Nestling cuckoos, *Cuculus canorus*, exploit hosts with begging calls that mimic a brood. *Proceedings of the Royal Society of London, Series B* **265**, 673-678.

Dawkins, M.S. & Guilford, T.C. 1991. The corruption of honest signalling. *Animal Behaviour* **41**, 865-873.

Dearborn, D.C. 1996. Video documentation of a brown-headed cowbird nestling ejecting an Indigo Bunting nestling from nest. *Condor* **98**, 645-649.

Dearborn, D.C. 1998. Begging behaviour and food acquisition by brown-headed cowbird nestlings. *Behavioral Ecology and Sociobiology* **43**, 259-270.

Dearborn, D.C. 1999. Brown-headed cowbird nestling vocalizations and the risk of nest predation. *The Auk* **116**, 448-457.

Dearborn, D.C., Anders, A.D., Thompson, F.R.III & Faaborg, J. 1998. Effects of cowbird parasitism on parental provisioning and nestling food acquisition and growth. *Condor* **100**, 326-334.

Dijak, W.D. & Thompson, F.R.III. 2000. Landscape and edge effects on the distribution of mammalian predators in Missouri. *Journal of Wildlife Management* **64**, 209-216.

Dunning, J.B.J. 1993. *CRC Handbook of Avian Body Masses*. Boca Raton: CRC Press.

Elliott, P.F. 1978. Cowbird parasitism in the Kansas tallgrass prairie. *The Auk* **95**, 161-167.

Emms, S.K. & Verbeek, N.A.M. 1991. Brood size, food provisioning, and chick growth in the pigeon guillemot *Cepphus columba*. *Condor* **93**, 943-951.

Ficken, M.S. 1965. Mouth colour of nestling passerines and its use in taxonomy. *Wilson Bulletin* **77**, 71-75.

Filliater, T.S. & Breitwisch, R. 1997. Nestling provisioning by the extremely dichromatic northern cardinal. *Wilson Bulletin* **109**, 145-153.

Fraga, R.M. 1979. Differences between nestlings and fledglings in screaming and bay-winged cowbirds. *Wilson Bulletin* **91**, 151-154.

Fraga, R.M. 1986. The bay-winged cowbird *Molothrus badius* and its brood parasites: interactions, coevolution and comparative efficiency. PhD thesis, University of California.

Fraga, R.M. 1996. Further evidence of parasitism of chopi blackbirds (*Gnorimopsar chopi*) by the specialized screaming cowbird (*Molothrus rufoaxillaris*). *Condor* **98**, 866-867.

Friedmann, H. 1929. *The Cowbirds: A Study in the Biology of Social Parasitism*. Springfield: C.C. Thomas.

Friedmann, H., Kiff, L.F. & Rothstein, S.I. 1977. A further contribution to knowledge of host relations of the parasitic cowbirds. *Smithsonian Contributions in Zoology* **235**, 1-75.

Gochfeld, M. 1979a. Begging by nestling shiny cowbirds: adaptive or maladaptive? *Living Bird* **17**, 41-50.

Gochfeld, M. 1979b. Brood parasite and host coevolution: interactions between shiny cowbirds and two species of meadowlarks. *American Naturalist* **113**, 855-870.

Godfray, H.C.J. 1995a. Evolutionary theory of parent-offspring conflict. *Nature* **376**, 133-138.

Godfray, H.C.J. 1995b. Signaling of need between parents and young: parent-offspring conflict and sibling rivalry. *American Naturalist* **146**, 1-24.

Grafen, A. 1990. Biological signals as handicaps. *Journal of Theoretical Biology* **144**, 517-546.

Hahn, D.C., Sedgwick, J.A., Painter, I.S. & Casna, N.J. 1999. A spatial and genetic analysis of cowbird host selection. In: *Research and Management of the Brown-Headed Cowbird in Western Landscapes* (Ed. by M.L. Morrison, L.S. Hall, S.K. Robinson, S.I. Rothstein, D.C. Hahn & T.D. Rich). Lawrence: Allen Press.

Hamilton, W.D. 1964. The genetical evolution of social behaviour. I & II. *Journal of Theoretical Biology* **7**, 1-52.
Harper, A.B. 1986. The evolution of begging: sibling competition and parent-offspring conflict. *American Naturalist* **128**, 99-114.
Haskell, D.G. 1994. Experimental evidence that nestling begging behaviour incurs a cost due to nest predation. *Proceedings of the Royal Society of London, Series B* **257**, 161-164.
Haskell, D.G. 1999. The effect of predation on begging-call evolution in nestling wood warblers. *Animal Behaviour* **57**, 893-901.
Hatch, S.A. 1983. Nestling growth relationships of brown-headed cowbirds and dickcissels. *Wilson Bulletin* **95**, 669-672.
Hebert, P.N. & Sealy, S.G. 1993. Hatching asynchrony in yellow warblers: a test of the hormonal hypothesis. *Ornis Scandinavica* **24**, 10-14.
Henderson, B.A. 1975. Role of the chick's begging in the regulation of parental feeding behaviour of *Larus glaucescens*. *Condor* **77**, 488-492.
Hofslund, P.B. 1957. Cowbird parasitism of the northern yellow-throat. *The Auk* **74**, 42-48.
Hudson, W.H. 1920. *Birds of La Plata.* New York: J.M. Dent.
Johnson, K.P. & Lanyon, S.M. 1999. Molecular systematics of the grackles and allies, and the effect of additional sequence (cyt *b* and ND2). *The Auk* **116**, 759-768.
Johnstone, R.A. & Grafen, A. 1992. The continuous Sir Philip Sidney Game: a simple model of biological signalling. *Journal of Theoretical Biology* **156**, 215-234.
Johnstone, R.A. & Grafen, A. 1993. Dishonesty and the handicap principle. *Animal Behaviour* **46**, 759-764.
Kattan, G.H. 1995. Mechanisms of short incubation period in brood-parasitic cowbirds. *The Auk* **112**, 335-342.
Kattan, G.H. 1996. Growth and provisioning of shiny cowbird and house wren host nestlings. *Journal of Field Ornithology* **67**, 434-441.
Kilner, R. 1995. When do canary parents respond to nestling signals of need? *Proceedings of the Royal Society of London, Series B* **260**, 343-348.
Kilner, R.M. & Davies, N.B. 1999. How selfish is a cuckoo chick? *Animal Behaviour* **58**, 797-808.
Kozlovic, D.R., Knapton, R.W. & Barlow, J.C. 1996. Unsuitability of the house finch as a host of the brown-headed cowbird. *Condor* **98**, 253-258.
Lack, D. 1954. *The Natural Regulation of Animal Numbers.* Oxford: Clarendon Press.
Lanyon, S.M. 1992. Interspecific brood parasitism in blackbirds (Icterinae): a phylogenetic perspective. *Science* **255**, 77-79.
Lee, W.-H. 1995. The ontogeny and developmental constraints of begging in red-winged blackbirds. MA thesis, Binghamton University.
Leonard, M.L. & Horn, A.G. 1998. Need and nestmates affect begging in tree swallows. *Behavioral Ecology and Sociobiology* **42**, 431-436.
Leonard, M.L., Teather, K.L., Horn, A.G., Koening, W.D. & Dickinson, J.L. 1994. Provisioning in western bluebirds is not related to offspring sex. *Behavioral Ecology* **5**, 455-459.
Lichtenstein, G. 1997. Begging behaviour and host exploitation in three species of parasitic cowbirds. PhD thesis, University of Cambridge.
Lichtenstein, G. 1998. Parasitism by shiny cowbirds of rufous-bellied thrushes. *Condor* **100**, 680-687.
Lichtenstein, G. 2001a. Selfish begging by screaming cowbirds, a mimetic brood parasite of the bay-winged cowbird. *Animal Behaviour* **61**, 1151-1158.
Lichtenstein, G. 2001b. Low success of shiny cowbird chicks parasitizing rufous-bellied thrushes: chick-chick competition or parental discrimination? *Animal Behaviour* **61**, 401-413.

Lichtenstein, G. & Sealy, S.G. 1998. Nestling competition, rather than supernormal stimulus, explains the success of parasitic brown-headed cowbird chicks in yellow warbler nests. *Proceedings of the Royal Society of London, Series B* **265**, 249-254.

Lorenzana, J.C. & Sealy, S.G. 1999. A meta-analysis of the impact of parasitism by the brown-headed cowbird on its hosts. In: *Research and Management of the Brown-Headed Cowbird in Western Landscapes* (Ed. by M.L. Morrison, L.S. Hall, S.K. Robinson, S.I. Rothstein, D.C. Hahn & T.D. Rich). Lawrence: Allen Press.

Lotem, A. 1993. Learning to recognize nestlings is maladaptive for cuckoo *Cuculus canorus* hosts. *Nature* **362**, 743-745.

Lotem, A. 1998. Manipulative begging calls by parasitic cuckoo chicks: why should true offspring not do the same? *Trends in Ecology and Evolution* **13**, 342-343.

Lowther, P.E. 1993. Brown-headed cowbird (*Molothrus ater*). In: *The Birds of North America, No. 47* (Ed. by A. Poole & F. Gill). Philadelphia & Washington: The Academy of Natural Sciences & The American Ornithologists' Union.

Manolis, T.D. 1982. Host relationships and reproductive strategies of the shiny cowbird in Trinidad and Tobago. Ph.D. thesis, University of Colorado.

Marvil, R.E. & Cruz, A. 1989. Impact of brown-headed cowbird parasitism on the reproductive success of the solitary vireo. *The Auk* **106**, 476-480.

Mason, P. 1980. Ecological and evolutionary aspects of host selection in cowbirds. PhD thesis, University of Texas.

Mason, P. 1987. Pair formation in cowbirds: evidence found for screaming but not shiny cowbirds. *Condor* **89**, 349-356.

Mayfield, H. 1960. *The Kirtland's Warbler*. Bloomfield Hills: Cranbrook Institute.

McMaster, D.G. & Sealy, S.G. 1998. Red-winged blackbirds (*Agelaius phoeniceus*) accept prematurely hatching brown-headed cowbirds (*Molothrus ater*). *Bird Behaviour* **12**, 67-70.

McMaster, D.G. & Sealy, S.G. 1999. Do brown-headed cowbird hatchlings alter adult yellow warbler behavior during the hatching period? *Journal of Field Ornithology* **70**, 365-373.

McRae, S.B., Weatherhead, P.J. & Montgomerie, R. 1993. American robin nestlings compete by jockeying for position. *Behavioral Ecology and Sociobiology* **33**, 101-106.

Mermoz, M.E. & Reboreda, J.C. 1994. Brood parasitism of the shiny cowbird, *Molothrus bonaeriensis*, on the brown-and-yellow marshbird, *Pseudoleistes virescens*. *Condor* **96**, 716-721.

Mermoz, M.E. & Reboreda, J.C. 1996. New host for a specialized brood parasite, the screaming cowbird. *Condor* **98**, 630-632.

Middleton, A.L.A. 1977. Effect of cowbird parasitism on American goldfinch nesting. *The Auk* **94**, 304-307.

Nice, M.M. 1937. Studies in the life history of the song sparrow, Part 1. *Transactions of the Linnean Society of New York* **4**, 1-247.

Nice, M.M. 1939. Observations on the behavior of a young cowbird. *Wilson Bulletin* **51**, 233-239.

Nolan, V.Jr. 1978. The ecology and behavior of the prairie warbler *Dendroica discolor*. *Ornithological Monographs* **26**, 1-595.

Nur, N. 1984. The consequences of brood size for breeding blue tit (*Parus caeruleus*). 2. Nestling weight, offspring survival, and optimal brood size. *Journal of Animal Ecology* **53**, 497-518.

O'Connor, R.J. 1978. Brood reduction in birds: selection for fratricide, infanticide and suicide? *Animal Behaviour* **26**, 79-96.

Ortega, C.P. 1998. *Cowbirds and Other Brood Parasites*. Tucson: University of Arizona Press.

Ortega, C.P. & Cruz, A. 1991. A comparative study of cowbird parasitism in yellow-headed blackbirds and red-winged blackbirds. *The Auk* **108**, 16-24.

Ortega, C.P. & Cruz, A. 1992. Differential growth patterns of nestling brown-headed cowbirds and yellow-headed blackbirds. *The Auk* **109**, 368-376.
Ortega, C.P., Ortega, J.C. & Cruz, A. 1994. Use of artificial brown-headed cowbird eggs as a potential management tool in deterring parasitism. *Journal of Wildlife Management* **58**, 488-492.
Parker, G.A., Mock, D.W. & Lamey, T.C. 1989. How selfish should stronger sibs be? *American Naturalist* **133**, 846-868.
Payne, R.B. 1977. The ecology of brood parasitism in birds. *Annual Review of Ecology and Systematics* **8**, 1-28.
Payne, R.B. 1991. Indigo bunting (*Passerina cyanea*). In: *Birds of North America, No. 4* (Ed. by A. Poole, P. Stettenheim & F. Gill). Philadelphia & Washington: The Academy of Natural Sciences & The American Ornithologists' Union.
Payne, R.B. & Payne, L.L. 1998. Brood parasitism by cowbirds: risks and effects on reproductive success and survival in indigo buntings. *Behavioral Ecology* **9**, 64-73.
Petit, L.J. 1991. Adaptive tolerance of cowbird parasitism by prothonotary warblers: a consequence of nest-site limitation? *Animal Behaviour* **41**, 425-432.
Price, K. & Ydenberg, R. 1995. Begging and provisioning in broods of asynchronously-hatched yellow-headed blackbird nestlings. *Behavioral Ecology and Sociobiology* **37**, 201-208.
Price, K., Harvey, H. & Ydenberg, R. 1996. Begging tactics of nestling yellow-headed blackbirds, *Xanthocephalus xanthocephalus*, in relation to need. *Animal Behaviour* **51**, 421-435.
Redondo, T. 1993. Exploitation of host mechanisms for parental care by avian brood parasites. *Etología* **3**, 235-297.
Redondo, T. & Castro, F. 1992a. The increase of predation risk with begging activity in broods of magpies *Pica pica*. *Ibis* **134**, 180-187.
Redondo, T. & Castro, F. 1992b. Signalling nutritional need by magpie nestlings. *Ethology* **92**, 193-204.
Robinson, S.K., Rothstein, S.I., Brittingham, M.C., Petit, L.J. & Grzybowski, J.A. 1995. Ecology and behaviour of cowbirds and their impact on host populations. In: *Ecology and Management of Neotropical Migratory Birds* (Ed. by T.E. Martin & D.M. Finch). New York: Oxford University Press.
Root, R.B. 1969. The behavior and reproductive success of the blue-gray gnatcatcher. *Condor* **71**, 16-31.
Rothstein, S.I. 1975. An experimental and teleonomic investigation of avian brood parasitism. *Condor* **77**, 250-271.
Rothstein, S.I. 1976. Cowbird parasitism of the cedar waxwing and its evolutionary implications. *The Auk* **93**, 498-509.
Rothstein, S.I. 1978. Geographical variation in the nestling coloration of parasitic cowbirds. *The Auk* **95**, 152-160.
Rothstein, S.I. 1990. A model system for coevolution: avian brood parasitism. *Annual Review of Ecology and Systematics* **21**, 481-508.
Sanz, J.J. 1997. Clutch size manipulation in the pied flycatcher: effects on nestling growth, parental care, and moult. *Journal of Avian Biology* **28**, 157-162.
Scott, D.M. & Lemon, R.E. 1996. Differential reproductive success of brown-headed cowbirds with Northern Cardinals and three other hosts. *Condor* **98**, 259-271.
Sealy, S.G. 1992. Removal of yellow warbler eggs in association with cowbird parasitism. *Condor* **94**, 40-54.
Sealy, S.G., Sánchez, J.E., Campos R.R.G. & Marin, M. 1997. Bronzed cowbird hosts: new records, trends in host use, and cost of parasitism. *Ornitologia Neotropical* **8**, 175-184.
Smith, H.G. & Montgomerie, R. 1991. Nestling American robins compete with siblings by begging. *Behavioral Ecology and Sociobiology* **29**, 307-312.

Smith, H.G., Kallander, H. & Nilsson, J.A. 1989. The trade-off between offspring number and quality in the great tit *Parus major*. *Journal of Animal Ecology* **58**, 383-402.

Smith, J.N.M. 1981. Cowbird parasitism, host fitness, and age of the host female in an island Song Sparrow population. *Condor* **83**, 152-161.

Teather, K.L. 1992. An experimental study of competition for food between male and female nestlings of the red-winged blackbird. *Behavioral Ecology and Sociobiology* **31**, 81-87.

Thompson, F.R.III, Dijak, W. & Burhans, D.E. 1999. Video identification of predators at songbird nests in old fields. *The Auk* **116**, 259-264.

Trine, C.L. 1998. Wood thrush population sinks and implications for the scale of regional conservation strategies. *Conservation Biology* **12**, 576-585.

Weatherhead, P.J. 1989. Sex ratios, host-specific reproductive success, and impact of brown-headed cowbirds. *The Auk* **106**, 358-366.

Wetherbee, D.K. & Wetherbee, N.S. 1961. Artificial incubation of eggs of various species and some attributes of neonates. *Bird Banding* **32**, 141-159.

Wiley, J.W. 1988. Host selection by the shiny cowbird. *Condor* **90**, 289-303.

Wolf, L. 1987. Host-parasite interactions of brown-headed cowbirds and dark-eyed juncos in Virginia USA. *Wilson Bulletin* **99**, 338-350.

Wyllie, I. 1981. *The Cuckoo*. New York: Universe Books.

Yasukawa, K. & Searcy, W. 1995. Red-winged blackbird (*Agelaius phoeniceus*). In: *The Birds of North America, No. 184* (Ed. by A. Poole & F. Gill). Philadelphia & Washington: The Academy of Natural Sciences & The American Ornithologists' Union.

20.

DISHONEST BEGGING AND HOST MANIPULATION BY *CLAMATOR* CUCKOOS

Tomas Redondo[1] & Jesus M. Zuñiga[2]
[1]*Estación Biológica de Doñana, CSIC, Apdo 1056, 41080 Sevilla, Spain (redondo@ebd.csic.es)*
[2]*Unidad de Producción y Experimentación Animal, CIC, Universidad de Granada, 18071 Granada, Spain (jmartinz@ugr.es)*

ABSTRACT

Brood parasites may be favoured over host nestlings due to variation in the honesty of their begging signals. Begging behaviour of great spotted cuckoo nestlings and their host magpie nestlings was recorded when controlling food need. Cuckoo begging effort was dishonest as an indicator of nutritional need, whilst magpie begging was not. Cuckoos begged for longer and emitted more calls at a higher rate irrespective of the degree of food deprivation, although in contrast to magpies, cuckoos ate food in relation to their need. Energetic and predation costs are unlikely to account for these differences. Differences in indirect inclusive fitness costs can explain the more intensive begging by cuckoos. Magpie parents given a choice favoured larger nestlings and those begging more intensively. Cuckoos obtained more food and a larger share than magpies of a similar size. Magpies therefore received less food in the presence of a cuckoo, and cuckoos received a similar share irrespective of their size. Lack of relatedness to their magpie hosts therefore allows cuckoos to exploit a set of adaptive rules in the host parents and manipulate them into providing the latter with preferential care.

INTRODUCTION

The great spotted cuckoo (*Clamator glandarius*) is an obligate brood parasite that uses magpies (*Pica pica*) as its major host. Unlike the European cuckoo (*Cuculus canorus*), great spotted cuckoo nestlings do not evict host

J. Wright and M.L. Leonard (eds.), The Evolution of Begging, 389–412.

eggs or nestlings after hatching, but severely depress the reproductive success of magpies by outcompeting host nestlings, which often starve to death. Magpie nestlings may occasionally survive (ca. 17% of successful parasitized nests), but they fledge at low masses and thus contribute little, if any, to their parents' reproductive output. Adult great spotted cuckoos are slightly smaller than magpies but grow faster and hatch two to three days earlier on average than magpies (Soler & Soler 1991). Field observations revealed that young cuckoos were not aggressive towards magpie nestlings. Instead, they appeared to monopolize the incoming food, and then precipitated the death of their emaciated nestmates by trampling and crowding them (Alvarez & Arias de Reyna 1974).

This chapter explores this apparent favouritism toward cuckoo nestlings by magpie parents, which may be confounded by factors that suggest alternative explanations. For instance, because cuckoos hatch earlier than magpie nestlings, and because across species nestlings beg more as they get older, more intense begging behaviour by cuckoos could simply be a side-effect of their older age (Redondo & Exposito 1990). Cuckoo nestlings may also have higher food requirements because of their larger size, faster growth (Soler & Soler 1991) or lower quality diet (Brooke & Davies 1989).

Differences in feeding success between cuckoos and magpies could simply be the result of competition, or due to traits other than begging, such as nestling number or gape morphology (Soler et al. 1995a,b). Previous experiments on broods containing a single cuckoo and several magpie nestlings have revealed that magpie parents feed cuckoo nestlings preferentially over magpie nestlings (Soler et al. 1995b). The distinctive appearance of cuckoo nestlings that are in the minority might allow them to receive more food, either because parents alternate between the type of nestling fed on successive visits, or because cuckoos provide a stronger stimulus following habituation to host young (Rothstein 1978).

In this chapter, we avoid these complications by comparing the begging behaviour of nestlings under controlled conditions and with similar nutritional need. We also tried to overcome the confounds of nestling number by giving magpie parents a choice between one nestling of each species, whilst controlling for growth, need and size, and studying nestlings at an early age when competitive interference between nestlings is poorly developed.

METHODS

Cuckoos and magpies were studied in Santa Fe (Granada, Spain), a population with remarkably high levels of parasitism (ca. 65%). Details of the study area can be found elsewhere (Zuñiga & Redondo 1992). Field data collected from this population between 1990 and 1998 were used to compute several parameters (e.g. survival and predation rates). Nests were inspected every other day during building and laying until clutch completion, and every day around hatching time, to determine nestling age precisely. Once broods had hatched, they were monitored every two days and matched to the treatments required for each experiment.

Experiment 1

To test whether cuckoos begged more than magpie nestlings for a comparable level of need, a laboratory experiment was performed with ten nestlings of each species of known age coming from different nests. In order to control for growth effects, nestling age was restricted to the day of maximum growth (8 days for cuckoos and 11 days for magpies), when daily mass gain (10-12g) and development for both species were most similar. Nestlings were collected near dusk the day before the experiment and not fed until the next morning in the laboratory. They were kept in individual nestboxes at 27°C. The feeding schedule involved transporting each nestling inside its box into a feeding chamber containing a stuffed adult magpie and a black glove that could be manipulated from behind a screen, and the recording equipment. Nestlings were stimulated to beg by moving the stuffed magpie and a hand inside the glove holding forceps to deliver the food. Nestlings were allowed to ingest *ad libitum* amounts of food (minced beef heart muscle) once every hour during 14 hours of artificial daylight. The amount of food consumed in each feeding session was measured by weighing food before and after feeding with an electronic precision balance (accuracy 0.01g). Nestlings were returned to their nests the following morning (ca. 36 hours being spent in the laboratory).

The degree of food deprivation was manipulated by modifying the above schedule with two short (0.5 hour) and two long (2.5 hour) intervals between feedings at randomly established times of the day. In this way we obtained three periods of food deprivation (0.5, 1 and 2.5 hours). Begging behaviour was recorded during the two feeding sessions following short and long deprivation intervals, plus two 1-hour sessions randomly chosen from the regular feeding schedule. During a feeding session, food was delivered to nestlings whenever they gaped and made begging movements and/or calls. However, we had to consider slightly different satiation criteria for the two species, which were: failing to beg or stopping begging in magpies; and failing to swallow two consecutive meals or throwing the food away in cuckoos (see Results). Nestlings were visually stimulated to beg without feeding for a minute or until they stopped begging. Immediately afterwards, they were again stimulated to beg, using a sound stimulus (voice) and fed *ad libitum* as usual until they met the satiation criteria (three consecutive begging failures, see below). The complete begging bout (the total amount of time begging without feeding plus time spent begging until satiation) was recorded. This measure combines both nestling willingness to be fed, plus their ability to maintain begging behaviour following a feeding event. It should also compensate for individual differences in nestling responses to begging rewards as a result of previous experience. Means per nestling were computed using two begging samples per inter-feed interval duration.

Begging calls were recorded through a condenser microphone (AKG 568 EB) attached to a Sony cassette recorder (WM D6C), and analysed in a real-time sound spectrograph (KAY 5500, Kay Elemetrics Corporation) with a transform size of 300 Hz. We examined the total duration of begging bouts, the duration and number of calls per bout and the total time spent calling. During the second half of the begging bout, we defined begging failures as a lack of begging response following a vocal stimulus before satiation (magpies especially had to be stimulated several times before receiving the first meal, then they often responded by begging again). Measurements of the acoustic intensity of calls (in dB SPL) were taken during 0.5 hour and 2.5 hour trials using a Brüel & Kjaer 2235 sonometer at a distance of 0.5m. Records consisted of peak values of sound intensity per call resulting from an automatic averaging of the whole frequency range (A setting). This procedure was chosen because background noise (which showed 11% variation between recording sessions) could be filtered from high frequency begging calls. The body posture of nestlings during begging was estimated as the highest score displayed during a begging record. Rank scores were as follows: (1) fails to beg, (2) resting on belly, tarsi flexed, (3) body stretched, tarsi extended, and (4) the same as (3) plus wing flapping (see Redondo & Castro 1992a). We assumed that a begging parameter (e.g. calling rate) varied 'honestly' with deprivation time if it both (i) correlated with increasing deprivation and (ii) at least two adjacent mean values (e.g. those corresponding to 0.5 and 1.0 hour, or 1.0 and 2.5 hours) were significantly different in that particular direction (Paired Wilcoxon test). Parameter values for cuckoos and magpies were compared using Mann-Whitney U tests.

Experiment 2

The differential response of magpie parents towards cuckoo and magpie nestlings was determined *via* choice experiments: one cuckoo-magpie set plus a control set containing two magpie nestlings in comparable size combinations. To create experimental two-nestling broods, we temporarily removed all brood contents from magpie nests containing nestlings of three to eight days old and replaced them with two experimental nestlings of two to six days old from a different nest. We chose this age range because differences in development (e.g. eyes opening) are less pronounced and nestlings show limited physical activity, which minimizes direct physical interference. Nestlings were weighed using an electronic precision balance (accuracy 0.01g), they were then left in the nest for three hours before being weighed again. On each occasion, we recorded the behaviour and position of nestlings. Relative Food Intake (RFI) was defined as mass gain by each nestling expressed as a percentage of initial body mass. Differences in RFI between nestlings were plotted as a function of relative mass asymmetry between nestlings, allowing detection of variation in RFI with regard to relative nestling size. An index of relative mass asymmetry was calculated as the difference in mass between nestlings divided by their average mass. This variable controls for existing biases in asymmetry caused by variation in absolute body mass. In addition, we computed food share (the percentage of total mass gain) and nestling mass ratios (dividing the mass of a focal nestling by the mass of its broodmate) as estimates of feeding success and relative size, respectively. We were careful to choose adequate combinations of nestling sizes to represent the whole range of values of relative mass asymmetry within the age span considered. Tests in which parents failed to deliver any food (i.e. negative or zero RFI) were excluded from analyses. Neither nestlings nor parents were tested more than once. There were no overall differences in the initial mass of cuckoos (mean ± SE: 24.2 ± 2.4g) and magpies (26.8 ± 3.7g) in the cuckoo-magpie set (Wilcoxon test, $P > 0.5$, $n = 29$ broods). The initial mass of cuckoos in those tests where the cuckoo was the larger nestling (35.4 ± 5.4g, $n = 17$) did not differ significantly from that of the larger magpie nestling in controls (29.6 ± 2.9g, $n = 16$; Mann-Whitney U test, $P > 0.6$). The initial mass of cuckoos in those tests where the cuckoo was the smaller nestling (14.6 ± 1.8g, $n = 12$) did not differ significantly from that of the smaller magpie nestling in controls (14.1 ± 1.2g, $n = 16$; Mann-Whitney U test, $P > 0.9$).

We chose nestlings with an average level of nutritional condition, i.e. those coming from broods containing between three and five nestlings (with two cuckoos at the most) and avoiding the smallest nestling in a brood.

However, six of the cuckoo-magpie tests involving nestlings of equal size deliberately included cuckoo nestlings showing extreme begging behaviour. Three such tests included a cuckoo nestling that was reared alone and another three included a cuckoo that was the smallest nestling in a brood containing at least three more cuckoos. Singleton cuckoo nestlings often failed to beg in the hand, while small cuckoo nestlings from multiply-parasitized broods begged most intensively. If begging was the main factor influencing food intake then such cases should make it especially clear.

Nonparametric statistical analyses were selected whenever possible, the main exception being analyses of covariance for comparisons of feeding success (food share, RFI and absolute mass gain) of experimental categories of nestlings (cuckoo plus magpie, magpie plus cuckoo and magpie plus magpie), after controlling for the effects of the covariate nestling size (body mass, size asymmetry and mass ratio). Requirements of normality (central distribution) and homogeneity of variance were not violated for both raw data and residuals in every case.

RESULTS

Experiment 1

In the laboratory, cuckoo nestlings begged for much longer and emitted more vocalizations, both in absolute terms and per unit time, than magpie nestlings irrespective of their degree of food deprivation. Time since the last feeding predictably affected the duration of begging bouts, the amount of calling per bout, the calling rate and the total number of begging calls emitted by magpie nestlings, while cuckoos showed no predictable variation in any of these parameters (Table 1). Postural scores followed a similar pattern. Cuckoos usually begged fully stretched and very seldom in resting postures. A conspicuous difference between the two species was the temporal pattern of calling (i.e. gaping). Magpie begging was discrete and well separated in time, with pauses during which nestlings neither gaped nor called. In contrast, cuckoos emitted a continuous quivering flow of calls with persistent gaping, accompanied by head-bowing movements. Acoustic intensity of calls was similar for both species and showed little variation in relation to need (although the high frequencies involved and the limited sensitivity of the apparatus may mask differences detectable with a more sensitive device).

Begging bouts of cuckoos were much longer than magpie begging bouts, and the former required less stimulation to complete a full bout before the behaviour ceased, suggesting a higher motivation for begging in cuckoos independent of need. This is depicted in Table 1, as the number of begging failures. Similarly, cuckoos accepted food on almost every occasion it was offered to them (97.3%, with the exceptions being only two instances by two different nestlings, n = 110 tests), while magpies were three times more likely to fail to eat any food (8.2% of 110 tests, five nestlings). At this and older ages, cuckoo nestlings begged for food in postures that evidently would have interfered with begging of a nestmate (had nestlings in the lab not been isolated), both by spreading their wings to full extension and pushing themselves forward. In natural nests this behaviour is likely to prevent magpies from placing themselves in nest locations closest to the parents, adding to the effect of the intensive begging behaviour of cuckoos.

Table 1. Begging in relation to nutritional need by cuckoo and magpie nestlings (means and SE in brackets).

	Time since the last feeding (hours)			
	0.5	1.0	2.5	P*
Magpies:				
Begging duration (s)	21.80 (2.37)	19.80 (1.51)	26.80 (1.94)	<0.01
Time calling (s)	5.50 (0.55)	8.10 (0.83)	9.60 (0.99)	<0.05
Begging calls /bout	9.50 (1.33)	12.80 (0.36)	16.00 (1.20)	<0.001
Calling rate (calls/s)	0.43 (0.05)	0.67 (0.04)	0.60 (0.03)	<0.05
Call intensity (dB SPL)	97.30 (1.81)	--	97.10 (1.29)	NS
Postural score	2.30 (0.17)	2.61 (0.14)	3.02 (0.11)	<0.05
No. of begging failures	8.00 (0.82)	6.20 (0.63)	6.00 (1.11)	NS
No. of meals eaten	2.60 (0.14)	2.59 (0.13)	3.76 (0.11)	NS
Cuckoos:				
Begging duration (s)	68.00 (9.67)	99.60 (17.08)	79.20 (9.56)	NS
Time calling (s)	34.90 (6.17)	46.70 (10.28)	35.20 (5.42)	NS
Begging calls /bout	114.30 (18.70)	169.30 (40.50)	139.30 (22.10)	NS
Calling rate (calls/s)	1.65 (0.09)	1.64 (0.18)	1.71 (0.11)	NS
Call intensity (dB SPL)	93.94 (2.34)	--	100.52 (1.85)	NS
Postural score	3.24 (0.07)	3.27 (0.11)	3.31 (0.06)	NS
No. of begging failures	0.30 (0.21)	0.40 (0.22)	0.20 (0.13)	NS
No. of meals eaten	3.02 (0.12)	4.05 (0.10)	9.50 (0.40)	<0.05

* Minimum tail probabilities (after Bonferroni corrections) in the comparison between deprivation intervals within species (Wilcoxon test). For all begging parameters except call intensity, cuckoos differ significantly from magpies at any level of food deprivation (Mann-Whitney U test, P<0.05).

Despite cuckoos being much smaller than magpies (cuckoos: 62.2 ± 3.95g; magpies: 106.0 ± 2.85g; Mann-Whitney U test, $P < 0.001$), nestlings of both species consumed similar cumulative amounts of food over the 14 hours (cuckoos: 43.1 ± 1.12g; magpies: 40.4 ± 1.68g; Mann-Whitney U test, $P > 0.9$). As a result, the cumulative relative food intake of cuckoos (62.0% of body mass ± 2.5g) was much higher than that of magpies (40.8% ± 1.1g; Mann-Whitney U test, $P < 0.001$). Cuckoos unexpectedly ate food in proportion to deprivation time, whilst magpies did not. This apparently anomalous result may be a consequence of honest begging (see below).

Compared to magpies, cuckoo nestlings never failed to beg when first stimulated, even if recently fed. Most between-interval variation in magpie begging (i.e. the key requirement for honesty) occurred within the first minute of each begging trial and before nestlings ate any food. Unlike cuckoos, which kept begging during and after being fed, the number of begging displays by magpies during the second minute was equivalent to the number of meals eaten (with minor errors due to feeding failures). Magpies terminated begging bouts almost independently of deprivation and after receiving a mean of 3.0 meals. The experimental protocol involved feeding nestlings in response to gaping, and so cuckoos were often fed without completely swallowing the food (80% of trials). Magpies seldom begged again before swallowing the previous meal (5%), but 8/10 cuckoos threw away mouthfuls of food after being fed several times just to beg again!

Experiment 2

When given a choice, magpie parents favoured the cuckoo. This occurred both in those tests where the cuckoo was the larger nestling (cuckoo: 14.1% RFI ± 1.76; magpie: 6.9% ± 1.86; Wilcoxon test, $P < 0.001$) and when the cuckoo was the smaller nestling (cuckoo: 10.9% ± 1.17; magpie: 7.8% ± 1.27; Wilcoxon test, $P < 0.001$). Overall, cuckoos had higher RFI than magpies in 86% of tests. Considering all the data, the two control magpie nestlings had similar RFI (Wilcoxon test, $P > 0.9$), although, the interaction between nestling size and RFI proved more complex, as follows.

Figure 1 shows the results from the control set of choice experiments concerning differential feeding success in relation to relative nestling size. When magpie nestlings were similar in size, parents fed them equally, but heavier magpie nestlings were preferentially fed over smaller ones when the asymmetry in nestling body mass exceeded a threshold value close to 0.7 (i.e. the large nestling was about 2.5 times larger than the small one).

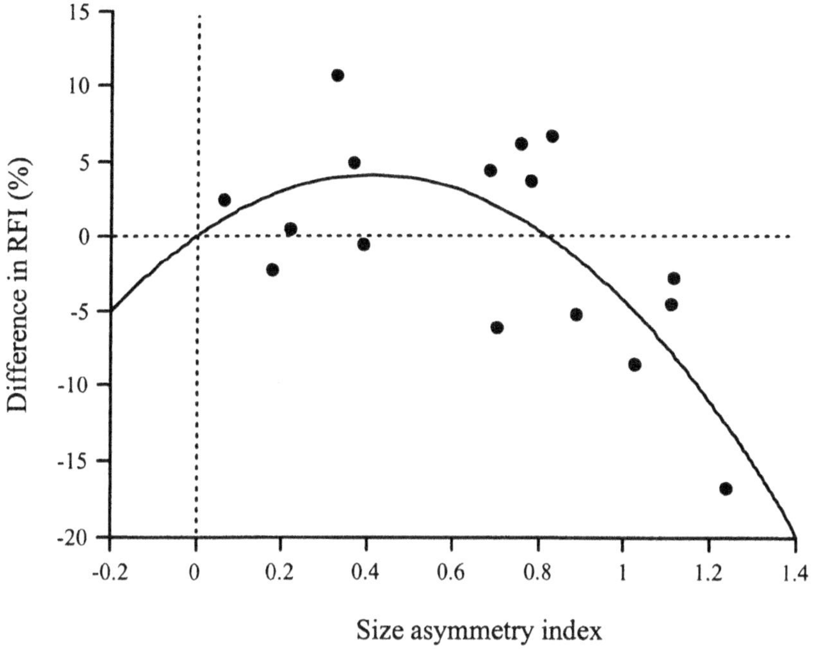

Figure 1. Results of the control set of experiments. Differences in RFI between magpie nestlings (small minus large) are plotted against the size asymmetry index (difference in mass between large minus small nestlings, divided by average mass). The data points have been fitted to a non-linear polynomial regression model ($y = 10.73x - 15.79x^3$; ANOVA, $P < 0.01$) subject to the following realistic restrictions: $y = 0$ when $x = 0$ (i.e. nestlings of equal size are fed the same); if so, y must be positive for small values of x (because parents must be allowed to feed both nestlings equal absolute amounts of food, hence the larger RFI for smaller nestlings, but y is continuous so that this must occur for values of x near zero).

Figure 2 shows an analogous result for the cuckoo-magpie set of choice experiments. Magpie parents always favoured cuckoo nestlings when they were larger or equal to magpie nestlings; the greater the mass asymmetry in favour of the cuckoo, the larger its food share. When smaller, cuckoo nestlings did better than a comparable magpie nestling by never being consistently disfavoured. Of the six magpie nestlings 2.5-3.0 times smaller than their magpie broodmate (Figure 1), five were not fed at all, while cuckoo nestlings in a similar situation never failed to be fed. There were no significant differences in RFI depending on whether magpie parents were caring for magpie nestlings, cuckoo nestlings, or both, prior to the test (ANCOVA, $P > 0.2$). Both in Figure 1 and 2, abscissa values outside the range shown are unrealistic under natural field conditions, so it does not matter if model curves approach infinity when x is much larger than zero.

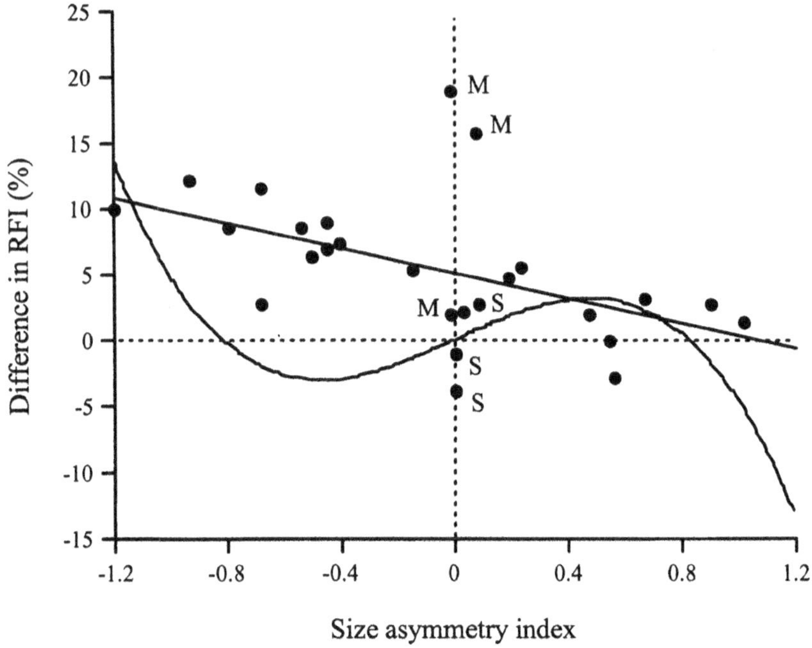

Figure 2. Results of the cuckoo-magpie set of choice experiments. Differences in RFI (cuckoo minus magpie) are plotted against the size asymmetry index (magpie minus cuckoo divided by their average), as in Figure 1. The non-linear regression function in the magpie test (see Figure 1) is shown for comparison. The data shown here fitted an exponential model ($y = 4.22e^{-1.031x}$; ANOVA, $P < 0.01$) better than a linear one. When cuckoos were smaller than magpies they were equivalent to the smaller magpie nestling in Figure 1 and x values are positive as in the control set. Conversely, negative values of x correspond to those tests where cuckoos were larger than magpies. Points marked with S refer to singleton cuckoo nestlings, and M to last-hatched cuckoo nestlings in multiply-parasitized broods.

As predicted, cuckoos showing extreme begging levels prior to the test showed the largest variation in feeding success for a given asymmetry. Two of the three cuckoos reared singly showed lower RFI than magpies, while two of the three small cuckoos coming from multiply-parasitized broods showed the highest RFI values (Figure 2).

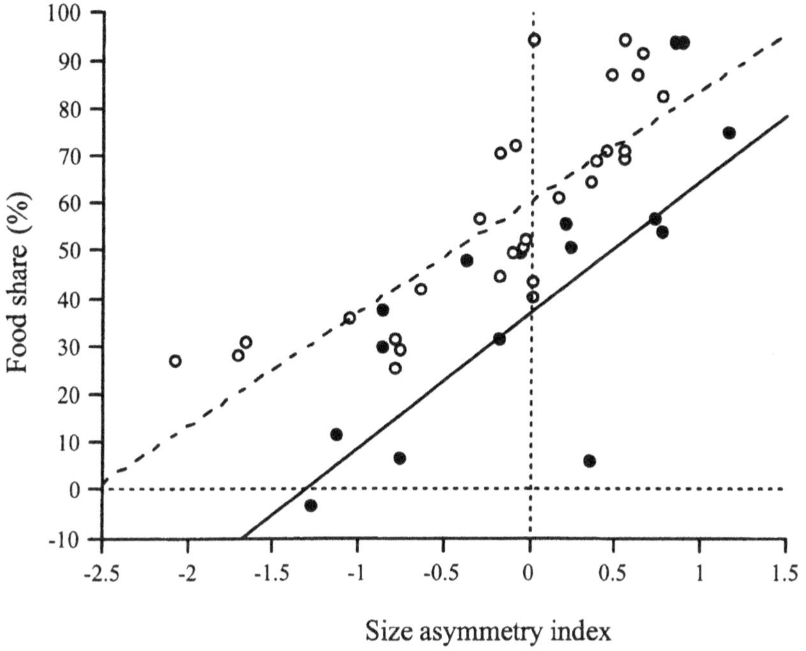

Figure 3. A comparison of linear regression lines for food share in relation to size asymmetry between cuckoo nestlings (open dots, dotted line) and comparable magpie nestlings (black dots, continuous line), when both are paired with a second magpie nestling of similar characteristics.

The higher competitive ability of cuckoos is demonstrated more clearly in Figure 3, where it is shown that cuckoos received more food than a comparable magpie nestling in similar circumstances. Relative size played a major role in the fraction of food obtained (linear regression: $F_{1,44} = 76.4$, $P < 0.001$). After controlling for the effect of nestling size relative to its nestmate, an ANCOVA showed that a cuckoo paired with a magpie nestling obtained a larger food share (adjusted mean: 65.3%) than a magpie nestling under the same circumstances (45.4%; $F_{1,44} = 17.0$, $P < 0.001$; SNK post-hoc comparisons, $P < 0.005$). Congruently, magpie nestling feeding success was more severely affected by the presence of a cuckoo nestmate as compared with the presence of a second magpie nestling. When sharing the nest with a cuckoo, a nestling obtained a much smaller food share (adjusted mean: 33.6%) than when sharing it with another magpie of identical relative size (49.9%; $F_{1,44} = 11.3$, $P < 0.005$, SNK, $P < 0.05$). However, the higher success of cuckoos relative to magpies was not merely a result of diverting a larger share of food to themselves, but actually because they received a larger absolute food intake (adjusted mean: 3.23g), as compared to a magpie nestmate (1.85g) and a control magpie nestling of similar size (2.17g; ANCOVA: $F_{1,44} = 3.84$, $P < 0.05$; Figure 4). As above, cuckoos had a higher RFI (adjusted mean: 12.32% ± 0.88) than magpie nestlings paired with a cuckoo (6.85%; $F_{2,73} = 8.1$, $P < 0.001$, SNK, $P < 0.010$).

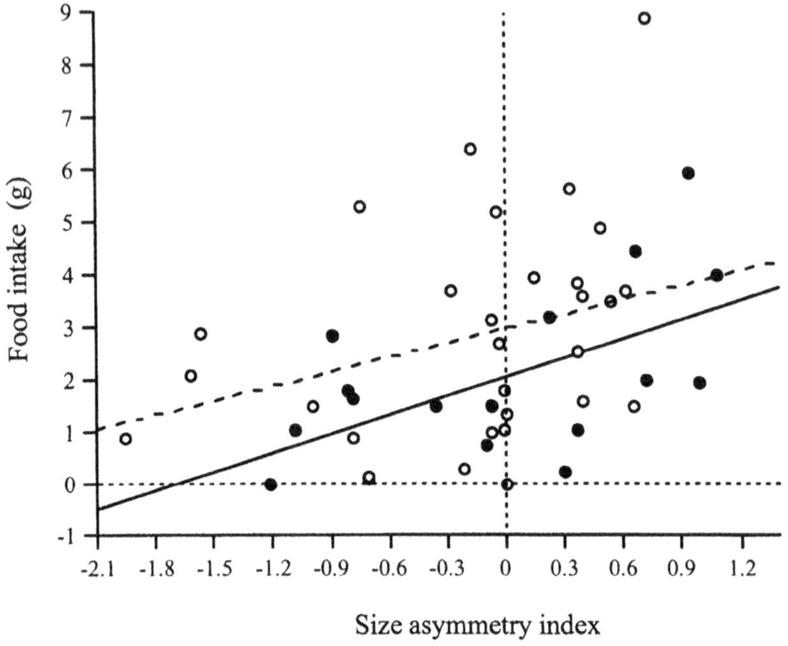

Figure 4. Regression lines for absolute food intake in relation to the size asymmetry of cuckoo nestlings (open dots, dotted line) and magpie nestlings (black dots, continuous line), when both are paired with a second magpie nestling of similar characteristics.

Considering absolute body mass of nestlings instead of relative size-asymmetry introduces a source of error into any comparisons. Despite this, cuckoos obtained a larger food share (adjusted mean: 63.3%) than both magpies of a similar size sharing the nest with them (34.8%; SNK, $P < 0.001$), and similar control-set magpies (49.0%; ANCOVA: $F_{2,73} = 11.2$, $P < 0.001$, SNK, $P < 0.050$; Figure 5).

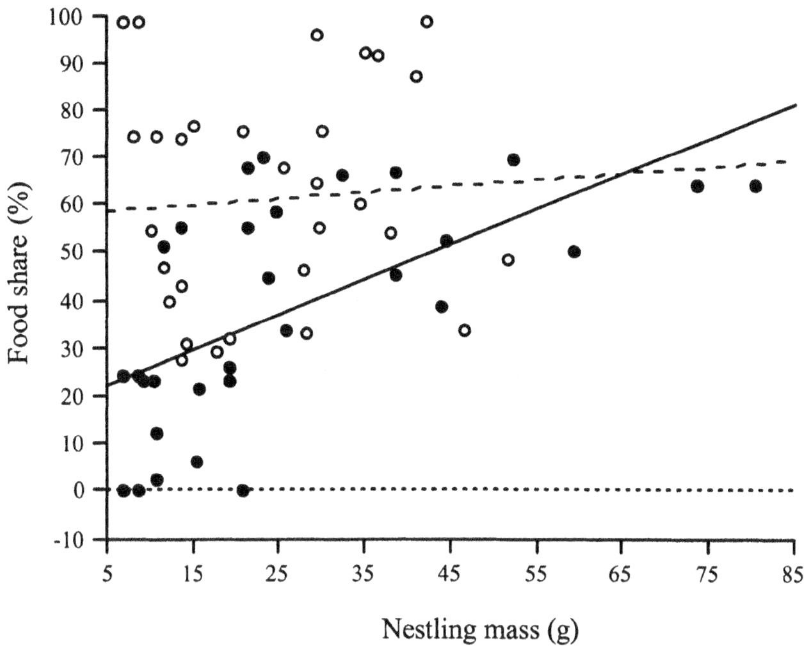

Figure 5. Food share obtained by a cuckoo nestling (open dots, dotted linear regression line) paired with a magpie nestling (black dots, continuous linear regression line) as a function of body mass.

It is interesting to note that, as one might expect from a size-dependent effect upon feeding success, magpie nestlings obtained a larger food share with increasing size. However, cuckoos were usually able to secure a similar (and often larger) share of food, independent of their size. Magpie nestlings did not differ in this respect, whether they shared the nest with a cuckoo or with a conspecific. This means that small cuckoo nestlings received about the same food share as larger ones, because cuckoos were able to compensate for any size differences. In summary, cuckoos received more food and a larger food share as compared to magpie nestlings of a similar absolute or relative size. Magpie nestlings paired with a cuckoo nestling obtained a smaller food share and RFI than magpie nestlings of similar absolute or relative size paired with another magpie nestling. Larger nestlings were fed more, but size effects were much less pronounced for cuckoos, which were fed independent of size.

Some of these results lend support to the conclusion that the higher feeding success of cuckoos was not the result of physical competitive interference between nestlings (i.e. larger nestlings preventing their smaller broodmates from begging at favoured nest locations and/or getting access to food). Instead, cuckoo begging success can be interpreted as the result of parental preferences. Assuming that the total amount of food provisioned by parents remained more or less constant across tests for a given brood mass, competitive interference should have manifested itself as negative covariance between mass gain values of both nestlings. This is because more competitive nestlings would have increased their food intake at the expense of their smaller broodmate, rather than as a result of preferential parental allocation. Total food mass delivered correlated positively with brood mass in both control (linear regression: $r^2 = 0.34$, $F_{1,15} = 8$, $P < 0.010$) and cuckoo-magpie tests ($r^2 = 0.66$, $F_{1,28} = 54.1$, $P < 0.001$). In control tests, no significant correlation was found between mass gains ($r = -0.15$, $df = 15$, $P > 0.5$), but in cuckoo-magpie tests the values of mass gain correlated positively ($r = 0.38$, $df = 28$, $P < 0.050$). Therefore, no nestling category gained mass exclusively at the expense of its broodmate. When parents delivered more food to mixed broods, both cuckoo and magpie nestlings received more food. Parents chose to feed cuckoo nestlings more, as evidenced by the increase in total food delivered to parasitized nests. Although cuckoo parasitism results in reduced fitness of host nestlings, the mechanism appears to be parental preference for cuckoo nestlings, not cuckoo aggression against host nestlings. In support of this, we never observed cuckoos adopting interfering postures during the age range covered by the experiment, as they usually do at older ages.

DISCUSSION

Do Cuckoos Beg More Because They Are in Greater Need?

Begging intensity can be defined as a complex variable incorporating postural and auditory components of begging co-varying with each other and with deprivation time (see R.M. Kilner this volume). In this sense, magpies begged more intensely with increasing need (see also Redondo & Castro 1992a), but was this the result of differential nutritional requirements between the species? Absolute energetic requirements of nestlings mainly depend upon growth and maintenance (e.g. metabolism and activity; O'Connor 1982). Therefore, because cuckoo and magpie nestlings in Experiment 1 grew at similar rates (ca. 10-12g per day, T. Redondo & J.

Zuñiga unpublished data), the total nutritional requirements must have been greater for the larger magpie nestlings. Recently, Soler et al. (1999) measured daily metabolic rate (from oxygen consumption), and the daily energy budgets of pre-fledging cuckoo and magpie nestlings were found to be surprisingly similar. In the laboratory, nestlings of both species were isolated and mostly inactive except while begging (i.e. at this age preening is not yet developed). The degree of plumage development and thermal insulation must also have been similar for both species, which operated at identical ambient temperatures. Therefore, differences in nestling energy budget between the two species must have been small, and unlikely to compensate for the 40% difference in body mass of magpie nestlings. Nestling assimilation efficiency is remarkably constant (ca. 70%) across species feeding on the same diet (O'Connor 1982). Cuckoo and magpie nestlings are adapted to an identical feeding regime at magpie nests (ca. one feeding per hour; Birkhead 1990), and should thus respond digestively in a similar way to variation in deprivation time. All of this suggests that any differences in need between nestlings were very unlikely to have been biased towards cuckoos. However, compared to magpies, begging by cuckoos was dishonest in the sense of being: (i) exaggerated (i.e. more intense for a similar level of need); and (ii) unreliable (i.e. not varying predictably in relation to deprivation time).

Direct Factors Limiting Escalation: Costs

Factors that limit exaggeration are those affecting the benefit/cost ratio of begging, including direct (energetic and predation) and indirect (inclusive fitness) costs. Unlike cuckoos, magpies refrained from exaggerated begging and consumed almost half the food intake of their cuckoo nestmates. This result demands explanation in terms of the factors limiting begging escalation in magpies, because they refrained from the dishonest begging that appeared effective for cuckoos in providing the reward of extra food.

Direct energetic costs of begging per unit time are unlikely to account for the more intense begging of cuckoo nestlings. Measurements of the energetic costs for both species have shown few differences in metabolic rate and energy utilization during begging, and these appear to contribute relatively little in terms of daily energetic expenditure (Soler et al. 1999; M.A. Chappell & G.C. Bachman this volume).

Direct predation costs are probably important (Haskell 1994; D.G. Haskell this volume), but it seems unlikely that they affect the two species of nestlings differentially. The most powerful predictor of predation risk is nest vulnerability (e.g. concealment, abundance of nearby predators), which depends on nest location and proximate habitat (Yahner & De Long 1992). There was no evident trend in the population studied for a differential habitat distribution of parasitized versus non-parasitized nests. If predation were a major determinant of begging intensity, we should expect: (1) a lower predation rate for cuckoos, so they could afford to beg more; (2) differences in predation to be more pronounced following enlarged brood sizes and increased age; and (3) differences in brood failure to be caused by predators other than adult cuckoos. The last prediction follows from the fact that female *Clamator* cuckoos routinely revisit nests and destroy the contents during incubation and early nestling periods (Soler et al. 1996).

We computed predation rates for broods of two to six nestlings containing either magpies or cuckoos exclusively during the first 20 days of life, excluding cases of total brood failure caused by humans. Between zero and ten days, predation rates were slightly higher for magpie (8/61, 13.1%) as compared to cuckoo broods (7/82, 8.5%), but this difference was nonsignificant (Fisher's exact probability test, $P = 0.540$). This nonsignificant difference diminished further between 10 and 20 days (7/115, 6.1%, and 9/215, 4.0%, respectively; Fisher's test, $P = 0.540$). Predated broods at both ages were not larger than non-predated broods, indeed predated broods were smaller in magpies when younger than 10 days (Mann-Whitney U test, $P < 0.010$). This brood size and age effect on predation is contrary to that expected from the second prediction, because larger and older broods will be noisier. In a different population of non-parasitized magpie nests but with high abundance and diversity of predators (Doñana National Park), noisier magpie broods have been shown to suffer higher predation and at earlier ages (Redondo & Castro 1992b). Further analyses of unpublished data from the Redondo and Castro (1992b) study reveals that predated broods younger than 20 days were significantly larger at the time of predation than non-predated ones. In the current study population at Santa Fe, most instances (87.5%) of predation in magpie broods younger than 10 days were restricted to broods of three nestlings or less, despite the fact that such broods contributed only a minority of the total (23.0%, Fisher's test, $P = 0.001$). In these broods, hatching success was low (41.1%), compared with 79.4% in non-predated broods. Egg breakage caused by adult cuckoos occurred in 100% of all-magpie broods predated before 10 days, as compared to 24.5% of non-predated ones. This strongly suggests that differential predation upon magpie broods was largely due to adult cuckoos. Summarizing this section, all-cuckoo broods were not predated any more frequently than all-magpie broods. The tendency for higher predation in magpie nests below 10 days of age was probably a consequence of adult cuckoos destroying nest contents in non-parasitized nests early in the nestling period.

Direct Factors Limiting Escalation: Benefits

Magpie and cuckoo nestlings may not differ greatly in the costs of begging, but they might differ in the benefits gained. Lotem (1998) suggested that exaggerated begging benefits the cuckoo more in species where it remains the sole occupant of the nest, than it does a host nestling which always shares the nest with several siblings. This is because extra food gained *via* begging is diluted among all broodmates, but the extra costs of begging are not. This comparison may not apply here, because both magpies and cuckoos share their nests and so they would accrue similar benefits (Redondo 1999). The present study also provides ample evidence against Lotem's (1998) suggestion. For example, singleton cuckoos actually begged the least, and larger magpie nestlings in non-parasitized broods begged less but obtained more food. This suggests similar benefits for both species.

Nestlings of both species would also benefit from maximizing body mass at independence. In many species, survival dramatically increases with fledgling mass, due to the combined effects of several factors related to foraging ability and social dominance (Garnett 1981; Richner et al. 1989; Magrath 1991). An adequate food supply at the nest can be the most powerful predictor of survival prior to breeding maturity (Spear & Nur 1994), which makes levels of food intake almost equivalent to lifetime reproductive value. In this system, there is evidence of size-biased mortality for fledglings of both species (Eden 1985; Soler et. al. 1994). For cuckoo fledglings, this is because they have to migrate to Africa, requiring fat reserves. Magpie nestlings suffer from size-biased mortality, both in the nest and before their first breeding season. The heaviest nestlings in successful magpie broods (already distinguishable within a few days of hatching) had a higher probability of fledging than their lighter siblings (0.83 versus 0.49, respectively, $n = 40$). And of the 10% of juveniles surviving to their first spring, more than 80% comprise the heaviest fledgling from their natal brood (T. Redondo unpublished data).

Indirect Inclusive Fitness Costs

We are led to the conclusion that the more intensive begging of cuckoos was a consequence of their lower (indeed zero) inclusive fitness costs. Small magpie nestlings obviously have an evolutionary interest in the survival prospects of their larger siblings, and may gain little from completely outcompeting them. In species where parents readjust feeding rates after brood reduction, the benefits to remaining nestlings of sibling starvation are small (Graves et al. 1984; Mock & Lamey 1991; Martins & Wright 1993), and this is likely to be the case here. Parent magpies not only feed young in relation to brood size (Redondo & Castro 1992b), but also according to brood mass. Nestlings may therefore benefit from the presence of nestmates, both directly (e.g. thermal insulation, post-fledging social support) and indirectly (kin selection), and this may select for tolerance towards smaller, lower quality siblings (Forbes & Ydenberg 1992; B. Glassey & S. Forbes this volume).

Field experiments show that magpie nestlings suffered from a lower food intake when sharing the brood with a cuckoo as compared to a magpie nestmate of a similar relative size. Any mutant magpie begging dishonestly would have had the same effect. On average, a magpie hatchling had a 0.81 probability of fledging in a successful brood if reared with other magpies (n = 73). However, this figure falls to 0.24 if one cuckoo was present in a brood of a similar size (n = 66). When two or more cuckoos were present, the probability of magpie nestling survival drops to 0.02 (n = 102). We computed survival probabilities attributed to nestling starvation before 10 days of age from synchronously hatched broods containing one cuckoo nestling. This would represent a reasonable estimate of the indirect cost of dishonest begging incurred by a hypothetical mutant magpie begging like a cuckoo (T. Redondo & J. Zuñiga unpublished data). Following parasitism by one cuckoo, a normal magpie nestling had a 0.47 probability of fledging, and a 0.45 probability of being accompanied by a sibling, as compared with figures of 0.85 and 0.83, respectively, in similar all-magpie broods. Among those magpie broods where at least two magpie siblings hatched, the probability that a nestling other than a focal survivor would fledge successfully (i.e. the probability of a sibling also leaving the nest) was 0.77, but the presence of a cuckoo lowered it to 0.36. Therefore, the presence of a single dishonest nestling of a similar age and size represents roughly a 0.50 reduction factor in nestling survival, without considering additional size-related post-fledging effects of insufficient food intake (Soler & Soler 1991).

Another experimental simulation of a mutation endowed with a more vigorous begging behaviour, not necessarily associated with brood parasitism or stronger physical competition, was performed by Alvarez et al. (1976), who placed jackdaw (*Corvus monedula*) nestlings in magpie broods (see also M. Soler this volume). Jackdaws beg more vigorously than magpie nestlings (louder and for longer, but less than cuckoos) and have an effect similar to that of cuckoos.

Three pieces of evidence lend additional support to the conclusion that begging honesty was mainly the consequence of indirect costs:

(1) Magpies begged more when in greater need, but the amount of food that they were willing to eat (i.e. hunger) depended upon deprivation time to a much lesser extent. Irrespective of need, magpie begging bouts were inhibited by feeding, and magpies 'prudently' stop begging after receiving a few meals. In contrast, cuckoos begged independently of need but consumed food in proportion to need, and failed to terminate begging in response to food.

(2) As in most birds, begging by magpies was expressed in discrete, well-defined units. This gives siblings the opportunity to be fed as nestlings cease to gape and vocalize between feedings. In contrast, cuckoos beg continuously, which is the rule among brood parasites (Redondo 1993; Davies et al. 1998).

(3) Unlike magpies, cuckoos stored food in their guts for at least 12 hours after eating, as evidenced by tracking barium-labelled food with Computerized Axial Tomography (Redondo 1993). This storing of food is obviously advantageous for individual cuckoos when prospective requirements (e.g. growth and thermoregulation) are great and a sustained, sufficient food supply is uncertain (e.g. due to variation in foraging conditions), and especially when energetic reserves are low, as in most altricial nestlings. This feature also suggests that food intake by cuckoos was determined by the available digestive capacity, whilst that of magpies was strictly under motivational control.

Therefore, in the absence of indirect costs, it is hard to explain why magpie nestlings refrain from escalating begging until their storage capacity is filled. Experimental evidence has shown that even minute supplemental increases in food intake early in the nestling period have enormous consequences for subsequent nestling survival (Graves et al. 1984), an effect that is also observed in magpies (Högstedt 1981; Hochachka & Boag 1987). The pattern of cuckoo and host nestling behaviour seen here is exactly what we might expect if costs do not depend upon how much a nestling begs *per se*, but instead upon the distribution of parental feedings. This is because the direct begging costs experienced by cuckoos should vary continuously in proportion to begging effort, whilst the honest begging of their hosts is also shaped by indirect fitness costs. Indeed, McCarty (1996) has suggested that honest begging itself could be maintained by such indirect inclusive fitness costs.

Do Cuckoos Beg More to Counteract Parental Neglect?

Experiments showed that magpies neglected the smallest nestling of two when its larger broodmate was roughly three times larger (equivalent to an index of asymmetry of one, Figure 1). There is a 0.70 probability (n = 326 parasitized broods) that cuckoo nestlings hatch out within an older, already parasitized brood. This indicates an adaptive value to exaggerated begging as a mechanism to prevent neglect by magpie parents if, as we suggest, exaggerated begging can compensate for the disadvantage of small relative size. Small magpie nestlings also begged more, but such begging was within the range of begging intensities for this species, and it was obviously not sufficient to prevent neglect (Figure 1). Manipulations of nestling begging and relative size showed that the more intense begging of small magpie nestlings was often insufficient to counteract parental favouritism towards larger nestlings, and progressively less so with increasing size differences (Redondo 1993). However, despite their extra begging, cuckoo nestlings still had a 0.11 probability of starving when hatching into broods containing older cuckoos (n = 716 nestlings in 199 successful broods).

Host Manipulation

Exaggerated begging by cuckoos elicited preferential feeding by magpie parents. We can be reasonably sure of this conclusion because we failed to find evidence for cuckoo feeding success being the consequence of direct physical competition with host nestmates, at the ages considered in the experiment. This explanation agrees with honest signalling models, in that preferential feeding of larger nestlings is a direct outcome of parental preferences rather than of the superior competitive ability of larger nestlings (Parker 1985; Mock & Parker 1997). Variation in feeding rates were, in part at least, tuned to differences in begging, even if other factors also contributed to the greater feeding success of cuckoos (e.g. a more conspicuous gape; Soler et al. 1995b).

Great spotted cuckoo nestlings exploited a set of magpie nestling-feeding rules that favoured nestlings of a large size and those begging intensively. A preference for larger nestlings is adaptive in fulfilling their higher absolute requirements due to a heavier mass and faster growth (during the earliest half period of exponential growth), but it may become amplified by directional selection in species where larger nestlings are more valuable, for instance those with size-biased survival, such as magpies (Haig 1990). It may also help in facilitating facultative brood reduction (selectively starving smaller nestlings), a trait that has been demonstrated to be adaptive in magpies (Husby 1986). Signalling theory accounts for the occurrence of dishonesty as part of an otherwise honest, stable signalling system, whenever a minority of signallers differ in advertising costs and thus can afford to emit higher-intensity signals when of a similar quality. Dishonest signals are consistently misinterpreted by receivers due to their inability to assess the difference in signaller category, thereby allowing the existence of stable manipulation (Johnstone & Grafen 1993). Begging by cuckoos and magpies fits well with this scenario.

FUTURE DIRECTIONS

Some of the evidence provided by this study is not conclusive and awaits further observations and experiments. This study has also generated many new hypotheses that require testing in the future. Here we enumerate the major points that should be addressed, at least in the short term.

Our conclusions regarding parental preferences and potential manipulation require observations at natural nests, to exclude the possibility of physical interference between nestlings as the causal factor explaining patterns of food allocation. Experiment 2 should perhaps have included video recordings of experimental broods in order to observe how parents and nestlings interacted.

More experimental data are required to quantify and compare the energetic cost of begging in cuckoos and magpies. In recent work (Rodríguez-Gironés et al. 2001) we assigned magpie nestlings to two treatments: group A was fed immediately after begging while group B was fed a similar amount of food but only after begging a lot. We detected significant differences in growth rates during the experimental period (three days): nestlings in group B grew at a lower rate after controlling for the effects of body mass and food intake. This is direct evidence for a negative effect of begging effort upon fitness (heavier nestlings surviving better to the next breeding season), which has gone undetected in studies measuring the oxygen consumption of begging as an estimate of its energetic cost (see M.A. Chappell & G.C. Bachman this volume). We failed to repeat the same experiment with cuckoo nestlings because they begged so much and independently of treatment. However, a new experimental protocol creating two treatment groups for cuckoos and magpies will test the hypothesis that cuckoos beg more because it is less costly than for magpies. Our results concerning differential predation costs for cuckoos and magpies should be interpreted with caution because the Santa Fe population studied is biased in two undesirable ways. First, it contains many cuckoos which destroy nest contents, but are not predators in a strict sense. Second, it lacks a sufficient number and diversity of predators because it is highly disturbed, agricultural land. Ideally, a study should be conducted on a non-parasitised population containing a high abundance and diversity of predators (e.g. Doñana National Park population, Redondo & Castro 1992a,b) in order to demonstrate that sound-guided predators are attracted to nests. Calls from magpie and cuckoo broods could be broadcast from natural rates and intensities from artificial nests to remove parental effects, such as differential nest guarding or defence. It would also be interesting to study how begging calls of both species propagate in the environment, to understand how they attenuate and degrade at varying distances from the source. Perhaps cuckoos beg at similar intensities to magpies but their calls degrade more easily and are less detectable (see D.G. Haskell this volume).

Studies on the digestive physiology of both species would be very desirable in order to understand the proximate causal factors underlying begging and satiation (see A.B. Clarke this volume; W.H. Karasov & J. Wright this volume). This information is crucial for the interpretation of the results of Experiment 1. There are several possible mechanisms (e.g. food mass or volume, number of meals eaten, caloric or nutrient content of food) that may cause hunger and satiation to vary between species. We would predict that cuckoos have a 'selfish' physiology; for example they can store food in their guts for several hours while magpies do not (Redondo 1993). We are currently addressing these questions.

Thus far, we have ignored the precise mechanisms determining which particular nestling is fed (e.g. position in the nest, relative height or proximity to a parent's beak) and the behavioural rules followed by parents in making a feeding decision. This requires video recordings of parent-offspring interactions under natural conditions, but would answer many issues arising from this study. For example, singleton cuckoos begged less than nestlings coming from multiple broods, and last-hatched nestlings begged more in multiply-parasitized broods. Is this evidence of cuckoos being honest concerning their nutritional need, or of their responsiveness to variation in the level of within-brood competition? Another example concerns the proximate causal factors underlying neglect of very small magpie nestlings. How do cuckoo nestlings under similar conditions manage to obtain a relative food intake similar to that ingested by their larger magpie nestmate?

Unfortunately, all our attempts to place video cameras, and even minute microphones (less than one cm in diameter), close to magpie nests have failed. The problem is that magpies are extremely wary of any strange device placed near their nests, and refrain from staying at the nest for as long as the device is present. We would like to conclude by taking this opportunity to ask readers for suggestions which could help us to solve this logistical obstacle.

REFERENCES

Alvarez, F. & Arias de Reyna, L. 1974. Mecanismos de parasitación por *Clamator glandarius* y defensa por *Pica pica*. *Doñana, Acta Vertebrata* **1**, 43-65.

Alvarez, F., Arias de Reyna, L. & Segura, M. 1976. Experimental brood parasitism of the magpie (*Pica pica*). *Animal Behaviour* **24**, 907-916.

Birkhead, T.R. 1990. *The Magpies*. London: T & A.D. Poyser.

Brooke, M. de L. & Davies, N.B. 1989. Provisioning of nestling cuckoos *Cuculus canorus* by reed warbler *Acrocephalus scirpaceus* hosts. *Ibis* **131**, 250-256.

Davies, N.B., Kilner, R.M. & Noble, D.G. 1998. Nestling cuckoos, *Cuculus canorus*, exploit hosts with begging calls that mimic a brood. *Proceedings of the Royal Society of London, Series B* **265**, 673-678.

Eden, S. F. 1985. Social organization and the dispersal of non-breeding magpies *Pica pica*. PhD Thesis, University of Sheffield.

Forbes, L.S. & Ydenberg, R.C. 1992. Sibling rivalry in a variable environment. *Theoretical Population Biology* **41**, 335-360.

Garnett, M.C. 1981. Body size, its heritability and influence on juvenile survival among great tits, *Parus major*. *Ibis* **123**, 31-41.

Graves, J., Whiten, A. & Henzi, P. 1984. Why does the herring gull lay three eggs? *Animal Behaviour* **32**, 798-805.

Haig, D. 1990. Brood reduction and optimal parental investment when offspring differ in quality. *American Naturalist* **136**, 550-566.

Haskell, D. 1994. Experimental evidence that nestling begging behaviour incurs a cost due to nest predation. *Proceedings of the Royal Society of London, Series B* **257**, 161-164.

Hochachka, W. & Boag, D. 1987. Food shortage for the black-billed magpie (*Pica pica*): an experiment using supplemental food. *Canadian Journal of Zoology* **65**, 1270-1274.

Högstedt, G. 1981. Effect of additional food on reproductive success in the magpie (*Pica pica*). *Journal of Animal Ecology* **50**, 219-229.

Husby, M. 1986. On the adaptive value of brood reduction in birds: experiments with the magpie (*Pica pica*). *Journal of Animal Ecology* **55**, 75-83.

Johnstone, R.A. & Grafen, A. 1993. Dishonesty and the handicap principle. *Animal Behaviour* **46**, 759-764.

Lotem, A. 1998. Manipulative begging calls by parasitic cuckoo chicks: why should true offspring not do the same? *Trends in Ecology and Evolution* **13**, 342-343.

Magrath, R.D. 1991. Nestling weight and juvenile survival in the blackbird, *Turdus merula*. *Journal of Animal Ecology* **60**, 335-351.

Martins, T.L.F. & Wright, J. 1993. On the cost of reproduction and the allocation of food between parent and young in the swift (*Apus apus*). *Behavioral Ecology* **4**, 213-223.

McCarty, J.P. 1996. The energetic cost of begging in nestling passerines. *The Auk* **113**, 178-188.

Mock, D.W. & Lamey, T.C. 1991. The role of brood size in regulating egret sibling aggression. *American Naturalist* **138**, 1015-1026.

Mock, D.W. & Parker, G.A. 1997. *The Evolution of Sibling Rivalry*. Oxford: Oxford University Press.

O'Connor, R.J. 1982. *The Growth and Development of Birds*. Chichester: John Wiley & Sons.

Parker, G.A. 1985. Models of parent-offspring conflict. V. Effects of the behaviour of the two parents. *Animal Behaviour* **33**, 519-533.

Redondo, T. 1993. Exploitation of host mechanisms for parental care by avian brood parasites. *Etología* **3**, 235-297.

Redondo, T. 1999. Manipulative begging by parasitic cuckoo nestlings and paradoxical host behaviour. *Trends in Ecology and Evolution* **14**, 107.

Redondo, T. & Castro, F. 1992a. Signalling of nutritional need by magpie nestlings. *Ethology* **92**, 193-204.

Redondo, T. & Castro, F. 1992b. The increase in risk of predation with begging activity in broods of magpies *Pica pica*. *Ibis* **134**, 180-187.

Redondo, T. & Exposito, F. 1990. Structural variations in the begging calls of nestling magpies *Pica pica* and their role in the development of adult voice. *Ethology* **84**, 307-318.

Richner, H., Schneiter, P. & Stirnimann, H. 1989. Life-history consequences of growth rate depression: an experimental study on carrion crows (*Corvus corone corone* L.). *Functional Ecology* **3**, 617-624.

Rodríguez-Gironés, M.A., Zuñiga, J.M. & Redondo, T. 2001. Effects of begging on growth rates of nestling chicks. *Behavioral Ecology* **12**, 269-274

Rothstein, S.I. 1978. Geographical variation in the nestling coloration of parasitic cowbirds. *The Auk* **95**, 152-160.

Soler, M. & Soler, J.J. 1991. Growth and development of great spotted cuckoos and their magpie host. *Condor* **93**, 49-54.

Soler, M., Palomino, J.J., Martinez, J.G. & Soler, J.J. 1994. Activity, survival, independence and migration of fledgling great spotted cuckoos. *Condor* **96**, 802-805.

Soler, M., Martinez, J.G. & Møller, A.P. 1995a. Chick recognition and acceptance: a weakness in magpies exploited by the parasitic great spotted cuckoo. *Behavioral Ecology and Sociobiology* **37**, 243-248.

Soler, M., Martinez, J.G., Soler, J.J. & Møller, A.P. 1995b Preferential allocation of food by magpies *Pica pica* to great spotted cuckoo *Clamator glandarius* chicks. *Behavioural Ecology and Sociobiology* **37**, 7-13.

Soler, M., Martinez, J.G. & Soler, J.J. 1996. Effects of brood parasitism by the great spotted cuckoo on the breeding success of the magpie host: an experimental study. *Ardeola* **43**, 87-96.

Soler, M., Soler, J.J., Martinez, J.G. & Moreno, J. 1999. Begging behaviour and its energetic cost in great spotted cuckoo and magpie host chicks. *Canadian Journal of Zoology* **77**, 1794-1800.

Spear, L.B. & Nur, N. 1994. Brood size, hatching order, and hatching date: effects of four life-history stages from hatching to recruitment in western gulls. *Journal of Animal Ecology* **63**, 283-298.

Yahner, R.H. & De Long, C.A. 1992. Avian predation and parasitism on artificial nests and eggs in two fragmented landscapes. *Wilson Bulletin* **104**, 162-168.

Zuñiga, J.M. & Redondo, T. 1992. No evidence for variable duration of sympatry between the great spotted cuckoo and its magpie host. *Nature* **359**, 410-411.

21.

BREEDING STRATEGY AND BEGGING INTENSITY: INFLUENCES ON FOOD DELIVERY BY PARENTS AND HOST SELECTION BY PARASITIC CUCKOOS

Manuel Soler
Departamento de Biología Animal y Ecología, Facultad de Ciencias, Universidad de Granada, E-18071 Granada, Spain (msoler@goliat.ugr.es)

ABSTRACT

Studies of begging behaviour and food provisioning have produced surprising results. Here, I present a new idea that may explain these results. That is, in species with brood reduction (brood-reducers) some nestlings starve, but in species that adjust clutch size (clutch-size adjusters) all nestlings typically survive to fledge. This suggests that parents adopting these strategies may follow different provisioning rules. In clutch-adjusters, parents tend to distribute food evenly among their nestlings, preferentially feeding young that are in poorer condition (i.e. begging at a higher intensity), but in brood-reducers parents selectively feed larger nestlings independently of begging intensity. These different rules may be important for brood parasites, which should be adapted to the breeding strategy of the host species. This model explains why eviction behaviour has evolved in some cuckoo species but not in others, and also why overlap in host use is very small between cuckoo species.

INTRODUCTION

Contradictory results in different studies of begging behaviour and food provisioning are very common. For example, Cotton et al. (1999) found that when hatching asynchrony was experimentally manipulated in starlings (*Sturnus vulgaris*) in asynchronous nests, older nestlings received

J. Wright and M.L. Leonard (eds.), The Evolution of Begging, 413–427.

significantly more food and grew faster than younger nestlings, even though they begged less. They concluded that these surprising results "violate the predictions of available theoretical models and, together with limitations in the universality of their assumptions, indicate that fundamental aspects of parent-offspring communication are not yet understood".

Here, I present a new hypothesis to explain why begging signals and the patterns of food delivery by parents vary across species. The idea is based on the suggestion that the breeding strategy of the species (brood reduction or adjustment of clutch size, see below) should be taken into consideration in order to understand the relationships between begging behaviour of nestlings and food delivery by parents. The key point is that in brood-reducers some nestlings starve, but in species that adjust clutch size all nestlings usually survive to fledge. This implies that parents with the two strategies do not follow the same rules when allocating food to their nestlings.

In the second part of this chapter, I analyse how this new idea may help to explain patterns of host selection by brood parasites. Avian brood parasites depend on other bird species (the hosts) for reproduction, since they lay their eggs in the nests of their hosts, which provide care for the eggs and nestlings of the brood parasite (Rothstein 1990). Host selection is a crucial decision, because the breeding success of brood parasites depends on the parental quality of the host species (Kleven et al. 1999) and of the host pair (Maynard Smith 1978; Krebs & Kacelnik 1991; Soler J.J. et al. 1995). If in clutch-size adjusters parents tend to distribute food evenly among their nestlings, but in brood-reducers parents selectively feed larger nestlings, brood parasites must be adapted to choose hosts of one breeding strategy. I present a model that explains host selection by cuckoos and the evolution of eviction behaviour based on the breeding strategy of hosts and the relative level of need of the parasite nestling. This model also enables one to make predictions about the fledging success of both host and parasite nestlings.

In the third part of the chapter, I suggest that begging intensity of host nestlings is one of the important factors affecting host selection by brood parasites. This idea is based upon the fact that there are important differences in the intensity of begging calls among species, mainly due to different selection pressures from predation. That is, nest predation may be responsible for maintaining some of the interspecific differences in the acoustic structure of begging calls (Haskell 1999; D.G. Haskell this volume). Thus, begging intensity and/or competitive ability of the nestlings in different host species should be an important factor influencing host choice by brood parasites, mainly in species where parasitic nestlings do not evict host nestlings. Some pieces of evidence are described showing that this is the case for great spotted cuckoos (*Clamator glandarius*) when selecting their corvid hosts.

BREEDING STRATEGIES

Two main breeding strategies have been described for parent birds: (1) direct adjustment of clutch size; or (2) laying more eggs and later, during the nestling period, adjusting the number of young to food supply (Lack 1947).

In most birds the extent of hatching asynchrony varies considerably with the range in hatching period, extending from a few hours in many passerine species to more than ten days in some parrot species (Stamps et al. 1985). Low levels of hatching asynchrony do not induce brood reduction, but increased levels provoke intense competition for food within broods, culminating in starvation of the last-hatched nestlings (Slagsvold et al. 1995; Stoleson & Beissinger 1995; Stenning 1996; but see Krebs et al. 1999 for an exception to the rule). Why parents preferentially provide food to some nestlings while leaving others unattended until starvation, remains controversial. The only point relevant here is that in clutch-size adjusters all hatched nestlings typically fledge, while brood-reducers lay more eggs and later adjust brood size to food supply provoking starvation of the smaller nestling(s). Clutch-size adjusters can be defined as those species where all nestlings fledge in more than 80% of the nests, and brood-reducers as those species where at least one nestling starves in more than 80% of the nests. The jackdaw (*Corvus monedula*) is a clear example of a brood-reducer. In this species, hatching success is very high, however, the mean percentage of eggs that produce fledglings is about 25% (Soler & Soler 1996). Jackdaws are never able to fledge more than four nestlings, even under experimental conditions of food supplementation (Soler & Soler 1996). Most passerines are clear examples of clutch-adjusters; for example, in the black wheatear (*Oenanthe leucura*) 99.8% of hatched eggs produce fledglings, and brood reduction occurs in only 1.2% of nests (two of 164 nests, calculated from Soler M. et al. 1995). In both cases they were the latest second clutch in two different years.

Most clutch-size adjusters are facultative brood-reducers when food is limited. However, this does not contradict my idea, it merely means that parents are able to make active decisions regarding food allocation, and in the case of food shortage they can avoid total brood loss by allocating more food to larger nestlings.

BEGGING BEHAVIOUR AND FOOD DELIVERY PATTERNS: GENERAL RULES AND EXCEPTIONS

A logical chain of events links begging signals and parental responses: nestling needs → begging signals → parental provisioning (Harper 1986;

Godfray 1991; Soler & Soler 1999). There are no published exceptions with respect to the first link of the chain (sibling competition has been shown to alter begging pattern, but only slightly with respect to the level of need, i.e. Muller & Smith 1978; Price & Ydenberg 1995).

With respect to the second link of the chain, a positive correlation between begging intensity and feeding rate, as well as the nestling's probability of receiving food, has been found in most studies (reviewed in Kilner & Johnstone 1997; Budden & Wright 2001). Theoretical models of the evolution of begging behaviour assume that parents decide the provisioning rate to the brood by simply responding to the total amount of begging emanating from their nests (Parker & Macnair 1979; Parker 1985; Godfray 1995a,b), and experimental results have supported this assumption (Muller & Smith 1978; Davies et al. 1998).

With respect to the delivery of food among the nestlings of a brood, it is assumed that parents should favour an even distribution of food only if young benefit equally from a unit of parental investment (Clutton-Brock 1991). Therefore, small nestlings in worse condition, which would derive greater fitness benefits per unit of parental investment, should be fed more than larger nestlings (Godfray 1991, 1995b; Kilner & Johnstone 1997). This point is singularly responsible for the frequently reported contradictory result that parents in some species in fact preferentially feed the larger nestlings, even when they beg at a lower rate (e.g. Cotton et al. 1999).

In agreement with the theoretical background described above, many studies have shown that hungry offspring beg more intensely and are more likely to be fed by their parents than satiated ones (reviewed in Kilner & Johnstone 1997; Budden & Wright 2001). However, in asynchronous broods, where theoretical models predict that parents should preferentially feed the young that are in poorer condition (the ones begging at higher intensity, usually the smaller nestlings, see above), the opposite is frequently observed: larger nestlings receive more feedings than smaller ones in spite of their lower begging rate compared to that of smaller nestlings (Rydén & Bengtsson 1980; Bengtsson & Rydén 1981, 1983; Fujioka 1985; Greig-Smith 1985; McGillivary & Levenson 1986; Hussell 1988; Martins & Wright 1993; Kilner 1995; Price et al. 1996; but for exceptions see Stamps et al. 1989; Krebs et al. 1999; E.A. Krebs this volume). First-hatched nestlings can outcompete smaller nestlings, provoking brood reduction and increasing their own probabilities of survival (Price & Ydenberg 1995; Slagsvold et al. 1995). Small siblings are fed less frequently which leads to reduced growth and higher mortality of last-hatched nestlings even when food is abundant (Slagsvold 1986; Magrath 1990; Stoleson & Beissinger 1995; Mock & Parker 1997). An extreme example is that provided by the hoopoe (*Upupa epops*). We have observed that before feeding small hungry nestlings that are begging quite aggressively, female parents first try to feed all of the bigger nestlings which are not begging by touching them several times with their beak (M. Martin-Vivaldi & M. Soler unpublished data).

There are many studies reporting this apparent contradiction to theoretical models (see references above), and most of them emphasize (frequently even in the title) that the study deals with an asynchronously hatching species.

THE NEW IDEA

As I have shown above, general predictions of current theoretical models are supported in species that adjust clutch size, but rarely in brood-reducing species. To explain these frequently reported contradictory results (see above), I suggest that the key question is the breeding strategy. The verbal argument is as follows. Some hatched nestlings starve in brood-reducers, whereas all hatched nestlings survive to fledge in species that directly adjust clutch size. This implies that parents using these two strategies do not follow the same rules when allocating food to nestlings. In species that adjust clutch size, parents tend to distribute resources evenly among their nestlings, preferentially feeding young that are in poorer condition (i.e. begging at higher intensity), and thus ensuring that all nestlings survive to fledging. On the other hand, parents of brood-reducers do not allocate food evenly among their offspring. In fact, since some young fledge in good condition while others starve, parents must be making active decisions and selectively feeding larger nestlings independently of begging intensity.

This does not imply that the communication system between parents and young does not work, since nestlings honestly advertise their need, as reported for species of both breeding strategies (see references above). The difference between previous models and that given here is with respect to the second link in the chain of events (nestling needs → begging signals → parent provisioning). Current models assume that parents provision young in relation to begging intensity, while my model states that this is true only in species that adjust clutch size. In species that brood reduce, parents base their decisions regarding food distribution not only on begging intensity, but also on other nestling characteristics usually related to their competitive ability (at least when differentiating between larger and smaller nestlings).

Parental investment theory, and therefore current thinking about begging, assumes that sometimes the nestling in worse condition may not represent the best investment for the parent per unit of care provided, because larger offspring represent a better absolute chance of survival to reproduction. This theoretical argument explains why larger nestlings are fed preferentially, but despite being correct this is not consistent with my idea. The theoretical argument starts off with the existence of a nestling in worse condition, but my idea explains why the nestling is in this condition. I suggest that nestlings are not fed according to their need, in spite of begging signals showing that physical condition is decreasing.

Thus, the most important difference between the two breeding strategies is that, contrary to expectation, in brood-reducers parents prefer to feed larger nestlings instead of hungrier ones. It is a key point that my model assumes that parents sometimes make active decisions regarding food allocation rather than feeding nestlings on the basis of begging intensity. Surely, in most cases of hatching asynchrony, parents decide on the allocation of food within the brood on the basis of nestling competition (Mock & Parker 1997). However, behavioural decisions made by parents when feeding nestlings have been emphasized in studies where exceptions to the general rules have been found (Rodríguez-Gironés et al. 1996; Lotem 1998; Cotton et al. 1999).

Of course, this new idea should be tested by performing a comparative study. Information is, however, needed from many different species on the number of nestlings starving per nest, the intensity of nestling begging and patterns of parental food delivery. Unfortunately, these data are frequently not available.

HOST SELECTION BY PARASITIC CUCKOOS IN RELATION TO BREEDING STRATEGY AND BEGGING INTENSITY

I will now deal with how the new idea described above can help to explain some aspects of the relationship between brood parasites and their hosts. With respect to host selection, several factors have been suggested to be important in the brood parasite's host choice, both at the species and the individual level (Payne 1977; Rothstein 1990). Here, I propose that the breeding strategy of the host and the begging intensity of host nestlings can influence host selection at the species level.

THE IMPORTANCE OF BREEDING STRATEGY IN HOST SPECIES - A MODEL

Considering that host species which adjust clutch size do not follow the same rules when distributing food within the brood as brood-reducing species (see above), brood parasites would be expected to preferentially choose hosts of only one strategy. However, another relevant point that should directly affect both host selection and the tactics of the brood-parasitic nestling, is the level of need of the parasitic nestling in relation to need of the host brood. This depends on both the relative size of the parasitic nestling with respect to that of the host nestlings (also see D.C. Dearborn & G. Lichtenstein this volume), and the feeding capacity of the host adults. The relative size of the parasite nestling compared to a host nestling may be considered as a good index of the level of need of the parasite in relation to that of the host brood. Therefore, both the breeding strategy of hosts and the relative need (relative body size) of the parasite nestling should affect the success of parasitic cuckoos (Figure 1). When the host parents tend to distribute food evenly, and the cuckoo nestling requires a quantity of food equivalent to that required by a whole brood of the host species, the only possibility of survival for the cuckoo nestling is to remain alone in the nest (Figure 1). This may be the reason why eviction behaviour has evolved in parasitic cuckoos, despite being a complex and costly adaptation (not only in terms of time and energy, but also in terms of the danger of the young cuckoo falling out of the nest during such effort; Wyllie 1981). If, however, host parents tend to allocate food preferentially to larger nestlings, it is less costly for parasitic cuckoos to compete for food with foster siblings (Figure 1). This is because parasitic nestlings are equipped with two adaptations that make them very effective in competing for food: they hatch earlier than the host nestlings as a consequence of a shorter incubation period (Payne 1977), and they have a higher begging rate (Soler et al. 1999).

The model enables us to make predictions about the fledging success of both hosts and parasites (Figure 1). Fledging success of host species of cuckoos that evict eggs or nestlings is 0% (Figure 1). Fledging success of host species of cuckoo that directly compete for food with host nestlings depends mainly on the size of the cuckoo nestling relative to the size of the host nestlings. This is because parents of brood-reducers preferentially feed larger nestlings (see references above). When the cuckoo nestling is of similar or larger size compared to a host nestling, the fledging success of the host species should be near 0%, because the cuckoo nestling outcompetes host nestlings for food (Figure 1). When host nestlings are larger than the parasitic nestling some of the host nestlings may survive. These predictions are supported by studies conducted on the great spotted cuckoo. Among parasitized magpies (*Pica pica*), only 0.6 host nestlings are fledged per nest, with most of the brood being outcompeted for food (Soler et al. 1996). However, when parasitizing the larger carrion crow (*Corvus corone*), 1.2 nestlings were fledged per brood; a value very similar to the 1.3 found in unparasitized nests (M. Soler, J.J. Soler, T. Pérez-Contreras & J.G. Martínez unpublished data).

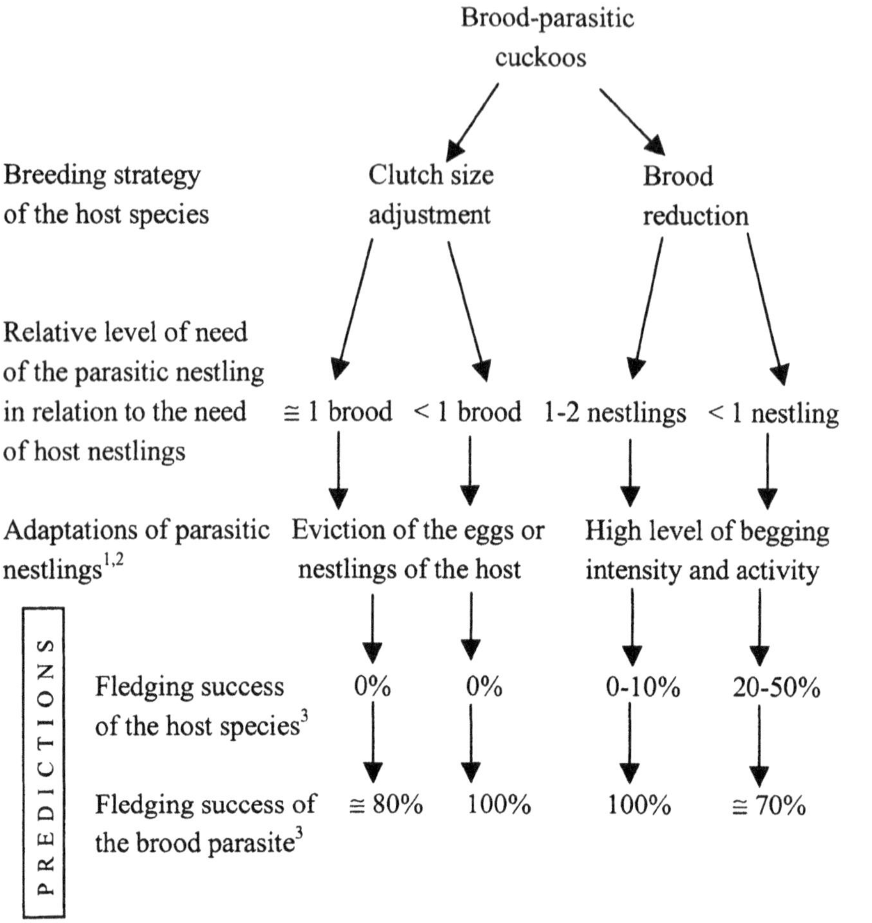

Figure 1. Model showing how the breeding strategy of the host species, and the relative level of need of the parasitic nestling in relation to the need of the host nestlings, can influence host selection and the adaptations developed by different species of brood-parasitic cuckoos. Some predictions are also specified. [1]In all cases a short incubation period will be advantageous for cuckoo nestlings. [2]A very high level of begging intensity should also be found in brood-parasitic cuckoo nestlings parasitizing host species that adjust clutch size, mainly when the relative level of need of the cuckoo nestling is approximately that of the entire host brood, because parasitic nestlings need their hosts to deliver food at maximum capacity. [3]Fledging success is the percentage of eggs hatched that result in nestlings fledged. Numbers here are approximate.

Fledging success of the parasite when the cuckoo nestling needs the same quantity of resources as does an entire host brood (i.e. when parasitizing small-sized hosts) should not reach 100% (Figure 1), because host parents of lower quality would be unable to raise a parasitic nestling that needs the same quantity of food as an entire host brood. This prediction is supported by the fact that cuckoo nestlings raised by the larger host species (great reed warblers, *Acrocephalus arundinaceus*) grow significantly faster and survive better than do those raised by the smaller one (reed warblers, *Acrocephalus scirpaceus*; Kleven et al. 1999).

If cuckoos parasitize brood-reducing species, and if the level of need of the parasite nestling is similar to that of one of the host nestlings in the brood, fledging success of the brood parasite should be about 100% (Figure 1). If, however, the cuckoo nestling is smaller than the host nestlings, fledging success of the parasite should not reach 100% (Figure 1). This is because some cuckoo nestlings will not hatch early enough to remain larger than the host nestlings, thereby losing any competitive advantage, which leads to starvation. These predictions are also supported by studies on the great spotted cuckoo parasitizing the magpie, where fledging success of the cuckoo is around 100% (Soler et al. 1996, 1998). When parasitizing the carrion crow, however, cuckoo fledging success is about 75% (M. Soler, J.J. Soler, T. Pérez-Contreras & J.G. Martínez unpublished data).

Finally, it could also be predicted that (1) if a cuckoo nestling adapted to parasitize species that adjust clutch size fails to remain alone in the nest, it will not survive, because it needs all the food that parents are able to carry to the nest. In addition, (2) a cuckoo nestling adapted to use a brood-reducing species as its host would not survive in the nest of a clutch-size adjusting species, because such foster parents would distribute food equally among nestlings independently of the begging activity of the larger cuckoo. I tried to test the first of these predictions by experimentally introducing cuckoo nestlings into three unparasitized nests of rufous-tail scrub robins (*Cercotrichas galactotes*), and by returning host nestlings to two nests where they had previously been evicted by cuckoos. These experiments were not successful, because cuckoo nestlings stopped begging and concentrated upon trying to evict the host nestlings (one of the cuckoos starved whilst attempting this costly task). Davies and Brooke (1988) found that, when a cuckoo nestling was experimentally placed within a brood of reed warblers, the cuckoo nestling received more food than any host nestling (4.5 feeds versus 3.5 feeds, recalculated from Davies & Brooke 1988), but the brood of reed warblers obtained more than 75% of the food provided by the parents. Considering that one cuckoo nestling needs much more than this, and about the same quantity of food as an entire brood of reed warblers (Brooke & Davies 1989), this result appears to support the prediction.

The second prediction was tested by experimentally introducing one great spotted cuckoo nestling into three black wheatear nests, each containing four host nestlings. Although cuckoo nestlings beg very aggressively at extremely high rates (Soler et al. 1999), and thereby usually outcompete their foster siblings in magpie nests (Soler et al. 1996, 1998), all the cuckoo nestlings in this experiment starved in black wheatear nests after three, four and four days, respectively. None of the black wheatear nestlings starved, although in all three nests they fledged with a lower mass than in control nests. The model presented here explains one long-standing question that remains an evolutionary enigma: why do the nestlings of some brood-parasitic cuckoo species evict host eggs or nestlings, while the parasitic nestlings of other species do not? This eviction behaviour is the most frequent strategy in parasitic cuckoo nestlings, although nestlings of some species have not developed this extremely specialized behaviour (M. Soler & J.J. Soler unpublished data).

THE IMPORTANCE OF BEGGING INTENSITY BY HOST NESTLINGS

Begging behaviour by brood-parasitic and host nestlings, and the problem of how parasitic nestlings deceive their foster parents, has recently received substantial attention from evolutionary biologists (Davies et al. 1998; Dearborn 1998; Lichtenstein & Sealy 1998; Kilner & Davies 1999; Kilner et al. 1999; D.C. Dearborn & G. Lichtenstein this volume). However, as far as I know, in the long list of factors suggested to be important in the brood parasite's host choice, nobody has suggested that the begging intensity of host nestlings is one of the important factors, and this is the idea that will now be addressed.

Comparative studies have shown that differences between species in begging calls may be considerable. For example, Briskie et al. (1994) reported a significant inverse correlation across species between the average degree of relatedness within broods and the amplitude of begging calls. However, one of the most important factors affecting differences among species in begging behaviour is the risk of predation. Begging is usually noisy and conspicuous and therefore could attract predators; this is the main reason why it is widely assumed that begging behaviour is costly for nestlings to produce (Redondo & Castro 1992; Haskell 1994; Leech & Leonard 1997; D.G. Haskell this volume). Considering this predation cost of begging, the auditory component of begging behaviour should vary greatly according to the vulnerability of nests (Haskell 1994, 1999). For example, nestlings of cavity-nesting birds should beg more intensely than do nestlings from species that build open nests (Redondo & Arias de Reyna 1988; Briskie et al. 1999; D.G. Haskell this volume).

Considering these differences among species, begging intensity and/or competitive ability of the nestlings in different host species should be an important factor influencing host choice by brood parasites, especially in those species where parasitic nestlings do not evict host nestlings. There is some evidence that this is the case for great spotted cuckoos when selecting their corvid hosts:

(1) The carrion crow is the secondary host of the great spotted cuckoo. Great spotted cuckoos appear to prefer to parasitize magpies because fledging success of the parasite is significantly higher in magpie nests, as a result of carrion crow nestlings being larger and of superior competitive ability compared to cuckoo nestlings (M. Soler, J.J. Soler, T. Pérez-Contreras & J.G. Martínez unpublished data). The larger carrion crow parents provide the cuckoo nestling with more food than it can eat, and there is thus always enough food for late-hatched host nestlings to grow larger and hence eventually outcompete the cuckoo nestling (M. Soler, J.J. Soler, T. Pérez-Contreras and J.G. Martínez unpublished data).

(2) Jackdaws are only sporadically parasitized (2.1%, n = 290, Soler 1990), and breeding success of the great spotted cuckoo parasitizing this host is very low. Only one of nine jackdaw nests (11.1%) which were parasitized fledged at least one cuckoo nestling (two of 11 eggs, M. Soler & J.J. Soler unpublished data).

(3) It was hypothesized that this low breeding success of great spotted cuckoos in jackdaw nests was the consequence of jackdaw nestlings begging at a higher level and competing more aggressively than magpie nestlings, and even more than cuckoo nestlings themselves. This may occur because jackdaws nest in holes and the risk of predation is lower than in open magpie nests. Therefore, contrary to the situation in magpie nests, jackdaw host nestlings should outcompete cuckoo nestlings.

(4) This hypothesis was tested in a series of cross-fostering experiments (M. Soler & J.J. Soler unpublished data). First, four jackdaw nests were experimentally parasitized with one cuckoo nestling, which was one or two days younger than the oldest host nestling. In all four cases the cuckoo nestling was outcompeted by the jackdaw nestlings. Second, eight magpie nests were parasitized with one jackdaw nestling of the same age or one day older than the oldest magpie nestling. All the jackdaw nestlings survived whilst the magpie nestlings survived in only three nests. That is, the experimental introduction of one jackdaw nestling into a magpie nest produced the same effect as brood parasitism by one cuckoo nestling. Third, five jackdaw nests were each parasitized with one magpie nestling, which was the same age or one day older than the oldest jackdaw nestling. All the magpie nestlings starved. These experiments clearly demonstrate that jackdaw nestlings with their aggressive begging behaviour are probably able

to outcompete great spotted cuckoo nestlings. In conclusion, begging behaviour and/or competitive ability of host nestlings should be an important factor influencing host choice by brood parasites.

FUTURE DIRECTIONS

This chapter covers areas that are not well studied within the same species, such as breeding strategy, begging intensity and factors affecting food delivery by parents. In fact, complete information is available for very few species. Thus, more empirical studies are needed in order to gather this type of information for as many species as possible. There is also a huge potential for experimental studies in this area. For example, my model predicts that the response of parents to a hungry nestling smaller than the rest of the brood will differ between clutch-size adjusters and brood-reducers. The former will feed it preferentially, whilst the latter will not. Host selection by brood parasites depends upon the breeding strategy of the host species, the relative level of need of parasitic nestlings in relation to host nestlings, as well as specific adaptations developed by particular species of brood-parasitic cuckoos. This approach therefore provides many possibilities for both descriptive and experimental studies in the future.

ACKNOWLEDGEMENTS

I am most grateful to Marty Leonard, Juan G. Martínez, Juan J. Soler and Jon Wright for their comments, which greatly improved this chapter. I am especially indebted to Juan J. Soler for his collaboration in obtaining part of the information presented in this chapter. Financial support was provided by the DGES PB97-1233-C02-02 research project.

REFERENCES

Bengtsson, H. & Rydén, O. 1981. Development of parent-young interaction in asynchronously hatched broods of altricial birds. *Zeitschrift für Tierpsychologie* **56**, 255-272.

Bengtsson, H. & Rydén, O. 1983. Parental feeding rate in relation to begging behaviour in asynchronously hatched broods of the great tit *Parus major*. *Behavioral Ecology and Sociobiology* **12**, 243-251.

Briskie, J.V., Naugler, C.Y. & Leech, S.M. 1994. Begging intensity of nestling birds varies with sibling relatedness. *Proceedings of the Royal Society of London, Series B* **258**, 73-78.

Briskie, J.V., Martin, P.R. & Martin, T.E. 1999. Nest predation and the evolution of nestling begging calls. *Proceedings of the Royal Society, Series B* **266**, 2153-2159.

Brooke, M. de L. & Davies, N.B. 1989. Provisioning of nestling cuckoos *Cuculus canorus* by reed warbler *Acrocephalus scirpaceus* hosts. *Ibis* **131**, 250-256.

Budden, A.E. & Wright, J. 2001. Begging in nestling birds. *Current Ornithology* **16**, 83-118.

Clutton-Brock, T.H. 1991. *The Evolution of Parental Care.* Princeton: Princeton University Press.

Cotton, P.A., Wright, J. & Kacelnik, A. 1999. Chick begging strategies in relation to brood hierarchies and hatching asynchrony. *American Naturalist* **153**, 412-420.

Davies, N.B. & Brooke, M. de L. 1988. Cuckoos versus reed warblers: adaptations and counteradaptations. *Animal Behaviour* **36**, 262-284.

Davies, N.B., Kilner, R.M. & Noble, D.G. 1998. Nestling cuckoos, *Cuculus canorus*, exploit hosts with begging calls that mimic a brood. *Proceedings of the Royal Society of London, Series B* **265**, 673-678.

Dearborn, D.C. 1998. Begging behavior and food acquisition by brown-headed cowbird nestlings. *Behavioral Ecology and Sociobiology* **43**, 259-270.

Fujioka, M. 1985. Sibling competition and siblicide in asynchronously-hatching broods of the cattle egret *Bubulcus ibis. Animal Behaviour* **33**, 1228-1242.

Godfray, H.C.J. 1991. Signalling of need by offspring to their parents. *Nature* **352**, 328-330.

Godfray H.C.J. 1995a. Evolutionary theory of parent-offspring conflict. *Nature* **376**, 133-138.

Godfray H.C.J. 1995b. Signalling of need between parents and young: parent-offspring conflict and sibling rivalry. *American Naturalist* **146**, 1-24.

Greig-Smith, P. 1985. Weight differences, brood reduction and sibling competition among nestling stonechats *Saxicola torquata* (Aves: Turdidae). *Journal of Zoology* **205**, 453-465.

Harper, A.B. 1986. The evolution of begging: sibling competition and parent-offspring conflict. *American Naturalist* **128**, 99-114.

Haskell, D. 1994. Experimental evidence that nestling begging behaviour incurs a cost due to nest predation. *Proceedings of the Royal Society of London, Series B* **257**, 161-164.

Haskell, D. 1999. The effect of predation on begging-call evolution in nestling wood warblers. *Animal Behaviour* **57**, 893-901.

Hussell, D.L.T. 1988. Supply and demand in tree swallow broods: a model of parent-offspring food-provisioning interactions in birds. *American Naturalist* **131**, 175-202.

Kilner, R.M. 1995. When do canary parents respond to nestling signals of need? *Proceedings of the Royal Society of London, Series B* **260**, 343-348.

Kilner, R.M. & Davies, N.B. 1999. How selfish is a cuckoo chick? *Animal Behaviour* **58**, 797-808.

Kilner, R.M. & Johnstone, R.A. 1997. Begging the question: are offspring solicitation behaviours signals of need? *Trends in Ecology and Evolution* **12**, 11-15.

Kilner, R.M., Noble, D.G. & Davies, N.B. 1999. Signals of need in parent-offspring communication and their exploitation by the common cuckoo. *Nature* **397**, 667-672.

Kleven, O., Moksnes, A., Røskaft, E. & Honza, M. 1999. Host species affects the growth rate of cuckoo (*Cuculus canorus*) nestlings. *Behavioral Ecology and Sociobiology* **47**, 41-46.

Krebs, E.A., Cunningham, R.B. & Donnelly, C.F. 1999. Complex patterns of food allocation in asynchronously hatching broods of crimson rosellas. *Animal Behaviour* **57**, 753-763.

Krebs, J.R. & Kacelnik, A. 1991. Decision-making. In: *Behavioural Ecology: An Evolutionary Approach.* (Ed. by J.R. Krebs & N.B. Davies). Oxford: Blackwell Scientific Publications.

Lack, D. 1947. The significance of clutch size. *Ibis* **89**, 302-352.

Leech, S.M. & Leonard, M.L. 1997. Begging and the risk of predation in nestling birds. *Behavioral Ecology* **8**, 644-646.

Lichtenstein, G. & Sealy, S.G. 1998. Nestling competition, rather than supernormal stimulus, explains the success of parasitic brown-headed cowbird nestlings in yellow warbler nests. *Proceedings of the Royal Society of London, Series B* **265**, 249-254.

Lotem, A. 1998. Brood reduction and begging behaviour in the swift *Apus apus*; no evidence that large nestlings restrict parental choice. *Ibis* **140**, 507-511.

Magrath, R.D. 1990. Hatching asynchrony in altricial birds. *Biological Reviews* **65**, 587-622.

Martins, T.L.F. & Wright, J. 1993. Brood reduction in response to manipulated brood sizes in the common swift (*Apus apus*). *Behavioral Ecology and Sociobiology* **32**, 61-70.

Maynard Smith, J. 1978. Optimization theory in evolution. *Annual Review of Ecology and Systematics* **9**, 31-36.

McGillivray, W.B. & Levenson, H. 1986. Distribution of food within broods of barn swallows. *Wilson Bulletin* **98**, 286-291.

Mock, D.W. & Parker, G.A. 1997. *The Evolution of Sibling Rivalry*. Oxford: Oxford University Press.

Muller, R.E. & Smith, D.G. 1978. Parent-offspring interactions in zebra finches. *The Auk* **95**, 485-495.

Parker, G.A. 1985. Models of parent-offspring conflict. V. Effects of the behaviour of the two parents. *Animal Behaviour* **33**, 519-533.

Parker, G.A. & Macnair, M.R. 1979. Models of parent-offspring conflict. Monogamy. *Animal Behaviour* **26**, 97-110.

Payne, R.B. 1977. The ecology of brood parasitism in birds. *Annual Review of Ecology and Systematics* **8**, 1-28.

Price, K. & Ydenberg, R. 1995. Begging and provisioning in broods of asynchronously-hatched yellow-headed blackbird nestlings. *Behavioral Ecology and Sociobiology* **37**, 201-208.

Price, K., Harvey, H. & Ydenberg, R. 1996. Begging tactics of nestling yellow-headed blackbirds, *Xanthocephalus xanthocephalus*, in relation to need. *Animal Behaviour* **51**, 421-435.

Redondo, T. & Arias de Reyna, L. 1988. Locatability of begging calls in nestling altricial birds. *Animal Behaviour* **36**, 653-661.

Redondo, T. & Castro, F. 1992. The increase in risk of predation with begging activity of magpies *Pica pica*. *Ibis* **134**, 180-187.

Rodríguez-Gironés, M.A., Cotton, P.A. & Kacelnik, A. 1996. The evolution of begging: signaling and sibling competition. *Proceedings of the National Academy of Sciences USA* **93**, 14637-14641.

Rothstein, S.I. 1990. A model system for coevolution: avian brood parasitism. *Annual Review of Ecology and Systematics* **21**, 481-508.

Rydén, O. & Bengtsson, H. 1980. Differential begging and locomotory behaviour by early and late hatched nestlings affecting the distribution of food in asynchronously hatched broods of altricial birds. *Zeitschrift für Tierpsychologie* **53**, 209-224.

Slagsvold, T. 1986. Asynchronous versus synchronous hatching in birds: experiments with the pied flycatcher. *Journal of Animal Ecology* **55**, 1115-1134.

Slagsvold, T., Amundsen, T. & Dale, S. 1995. Costs and benefits of hatching asynchrony in blue tits *Parus caeruleus*. *Journal of Animal Ecology* **64**, 563-578.

Soler, J.J., Soler, M., Møller, A.P. & Martínez, J.G. 1995. Does the great spotted cuckoo choose magpie hosts according to their parenting ability? *Behavioral Ecology and Sociobiology* **36**, 201-206.

Soler, M. 1990. Relationships between the great spotted cuckoo *Clamator glandarius* and its magpie host in a recently colonized area. *Ornis Scandinavica* **21**, 212-223.

Soler, M. & Soler, J.J. 1996. Effects of experimental food provisioning on reproduction in the Jackdaw *Corvus monedula*, a semi-colonial species. *Ibis* **138**, 377-383.

Soler, M., Moreno, J., Møller, A.P., Lindén, M. & Soler, J.J. 1995. Determinants of reproductive success in a Mediterranean multi-brooded passerine: the black wheatear *Oenanthe leucura*. *Journal für Ornithologie* **136**, 17-27.

Soler, M. & Soler, J.J. 1999. The cuckoo chick tricks their reed warbler foster parents, but what about other host species? *Trends in Ecology and Evolution* **14**, 294-295.

Soler, M., Martínez, J.G. & Soler, J.J. 1996. Effects of brood parasitism by the great spotted cuckoo on the breeding success of the magpie host: an experimental study. *Ardeola* **43**, 87-96.

Soler, M., Soler, J.J. & Martínez, J.G. 1998. Duration of sympatry and coevolution between the great spotted cuckoo (*Clamator glandarius*) and its primary host, the magpie (*Pica pica*). In: *Parasitic Birds and Their Hosts: Studies in Coevolution* (Ed by S.I. Rothstein & S.K. Robinson). New York: Oxford University Press.

Soler, M., Soler, J.J., Martínez, J.G. & Moreno, J. 1999. Begging behaviour and its energetic cost in great spotted cuckoo and magpie host chicks. *Canadian Journal of Zoology* **77**, 1794-1800.

Stamps, J., Clark, A.B., Arrowood, P. & Kus, B. 1985. Parent-offspring conflict in budgerigars. *Behaviour* **94**, 1-40.

Stamps, J.A., Clark, A., Arrowood, P. & Kus, B. 1989. Begging behaviour in budgerigars. *Ethology* **81**, 177-192.

Stenning, M.J. 1996. Hatching asynchrony, brood reduction and other rapidly reproducing hypotheses. *Trends in Ecology and Evolution* **11**, 243-246.

Stoleson, S.H. & Beissinger, S.R. 1995. Hatching asynchrony and the onset of incubation in birds, revisited: when is the critical period? *Current Ornithology* **12**, 191-270.

Wyllie, I. 1981. *The Cuckoo*. London: Batsford.

22.

BEGGING FOR PARENTAL CARE FROM ANOTHER SPECIES: SPECIALIZATION AND GENERALIZATION IN BROOD-PARASITIC FINCHES

Robert B. Payne & Laura L. Payne
Museum of Zoology and Department of Biology, University of Michigan, Ann Arbor, MI 48109-1079, USA (rbpayne@umich.edu)

ABSTRACT

African indigobirds (*Vidua* species) are species-specific brood parasites of estrildid finches. Although the mouth patterns of nestlings mimic their host nestlings, the begging calls of young indigobirds are not host-specific, and in only some species do they resemble begging calls of the host. Adult male indigobirds mimic calls and songs of their host species. Their song incorporates two kinds of begging call, an innate call like that used by nestling indigobirds, and a second learned one when males imprint and then mimic the foster species' begging calls in male song. We recorded young and adult indigobirds in the field, and the begging calls of young and adult song mimics reared under alternative foster species. The innate begging calls in all indigobird species matched the begging calls of only certain firefinch (*Lagonosticta*) host species, even in indigobirds that normally parasitized other hosts. This innate call is used by nestlings to gain parental care. Both kinds of begging calls are used by adults in mate choice. Host-specific begging call in mimicry songs of adult male indigobirds would allow females to assess whether males were reared by their own foster species.

INTRODUCTION

Brood-parasitic birds need the parental care of other bird species to rear their young. In some, the generalist brood parasites, the nestlings give non-

J. Wright and M.L. Leonard (eds.), The Evolution of Begging, 429–449.

specific begging signals to gain the parental care of several host species and their begging calls do not closely resemble the calls of their host young (Payne 1997, 1998a; Davies 2000). Other brood parasites are host specialists, and their begging signals may mimic the specific begging calls of their host species (Payne & Payne 1998). The most species-specific brood parasites are the African finches; the indigobirds and whydahs of the genus *Vidua* (Nicolai 1964; Payne 1997). All 19 species are brood-parasitic and do not rear their own young. Most of these finch species are host-specific; that is, each parasitizes a single species of host, a finch in the family Estrildidae. Most parasitic finches have nestlings with mouth patterns and colours that mimic the mouths of their host species' nestlings. In addition, the parasitic finches have been reported to have begging calls like those of their host species' young. In most cases, however, the begging calls were not recorded from the young brood parasites while they were in the care of their foster parents. Rather, the begging calls of parasitic finches were recorded in the songs of adult males, which mimic both the songs of their host species and the calls of the adult hosts and their young (Nicolai 1964, 1973).

In experiments on the development of mimicry songs in indigobirds, we have determined that these begging calls, as well as the other mimicry calls and songs, are learned as a result of imprinting of the young indigobird upon its foster parent (Payne et al. 1998). A young indigobird that is reared by its normal host species develops the calls and songs of the species that reared it, whereas a young indigobird reared by an experimental foster species develops the calls and songs of the experimental foster species. In addition, in at least one species of indigobird, the begging calls of the young bird itself are different from the begging calls of the young of its host species (Payne et al. 1998). Because these calls in the songs of the adult males are learned, the calls in their songs are not necessarily the same begging calls that a young brood parasite uses to gain care from its foster parents.

We suspect that the begging calls of brood-parasitic finches involve a more complex behaviour than was previously thought. First, these finches are usually species-specific in their brood parasitism, yet they have shifted from time to time to parasitize new host species. The evidence for host switching includes field observations of male indigobirds that mimic the songs of a novel host (Payne et al. in press), and molecular genetic studies that show multiple colonizations of a host species (Klein & Payne 1998). If the begging calls are specialized to mimic one host, then the specialized behaviour which evolved to elicit parental care from the old host species might put a specialist at a disadvantage, relative to a generalist, when opportunities arise to parasitize a new host species. Yet, if the begging calls do not elicit parental care from the host species, the young indigobird will not be reared. In addition, the begging calls are used by these birds during two different stages in their lives, once when they are dependent young, and later when they sing and mate. The begging calls that appear in the songs of the adult males may include both the calls that the young bird has learned as a result of imprinting on its foster parents, as well as its own begging calls. This idea of two origins of begging calls in the male song repertoire was first suggested by Nicolai (1973). Here, we develop this idea and trace both kinds of calls in a comparison of different components of indigobird song.

In this chapter, we review the observations and experiments on vocal mimicry in brood-parasitic finches, and describe and compare the begging calls of young indigobirds and their host species to determine the specificity of matching in parasite and host. We compare these begging calls to calls given by male indigobirds in adult song. Finally, we review the importance of vocal learning and imprinting in the development of calls, and the evolutionary significance of vocal signals in both parental care and mate choice in these brood parasites.

Begging Behaviour and Parental Care in Finches

Estrildid nestlings and fledglings crouch, hold the mouth open and twist and wave the head from side to side. The parent then inserts its bill into the mouth of the young and regurgitates seeds into the crop. This unique begging behaviour is thought to restrict successful parental care of *Vidua* to estrildid finches, as these brood parasites beg and are fed in the same way.

In addition, the specificity of the brood parasite - host association is the result of visual signals of the begging young. Nestling estrildids have species-characteristic markings and colours in their mouths, which they display when they beg for parental care. In parasitized broods, *Vidua* nestlings are reared together with the host nestlings, and most species mimic the mouth colours of their host nestlings (Nicolai 1964, 1969, 1974, 1989; Payne 1973a, 1982, 1997, 1998a,b). Although their begging behaviour is similar, estrildid species differ in nestling mouth colour, skin colour, colour and density of the natal down, size and details of posture and begging calls (Kunkel 1959; Immelmann 1962, 1982; Nicolai 1964, 1967, 1974; Goodwin 1982) and all these traits may affect parental care.

Nestling Mimicry

The brood-parasitic indigobirds are specialized to match their host species in the mouth coloration of their young. The mouth colours of nestling village indigobirds (*Vidua chalybeata*) are like the mouth colours of nestlings of their host species, red-billed firefinches (*Lagonosticta senegala*), with a gape marked by a dorsal and ventral pair of swollen white papillae with a blue base, and a yellow palate with black spots (Nicolai 1964; Payne 1973a). Nestling purple indigobirds (*V. purpurascens*) are like nestlings of their host species, Jameson's firefinch (*L. rhodopareia*), with a broad pinkish-violet oral flange and narrow blue band between two small white papillae and a pink palate with black spots. Another pattern occurs in the bar-breasted firefinch indigobird (*V. wilsoni*); like young of their host species, the bar-breasted firefinch (*L. rufopicta*), the gape has a swollen oral flange, white to light bluish in colour and a pink palate with black spots (Payne 1982, 1996). In contrast to these species, in the goldbreast indigobird (*V. raricola*) and the quail-finch indigobird (*V. nigeriae*) the mouth pattern does not match the nestlings of their host species (Payne & Payne 1994), and in some other indigobirds the mouth pattern is unknown.

In contrast to their mouth colours, the begging calls have been described in nestling indigobirds of only one species, the village indigobird, and these begging calls differ from those of its host species (Payne et al. 1998).

Adult Song Mimicry

Nicolai (1964) proposed that young whydahs form an association with their host species through behavioural imprinting. This hypothesis has guided our experimental work with the indigobirds. We find that adult males mimic songs and calls of their foster species, and this behaviour is learned by the young during their period of foster parental care, even when the foster species is not the natural host. When the adult male sings, he advertises that he was reared by this foster species. Females are attracted to songs of males that mimic the same species that reared the females. The importance of mimicry songs is that females mate with males that were reared by the same host species. That is, their offspring will have the behaviours and the mouth colours and patterns that elicit parental care from the breeding host. Mate choice by female indigobirds is determined by mimicry songs of the males, and these songs have been elaborated through a process of sexual selection (Nicolai 1964; Payne 1973a,b, 1983, 1990, 1997; Payne et al. 1998, 2000a).

Vocal mimicry in adult males includes calls like the begging calls of the host young. Begging calls that are specific to the host species may be an asset in mate choice. Adult male song also includes calls like the begging call of the young indigobird (Payne et al. 1998). Not only does song mimicry advertise the male's early experience with the foster species' songs, it also provides us, as well as the females, with copies of the indigobird's own begging calls and his foster species' begging calls.

Begging for Parental Care and Begging for Sex

We present a model to account for the function and development of begging calls in the indigobirds. First, we propose that there are two begging calls; one used by the young to attract parental care, and the other used by adult males to attract a mate. Second, we propose that begging calls of a young nestling are innate and develop normally whether or not it experiences parental care from its normal host species or hears the begging calls of its normal host species. We also suggest that the mimicry songs of an adult male include this innate call, and also include begging calls that he copies from other male indigobirds or from the young of his foster species.

We used the variation in begging calls in our field observations and imprinting experiments to test two hypotheses; one concerned with the function of begging calls in parental care, and the second with their function in mate choice. The first hypothesis leads to a prediction of generalized begging calls in nestlings, while the second leads to a prediction of host-specific begging calls in adult males in all the indigobird species.

Hypothesis 1. Begging calls attract foster-parental care. If the begging calls of host nestlings are the same across species, then we would not expect to see covariation in the calls of their brood parasites. If, however, begging calls differ among the host species, then we expect the following: if nestlings are fed only when they have begging signals of the host young, we predict the begging calls of brood-parasitic young to match those of the host young. Because indigobird lineages occasionally have shifted between host species in natural conditions, we predict that the begging calls of their young are not absolutely host-specific, but are general enough to elicit parental care in more than one host species. We also predict that begging calls in young indigobirds match the host species that indigobirds have been associated with for a long time, but do not match more recent host species.

Hypothesis 2. Mimicry begging calls attract a mate. Sexual selection theory predicts that a male with more potent sexual signals will be more successful in attracting a mate than a male with fewer such signals (Payne 1973b, 1983; Andersson 1994). Indigobird males learn the songs of their host species through imprinting on their host species, and then give these songs as adults, while females are attracted to males that mimic the same kind of birds that reared them. The hypothesis predicts that a male includes mimicry of host nestling begging calls in his songs. It also predicts that a male has more than one kind of begging call in his songs, both the generalized parasite begging call that he had when he was young and a specific begging call that he learns, because together these are cues to the female that could affect her choice of a mate.

Sources of Information: What We Did and How We Did It

We compared the begging calls of estrildid host young with begging calls of indigobird young and calls in the songs of adult male indigobirds. We observed the estrildids and the ten species of indigobirds and recorded their calls and songs in the field. For indigobirds we recorded between 400 and 60,000 songs for each species (Payne 1985, 1996, 1998a,b). For estrildid finches we recorded young in the field, and we bred several species in captivity to record the begging calls given by nestlings and fledglings as their parents approached and fed them. We also recorded the begging calls of young village indigobirds that we bred and had reared by their normal host, the red-billed firefinch, as well as by other experimental foster species.

We did not find the young of most species of indigobirds in the field, so we could not record their begging calls directly. Instead, we used the observation that adult male indigobirds incorporate learned begging calls and calls like those they gave as begging young into their song repertoire (Payne et al. 1998), to help us distinguish between learned and innate begging calls in the other indigobird species.

To record songs and calls we used a Uher L tape recorder and Uher M514-517 microphone, or a Sony TC-D5M recorder and Sennheiser ME40 microphone, in a parabolic reflector or placed near the nest. For each bird we printed and examined all songs and calls with a Kay Elemetrics 'Vibralyzer' 7029A, or examined them on screen with a Kay Elemetrics DSP-5500 Sonagraph or used a PAR-4512 real-time spectrum analyser to print continuous 35-mm film audiospectrograms (Payne 1985). For songs and calls with fundamentals above 8 kHz, we re-examined the vocalizations on the DSP-5500 using a 0-16 kHz range, and a Kay-5509 printer.

We used the following criteria to estimate the developmental sources of calls in young and adult male indigobirds. First, we compared begging calls of nestlings that lacked experience with young of their normal host species with those of nestlings that had the experience. Second, we compared the calls in songs of adult males with the begging calls of young of their normal host species. If the adult calls are the same, then they could have been retained by the young indigobird itself, or learned from an adult male indigobird or the host young. On the other hand, when a call occurs in adult song and differs from the begging calls of young of the song-mimicked species, then the call corresponds to the begging call the adult had as a young bird. Our criterion of whether a call in an adult male indigobird was like the call in the host young was a visual match-to-sample with audiospectrograms (as in an earlier study of song mimicry, Payne et al. 2000b), using features of shape, frequency and time, and allowing for the variation that we observed when we recorded begging calls in the young. In no case did a male indigobird have calls like two or more host species, where the hosts had different calls (some host species have very similar begging calls, as in other indigobird species and their firefinch hosts). For figures in this chapter, we used the clearest examples for hosts and the most similar examples for brood parasites. Where a call matched the known begging call of the host young for species with no recorded examples of begging calls, we assumed similar calls to those of their closest relatives. Where calls in adult song matched begging calls in young village indigobirds, we recognized them as begging calls in the indigobird.

Begging Calls of Indigobirds and Their Host Species

First, we present the results of observations and experiments on indigobird species for which we recorded begging calls of the young for both the parasite and its host. Next, we compare the begging calls of the firefinch host species with begging calls that are similar across species. Then, we compare the begging calls of the host species with begging calls that are unique for their species, first for certain firefinch species and then for other estrildid species that are associated with a brood-parasitic indigobird. For these later cases we have no begging calls recorded from the young parasites, but we were able to distinguish the begging calls in songs of an adult male indigobird.

Village Indigobirds, Their Firefinch Host and Other Normal and Experimental Foster Species

Begging calls of young red-billed firefinches are complex. One trace in the note rises to peak at 5.5 kHz; another trace rises and meets the first. The note repeats at 5-12 times per second, and more rapidly as the parents approach and feed the young. Older nestlings and fledglings give a second begging call; a two-part whistle, alternated with the first call (Figure 1a,b). Young firefinches give whistle calls when the parents are away from the group, and short calls when they are nearby, with fledglings giving both. As young become independent of their parents, the double note changes into the single trace of the adult contact call, a single whistle ‘pea’ (Payne 1973a).

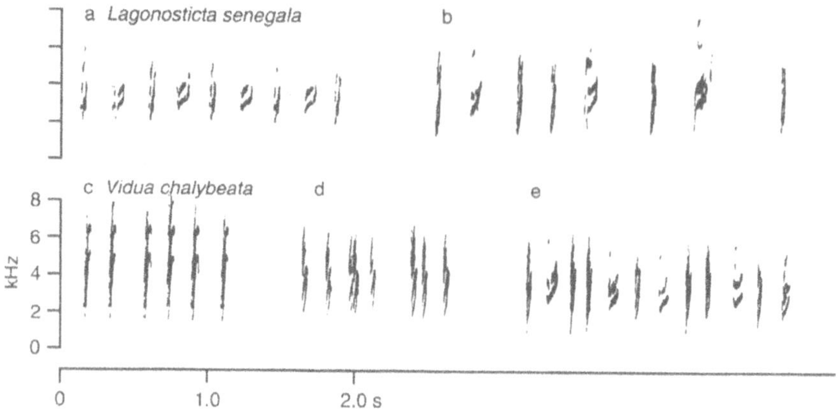

Figure 1. (a) Begging calls of a fledgling red-billed firefinch) fed by its parents (captive); (b) begging calls of a fledgling firefinch fed by an adult (Maun, Botswana); (c) begging calls of a fledgling village indigobird) in a family group of firefinches; (d-e) adult male village indigobird mimicry of fledged begging firefinches (c-e, Lochinvar National Park, Zambia).

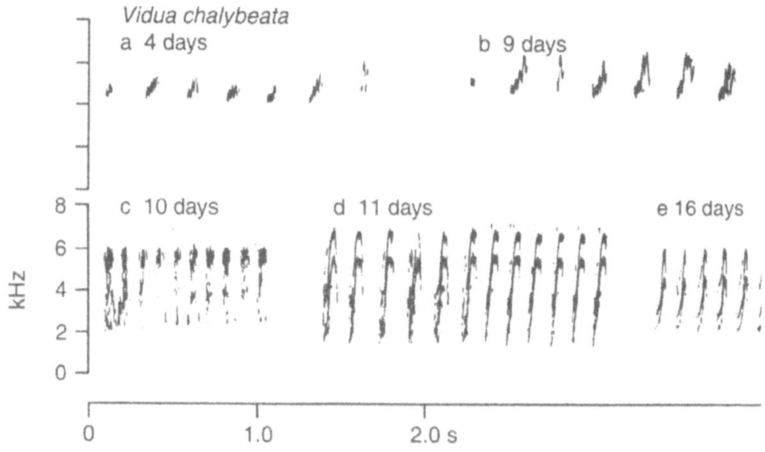

Figure 2. Begging calls of nestling village indigobirds (*V. chalybeata*) reared by alternative foster species: (a), (b) and (d) in nest of Jameson's firefinch at 4, 9 and 11 days after hatching; (c) begging calls at 10 days, in nest of brown firefinch (Kazungula Island, Zambezi River); and (e) begging calls at 16 days, in nest of goldbreast.

Begging calls of young village indigobirds are simple repeated notes that peak at about 5 kHz with an overtone at about 5.8 kHz (Figure 1c,d). The call is heard on hatching day. By day 4 it is a series of rising notes 4.5-6 kHz (Figure 2a). By day 9 the notes rise then fall (Figure 2b) and by day 11 the calls have the same form as in a fledged bird (Figure 2c,d).

The species differ by day 8, when an indigobird repeats a single note and a firefinch alternates whistled and rising notes. After fledging, the firefinch whistled call becomes longer, while the indigobird repeats the single note through its period of parental care. The same calls appear in songs of adult male indigobirds (Payne et al. 1998), as do calls like the calls of fledged firefinches (Figure 1e).

Development of begging calls in a young indigobird is independent of whether it is reared with a firefinch nestmate. In a mixed-species brood in Nigeria, the indigobirds gave regular short calls and the two firefinches gave irregular short and long calls. Begging calls in broods where an indigobird is the only young (Figure 2c) and in mixed-species broods (Figure 1c) are the same, so the presence of a young firefinch does not affect the form of the begging calls. Begging calls are the same in an indigobird reared by an experimental foster species, Bengalese finch (*Lonchura striata*), even when reared together with a young Bengalese (Payne et al. 1998). In other cases, young indigobirds reared by Jameson's firefinch, brown firefinch (*L. nitidula*) and goldbreast (*Amandava subflava*) all had the same simple begging calls (Figure 2a-e) as young reared by their normal host, the red-billed firefinch (Figure 1c; Payne et al. 1998).

Other Indigobird Species and Their Firefinch Hosts

Begging calls of bar-breasted firefinches have a peaked note with an overtone and a drop in pitch (Figure 3a,b). Calls in mimicry songs of their brood parasite, the bar-breasted firefinch indigobird, are similar (Figure 3c). A fledged bar-breasted firefinch indigobird attended by an adult bar-breasted firefinch at Assop in Nigeria, also gave begging calls like the host young in the same brood. The southern African allospecies of this firefinch is the brown firefinch, whose song is the same as the bar-breasted firefinch (Payne 1982). In a population on the Zambezi River from Kazungula to Victoria Falls, southern African village indigobird mimics of brown firefinches have songs with calls like begging calls of bar-breasted firefinches (Payne et al. in press), with long whistles like begging red-billed firefinches, but with notes slurring downward (Figure 3d).

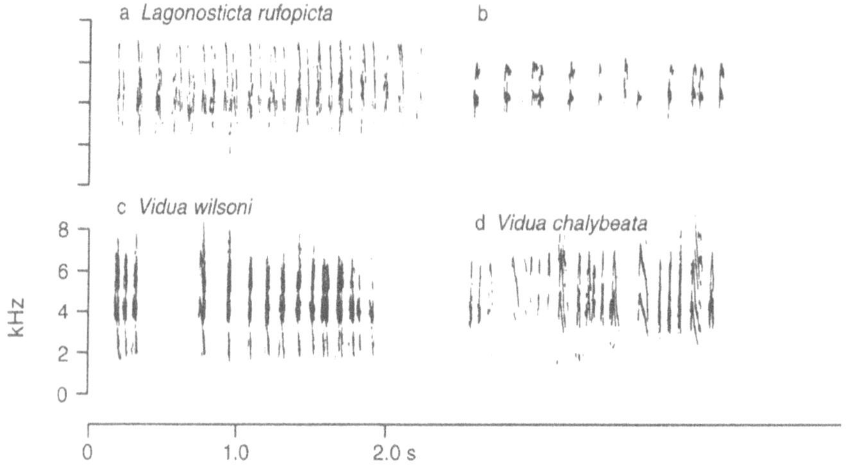

Figure 3. (a) Begging calls of a brood with two nestling bar-breasted firefinches and two nestling bar-breasted firefinch indigobirds (Bukuru, Nigeria); (b) begging calls of a fledged bar-breasted firefinch with an adult that fed it (Meng, Cameroon); (c, d) begging calls in song of adult male indigobirds, (c) bar-breasted firefinch indigobird (song mimics bar-breasted firefinch, Vom, Nigeria); and (d) village indigobird (song mimics brown firefinch, Kazungula, Zambezi River).

Begging calls of other firefinch species differ from those of the red-billed and bar-breasted firefinch. Young Jameson's, African (*L. rubricata*) and Mali (*L. virata*) firefinches have short notes in which an element sweeps upward in pitch to peak at 5.8 kHz then drops in pitch; these notes are repeated at 14 per second (Figure 4a-d). The structure of the begging call varies with age (Figure 4a,b), hunger and the approach of the parent (not shown). Although begging calls of young indigobirds under the care of these firefinches were not recorded, the corresponding calls in the song of the purple indigobird, dusky indigobird (*V. funerea*) and Cameroon indigobird (*V. camerunensis*) that mimic these firefinches' songs are like the begging calls of their firefinch hosts (compare Figures 4a,b and 5a; Figures 4c and 5b; Figures 4d and Figures 4g and 5g). In addition, the begging calls in the song of the different indigobird mimics and species all were like each other (Figure 5a-i) and these were like the begging calls of young village indigobirds (Figure 2). Calls in the song of the Jos Plateau indigobird (*V. maryae*; not shown, see Payne 1998b) that mimic songs of rock firefinches (*L. sanguinodorsalis*) are the same as in these other indigobirds. All these indigobirds have innate begging calls like the young of the firefinch species complex that includes Jameson's firefinch, African firefinch and Mali firefinch.

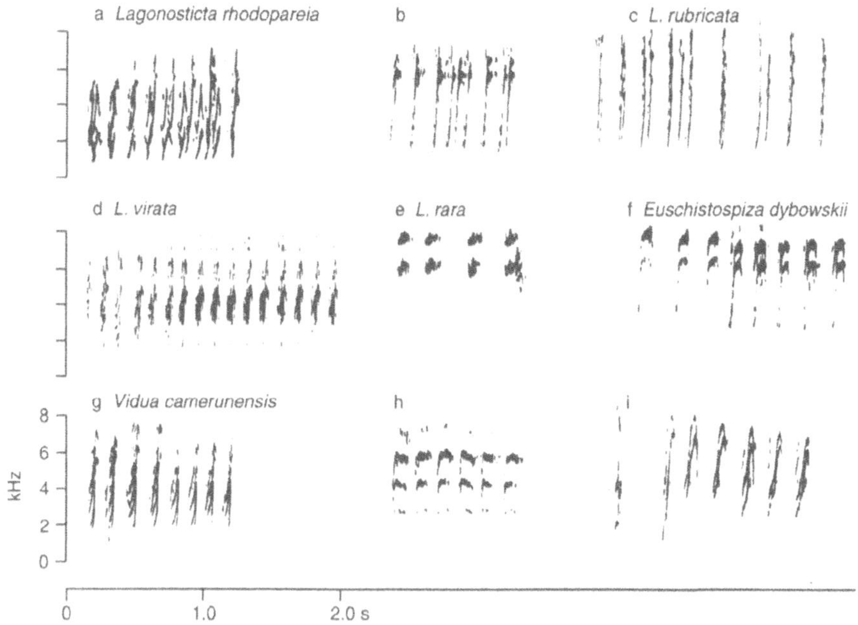

Figure 4. (a) Begging calls of two nestling Jameson's firefinches at 12 days of age, captive; and (b) calls of the same birds at fledging. Begging calls of captive host fledglings: (c) African firefinch; (d) Mali firefinch; (e) black-bellied firefinch; and (f) Dybowski's twinspot. Begging calls in song of adult male Cameroon indigobirds (g-i): (g) song mimics Mali firefinch (Tienfala, Mali); (h) song mimics black-bellied firefinch (Banyo, Cameroon); and (i) song mimics Dybowski's twinspot, captive.

Certain firefinches have different begging calls. Black-bellied firefinch (*L. rara*) begging calls are notes with an overtone (Figure 4e). When the parent approaches and feeds the young, the notes are shorter, higher in pitch and have a different shape (not shown). Some calls in the song of adult Cameroon indigobirds that mimic black-bellied firefinch songs are like the begging calls of this young firefinch (Figure 4g). Other calls in the song (Figure 5h) are like the begging calls of young village indigobirds (Figure 2). Some calls in the song of barka indigobirds (*V. larvaticola*) that mimic the songs of its host species, the black-faced firefinch (*L. larvata*; Figure 5d) are like the begging calls of young village indigobirds (Figures 1c, 2), as are calls in the song of adult village indigobirds that mimic the song of brown firefinches (Figure 5e). We have not recorded the begging calls of young black-faced firefinches or brown firefinches.

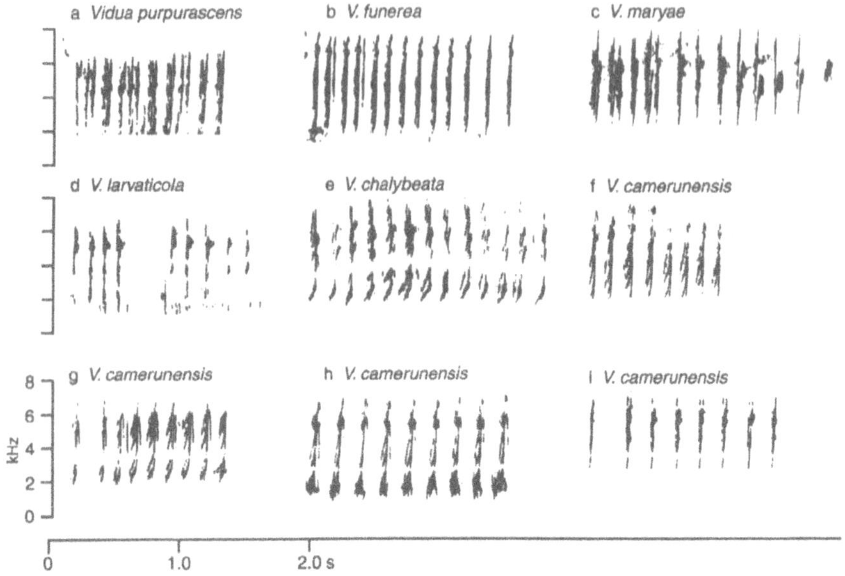

Figure 5. Begging calls in song of adult male indigobirds: (a) purple indigobird (song mimics Jameson's firefinch, Lochinvar, Zambia); (b) dusky indigobird (song mimics African firefinch, Tzaneen, Transvaal); (c) Jos Plateau indigobird (song mimics rock firefinch, Panshanu, Nigeria); (d) barka indigobird (song mimics black-faced firefinch, captive); (e) village indigobird (song mimics brown firefinch, Mosi-oa-Tunya, Zambia); (f-i) Cameroon indigobirds: (f) song mimics Mali firefinch (Forêt de Tienfala, Mali); (g) song mimics African firefinch (Tibati, Cameroon); (h) song mimics black-bellied firefinch (Lélouma, Guinea); and (i) song mimics Dybowski's twinspot (Dalaba, Guinea).

Indigobirds and Non-Firefinch Hosts

Begging calls of Peters's twinspot (*Hypargos niveoguttatus*) are high-pitched notes that drop in pitch and have peak amplitudes at 7.6 and 8.1 kHz (Figure 6a,b). In contrast, songs of Peters's twinspot indigobird (*V. codringtoni*) include upslurred notes (Figure 6c) like the begging calls of African firefinches, Jameson's firefinches and Mali firefinches (Figure 4a-d) and like phrases in the mimicry songs of other indigobirds (Figures 4h,i and 5a-h). Peters's twinspot indigobird mimicry songs (Payne et al. 1992) have calls like begging calls of young firefinches and like young village indigobirds, with peak amplitudes well below 7 kHz. Male Peters's twinspot indigobirds also give high-pitched calls at 8 kHz, like begging calls of young twinspots with a double peak amplitude at 7.6 and 8.1 kHz (Figure 6d).

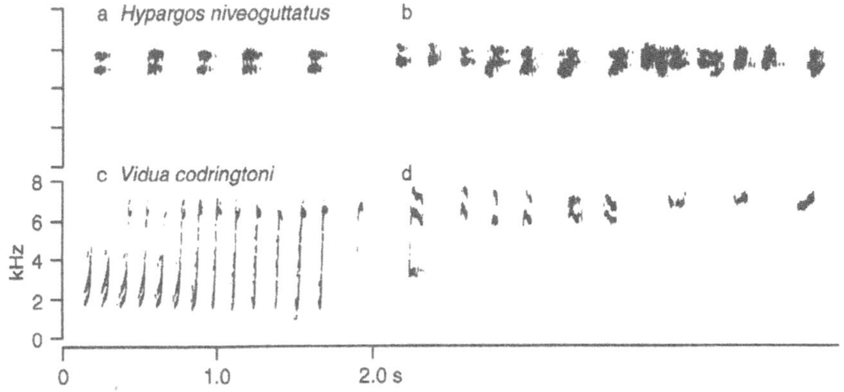

Figure 6. Begging calls of fledged Peters's twinspot, captive: (a) 4 and (b) 8 days out of nest; and (c, d) begging calls in adult male song of Peters's twinspot indigobirds (Thondwe, Malawi).

Begging calls of Dybowski's twinspot (*Euschistospiza dybowskii*) have an overtone with a double peak amplitude at 7.1 and 8.2 kHz (Figure 4f). Some calls in the song of male Cameroon indigobirds that mimic this twinspot (Figure 5i; Payne & Payne 1995) are like the calls of firefinch mimics (Figure 5a-h) and are like the begging calls of young village indigobirds (Figure 2), while others (Figure 4i) are like begging calls of young twinspots (Figure 4f). Begging calls (not shown) of captive-bred nestling Cameroon indigobirds (their father mimicked the twinspot songs) were like the begging calls of young village indigobirds. We lack begging calls of the brown twinspot (*Clytospiza monteiri*). Male Cameroon indigobirds that mimic brown twinspots (not shown, but see Payne & Payne 1994) have phrases like those of firefinch mimics and young village indigobirds (Figures 1c and 2).

Begging calls of goldbreasts are high-pitched 'chee-chee' notes at 6-8 kHz that form zig-zag 'M' or 'W' traces in audiospectrograms. The modulations change into inverted 'U' and 'V' traces when young are approached by the parents (Figure 7a). These calls are conspicuous in songs of males of the goldbreast indigobird (Figure 7b), while other calls (Figure 7c) of these males are like the begging calls of young village indigobirds (Figures 1c and 2). Like the fledgling goldbreast, the calls occur in a transitional series in the adult male's mimicry song.

Begging calls of quail-finch (*Ortygospiza atricollis*) are low in pitch and have an overtone, the low trace at 2.2 kHz and the higher at 3.8 kHz; the phrase is repeated 8-10 times a second (Figure 8a,b), and faster when a young is approached and fed. Adult male quail-finch indigobirds that mimic songs and calls of quail-finches have calls with the same acoustic structure as quail-finch begging calls (Figure 8c), while other calls (Figure 8d) are like the begging calls of young village indigobirds.

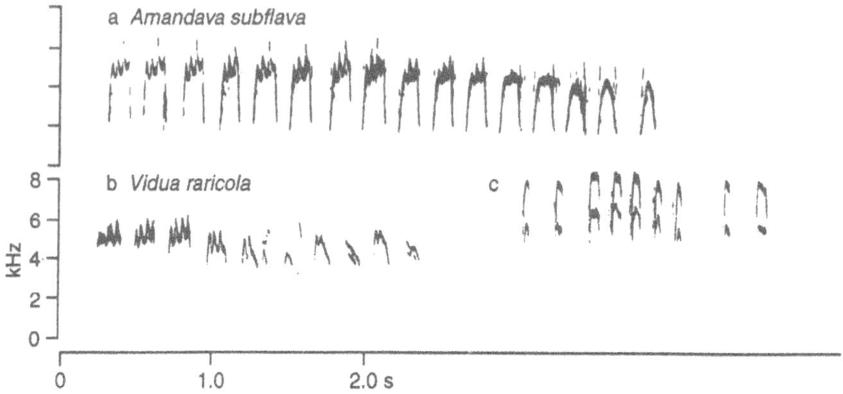

Figure 7. (a, b) Begging calls of fledged goldbreast, captive; and (b, c) begging calls in song of adult male goldbreast indigobird (b, Tibati, Cameroon; c, Kabala, Sierra Leone).

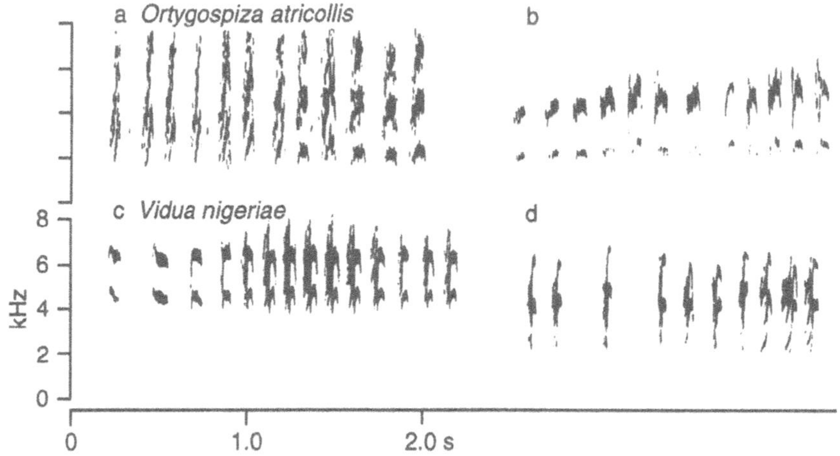

Figure 8. (a, b) Begging calls of fledged quail-finch, captive; and (c, d) begging calls in song of adult male quail-finch indigobird (Garoua, Cameroon).

DISCUSSION

We recorded begging calls from young of ten of the 13 host species known in natural conditions, from parasitized broods in the nest or after fledging or from song mimicry of adult indigobirds. In the young indigobirds that we recorded, none matched exactly the begging calls of the host young, although we did not record calls of the young indigobirds themselves with all host species. In all ten species of indigobirds, adult males gave songs with calls like the begging calls of young village indigobirds, and these are like the begging calls of certain firefinches (African, Jameson's, Mali). Adult male song also includes calls like the begging calls of their host species where these differed from calls of the young indigobirds (Table 1).

Innate begging calls like those of a young village indigobird appear in songs of adult male village indigobirds and other indigobird species. The short, repeated notes that slur upwards and have a double peak around 5.6 and 6 kHz, in young of species that were recorded as young and in adult song, suggest that nestling begging calls may be the same in all species of indigobirds. In village indigobirds that we reared with experimental foster species, the young indigobirds give this begging call as nestlings and fledglings, and they give the same call in their song as adults (Payne et al. 1998). The call in adult song develops normally even without hearing it from another bird (an adult indigobird, a young indigobird or young firefinch). All species of indigobirds have begging calls like those of certain firefinches (African, Jameson's and Mali). We suspect that this is recognized as a call for parental care from the foster parents in all firefinch species, and it elicits parental care in host species whose young have begging calls that differ (other firefinches, twinspots, goldbreast, quail-finch). These observations support the predictions of a generalized begging call in brood-parasitic finches (Hypothesis 1).

Songs of adult male indigobirds also mimic the species-specific calls of the begging young of their host species. These include the alternating short and long elements of begging calls in red-billed firefinches by village indigobirds, the high-pitched calls of Peters's twinspot by its indigobird, the 'M' and 'W' elements in goldbreast by goldbreast indigobirds and the overtones and harmonics in quail-finch by quail-finch indigobirds. In some indigobirds the young are known to be mouth mimics of their host species, and in others (goldbreast indigobird, quail-finch indigobird) they are not (Payne & Payne 1994). The presence in indigobird song of some calls like the begging calls of their host species supports the prediction of specialized mimetic begging calls in mate choice (Hypothesis 2).

Table 1. Begging calls of indigobird brood parasites[a] and estrildid host species.

				Parasite Adult Song	
Indigobird	Host or song model	Host young	Parasite young	Innate calls	Mimicry beg calls
Village	Red-billed firefinch	+	+	+	+
Village	Brown firefinch	-	+	+	+
Bar-breasted firefinch	Bar-breasted firefinch	+	+	+	+
Purple	Jameson's firefinch	+	-	+	+*
Dusky	African firefinch	+	-	+	+*
Barka	Black-faced firefinch	-	-	+	+
Jos Plateau	Rock firefinch	-	-	+	+
Cameroon	African firefinch	+	-	+	+*
Cameroon	Black-bellied firefinch	+	-	+	+*
Cameroon	Mali firefinch	+	-	+	+*
Cameroon	Dybowski's twinspot	+	+	+	+
Cameroon	Brown twinspot	-	-	+	?
Peters's twinspot	Peters's twinspot	+	-	+	+
Goldbreast	Goldbreast	+	-	+	+
Quail-finch	Quail-finch	+	-	+	+

[a] includes indigobirds reared in natural conditions (see text for experimentally reared birds); + = calls recorded in this study; - = calls not recorded in this study; * = begging calls of host young and begging calls of indigobird young not distinguishable; ? = not recognized in the sample recorded, but may be present.

In our experiments, male village indigobirds that were reared by an alternative foster species do not mimic the begging calls of their foster species, even when they were reared with foster nestmates. On the other hand, males imitate the songs and begging calls of their foster species heard after independence from their foster parents. More often, males copy songs and calls of their foster species directly from other adult male indigobirds that mimic the same species. This involves two stages in the development of behaviours that are affected by their foster care. First, males imprint on their foster parents and second, they learn details of their songs selectively from other birds that have calls and songs like those of their foster parents (Payne et al. 1998). Begging calls of foster species are acquired when a male copies the mimicry songs of other adult indigobirds. A male acquires the calls in his adult song as long as he hears an indigobird male sing these (Payne et al. 1998). While males usually learn the calls from other adult males that mimic songs of the same foster species, at some point after a successful host switch by a laying female indigobird, an imprinted male would learn the begging calls directly from a brood of his foster species. These calls would become established in the mimicry songs of the male and his cultural descendants; that is, the other male indigobirds that copy his songs.

Our observations show that begging calls of most species of indigobirds mimic the begging calls of the host species only in the mimicry song of the adults, and not in the young indigobirds. Nicolai's (1964) evidence for host-specific begging-call mimicry came from the calls that he recorded in adult male song, and not in the young themselves, and this led to a misleading inference about co-speciation and coevolution. Our observations in the indigobirds are consistent with Nicolai's observations of the development of begging calls in the whydahs. First, adult male straw-tailed whydahs (*V. fischeri*) give begging calls in their song (Nicolai 1964). Nicolai proposed that these calls are innate and recapitulate the behaviour of a male when it was a begging nestling. He took whydah eggs and nestlings from nests of their normal host, purple grenadier (*Granatina ianthinogaster*), and reared the young whydahs with Bengalese finch foster parents. Whydahs that had never heard their normal host species gave normal begging calls in their adult song like begging calls of young whydahs, so the development of this call is innate (Nicolai 1973). Second, young nestling paradise whydahs (*V. paradisaea*) give begging calls somewhat like the calls of their, the host melba finch (*Pytilia melba*), but as they become older the calls of whydahs diverge from those of melbas and differ by fledging age (Nicolai 1969). These early begging calls of young whydahs, as in young indigobirds, differ from their host species' calls and the calls develop without imitation of their foster species.

Functional Significance of Begging Calls in Nestling and Adult Indigobirds

Because nesting estrildids can rear young that do not look and sound like their own young, it is not necessary for a young finch to have the begging calls and visual signals of its foster species in order to receive parental care from the foster adults. It is known that some estrildids will rear nestlings that are not their own species' young (Immelmann et al. 1965), and we have repeated this work in greater detail. In our observations, the location of the young in the nest and the generalized features of estrildid begging behaviour are sufficient to elicit parental care. Nevertheless, host-mimetic indigobird nestlings were more successful than were non-mimetic estrildids in being reared by red-billed firefinches (Payne et al. 2001).

The observation that all indigobird species have an adult song with calls like the begging calls of young in certain host firefinches suggests that this common begging call is sufficient for parental care by other estrildid finches. Because the begging calls are the same, an indigobird that parasitizes one of them could elicit parental care from other estrildid finches, not only from its normal host species. Molecular genetic comparisons indicate that indigobirds in fact have switched from host to host in natural conditions (Klein & Payne 1998), and their generalized nestling begging calls may provide a mechanism which allows this switch.

On the other hand, the begging calls in the song of adult male indigobirds can influence females in their choice of a mate. In indigobirds, begging calls comprise up to 20% of the mimetic phrases and are a conspicuous part of mimicry song (Payne 1973a, 1979, 1983, 1985). Imprinting and learning the calls and songs of foster parents allows indigobirds to track the behaviour of a new host species across generations when they undergo a host shift, and to include these new calls and songs in their own mimicry songs. Males that include in their songs the begging calls of the species that reared them may do better than males without these calls in attracting a mate (Payne & Payne 1977, 1997; Payne et al. 1998, 2000a).

In summary, we find that all species of brood-parasitic indigobirds have two kinds of begging call, one that is innate in the young birds, and another in the song of adult males, which is copied from other indigobirds that mimic the same host species upon which the adult male has imprinted. The first of these calls is used in parental care, whereas both kinds of begging call given by a singing male may be used in mate choice.

FUTURE DIRECTIONS

Directions for future work might include: (1) fieldwork in Africa to find nestling indigobirds and record their begging calls, and to record singing males to determine the rates of host switching in different regions; and (2) experiments to rear the young indigobirds in the nests of foster species that are not their normal hosts, to test whether their begging calls as young and their calls in adult songs differ with their early experience. Finally, (3) we are using molecular genetics to determine evolutionary relationships. A prediction from the evolutionary history of host switches and rapid speciation, and the changes in paleoclimate (Roche 1991) and grain cultivation (National Research Council 1996) that accompanied these events, is that ancestral indigobirds parasitized the species complex of the African firefinch, whose begging calls the indigobirds closely match. In addition, indigobirds that parasitize species whose begging calls they do not mimic (e.g. indigobirds with red-billed firefinches, twinspots, goldbreast and quail-finch) were derived from indigobirds that parasitize the African firefinch species complex. Our preliminary results suggest that the initial separation of indigobirds and the other brood-parasitic whydahs (Klein & Payne 1998) occurred much earlier in the evolutionary past than the coalescence times of existing indigobird species. The innate begging calls perhaps evolved to match the begging calls of a host in the remote past, before other indigobird lineages became extinct. If so, then the behaviour of parental care provides us with a window on early evolutionary time in addition to that we can acquire through molecular genetics.

ACKNOWLEDGEMENTS

Clive Barlow, Ian Hinze, Kit Hustler and Jürgen Nicolai recorded begging calls and songs of certain finches. Jean Woods was helpful in aviary research. For comments on the manuscript we thank Sal Cerchio, David Lahti, Marty Leonard, Alec Lindsay, Jon Wright and an anonymous reviewer. Research was supported by the National Science Foundation and the University of Michigan Museum of Zoology.

REFERENCES

Andersson, M. 1994. *Sexual Selection*. Princeton: Princeton University Press.

Davies, N.B. 2000. *Cuckoos, Cowbirds and Other Cheats*. London: T. & A.D. Poyser.

Goodwin, D. 1982. *Estrildid Finches of the World*. London: British Museum of Natural History.

Immelmann, K. 1962. Beiträge zu einer vergleichender biologie australischer Prachtfinken (Spermestidae). *Zoologische Jahrbücher Abteilung für Systematik, Geographe und Biologie der Tiere* **90**, 1-196.

Immelmann, K. 1982. *Australian Finches in Bush and Aviary*. Sydney: Angus and-Robertson.

Immelmann, K., Steinbacher, J. & Wolters, H.E. 1965. *Prachtfinken. Vol. 1: Astrilde. 2nd Edtn.* Aachen: Verlag Hans Limberg.

Klein, N.K. & Payne, R.B. 1998. Evolutionary associations of brood parasitic finches (*Vidua*) and their host species: analyses of mitochondrial restriction sites. *Evolution* **52**, 299-315.

Kunkel, P. 1959. Zum verhalten einiger Prachtfinken (Estrildinae). *Zeitschrift für Tierpsychologie* **16**, 302-350.

National Research Council. 1996. Fonio (acha). In: *The Lost Crops of Africa. Vol. 1: Grains.* Washington: National Academy Press.

Nicolai, J. 1964. Der brutparasitismus der Viduinae als ethologisches problem. *Zeitschrift für Tierpsychologie* **21**, 129-204.

Nicolai, J. 1967. *Vogelhaltung—Vogelpflege. 2nd Edtn.* Stuttgart: Franckh'sche Verlag.

Nicolai, J. 1969. Beobachtungen an Paradieswitwen (*Steganura paradisaea* L., *Steganura obtusa* Chapin) und der Strohwitwe (*Tetraenura fischeri* Reichenow) in Ostafrika. *Journal für Ornithologie* **110**, 421-447.

Nicolai, J. 1973. Das lernprogramm in der gesangsausbildung der Strohwitwe *Tetraenura fischeri* Reichenow. *Zeitschrift für Tierpsychologie* **32**, 113-138.

Nicolai, J. 1974. Mimicry in parasitic birds. *Scientific American* **231**, 92-98.

Nicolai, J. 1989. Brutparasitismus der Glanzwitwe (*Vidua hypocherina*). *Journal für Ornithologie* **130**, 423-434.

Payne, R.B. 1973a. Behavior, mimetic songs and song dialects, and relationships of the parasitic indigobirds (*Vidua*) of Africa. *Ornithological Monographs* **11**, 1-333.

Payne, R.B. 1973b. Vocal mimicry of the paradise whydahs (*Vidua*) and response of female whydahs to the songs of their hosts (*Pytilia*) and their mimics. *Animal Behaviour* **21**, 762-771.

Payne, R.B. 1979. Song structure, behaviour, and sequence of song types in a population of village indigobirds, *Vidua chalybeata. Animal Behaviour* **27**, 997-1013.

Payne, R.B. 1982. Species limits in the indigobirds (Ploceidae, *Vidua*) of West Africa: mouth mimicry, song mimicry, and description of new species. *Miscellaneous Publications of the University of Michigan Museum of Zoology* **162**, 1-96.

Payne, R.B. 1983. Bird songs, sexual selection, and female mating strategies. In: *Social Behavior of Female Vertebrates* (Ed. by S.K. Wasser). New York: Academic Press.

Payne, R.B. 1985. Behavioral continuity and change in local song populations of village indigobirds *Vidua chalybeata. Zeitschrift für Tierpsychologie* **70**, 1-44.

Payne, R.B. 1990. Song mimicry by the village indigobird (*Vidua chalybeata*) of the red-billed firefinch (*Lagonosticta senegala). Vogelwarte* **35**, 321-328.

Payne, R.B. 1996. Field identification of the indigobirds. *Bulletin of the African Bird Club* **3**, 14-25.

Payne, R.B. 1997. Avian brood parasitism. In: *Host-Parasite Evolution: General Principles and Avian Models* (Ed. by D.H. Clayton & J. Moore). Oxford: Oxford University Press.

Payne, R.B. 1998a. Brood parasitism in birds: strangers in the nest. *BioScience* **48**, 377-386.

Payne, R.B. 1998b. A new species of firefinch *Lagonosticta* from northern Nigeria, and its association with the Jos Plateau indigobird *Vidua maryae. Ibis* **140**, 368-381.

Payne, R.B. & Payne, K. 1977. Social organization and mating success in local song populations of village indigobirds, *Vidua chalybeata. Zeitschrift für Tierpsychologie* **45**, 113-173.

Payne, R.B. & Payne, L.L. 1994. Song mimicry and species associations of west African indigobirds *Vidua* with quail-finch *Ortygospiza atricollis*, goldbreast *Amandava subflava* and brown twinspot *Clytospiza monteiri. Ibis* **136**, 291-304.

Payne, R.B. & Payne, L.L. 1995. Song mimicry and association of brood-parasitic indigobirds (*Vidua*) with Dybowski's twinspot (*Euschistospiza dybowskii*). *The Auk* **112**, 649-658.

Payne, R.B. & Payne, L.L. 1997. Field observations, experimental design, and the time and place of learning in bird songs. In: *Social Influences on Vocal Development* (Ed. by C. Snowdon & M. Hausberger). Cambridge: Cambridge University Press.

Payne, R.B. & Payne, L.L. 1998. Nestling eviction and vocal begging behaviors in Australian glossy cuckoos *Chrysococcyx basalis* and *C. lucidus*. In: *Parasitic Birds and Their Hosts: Studies in Coevolution* (Ed. by S.I. Rothstein & S.K. Robinson). Oxford: Oxford University Press.

Payne, R.B., Payne, L.L. & Nhlane, M.E.D. 1992. Song mimicry and species status of the green widowfinch *Vidua codringtoni*. *Ostrich* **63**, 86-97.

Payne, R.B., Payne, L.L. & Woods, J.L. 1998. Song learning in brood parasitic indigobirds *Vidua chalybeata*: song mimicry of the host species. *Animal Behaviour* **55**, 1537-1553.

Payne, R.B., Payne, L.L., Woods, J.L. & Sorenson, M.D. 2000a. Imprinting and the origin of parasite-host species associations in brood parasitic indigobirds *Vidua chalybeata*. *Animal Behaviour* **59**, 69-81.

Payne, R.B., Woods, J.L., Siddall, M. & Parr, C.S. 2000b. Randomization analyses: mimicry, geographic variation and cultural evolution of song in brood-parasitic straw-tailed whydahs, *Vidua fischeri*. *Ethology* **106**, 261-282.

Payne, R.B., Woods, J.L. & Payne, L.L. 2001. Parental care in estrildid finches: experimental tests of a model of *Vidua* brood parasitism. *Animal Behaviour* **62**, 473-483.

Payne, R.B, Hustler, K., Stjernstedt, R., Sefc, K. & Sorenson, M.D. in press. Behavioural and genetic evidence of a recent population switch to a novel host species in brood parasitic indigobirds *Vidua chalybeata*. *Ibis*.

Roche, E. 1991. Evolution des paléoenvironnements en Afrique centrale et orientale au Pléistocène supérieur et à l'Holocène. Influences climatiques et anthropiques. *Bulletin de la Société Géographique de Liège* **27**, 187-208.

STATISTICAL APPROACHES

23.

LOGISTIC REGRESSION AND THE ANALYSIS OF BEGGING AND PARENTAL PROVISIONING

Daniela S. Monk
School of Biological Sciences, Washington State University, Pullman WA 99164, USA (dmonk@wsu.edu)

ABSTRACT

The statistical analysis of parental responses associated with complex offspring begging signals is challenging. The outcome is frequently dichotomous – one nestling receives food first while all others do not, all nestmates potentially affect the outcome, and begging signals are comprised of a variety of cues. Logistic regression is a method that is ideally suited to analyse parental responses (care / no care) relative to offspring cues. I first illustrate the use of simple, univariate logistic regression to examine whether male or female offspring are preferentially fed by their parents, when taking the sex ratio of the entire brood into account. Next, I introduce a model (called *Arena*) which is specifically designed to analyse the probability of a nestling receiving food given multiple explanatory variables (multivariate), while integrating information from all interacting nestlings (multinomial). Overall, logistic regression models are a statistically rigorous and flexible method to analyse complex parental care data.

INTRODUCTION

As previous chapters of this book illustrate, the dynamics in a nest during feeding events are very complex. Frequently, when a parent bird returns to its nest to feed its young it is faced by multiple offspring that are begging to various degrees, that are moving about, that are interacting with one another in more or less aggressive ways and that all need to be fed at some point. Thus, factors such as loudness or intensity with which a nestling begs, its position in the nest, or some other visual or vocal trait reflecting the

J. Wright and M.L. Leonard (eds.), The Evolution of Begging, 451–471.

nestling's state may influence parental feeding decisions. For close to half a century questions regarding the whys and hows of parental responses to begging offspring have been explored empirically, as well as theoretically (Lack 1968; Clutton-Brock 1991). It is now recognized that parental responses to begging signals are likely to be a function of honest signals given by the young as an index of their state, as well as a function of conflicts of interests between parents, conflicts of interest between parents and offspring, and competition between siblings (reviewed by R.A. Johnstone & H.C.J. Godfray this volume). However, statistical analysis of nestling begging behaviour and associated parental care is not straightforward. Statistical analysis needs to simultaneously include multiple variables associated with complex parent-offspring interactions and with multiple cues from several interacting nestlings. In addition, studies on begging behaviour frequently have to deal with small sample sizes and unknown sampling distributions which limit available statistical options (S. Forbes this volume).

Traditional attempts to deal with the complexity of behaviour typically involve the simplification and comparison of grouped data using techniques that assume normal distributions. For example, some authors combine information from individual nestlings into categories and compare mean values of those categories (e.g. Teather 1992; Price & Ydenberg 1995; Cotton et al. 1996; Leonard & Horn 1996, 1998; Smiseth et al. 1998). In other cases researchers focus behavioural observations on one nestling, and examine just the values of those focal nestlings (e.g. Leonard et al. 1994; Kacelnik et al. 1995). Neither of these methods incorporate the behaviour of each of the interacting offspring that is not fed. Also, these techniques usually accumulate averages over multiple events, and lose the information contained in each feeding event. In addition, authors need to use nonparametric techniques when the sampling distribution cannot be assumed to be normal (e.g. Stamps et al. 1987; Gowaty & Droge 1990). Unfortunately, use of nonparametric tests in situations where all assumptions of parametric tests are met will result in a waste of data (Daniel 1987).

Despite the complexity of parental responses to begging signals, there is a common denominator: the outcome of parent-offspring interactions can be viewed as dichotomous. A parent returning to the nest will feed only one of many nestlings, or will feed one first or most. The analysis of a dichotomous (or, more generally, discrete) dependent variable, such as success / failure or fed / not fed, requires special treatment because the errors associated with such variables are usually not normally distributed. Traditional analyses are not designed to handle the dichotomy of the outcomes and their associated binomial distribution of errors. Logistic regression, on the other hand, is ideally suited for such analyses because the dichotomy of outcomes is formally taken into account by using the logit (log-odds) transformation (Agresti 1990; Trexler & Travis 1993).

In this chapter, I will discuss logistic regression and describe the advantages of this kind of analysis. I will then illustrate the application of logistic regression approaches to situations involving parental responses to their young. The first example uses a simple logistic regression approach to examine the relationship between one independent variable (sex ratio of the brood) and whether male or female offspring are disproportionately fed by their parents. The second example illustrates a more complex situation and the application of a model related to logistic regression (*Arena* model). Using the *Arena* model, I will examine the attributes of a nestling that are associated with its success in receiving food, given that there are several nestlings in the nest and many of them are begging. This model determines the relative importance of nestling attributes in predicting the probability of receiving food. I will conclude with a discussion of the value of models related to logistic regression for analysis of nestling begging.

WHAT IS LOGISTIC REGRESSION?

Logistic regression is a mathematical modelling approach that describes the relationship between one or several independent variables (x_1, x_2,..., x_p) and a dichotomous dependent variable, Y, typically labelled 'success' and 'failure', and represented by numerical values 1 and 0 (Hosmer & Lemeshow 1989; Dobson 1990; Kleinbaum 1994; Flury 1997). Logistic regression analysis, however, does not just test for this relationship - it estimates the probability of something happening (the success) as a function of one or several independent variables.

Logistic regression is one special case of generalized linear models (GLM); linear regression, ANOVAs, ANCOVAs, log-linear analysis and multinomial response models are other special cases of GLM (McCullagh & Nelder 1989; Agresti 1990; Dobson 1990). As with other GLM, logistic regression is specified by three components. First, the observations of the dependent variable, Y, are from an exponential distribution; the Binomial. Second, there is a set of independent variables (x_1, x_2, etc.) and a set of associated, unknown parameters (α, β, etc.). Third, the relationship between the dependent variable and the set of independent variables is defined by the link function, which is a monotonic differentiable function. In logistic regression the link function is the logit function.

In more formal terms, logistic regression models relate the success probability, $\pi(x)$, which is approximated by the proportion of successes, to the independent (explanatory) variable according to the logistic regression function:

$$\pi(x) = \exp(\alpha + \beta x) / (1 + \exp(\alpha + \beta x)) \quad (1)$$

(Hosmer & Lemeshow 1989; Agresti 1990; Flury 1997). In this function, $\pi(x)$ represents $E(Y|x)$, the expected value of Y given the value of x. The parameters α and β determine the exact shape of the curve that relates the probability of success to the independent variable. It is customary to transform the logistic regression function by solving for: $\alpha + \beta x$. This is called the logit or log-odds transformation, such that:

$$\text{logit } \pi(x) = \ln[\pi(x)/(1-\pi(x))] = \alpha + \beta x \quad (2)$$

This equation expresses the basic assumption of the logistic regression model: the logit of the success probability is a linear function of the dependent variable(s) (Flury 1997). The logistic regression model can be generalized for the case with multiple independent variables each with an associated β parameter, in this case the exponent in equation (1) would be $\alpha + \beta_1 x_1 + \beta_2 x_2 + \ldots + \beta_p x_p$ (Hosmer & Lemeshow 1989). Logistic regression can also be generalized for a polychotomous (= polytomous) dependent variable (Hosmer & Lemeshow 1989; Agresti 1990), in this case the outcome variable, Y, has more than two categories, for example, three outcome categories that may be coded as 0, 1 or 2.

To obtain estimates for the unknown α and β parameters and their standard errors the preferred method used in logistic regression is maximum likelihood estimation (Hosmer & Lemeshow 1989; Dobson 1990; Kleinbaum 1994; Flury 1997). Usually the parameter estimates have to be obtained numerically by an iterative procedure, for example the Newton-Raphson method (Agresti 1990; Dobson 1990). When using maximum likelihood estimation, hypothesis testing is frequently carried out by the likelihood ratio test, the Wald test or the score test (Kleinbaum 1994).

Several computer programs are available for performing logistic regression, including: program PLR in BMDP, procedure PROC GLM in SAS, option REGRESSION in SPSS, and programs in SYSTAT and GLIM.

A UNIVARIATE LOGISTIC REGRESSION ANALYSIS OF PROVISIONING TO MALE AND FEMALE YOUNG

The first example of an application of logistic regression describes a univariate situation, in which the effect of only one independent variable is examined. I will examine whether male or female offspring are preferentially fed by their parents, when taking the sex ratio of the entire brood into account. In sexually size dimorphic species in which the energy requirements of young of the larger sex are higher than that of the smaller sex, parents can be expected to feed the larger sex more often. For example, in the red-winged blackbird (*Agelaius phoeniceus*) in which males are larger, Teather (1992) showed that male offspring were offered more food by parents than were female offspring. Similarly, in crimson rosellas (*Platycercus elegans*) in which males are also larger, Krebs et al. (1999) showed that first- and middle-hatched male nestlings were fed more than last-hatched males, and more than all female nestlings. For sexually size monomorphic species, on the other hand, male and female nestlings are generally not expected to receive different amounts of feeding by parents. This prediction was supported in a study on budgerigars (*Melopsittacus undulatus*), in which Stamps et al. (1987) found no difference between the rates at which fathers or mothers fed male and female nestlings. A study on western bluebirds (*Sialia mexicanus*) also showed that there was no difference between the feeding frequency to sons versus daughters for either parent (Leonard et al. 1994). On the other hand, Gowaty and Droge (1990) showed that in eastern bluebirds (*Sialia sialis*), fathers fed their daughters significantly more than they fed their sons.

In all but one of the studies described above, the authors used nonparametric comparisons between paired groups. Groups of male nestlings and female nestlings were compared with respect to feeding frequencies by parents. However, because the outcome variable for the analysis of food distribution among male and female offspring can be considered as binomial (e.g. a male (or female) offspring is either fed or not fed), the use of logistic regression is statistically appropriate. In addition to comparing whether male nestlings as a group receive more food than females, logistic regression can address the question of whether the sex of the offspring is a determining factor of feeding success, and what effect independent variables, such as brood sex ratio, may have on the outcome.

I studied parental behaviour of mountain bluebirds (*Sialia currucoides*) between 1992 and 1995 in the Colorado Rockies, near the Mountain Research Station of the University of Colorado, USA. Mountain bluebirds, along with western bluebirds and eastern bluebirds are the only members of this North American thrush family. I was particularly interested in whether there was sex-biased provisioning in mountain bluebirds because, as just described, its two congeners produced contrasting results (Gowaty & Droge 1990; Leonard et al. 1994).

Since bluebirds are secondary cavity-nesters and readily adopt nestboxes, I was able to videotape 24 mountain bluebird families (for details see Monk

1999). To record activities in a nest, a video camera was placed into a lateral box attached to the nesting box for two hours in the morning. Nestlings were individually marked and videotapes were analysed using a stop-frame analyser. For each nest, I determined the sex ratio, the proportion of successes (i.e. number of times a male offspring was fed) and the proportion of failures (i.e. number of times a female offspring was fed).

Using a simple logistic regression, I tested whether parents fed male offspring more than female offspring at any given sex ratio of the brood. This model is based on the situation in which a parent bird returns to its nest and feeds one of several nestlings. The parent may feed both sexes equally, or feed males unconditionally more than females, or only feed males more when males are more common in the nest, or only when males are rare in the nest. The logistic regression model to address this situation describes the probability of a son being fed, the success probability π (x), as a function of the independent variable x and the sex ratio of the brood. Thus, as in equation (2), logit π (x) = $\alpha + \beta$ x. In this logistic regression model, specific values of the model estimates (α, β) describe specific provisioning patterns to male or female offspring. By logit transforming the independent variable, sex ratio, I could distinguish between the various outcomes with just one analysis (Figure 1). The specific logistic regression model I used was:

$$\text{logit } \pi \text{ (x)} = \text{logit (prob. son being fed)} = \alpha + \beta \text{ [logit (sex ratio)]} \quad (3)$$

In this model, α is an indicator of how much male offspring are fed. Values of α greater than zero indicate that males receive more food than expected based on their representation in the brood. β describes the influence of the sex ratio of the brood on the probability of a male receiving food. Thus, β is an indicator of how much the more common sex is fed.

The predictions of this model can be illustrated graphically (Figure 1). When both male and female offspring are equally likely to receive food, then the model estimates will be $\alpha = 0$ and $\beta = 1$, and the data will fall along the diagonal, dashed line. In this situation, the probability of one of the males being fed will equal the proportion of males in the brood. If males are being fed disproportionately more often than females given their proportion in the brood, then $\alpha > 0$ and $\beta = 1$, and the data points will lie above the dashed line (Figure 1a). Conversely, if females are being fed more often than males given their proportion in the brood, then the data points will fall below the dashed line (Figure 1b). If, on the other hand, at every sex ratio the common sex is fed more often, the data points will fall below the diagonal line for sex ratios smaller than 50% and above the diagonal line for sex ratios greater than 50% (more common category wins, $\alpha = 0$ and $\beta > 1$; Figure 1c). When the rarer sex of nestling receives more food at each sex ratio, the data will lie above the diagonal line for sex ratios smaller than 50% and below the diagonal for higher sex ratios (affirmative action; Figure 1d).

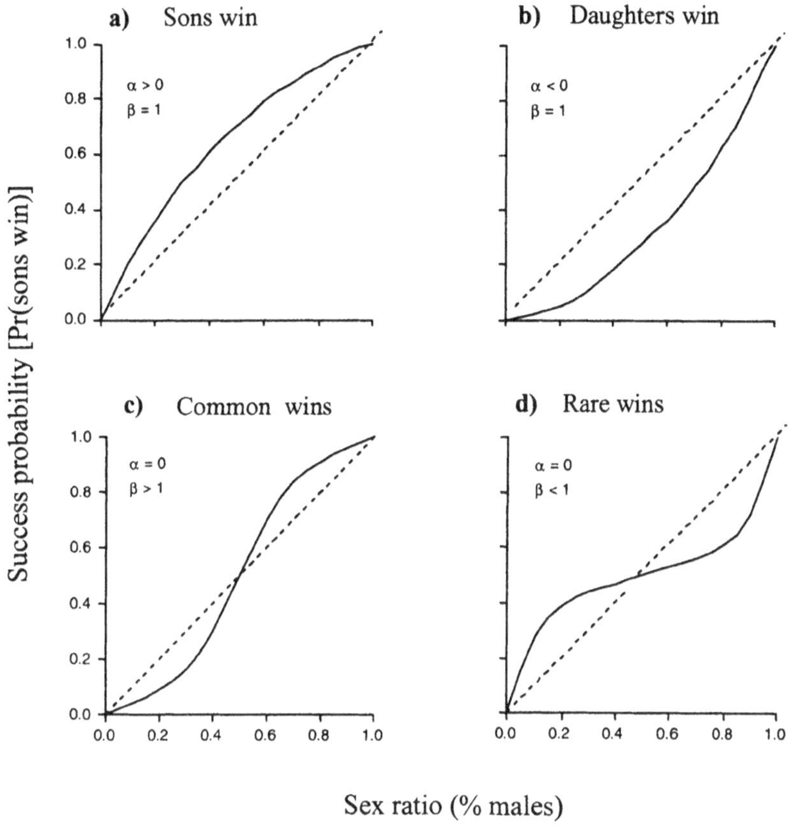

Figure 1. A graphical illustration of the predictions of the logistic regression model used to evaluate parental care to individual male and female nestlings. The parameter α is an indicator of how much food males receive and the parameter β is an indicator of how much the sex that is more common in the brood receives. The lines represent simulated data for various scenarios. The dashed line represents the situation where each sex receives food in proportion to its abundance in the nest (i.e. male and female nestlings receive the same amount). In (a) males receive more feedings, in (b) females receive more feedings, in (c) the sex that is more common in the brood receives more (i.e. daughters in female-biased broods and sons in male-biased broods), and in (d) the sex that is rarer receives more feedings (i.e. sons in female-biased broods and daughters in male-biased broods).

I used the Wald statistic to test whether the coefficients (α and β) differ significantly from zero. The Wald test is usually performed when there are only one or two parameters being tested. Under the hypothesis that an individual coefficient is zero (H_0: $\alpha = 0$; H_0: $\beta = 0$), the square of the Wald statistic will follow approximately a chi-square distribution, with one degree of freedom (Kleinbaum 1994).

Analysis of data on mountain bluebirds using the logistic regression model outlined above showed that parents did not discriminate among nestlings based on the sex of the young ($\alpha = -0.052$ (± 0.07), W = 0.62, P = 0.40 and $\beta = 1.01$ (± 0.07), W = 183.1, $P < 0.001$). Parents fed males and females in proportion to their representation in the nest (Figure 2).

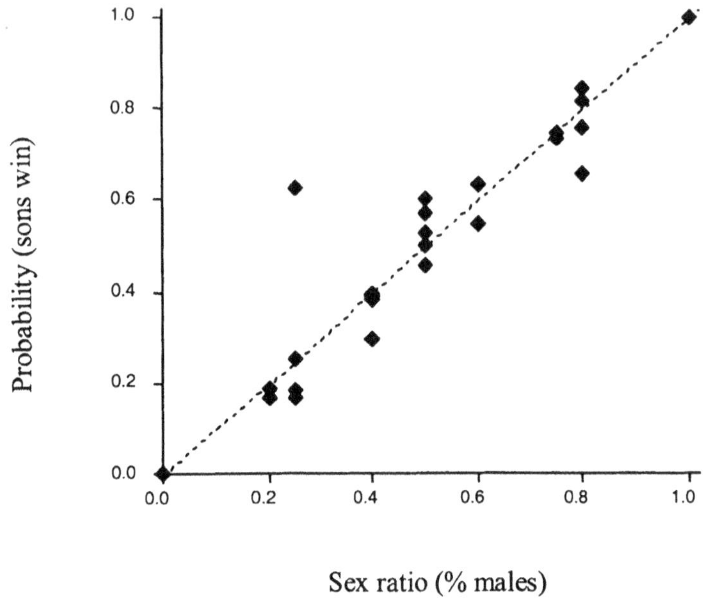

Figure 2. A scatter-plot of the probability of food deliveries by mountain bluebird parents to male offspring with respect to brood sex ratio. n = 24 broods. Logistic regression parameter estimates are α = -0.052 and β = 1.01, indicating, respectively, that sons do not receive more food, and that each sex receives food in proportion to its representation in the brood.

Thus, the result from mountain bluebirds is in accord with the result in western bluebirds (Leonard et al. 1994). No difference in feeding of male and female offspring is, however, somewhat counter to expectation, because the primary sex ratio of this population is female-biased and male parents increase their provisioning rate to older male-biased broods (Monk 1999).

THE ANALYSIS OF MULTIPLE BEGGING TRAITS AND MULTIPLE OFFSPRING

The second example of an application of logistic regression is a test to determine which traits of individual nestlings influence the feeding decisions of parents, as well as the relative importance of these traits. This model is also based on situations in which a parent bird returns to its nest and feeds one of several nestlings. The nestlings are begging and interacting with one another. This scenario is quite complex, and I think there are three factors that contribute heavily to this complexity. First, there are multiple interacting nestlings, and the characteristics of the one receiving food as well as the characteristics of the ones not receiving food are likely to influence the outcome. Second, parent birds are likely to have a variety of cues upon which they base their provisioning decisions. Third, nestling provisioning is a dynamic, sequential process, and parents may integrate information over many successive events.

Traditional analyses of success-failure situations frequently compare winners with losers, or otherwise compare groups. This generalization also holds for analyses of food distribution to nestling birds. For example, Leonard and Horn (1996) tested how successful nestlings (i.e. those fed) compared to unsuccessful nestlings (those not fed) by using repeated measures ANOVA. These authors found that those tree swallow (*Tachycineta bicolor*) nestlings which received food from their parents initiated begging sooner, reached higher and were closer to the entrance of the box, compared with those nestlings that were not fed. Similarly, Price and Ydenberg (1995) used a Wilcoxon matched pairs test to compare values for the fed nestling with the mean for the unfed nestlings combined per visit. They found during a pre-experiment period, that yellow-headed blackbird (*Xanthocephalus xanthocephalus*) nestlings that were fed by their parents were those that begged first, begged more intensely and begged more loudly. An alternative approach to grouping by winners versus losers is to group nestlings by trait. For example, the probability that nestlings that are satiated are fed can be compared to the probability that hungry nestlings are fed. Leonard and Horn (1998) compared broods of nestlings that were deprived with broods that were fed, and found that hungry broods begged more and that parental feeding frequency to hungry broods was higher. Because these approaches compare similarities and differences between groups based on the shared outcome, they can be considered retrospective.

The *Arena* Model

Together with Stefania Bartoletti, from the Institute of Statistics at the University of Rome, I developed a model related to logistic regression to analyse complex processes such as food distribution among nestlings. We called it the *Arena* model because it is applicable to a variety of social circumstances in which individuals interact within an arena-type situation, such as nests, leks and experimental choice-arenas (see Monk 1999). The *Arena* model evaluates the probability of obtaining food for a given nestling, depending on the characteristics of all the interacting individuals.

In the *Arena* model, the probability that individual j receives food (or is chosen) at time t depends on the 'total character stimulus' with which that nestling begs for food (λ_j) in relation to the total character stimulus of all individuals, 1 through k, in that arena ($\Sigma\ \lambda_h$). The total character stimulus is a linear function of the various explanatory variables ($x_1, x_2, \ldots x_p$), each weighted by a parameter ($\beta_1, \beta_2, \ldots \beta_p$):

$$\pi j(x) = \Pr(\text{ nestling j is fed }) = \lambda_j \Big/ \sum_{h=1}^{k} \lambda_h$$

$$\text{with } \lambda_j = \exp(\beta_1 x_{j1} + \beta_2 x_{j2} + \ldots + \beta_p x_{jp}) \qquad (4)$$

Thus, each explanatory (independent) variable has an associated β parameter. Notice the similarity between the *Arena* model and equation (1). In our model, as generally is the case for logistic regression models, each β parameter and its standard error are estimated using the method of maximum likelihood, and the numerical values are obtained by the Newton-Raphson algorithm (Hosmer & Lemeshow 1989; Agresti 1990; Dobson 1990; Flury 1997). The β estimates indicate which trait is important in predicting which nestling succeeds. The standard error indicates the stability of the β estimate. We designed the *Arena* model so that the probability of success depends on the characteristics of all individuals in the nest, because when a parent feeds, its choice is likely to depend upon a comparison of characteristics of the fed nestling with the characteristics of all the other nestlings it did not feed.

The *Arena* model is based on principles of multinomial (polychotomous = polytomous) logistic regression. For a description of polychotomous logistic regression see Hosmer and Lemeshow (1989). The *Arena* model differs from the polychotomous approach in three significant ways (Bartoletti 1998; Monk 1999; S. Bartoletti & D.S. Monk unpublished data). First, the polychotomous approach uses a separate parameter vector for each category of the outcome variable. In contrast, the *Arena* model provides us with the same parameter vector β for all categories. We solved for one parameter vector, because we were interested in obtaining the weightings for the explanatory variables that were of high predictive value for all nestlings. Second, polychotomous regression requires a reference group whereas the *Arena* model does not. And thirdly, polychotomous regression has an intercept term α. In order to make our model identifiable (i.e. the estimation process yields exactly one value for each of the parameters), the *Arena* model has no intercept term. The *Arena* model is similar to the proportional-hazards model by Cox (1972); and in econometrics there is an equivalent approach to the *Arena* model, the conditional logit analysis of qualitative choice behaviour by McFadden (1974).

By analysing nestling feeding events with logistic regression methods we are using what can be considered a prospective approach, because logistic regression develops a model that enables the prediction of which player is likely to win or lose. A logistic model is a kind of forecast: if we know the parameters α, β, etc., and we have values for the independent variables, we can calculate the probability that a particular individual will be successful. In a way, a logistic regression model is an attempt to determine the rule of thumb a parent bird might use to decide which nestling to feed.

Model Building

The best model for predicting success is a model that includes only significant explanatory variables. Variables that have no detectable effect on the response should generally be excluded from the model (McCullagh & Nelder 1989). To obtain such a reduced, predictive model, there are several procedures that enable selection or deletion of independent variables, including forward selection, backward elimination and stepwise regression (Hosmer & Lemeshow 1989; McCullagh & Nelder 1989; Kleinbaum 1994). For the *Arena* model we used a backward elimination process in combination with an evaluation of the stability of each parameter estimate, and an evaluation of the quality of each estimate with respect to how well it predicted a successful outcome (Monk 1999).

The goal of variable selection procedures is to find the model with the fewest variables that nevertheless is highly predictive. In multivariate logistic regression, each model has an associated log likelihood statistic ($-2 \ln \hat{L}$) which is based on the maximized likelihood value ($\hat{L}$), a numerical value of the likelihood function when the maximum likelihood estimates are substituted for their corresponding parameter values (Kleinbaum 1994). The ratio of the maximized likelihood values of two models, one with and one without a particular independent variable in the equation, is called the likelihood ratio (Kleinbaum 1994). The likelihood ratio is used in the calculation of the likelihood ratio test or deviance to test for the magnitude of the difference between two models. This test statistic has approximately a chi-square distribution with degrees of freedom (df) equal to the difference between the number of parameters in the two models, i.e. the number of parameters β_p set to zero (Hosmer & Lemeshow 1989; McCullagh & Nelder 1989; Agresti 1990; Trexler & Travis 1993; Kleinbaum 1994). Thus, the degrees of freedom for model fitting are affected only by the number of parameters that are excluded from the full model, and not by the number of data points. The smaller the P-value associated with the likelihood ratio or deviance statistic, the more important the contribution of the parameter. Such parameters are candidates to be retained in the reduced, final model.

For the *Arena* model, the P-values were primarily used to rank parameter estimates according to their importance to the predictive capability of the model. We used these rankings in the backward elimination procedure for variable selection. In our procedure we first eliminated one variable at a time, then two variables at a time, followed by dropping several variables at once and evaluated the contribution to the model at each step (Monk 1999). We set a guideline cut-off point for inclusion at $\alpha = 0.05$. A P-value smaller than $P = 0.05$ indicated that the parameter(s) was likely to be of considerable importance in the model. However, parameters associated with P-values greater than $P = 0.05$ still remained under consideration. A parameter was retained in the final model if the estimate was stable and if the parameter helped predict the outcome. A β parameter estimate was considered stable if the estimate $\pm$ 2 SE did not include the value 0. We evaluated how good each parameter was at helping to predict the outcome by calculating how often the model containing only that parameter correctly identified the winning nestling, assuming that the winner would be correctly identified due to chance alone 25% of the time (i.e. success probability $\pi_j(x) > 0.25$, one out of four nestlings).

Interpretation of Parameter Estimates

Once the independent variables that contribute to the predictive capability of the model have been identified, the interpretation of the numeric value of β parameter estimates in the *Arena* model is straightforward. A positive β parameter estimate indicates that higher values of the measured variable are associated with a greater probability of being fed. Conversely, a negative β parameter estimate indicates that lower values of the measured variable are associated with a greater probability of being fed.

Begging Traits of Mountain Bluebird Nestlings

I was interested in determining the characteristics of nestling mountain bluebirds that are relevant for predicting which one receives food from a parent. I also wanted to determine the relative importance of these predictor variables (see Monk 1999). In this study a total of 29 families were monitored, each with four or five nestlings. I video recorded activities at the nest for two hours in the morning when nestlings were 9-12 days post-hatch (hatching = day 1), when they were 13-16 days old, and shortly before fledging at 17-20 days old. Mountain bluebird nestlings generally fledge when they are 18 to 20 days post-hatch. Nestlings were individually marked. From frame-by-frame analyses I extracted information on nestling characteristics (nestling behaviours and other traits) associated with feeding events (Table 1; see Monk 1999 for a list of variables).

Table 1. Summary table of results obtained by applying the *Arena* model to data on food distribution to mountain bluebird nestlings. Results are given for the reduced models.

Age of nestlings	9 - 12 Days		13 - 16 Days		17 - 20 Days	
Sex of parent	Female	Male	Female	Male	Female	Male
Begging intensity	+	+	+	+	+	+
Relative position	-	-	-	-	-	-
Begging order			-	-	-	-
Absolute position						-
Sex				-		
Mass					+	
Previously fed						

The sign (positive or negative) of the maximum likelihood β parameter estimates, rather than the numerical value, for the model is given. The size of the sign reflects the relative importance of that variable to the predictive capability of the reduced model. Begging intensity (0 = lowest); relative position (1 = closest); begging order (1 = first to beg); absolute position (1 = closest); sex (M = 1, F = 0); mass = ln (weight); previously fed (1= fed at previous feeding, 0 = not fed).

Results obtained from the *Arena* model revealed that when nestlings were young only two variables were important predictors of feeding success.

Those nestlings that begged more intensely, and those that were closer to the feeding parent (after the parent had hopped into the nestbox) were more likely to receive food (Table 1). The other β parameter estimates and associated variables were either unstable, and/or did not add much to the predictive capability of the model. As nestlings grew older, more variables became necessary to predict which nestling was successful at receiving food. For all older nestlings (Table 1), the order of initiation of begging was an important predictor of success at obtaining food, but generally less important than begging intensity and position relative to the feeding parent. A nestling that begged vigorously, was close to the feeding parent, and was also the first to beg was more likely to receive food. For 13-16 day old nestlings, the sex of the offspring was a relevant predictor of being fed by male parents, but not female parents. Female offspring were more likely to receive food from male parents. Finally, when nestlings were 17-20 days old and were soon ready to fledge, four variables contributed considerably to the predictive capability of the model. Both parents paid most attention to the intensity at which the nestlings begged and their relative position, favouring those nestlings that begged more intensely and were closer to them. Both paid some attention to which nestling(s) started begging first. In addition, but of less importance, heavier offspring were more likely to receive food from the female parent, and nestlings that positioned themselves at the entrance were more likely to receive food from the male parent (Table 1).

Strengths and Weaknesses of the Arena Model and Other Logistic Regression Approaches

There are several properties of the *Arena* model that make it valuable for the study of nestling begging behaviour. Many of these properties are shared by other multivariate and multinomial logistic regression analyses in general. First, as a logistic-regression-based analysis, the *Arena* model is specifically designed to deal with dichotomous success / failure (fed / not fed) data, and thus the distribution of errors is known and specified properly. Second, because it is a multinomial approach, the *Arena* model incorporates the attributes of all interactants in each event simultaneously. Information from nestlings that were fed as well as those not fed is incorporated in the result, and therefore the full social context during nestling feeding is reflected in the analysis. Third, the *Arena* model is a multivariate model. As such, it includes all specified independent variables (traits) in one analysis. This enables comparison of the relative importance of these traits within a single analysis. These attributes of the *Arena* model contrast with traditional methods of analysing nestling interactions. Traditional models are generally not specifically designed to deal with dichotomous outcome variables. Also, traditional methods frequently do not simultaneously incorporate the attributes of all interacting individuals. Even though each nestling affects every outcome, traits of the nestlings are frequently grouped into categories. In addition, traditional approaches are often not able to incorporate all independent variables into one analysis, but instead require multiple tests.

Despite this array of advantages of the *Arena* model, there are aspects of the model that require consideration, including data selection as well as model building.

Data Selection

The scope of inference and conclusions drawn from any analysis depend on proper replication (Hurlbert 1984; Heffner et al. 1996). Parental care studies are often constrained by the use of multiple events from non-independent individuals or the same individual. Statistical analyses may lead to pseudoreplication if the degrees of freedom are equated with the number of events, including some that are non-independent. Evaluation of significance is thus troublesome. The model fitting procedure in the *Arena* model uses the likelihood ratio test. As mentioned already, in this test the degrees of freedom for model fitting are affected by the number of parameters that are excluded from the full model, and not by the number of data points. Therefore, it is possible to include observations of all feeding events in the data set without inflating the degrees of freedom.

To show that statistical conclusions obtained by using the *Arena* model are robust to the use of repeated measurements on the same family, I used data from a subset of eight female mountain bluebirds feeding their nestlings at 9-12 days of age (Table 2). In the first data set, I included five feedings for each female (a total of 40 feeding events). In the second data set, I repeated the exact data twice, thus for each female there were now ten feeding events (total 80 events).

Doubling the data did not impact the statistical inferences obtained from the *Arena* model (Table 2). There were only slight changes in the β parameter estimates, and the changes had no bearing on the conclusions. Obtaining robust parameter estimates is one of the strengths of the *Arena* model. The standard error for each estimate decreased when the data were doubled. This makes sense, because the maximum likelihood procedure uses the same information twice to attain estimates for the β parameters, the estimates become more stable and the standard errors become smaller. The value of the likelihood ratio statistic (LRS) also changed. The LRS associated with very predictive variables increased in magnitude, and thus the P-value decreased for those parameters. For less informative variables the LRS changed little, or even decreased in magnitude.

Table 2. Summary table of β parameter estimates and their standard errors for the full model, the likelihood ratio test statistic (LRS) and associated P-values for the comparison between the full model and the model in which that parameter was omitted, and percentage of events in which a reduced model including only that parameter correctly predicted the winner. Results were obtained by applying the *Arena* model to data on food distribution to nestlings using 40 feeding events and 80 feeding events, the latter obtained by repeating the 40 feeding events twice.

	40 Feedings				80 Feedings			
	Estimate ± SE	LRS	P-value	%correct	Estimate ± SE	LRS	P-value	%correct
Begging intensity	1.37 ± 0.43	13.30	< 0.000	63	1.74 ± 0.36	17.70	< 0.000	64
Relative position	- 0.66 ± 0.32	4.37	0.04	60	- 0.97 ± 0.27	9.95	0.002	58
Begging order	- 0.06 ± 0.27	0.05	0.83	50	- 0.02 ± 0.20	0.20	0.66	54
Absolute position	- 0.16 ± 0.11	2.40	0.12	40	- 0.20 ± 0.08	0.56	0.45	36
Sex	0.44 ± 0.44	0.98	0.32	18	0.52 ± 0.34	2.34	0.13	18
Mass	0.24 ± 0.14	3.16	0.08	25	0.42 ± 0.11	1.82	0.18	36
Previously fed	0.21 ± 0.45	0.20	0.65	10	0.29 ± 0.35	0.09	0.76	8

Results from both data sets showed that female parents were most likely to feed those nestlings that begged with greatest intensity, and female parents were more likely to feed a nestling close to them. For both data sets, omitting the variable 'begging intensity' made a highly significant difference to the predictive capability of the model ($P < 0.0001$). Omitting the variable 'relative position' made a significant difference as well ($P \leq 0.04$). Even though some of the other variables had low P-values, they were either unstable or did not predict successful feeding events very well, and thus were excluded from the final model.

As I have shown, using multiple non-independent events does not affect statistical inference, but it is possible that the outcomes of sequential events are correlated. In the application of the *Arena* model to food distribution among begging nestlings, I used sequential feeding events. To test for correlations, I included a variable called 'previously fed' in this example, as well as in all other analyses of food distribution to nestling bluebirds. This variable tested whether the outcome of the previous feeding event was predictive of the current outcome. In all analyses this variable showed that, given other explanatory variables, the previous feeding event was irrelevant to the model.

Another potential problem with using non-independent events (i.e. repeated measures from the same family) is the scope of generalization that is desired, and the extent of this problem is question-dependent. For any statistical approach, the data need to be collected from an appropriate sample of families, populations, species or group of species, depending on the question. Other non-logistic-regression analyses may provide false confidence about the scope of generalization, since they are not based on an underlying distribution appropriate to dichotomous outcome variables. Obviously, the result of a study with hundreds of events from a single family is inappropriate for generalizations outside that family. In order to generalize over the entire population of bluebirds studied, I made sure that the design was balanced and that the sample was representative. The *Arena* model is not sensitive to whether a particular data set was obtained from one family or many families. In other words, 40 feeding events from one family result in the same parameter estimates and standard errors as would the same 40 events if taken from ten families. In order, therefore, to feel confident about making generalizations regarding the bluebird population, I included as many families as possible in the analysis. The *Arena* model is also not very sensitive to data outliers (see Monk 1999). Nonetheless, I insured that no family was over-represented.

Model Building

As part of the procedure to fit a model with the greatest predictive capability, but with the fewest variables, the *Arena* model uses a backward elimination procedure. Any model fitting procedure for inclusion and/or exclusion of variables should be performed in a systematic fashion and the resulting model should be viewed with caution. Because the selection process identifies variables as important based solely on statistical grounds, the final model should be evaluated in terms of biological relevance (Hosmer & Lemeshow 1989). Kleinbaum (1994) recommends strategies for systematic procedures of variable specification and for selecting a hierarchically and well-formulated model, and Hosmer and Lemeshow (1989) illustrate several stepwise selection procedures. Below, I will outline some of the reasons for caution when variables are selected or deleted from a model using any selection procedure.

First, specifying a significance level (α level) in a variable selection procedure determines how stringent the selection criteria are, and thus, whether a variable will be included in the final model or not. For the *Arena* model we chose to use $\alpha = 0.05$ as the significance level. Some authors argue that the choice of $\alpha = 0.05$ may be too stringent, often excluding important variables from the model (Hosmer & Lemeshow 1989). On the other hand, retaining too many variables in the model might result in overfitting. That is, obtaining a model that includes most of the variables and therefore fits the data very well, yet aids little in their explanation and is less likely to generalize widely (Hosmer & Lemeshow 1989).

Second, for any multivariate regression technique (linear and logistic), a close functional relationship between two or more independent variables (collinearity) poses difficulty, because one of the collinear variables will arbitrarily be eliminated by the selection procedure (Hosmer & Lemeshow 1989; Dobson 1990). Experimental manipulation is the only certain method of teasing apart the influence of collinear variables. In logistic regression models, however, collinearity between variables tends to manifest itself *via* aberrantly large estimated standard errors and sometimes *via* a large estimated coefficient (Hosmer & Lemeshow 1989), and thus can be detected by the investigator. In addition, a variable selection procedure that includes steps where several independent variables are added to or eliminated from the model simultaneously (as was done in the *Arena* model) can help avoid the elimination of a biologically significant variable. One of two variables that are important and collinear will be nonsignificant individually, but highly significant when dropped in combination.

A final word of caution should be given here regarding multiple testing. When selecting or eliminating variables from a model, many tests of significance are typically performed. When several tests are performed on the same data set, there is a certain likelihood that a statistically significant result is obtained even though there is no real association in the data (Kleinbaum 1994). Thus, the process of variable selection may result in a model that contains more variables than necessary. Unfortunately, to my knowledge there is no explicit method available for adjusting for multiple testing in logistic regression (see also Kleinbaum 1994). However, a systematically applied method of model building, which includes evaluation from a biological standpoint, will guard against inclusion of variables that do not contribute to the model.

In my application of the *Arena* model I addressed the above model building concerns by using a backward elimination procedure that included simultaneous elimination of groups of variables to test for collinearity and to test for inclusion of non-relevant variables. I also evaluated each parameter estimate for its stability and its capability of predicting a successful outcome.

Finally, I visually inspected plots of standardized residuals. Plots of standardized residuals, against each of the independent variables, can help determine whether a particular model adequately describes the effect of the variable (Dobson 1990). Similarly, residual plots against fitted values can help determine the adequacy of the model in predicting the outcome (Dobson 1990). The absence of any apparent pattern in the plot indicates that the model is appropriate. A plot of the standardized residuals depicting curvilinearity, or some other systematic pattern, would suggest that additional or alternative terms should be included in the model to describe the effect of the independent variable (Dobson 1990).

FUTURE DIRECTIONS

Theory and questions about offspring signals and parental responses increasingly rely upon a more detailed understanding of the complexity associated with begging signals. Concomitantly, statistical analyses of begging and parental provisioning need to incorporate more of the complexity found in these social interactions. Statistical models based upon logistic regression can deal with such data on parental responses to begging. The development of the *Arena* model, which is a multivariate and multinomial extension of logistic regression, provides a tool that may lead to insight into the relative role of begging characteristics. For example, the analysis across developmental stages in mountain bluebirds demonstrates that there is the potential for interesting dynamics between nestling interests and parental interests (see Monk 1999; R.M. Kilner this volume). When nestlings are young, conflicts of interest between parents and young did not appear to be very strong. Both male and female adults fed the young that begged most vigorously. This variable is likely to be an indicator of nestling need (e.g. Redondo & Castro 1992; Price & Ydenberg 1995; Leonard & Horn 1998). As young grow older, they may be challenging parental interests to some degree, as well as competing with one another for access to food. At older ages, parents may have had to rely on attributes such as nestling mass or sex to distribute food, in order not to be manipulated. In addition, nestlings appeared to be competing with each other for position (absolute position in the nestbox was a significant factor), and thus may have been manipulating food distribution.

Logistic regression and related approaches can be an important tool for addressing the complexity of begging signals. I hope that the models described in this chapter serve as an impetus to further development of models for the analysis of parent-offspring interactions. For example, in situations in which a parent bird divides food among several young during each visit, a polychotomous (as opposed to dichotomous) logistic regression model could be developed. Such a model would enable the prediction of which nestling in a brood of young would be fed first, which second, and so on, given a set of observations taken on the entire brood.

ACKNOWLEDGEMENTS

I would like to dedicate this chapter to the memory of Berhard D. Flury, professor of mathematics at Indiana University, USA and of statistics at the Université de Neuchâtel, Switzerland. He died unexpectedly in a mountaineering accident in 1999, and will be missed by many of his colleagues and students. Without his expertise and enthusiastic guidance these models would not have been developed.

My thanks go to Bernhard Flury and Stefania Bartoletti for a productive collaboration, and to Mark Dybdahl, Ellen Ketterson, Marty Leonard, Jon Wright and two anonymous reviewers for helpful comments on earlier versions of this manuscript. I gratefully acknowledge financial support during the development of these statistical models from CISAB at Indiana University and NSF grant IBN 9412288.

The *Arena* model is not a readily available computer program. It was encoded by Stefania Bartoletti in GAUSS. For further information on the *Arena* model please refer to Bartoletti (1998) and Monk (1999) or contact the author of this chapter.

REFERENCES

Agresti, A. 1990. *Categorical Data Analysis*. New York: John Wiley & Sons.

Bartoletti, S. 1998. Stima statistica di misture con covariate. PhD thesis, Universita' La Sapienza.

Clutton-Brock, T.H. 1991. *The Evolution of Parental Care*. Princeton: Princeton University Press.

Cotton, P.A., Kacelnik, A. & Wright, J. 1996. Chick begging as a signal: are nestlings honest? *Behavioral Ecology and Sociobiology* **7**, 178-182.

Cox, D.R. 1972. Regression models and life tables. *Journal of the Royal Statistical Society, Series B* **74**, 187-220.

Daniel, W.W. 1987. *Biostatistics: A Foundation for Analysis in the Health Sciences, 4th Edtn*. New York: John Wiley & Sons.

Dobson, A.J. 1990. *An Introduction to Generalised Linear Models*. London: Chapman and Hall.

Flury, B.D. 1997. *A First Course in Multivariate Statistics*. New York: Springer.

Gowaty, P.A. & Droge, D.L. 1990. Sex ratio conflict and the evolution of sex-biased provisioning in birds. *Acta XX Congressus Internationalis Ornithologici* **II**, 932-945.

Heffner, R.A., Butler, M.J.IV. & Reilly, C.K. 1996. Pseudoreplication revisited. *Ecology* **77**, 2558-2562.

Hosmer, D.W. & Lemeshow, S. 1989. *Applied Logistic Regression*. New York: John Wiley & Sons.

Hurlbert, S.H. 1984. Pseudoreplication and the design of ecological field experiments. *Ecological Monographs* **54**, 187-211.

Kacelnik, A., Cotton, P.A., Stirling, L. & Wright, J. 1995. Food allocation among nestling starlings: sibling competition and the scope of parental choice. *Proceedings of the Royal Society of London, Series B* **259**, 259-263.

Kleinbaum, D.G. 1994. *Logistic Regression: A Self-Learning Text*. New York: Springer.

Krebs, E.A., Cunningham, R.B. & Donnelly, C.F. 1999. Complex patterns of food allocation in asynchronously hatched broods of crimson rosellas. *Animal Behaviour* **57**, 753-763.

Lack, D. 1968. *Ecological Adaptations for Breeding in Birds*. London: Methuen & Co.

Leonard, M. & Horn, A. 1996. Provisioning rules in tree swallows. *Behavioral Ecology and Sociobiology* **38**, 341-347.

Leonard, M.L. & Horn, A.G. 1998. Need and nestmates affect begging in tree swallows. *Behavioral Ecology and Sociobiology* **42**, 431-436.

Leonard, M.L., Teather, K.L, Horn, A.G., Koenig, W.D. & Dickinson, J.L. 1994. Provisioning in western bluebirds is not related to offspring sex. *Behavioral Ecology* **5**, 455-459.

McCullagh, P. & Nelder, J.A. 1989. *Generalised Linear Models, 2nd Edtn*. London: Chapman & Hall.

McFadden, D. 1974. Conditional logit analysis of qualitative choice behavior. In: *Frontiers In Econometrics* (Ed. by P. Zarembka). New York: Academic Press.

Monk, D.S. 1999. Differential allocation of parental care in free-ranging mountain bluebirds. PhD thesis, Indiana University.

Price, K. & Ydenberg, R. 1995. Begging and provisioning in broods of asynchronously-hatched yellow-headed blackbird nestlings. *Behavioral Ecology and Sociobiology* **37**, 201-208.

Redondo, T. & Castro, F. 1992. Signalling of nutritional need by magpie nestlings. *Ethology* **92**, 193-204.

Smiseth, P.T., Amundsen, T. & Hansen, L.T.T. 1998. Do males and females differ in the feeding of large and small siblings? An experiment with the bluethroat. *Behavioral Ecology and Sociobiology* **42**, 321-328.

Stamps, J., Clark, A., Kus, B. & Arrowood, P. 1987. The effects of parent and offspring gender on food allocation in budgerigars. *Behaviour* **101**, 177-199.

Teather, K.L. 1992. An experimental study of competition for food between male and female nestlings of the red-winged blackbird. *Behavioral Ecology and Sociobiology* **31**, 81-87.

Trexler, J.C. & Travis, J. 1993. Nontraditional regression analyses. *Ecology* **74**, 1629-1637.

24.

STATISTICAL CHALLENGES IN THE STUDY OF NESTLING BEGGING

Scott Forbes
Department of Biology, University of Winnipeg, Winnipeg MB R3B 2E9, Canada (s.forbes@uwinnipeg.ca)

ABSTRACT

The study of avian begging poses formidable statistical challenges. Most work is field-based, limiting control over an array of environmental variables such as unpredictable weather and feeding conditions and predators with a plain disrespect for experimental design. Small sample sizes and unconventional statistical distributions often leave one without shelter from the central limit theorem, and present stumbling blocks for parametric methods. Nonparametric methods present workable solutions, but randomization methods are more powerful, though error prone when samples are small and variance large. Broods of nestling birds might have been drawn from Stuart Hurlbert's worst nightmare: behavioural variables are highly correlated in often subtle ways and differences within and across broods are frequently large and little understood. The price of studying avian begging is eternal vigilance against pseudoreplication, working around low power, dealing with multicollinearity and statistical improvisation. I suggest techniques to draw stronger inferences from often imperfect data sets, including rank-based indices.

INTRODUCTION

Interest in nestling begging has burgeoned in recent years, and so, too, has our awareness of the complexity of this behaviour. Nestlings, for example, express behaviour in the context of an evolutionary game. What one individual in a brood does potentially affects everyone else. Thus, events are dependent. This problem is compounded by the common behavioural and

J. Wright and M.L. Leonard (eds.), The Evolution of Begging, 473–491.

nutritional histories of broodmates. Thus, there is potential for pseudoreplication. Nestlings within broods are influenced by an array of environmental variables such as rainfall, temperature and food availability, that vary at times, ranging from second-to-second to year-to-year. Thus, distributions are not stationary. Add to this the age-related dynamics of maturation of sensory, motor and homeothermic abilities and it is clear that the study of nestling begging is a veritable minefield for the correct application of statistical models. In this chapter I shall do two things. First, I shall review briefly the statistical methods used in the field to date, and examine the limitations of these data sets. Second, I shall review four main statistical problems associated with begging studies and their potential solutions: (1) small sample statistics; (2) the role of measurement error and multicollinearity; (3) pseudoreplication and (4) analysis of begging indices.

Survey of the Begging Literature

I reviewed 50 journal articles where the chief focus of the study was nestling begging. A listing of the papers reviewed is presented below Table 1. Though not exhaustive, I believe this review is sufficiently comprehensive to be representative of the field as of this writing (October 2000). For each paper I compiled an inventory of the statistical methods used, as well as the sample sizes (n). Where more than one experiment/analysis was repeated in a study, I included separate values for n. As analyses within and sometimes even across studies are not independent, I confess an element of pseudoreplication.

Statistical Models

As expected, a wide array of statistical methods has been adopted by workers in the field (Table 1). Nonparametric methods, most notably the Wilcoxon and Mann-Whitney U tests, are popular in the field, and are associated with ranked data that are commonly gathered when begging indices are used. A few unconventional methods were also applied such as testing for multicollinearity (Price et al. 1996), and circular statistics for examining the location of parental arrivals at nests (Kölliker et al. 1998). Equally interesting are the methods that were not used. No study made use of permutation methods, and only five (Christe et al. 1996; Burford et al. 1998; Dearborn 1998; Smiseth et al. 1998; Leonard et al. 2000) reported the power of nonsignificant results, even though positive conclusions were often drawn from negative results.

Table 1. Survey of statistical methods used in 50 begging papers; n = number of papers in which a method is used.

STATISTICAL METHODS	n
Parametric Methods	
ANOVA	14
Repeated measures ANOVA	13
Student's t-test	13
Correlation	10
Paired t-test	9
Regression	8
ANCOVA	4
Logistic regression	1
Factor analysis	1
Nonparametric Methods	
Wilcoxon test	23
Mann-Whitney U test	13
Spearman correlation	13
Chi-square test	12
Kruskal-Wallis ANOVA	6
Sign test	6
Friedman ANOVA	5
Binomial test	4
G-test	4
Median test	1
Other Methods	
Power analysis	5
Meta-analysis	1
Testing for multicollinearity	1
Circular statistics	1
Randomization / permutation methods	0

Papers reviewed for survey: Abraham & Evans 1999; Bengtsson & Rydén 1981, 1983; Briskie et al. 1994; Burford et al. 1998; Bustamante et al. 1992; Cash & Evans 1986; Choi & Bakken 1990; Christe et al. 1996; Clark & Lee 1998; Cotton et al. 1996; Dearborn 1998; Drummond & Osorno 1992; Evans 1992; Forbes 1991; Fujioka 1985; Göttlander 1987; Hofstetter & Ritchison 1998; Hussell 1988; Iacovides & Evans 1998; Kilner 1995; Kilner & Davies 1998; Kölliker et al. 1998; Krebs & Magrath 2000; Leech & Leonard 1996, 1997; Leonard & Horn 1996, 1998; Leonard et al. 2000; Lichtenstein & Sealy 1998; Lotem 1998a,b,c; Mock & Ploger 1987; Mondloch 1995; Ottosson et al. 1997; Ploger & Mock 1986; Price & Ydenberg 1995; Price et al. 1996; Price 1998; Redondo & Castro 1992; Roulin et al. 2000; Rydén & Bengtsson 1980; Saino et al. 2000; Smiseth et al. 1998; Smith & Montgomerie 1991; Stamps et al. 1987, 1989; Teather 1992; Whittingham & Robertson 1993.

Sample Sizes in Begging Studies

The analysis of sample sizes was similarly revealing (Figure 1). Sample sizes ranged from a minimum of one to a maximum of 256. Most studies were based on 15 or fewer broods and the modal sample size was six to ten (Figure 1). Thus, it is apparent that this is a field that lies primarily within the domain of small sample statistics.

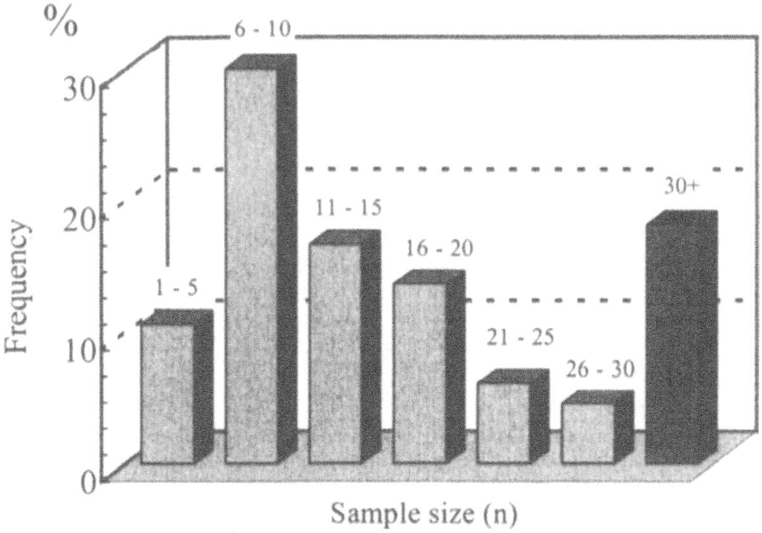

Figure 1. Frequency of sample sizes (number of broods) in the survey of begging literature. Range of sample sizes is shown above bars. Studies with sample sizes greater than 30 are pooled in the final histogram bar. This figure is based upon the same papers as listed below Table 1.

The Challenge of Small Sample Statistics

"Nonparametric statistical tests were used for analyses involving behavioural variables, since sample sizes were generally too small to determine if assumptions of equal variance and normality applied."

Stamps et al. 1987

Normality Assumption

When n is very small, testing for normality is largely pointless. The power of such tests is inescapably low and one will almost never reject the null hypothesis at conventional levels of alpha. If one does proceed with a parametric test after obtaining a negative result, one does so as an article of faith, not on the basis of a well-founded statistical inference. How is one to proceed? The approach of Stamps et al. (1987) can hardly be faulted. They accept the limitations of small data sets, and turn to the appropriate nonparametric models. Another possible solution involves permutation methods that are performed on a computer. As the nonparametric methods are well known to workers in the field of nestling begging, but permutation methods, evidently, are not (Table 1), I provide a brief primer on the latter below.

Hypothesis Testing Using Permutation Methods

These are often referred to as computer intensive statistics and include the techniques of exact and approximate randomization, and bootstrapping. They have been known for a long time, dating back to Fisher (1935), but were relatively little used until desktop computing arrived. In essence, they use brute force instead of mathematical elegance to produce results. And produce results they do. They are at least as powerful as their parametric analogues, require far fewer assumptions, and (perhaps most importantly) are conceptually simple.

There are two basic approaches to hypothesis testing with permutation methods. The most rigorous is exact randomization. Under conventional hypothesis testing, we obtain a test statistic and want to know what is the probability, P, of obtaining this result if the null hypothesis is true. To do so we must specify a known sampling distribution for the test statistic (e.g. F or t or χ^2). Permutation methods eliminate this assumption-laden step by generating a customized null distribution from the sample data.

Let us consider a simple example for the purposes of illustration. We have gathered two samples of data, with sample sizes n_1 and n_2 for comparison using a conventional Student's t-test. We compute a t-statistic, and instead of using the standard method of estimating P from the Student's-t distribution, we use exact randomization instead. The method is astonishingly simple. One computes the total number of possible arrangements of the sample data into two samples of size n_1 and n_2. Call this number T. One then reviews all of these possible arrangements of the data, computes for each a value of the test statistic (which we shall call a pseudostatistic, t'), and compares the pseudostatistic to the observed test statistic. The number of cases where $t' \geq t$ we shall call c. The probability of our observed test statistic is simply the proportion of the total number of permutations of the data set that yield a result as, or more, extreme than that we observed, which is just $P = c / T$.

This method is conceptually simple and easy to use, except for one small drawback. It takes a long time to compute an exhaustive list of the permutations of the data set when sample sizes get reasonably large. This constraint is temporary, as rapid advances in computing speed and the development of more efficient algorithms for computing all permutations will soon render exact randomization feasible for most data sets. Until then we can use the second permutation method, that is for all intents and purposes, just as good: approximate randomization.

Approximate randomization uses a slightly different method to compute P, but again bypasses all the messy distributional assumptions of parametric statistics. Instead of computing an exhaustive list of the possible permutations of the data set, one simply samples from this distribution a large number of times, and determines, just as in exact randomization, the proportion of cases where the computed pseudostatistic meets or exceeds the observed test statistic. This technique is also very simple. One begins by creating an array on the computer. For our Student's t-test, we could do this by creating a single column of data N rows in length, where $N = n_1 + n_2$. The first n_1 rows we shall call sample one; the remaining rows in the column will be sample two. As before we shall compute our Student's t-statistic by the usual method.

Approximate randomization simply involves random shuffling of the data in our array. Each time we do this, we will generate a new permutation of the data set (a new set of values will appear in the first n_1 rows of our column, as will a new set of values in the last n_2 rows). We use this shuffled data to compute a pseudostatistic, t' and compare this to the observed test statistic. A running tally of the number of times the value of the pseudostatistic meets or exceeds the observed test statistic is kept, as the process of shuffling and comparison of pseudostatistic to observed test statistic is repeated many (≥ 5000) times. As with exact randomization, P is simply the number of shuffles where $t' \geq t$. What could be simpler than that?

In fact, the method can be even simpler. In the example above, a conventional test statistic was calculated, but P was estimated by randomization. As with all parametric statistics, there are assumptions about the distributional properties of the data. For example, we customarily assume homogenous variance. This is not an assumption of direct interest to biologists, but rather is made so we can specify the null distribution of the test statistic. Indeed, we may not care at all about the properties of the variance, and be interested solely in whether the sample means differ.

Permutation methods allow us to specify a more precise null hypothesis. Instead of using a t-statistic (where the difference in sample means appears in the numerator of the calculation formulae, and sample variance appears in the denominator), we can calculate the simpler test statistic, the difference in sample means, d:

$$d = \bar{x}_1 - \bar{x}_2 \tag{1}$$

where x_1 and x_2 are the means of samples one and two respectively. For a two-tailed test we would use the absolute value of the right hand side of equation 1: for a one-tailed test where we hypothesized that $x_2 > x_1$ we would reverse the order of the right hand side of equation 1. To compute P we would proceed as described above, except that in each permutation of the data set, we would compute a pseudostatistic, d', based on the difference in sample means, instead of a pseudostatistic, t', based upon t.

An attractive property of these so-called computer intensive methods is that they can be easily adapted to the odd situations. If one can specify a clear expectation under the null hypothesis, one can generally design a custom test statistic that can be evaluated by approximate randomization. Another attractive property is that approximate randomization can be used in conjunction with conventional parametric (or nonparametric methods, though one wonders why you would bother) test statistics when one is unsure of whether the assumptions of normality and homogeneity of variance have been met. Those are only worrisome in that one needs to make those assumptions to specify the null distribution of the test statistic. If you generate a custom null distribution from the observed data, these assumptions become irrelevant.

One caveat about approximate randomisation. If one has very small samples (i.e. single digits), and very unequal variances, the estimated P values are unreliable (see discussion and references in Hayes 2000); approximate randomization will yield too many Type I errors. In truth, however, when one has the bad fortune to obtain data of this type, statistical methods of any sort are not going to help much anyway. For readable general discussions of the use of exact and approximate randomization, see Noreen (1989), Manly (1991) and Edgington (1995). Adams and Anthony (1996) describe how randomization methods can be used in behavioural studies. See Hayes (2000) for a recent discussion of the limitations of randomization methods in behavioural studies, and references to statistical packages that perform permutation-based statistics.

Bootstrapping

The bootstrap is another permutation method used to estimate population parameters from sample data. Under conventional statistical methods, we compute interval estimates of population parameters by specifying the exact form of sampling distributions (e.g. t or Z distribution). As noted above, to do so, we make certain, sometimes untestable assumptions, about the underlying distribution from which the sample was drawn. Efron (1979) suggested a novel and logical approach to the problem (see also Efron & Tibshirani 1991, 1993). Since the sample represents all that we know about the population, why not act as though the sample is the population? Bootstrap sampling works by taking random samples with replacement from the original sample. Each time we compute a bootstrap sample, b', from the original sample, b_0, we compute a bootstrap estimate of our test statistic, t.

Computing bootstrap confidence intervals for t is relatively straightforward. We simply generate a large number of bootstrap estimates (5000 seems to be an agreed minimum), to derive a bootstrap distribution of t. Here the confidence intervals are derived empirically. One sorts the obtained estimates of t in order, and then takes the values at the 97.5% and 2.5% percentiles as the confidence limits for a 95% confidence interval.

There is again one caveat about bootstrap confidence intervals. When the sample size is very small, the 95% bootstrap confidence interval can miss the true population mean more than 95% of the time. The reason is this: when one draws the first datum from the underlying population, that point is going to be either above or below the true population mean (unless it is exactly the mean); let us say it is above. The second data point may also come from the upper half of the distribution. Now, if we stop right here, and compute our bootstrap confidence interval, the bootstrap estimates would all fall in the range between the two data points, both of which lie above the true population mean. In practice, the risk of missing the true population mean with even modest sample sizes (double digits) is quite low and is not a serious concern; but for very small sample sizes - worry.

Measurement Error and Multicollinearity

The importance of measurement error grows when sample sizes are small. If the effect were only to add white noise, it would be a forgiveable sin, but the problem is more worrisome. First, measurement error tends to bias results towards a verdict of no effect. For example, in simple linear regression, measurement error in X tends to spread the data horizontally. As a direct consequence, the slope of the fitted regression is biased downward, thus underestimating a real effect if one is present.

Second, and perhaps even more importantly, some variables are more easily measured than others. This becomes especially important when facing the spectre of multicollinearity, the sticky problem of correlation(s) among independent variables. Consider the case of multiple linear regression. Where substantial correlations exist among the X variables, the 'best-fit' model is suspect, and the parameter estimates unstable. Two variables, X_1 and X_2 may explain essentially the same variation in the dependent variate, Y. Typically, when both are included in a model, one, say X_1, will register as explaining much more variation than the other, X_2. When X_1 is deleted from the model, however, X_2 registers as explaining about the same variation as X_1. The incorrect conclusion is that X_2 is unimportant while X_1 is important. The difference between the two may be trivial and the direction of causality may even run in the opposite direction. An illustration of the problem might be helpful.

Richard Herrnstein and Charles Murray (1994) relied upon a single statistical technique in their infamous 'The Bell Curve' multiple logistic regression. A variety of indicators of social performance were dependent variates (e.g. poverty, unemployment, dropout rates, average income). Two variables, one essentially a measure of IQ, the other a measure of environment, the Socioeconomic Status Index (SES), were featured as independent variates. Herrnstein and Murray attempted to determine whether genes (innate cognitive ability) or environment (SES) were stronger predictors of social performance and used multiple regression results as their statistical tool. Quite uniformly they found that cognitive ability explained more variation than SES in their measures of social performance. Their conclusion - most variation in social performance stems from innate, not extrinsic causes. In light of this conclusion, they recommended controversial changes to social policy. For example, why bother teaching children incapable of learning? Was this conclusion justified?

Imagine, for the sake of argument, that variation in IQ is determined in large part by early social environment. That is, the two independent variates in their regression models are highly correlated. As a technical matter IQ, whatever its meaning, is easily and well measured. Socioeconomic status is measured as an index, and compared with IQ, is poorly measured and is an extremely fuzzy variable (those who use begging indices should take note!). Thus, we have two variables, one precisely measured, the other not, and both explain much of the same variation. When the two are considered in the same multiple regression model, the well measured variable is almost certain to explain more variation than the poorly measured variable, even if it is the latter that is truly causal. It is entirely possible that all of Herrnstein and

Murray's results are explained by this statistical artefact alone! Statistical models such as multiple regression are unreliable guides in the face of multicollinearity. To get to the truth, one must arrange for orthogonal contrasts, usually by experimentation.

How does this relate to begging studies? Again, a simple example is helpful. The duration and intensity of begging are likely to be correlated variables in many cases, and for the sake of argument, let us assume that they are, and let us further assume that it is intensity, *per se*, that matters most to parents. The duration of begging is measured relatively easily and with almost no error. Such is not the case for begging intensity, which is generally measured as an index. Such indices are at best very coarse, and at worst wildly inaccurate. How a parent perceives and responds to intensity may bear only a weak relationship to a four-point scale. If we analyse parental behaviour in relation to begging duration and intensity in either univariate or multivariate analyses, we may well find that duration *per se* is the better predictor. For example, in a backwards stepwise regression, the biologically more relevant variable, begging intensity, is likely to be deleted first. And if we take these results at face value, we would erroneously conclude that begging duration is the more important variable. The same effect will likely show up in univariate analyses; the P value for begging duration is likely to be lower than that for begging intensity.

What can one do about the problem? First, one can attempt to diminish measurement error as much as possible by careful measurement. Second, simple awareness of the problem is half the battle. One must always be vigilant for spurious effects that are due to differences in data quality, not genuine biological differences. Modest differences in P values are not grounds for firm conclusions about the relative importance of two variables when they differ in quality of measurement.

Model Fitting

Students of begging behaviour face systems of intertwined variables. A major challenge is to unravel the causal threads. This requires multivariate approaches. Multiple regression, again, is an exemplar in this discussion, being closely related to a family of linear models. Which independent variates are important? Which are not? How does one find the best model? The last question is hard to answer. The converse, how not to, is easily answered: stepwise regression. It is a terrible choice for model fitting. The sequence in which variables are entered/eliminated can have profound effects on the resultant model, and this order may be due to tiny and insignificant differences among independent variates. The best one can hope for is that the resulting model is potentially useful. The more subtle features of a system may be missed entirely, or otherwise misdiagnosed.

There are better methods for model fitting that trade off goodness of fit (as estimated by the residual mean square) against the number of parameters. Models that explain much of the variation with few parameters are good; models that explain little variation with many parameters are bad. There are three often-used criteria for fitting linear models: Mallow's Cp statistic, Schwarz's Bayesian Information Criterion (BIC) and Akaike's Information Criterion (AIC). The approach is similar for all three, though the exact computations differ, the latter two relying upon maximum likelihood estimation. One proceeds by estimating the value of one of these statistics for all possible models (with k independent variates, there are 2^k models to fit), and then choosing the optimum model(s). By looking at many possibilities, one often sees that quite different models obtain similar explanatory power. This result means that more work is needed to discriminate between the alternatives. For further reading on model fitting techniques see Akaike (1974), Schwarz (1978) and especially Burham and Anderson (1998) for a recent review.

The Power and the Glory

Perhaps the greatest sin that I detected in my reading of the begging literature was the temptation to overinterpret modest data sets. One cannot reasonably expect to fit complex statistical models to modest data sets and produce definitive results. The behaviour we study is noisy (in the statistical as well as auditory sense) and therefore large data sets are required to tease apart interactions and subtle effects. The failure to detect such effects is not evidence of their absence; if one concludes that a negative result equates to the absence of an effect, then one has risen several notches on our scale of sin - into the 'mortal' range. The method for assessing the strength of inference one can draw from one's data is straightforward: assess statistical power. Power is the probability of correctly detecting a genuine effect if it is present. If power is low, your data are uninformative. That is, they can neither confirm nor deny the presence of an effect. Thus, one must remain agnostic about one's data. Such data correctly belong in the file drawer, though surprisingly often appear in print as evidence of 'no effect'.

In most realms of biology, alpha (α) = 0.05 has taken on near-sacred meaning. The tradition stems from the pre-computer era when computing exact probabilities was impractical. With a conventional level of significance arose a simple decision rule: reject the null hypothesis if P is less than alpha; do not reject the null hypothesis otherwise. Somewhere along the way, biologists transmuted this rule into something quite different: P less than alpha equals an effect; P greater than alpha equals no effect. This subtle change has caused no end of trouble. P greater than alpha does not mean no effect, and we should not act as though it does. Why? Type II errors; which are especially common when sample sizes are small. This seemingly obvious fact is ignored surprisingly often. Many journals still require that nonsignificant results be reported as NS, as though a P value of 0.51 is the same as 0.051. Even the enlightened journal *Animal Behaviour* does not insist upon routine power analysis, only asking for power calculations when samples are small (evidently not recognizing that, for some statistical tests, a sample of 500 may still have low power). Power should be reported for all negative results when possible, end of story (note: power calculations, sadly, are not routinely available for many nonparametric and multivariate statistics).

Moreover, common sense should return to data analysis. I can best illustrate this with two examples from papers that I have been asked to review recently. In the first, a series of similar tests were performed on the effects of an experimental manipulation upon broods of different ages. Ten tests were performed, all of which had results in the same direction, and of similar magnitude, though sample sizes and variability differed somewhat. Four of the tests registered as nonsignificant, with P values slightly above 0.05: the other six tests were significant at $\alpha = 0.05$. The authors ignored the obvious fact that all the data showed an identical pattern with only slight variations that rendered some results, technically, nonsignificant and argued that there was no effect of the experiment in the four cases with $P > 0.05$. They then attempted to concoct an explanation for this odd pattern. The only thing odd was the authors' interpretation, arguing against what the data, especially the point estimates, were telling them. Their sin was to use a literal interpretation of P to build a nonsensical biological argument.

A second example also helps to illustrate the level of innumeracy that is extant among biologists today. Here the growth of hatchlings from first- and last-laid eggs was being correlated with egg mass. The correlations reported were nearly identical, 0.48 versus 0.42, although the sample sizes differed slightly. For one, the P value was 0.043 (larger eggs), and 0.051 (smaller eggs) for the other with a slightly smaller sample size. Naturally, the authors argued that a significant effect did exist for large eggs, but no such effect existed for small eggs! They then proceeded to attempt to explain why. Again, a literal interpretation of P led to a nonsensical conclusion. For those who are tempted to agree with the authors, the correct conclusion is that an almost identical effect exists for large and small eggs, and that a difference between 0.043 and 0.051 is essentially meaningless (see Stoehr 1999 for a similar example).

If these were isolated cases, there would be little cause for worry. But they are not. All too often positive conclusions are drawn from negative results without justification (i.e. a demonstration that a genuine effect could have been detected). I found many such examples in my survey of the begging literature. I will not embarrass the offenders, mainly because I would have to include myself in the list. I will, however, applaud the authors of the five papers (10% of the surveyed papers) that did make use of power analysis (see above) for setting an example for others to follow.

Routine power analysis would greatly enhance the interpretation of negative results. *Post hoc*, one can compute the power of negative results by acting as though the point estimates are the true population estimates (see Cohen 1988); even better, one can compute the detectable effect size given the sample sizes and sample variances obtained. Best of all, one can use power *a priori* in experimental design. Doing so can help to ensure that one does not waste one's scarce resources (e.g. time, grant funds) on a futile, low power research program. Equally, one can use *a priori* power analysis to ensure that one does not use too many study subjects, which is important for obvious ethical reasons.

For those wishing to investigate power analysis further, Cohen (1988) is the canonical reference. For more readable discussions of power, see Dixon and Massey (1983), Kraemer and Thiemann (1987), Peterman (1990) and Zar (1996). See Thomas and Juanes (1996) for discussion of the importance of power analysis in the study of animal behaviour and Engeman and Shumake (1993) for a discussion of the animal welfare implications of statistical power. Finally, Thomas and Krebs (1997) review power analysis software packages.

Pseudoreplication

A major statistical impediment to the study of begging is the problem of non-independence of the data. The bad news is that nestlings share a common micro-environment and nutritional history, and respond to similar external stimuli, and even each other. Therefore, they can hardly be treated as independent data points. The good news is that most workers are well aware of this problem and take the appropriate steps. For example, they use brood means or use the appropriate ANOVA designs to control for brood effects; paired t-tests and matched pairs Wilcoxon tests. The fear of being cited in a Hurlbert-style paper appears to have largely eliminated what at first glance appears to be a major impediment to the study of nestling begging.

Begging Indices and Measurement Scales

Begging indices are used routinely in the begging literature as a measure of begging intensity (17 of the 50 studies listed in Table 1 reported some form of begging index). These are familiar to most, usually being built on a four- to six-point scale e.g. 0 = no gaping; 1 = gaping; 2 = gaping + neck stretching; 3 = gaping + neck stretching + standing and 4 = gaping + neck stretching + standing + wing flapping.

Most researchers have analysed these in an uncontroversial manner using nonparametric statistics (e.g. Kruskal-Wallis ANOVA, Wilcoxon or Mann-Whitney U tests). The problem of analysis that they present is that they are measured on ordinal, not interval scales. Can parametric statistics be used for such data? This is a topic of considerable controversy.

The issue of measurement scales versus statistical technique has long been debated in the psychology literature, beginning (but certainly not ending) with Stevens (1946). To put the issue in a nutshell, some take the view that the issue of what the numbers mean is irrelevant to which statistical test is used. All that matters is whether the mathematical assumptions (e.g. normality, homogeneity of variance) hold; others disagree. A diversity of opinion exists over this problem and the debate, particularly heated among psychologists, will not be resolved here. My own view is that as a technical matter, the meaning of the numbers is important, and ignoring this could lead to biased or erroneous inferences. At a practical level, however, in most cases it probably makes little difference, but when can we be sure? Thus, I side with those who argue against the use of parametric methods for ordinal data.

The crux of the problem is that parametric statistics are based upon the location of parameters, in particular the sample means. It is really a problem of how you interpret the currency of measurement of a variable. On an interval scale, a difference between 0 and 1 should mean the same thing as a difference between 3 and 4. There is reason to view this assumption with healthy scepticism when it comes to begging indices. A difference between a 0 and a 1 (no gaping versus gaping), which may be biologically trivial in terms of energy expended or parental response, is likely to be quite different from the difference between a score of 3 and 4. Yet, within a parametric model, both register as identical one-unit differences. An ordinal variable, thus, tells us that two behaviours are different, and the order. An interval scale additionally tells us how different.

Why should we care about this? Categorical variables can usually be transformed to make them approximately normal to satisfy distributional assumptions if needed. The problem is one of interpretation, and potentially bias. In terms of parental response, a score of 0 and 1 might be approximately accurate, but a 3 might be a 10 and a 4 really a 20. What we need to know is how the index maps onto the underlying variable of interest. For example, imagine that we are interested in the energetic cost of begging intensity: the difference between a score of 0 and 1 is likely to be negligible, whereas the difference between a 3 and 4 may represent a substantial caloric difference. Both differences count as the same under our index though. Perhaps the greatest difference is between a 2 and 3. An index potentially introduces an unknown bias into the results.

The obvious solution is to calibrate the index by measuring the underlying variable directly. If energetic cost is the variable of interest, then that should be measured under controlled conditions - even an approximate calibration would be better than the raw index. If it is the relationship between begging intensity and parental response that is of interest, one might perform the equivalent of a behavioural 'titration' experiment. Measure directly the level of parental response in relation to the begging intensity, and for subsequent analyses that rely upon begging intensity, substitute a calibrated index. Alternatively, one might search for variables (e.g. the duration of begging) that might be highly correlated with intensity, and more appropriate for parametric analyses. Or one might resort to the argument that it does not matter where the numbers come from and just use parametric statistics on ranks. That is certainly the easiest solution.

One overlooked approach to begging indices, of course, is to make use of the powerful techniques available for the analysis of categorical data (see D. Monk this volume, for a discussion of the use of logistic regression in begging studies).

For discussion of the controversy concerning the analysis of behavioural indices see Heerman and Braskamp (1970), Gaito (1980, 1987), Michell (1987), Marcus-Roberts and Roberts (1988), Stine (1989), Hunter and May (1994) and Lovie (1997). See Bishop et al. (1975) and Long (1997) for the analysis of categorical data.

FUTURE DIRECTIONS

From a statistical perspective, the study of nestling begging behaviour is no easy task. The easiest way to avoid trouble is good experimental design that allows one to ask clean, unambiguous questions. For logistical reasons, it seems likely that nestling begging behaviour will remain in the domain of small sample statistics, and as such will necessarily involve compromises in the choice of statistical models used. Nonparametric statistics are relatively safe methods, though they sacrifice power, especially important when n is small. Moreover, the nonparametric models used today in animal behaviour studies are primitive when compared with parametric linear models that allow for easy investigation of the interactions between variables. My recent reading of the statistical literature indicates that nonparametric analogues of these methods are available (particularly more complex ANOVA designs), but have not yet made their way into widespread use; someone will need to translate between the theoreticians and we the end-users. At the same time, powerful techniques for the analysis of categorical data (e.g. the equivalents of regression and ANOVA) have long been available (e.g. Bishop et al. 1975) and yet, are not widely used in behavioural research. I suspect this is because others also find this literature virtually impenetrable unless one has an advanced degree in mathematics or statistics (see Long 1997 for an exception). Permutation methods seem to offer the greatest potential to solve our difficulty of verifying the assumptions of normal theory methods, by using randomization to compute P within classical parametric models. My advice? Keep your friends close; keep your statistician closer.

ACKNOWLEDGEMENTS

I thank Jim Clark, of the University of Winnipeg Psychology Department, for introducing me to the statistical thinking of psychologists. Jonathan Newman, Marty Leonard, Jon Wright and an anonymous reviewer provided helpful comments on the manuscript. This research was supported by an NSERC research grant.

REFERENCES

Abraham, C.L. & Evans, R.M. 1999. Metabolic costs of heat solicitation calls in relation to thermal need in embryos of American white pelicans. *Animal Behaviour* **57**, 967-975.

Adams, D.C. & Anthony, C.D. 1996. Using randomisation techniques to analyse behavioural data. *Animal Behaviour* **51**, 733-738.

Akaike, H. 1974. A new look at statistical model identification. *IEEE Transactions on Automatic Control* **19**, 716-723.

Bengtsson, H. & Rydén, O. 1981. Development of parent-young interaction in asynchronously hatched broods of altricial birds. *Zeitschrift für Tierpsychologie* **56**, 255-272.

Bengtsson, H. & Rydén, O. 1983. Parental feeding rate in relation to begging behavior in asynchronously hatched broods of the great tit *Parus major*. *Behavioral Ecology and Sociobiology* **12**, 243-251.

Bishop, Y.M.M., Fienberg, S.E. & Holland, P.W. 1975. *Discrete Multivariate Analysis: Theory and Practice.* Cambridge: MIT Press.

Briskie, J.V., Naugler, C.T. & Leech, S.M. 1994. Begging intensity of nestling birds varies with sibling relatedness. *Proceedings of the Royal Society of London, Series B* **258**, 73-78.

Burford, J.E., Friedrich, T.J. & Yasukawa, K. 1998. Response to playback of nestling begging in the red-winged blackbird, *Agelaius phoeniceus*. *Animal Behaviour* **56**, 555-561.

Burnham, K.P. & Anderson, D.R. 1998. *Model Selection and Inference: A Practical Information-Theoretic Approach.* New York: Springer-Verlag.

Bustamante, J., Cuervo, J.J. & Moreno, J. 1992. The function of feeding chases in the chinstrap penguin, *Pygoscelis antarctica*. *Animal Behaviour* **44**, 753-759.

Cash, K.J. & Evans, R.M. 1986. The occurrence, context and functional significance of aggressive begging behaviours in young American white pelicans. *Behaviour* **102**, 119-128.

Choi, I.-H. & Bakken, G.S. 1990. Begging response in nestling red-winged blackbirds (*Agelaius phoeniceus*): effect of body temperature. *Physiological Zoology* **63**, 965-986.

Christe, P., Richner, H. & Oppliger, A. 1996. Begging, food provisioning, and nestling competition in great tit broods infested with ectoparasites. *Behavioral Ecology* **7**, 127-131.

Clark, A.B. & Lee, W.-H. 1998. Red-winged blackbird females fail to increase feeding in response to begging call playbacks. *Animal Behaviour* **56**, 563-570.

Cohen, J. 1988. *Statistical Power Analysis for the Behavioral Sciences, 2nd Edtn.* Hillsdale: Lawrence Erlbaum.

Cotton, P.A., Kacelnik, A. & Wright, J. 1996. Chick begging as a signal: are nestlings honest? *Behavioral Ecology* **7**, 178-182.

Dearborn, D.C. 1998. Begging behavior and food acquisition by brown-headed cowbird nestlings. *Behavioral Ecology and Sociobiology* **43**, 259-270.

Dixon, W.J. & Massey, F.J. Jr. 1983. *Introduction to Statistical Analysis, 4th Edtn.* New York: McGraw-Hill.

Drummond, H. & Osorno, J.L. 1992. Training siblings to be submissive losers: dominance between booby nestlings. *Animal Behaviour* **44**, 881-893.

Edgington, E.S. 1995. *Randomisation Tests, 3rd Edtn.* New York: Marcel Dekker.

Efron, B. 1979. Computers and the theory of statistics: thinking the unthinkable. *SIAM Review* **2**, 460-480.

Efron, B. & Tibshirani, R. 1991. Statistical data analysis in the computer age. *Science* **253**, 390-395.

Efron, B. & Tibshirani, R. 1993. *An Introduction to the Bootstrap.* New York: Chapman & Hall.

Engeman, R.M. & Shumake, S.A. 1993. Animal welfare and the statistical consultant. *American Statistician* **47**, 229-233.

Evans, R.M. 1992. Embryonic and neonatal vocal elicitation of parental brooding and feeding responses in American white pelicans. *Animal Behaviour* **44**, 667-675.

Fisher, R.A. 1935. *The Design of Experiments.* Edinburgh: Oliver & Boyd.

Forbes, L.S. 1991. Hunger and food allocation among nestlings of facultatively siblicidal ospreys. *Behavioral Ecology and Sociobiology* **29**, 189-195.

Fujioka, M. 1985. Sibling competition and siblicide in asynchronously-hatching broods of the cattle egret *Bubulcus ibis*. *Animal Behaviour* **33**, 1228-1242.

Gaito, J. 1980. Measurement scales and statistics: resurgence of an old misconception. *Psychological Bulletin* **87**, 564-567.

Gaito, J. 1987. Some issues in the measurement-statistics controversy. *Canadian Psychology* **27**, 63-68.

Göttlander, K. 1987. Parental feeding behaviour and sibling competition in the pied flycatcher *Ficedula hypoleuca*. *Ornis Scandinavica* **18**, 269-276.

Hayes, A.F. 2000. Randomisation tests and the equality of variance assumption when comparing group means. *Animal Behaviour* **59**, 653-656.

Heerman, E.F. & Braskamp, L.A. 1970. *Readings in Statistics for the Behavioral Sciences.* London: Prentice-Hall.

Herrnstein, R.J. & Murray, C. 1994. *The Bell Curve.* New York: Free Press Paperbacks.

Hofstetter, S.H. & Ritchison, G. 1998. The begging behavior of nestling eastern screech owls. *Wilson Bulletin* **110**, 86-92.

Hunter, M.A. & May, R.B. 1994. Some myths concerning parametric and nonparametric tests. *Canadian Psychology* **34**, 384-389.

Hussell, D.J.T. 1988. Supply and demand in tree swallow broods: a model of parent-offspring food-provisioning interactions in birds. *American Naturalist* **131**, 175-202.

Iacovides, S. & Evans, R.M. 1998. Begging as graded signals of need for food in young ring-billed gulls. *Animal Behaviour* **56**, 79-85.

Kilner, R. 1995. When do canary parents respond to nestling signals of need? *Proceedings of the Royal Society of London, Series B* **260**, 343-348.

Kilner, R. & Davies, N.B. 1998. Nestling mouth colour: ecological correlates of a begging signal. *Animal Behaviour* **56**, 705-712.

Kölliker, M., Richner, H., Werner, I. & Heeb, P. 1998. Begging signals and biparental care: nestling choice between parental feeding locations. *Animal Behaviour* **55**, 215-222.

Kraemer, H.C. & Thiemann, S. 1987. *How Many Subjects? Statistical Power Analysis in Research.* Newbury Park: Sage Publications Limited.

Krebs, E.A. & Magrath, R.D. 2000. Food allocation in crimson rosella broods: parents differ in their responses to chick hunger. *Animal Behaviour* **59**, 739-751.

Leech, S.M. & Leonard, M.L. 1996. Is there an energetic cost to begging in nestling tree swallows (*Tachycineta bicolor*)? *Proceedings of the Royal Society of London, Series B* **263**, 983-987.

Leech, S.M. & Leonard, M.L. 1997. Begging and the risk of predation in nestling birds. *Behavioral Ecology* **8**, 644-646.

Leonard, M.L. & Horn, A.G. 1996. Provisioning rules in tree swallows. *Behavioral Ecology and Sociobiology* **38**, 341-347.

Leonard, M.L. & Horn, A.G. 1998. Need and nestmates affect begging in tree swallows. *Behavioral Ecology and Sociobiology* **42**, 431-436.

Leonard, M.L., Horn, A.G., Gozna, A. & Ramen, S. 2000. Brood size and begging intensity in nestling birds. *Behavioral Ecology* **11**, 196-201.

Lichtenstein, G. & Sealy, S.G. 1998. Nestling competition, rather than supernormal stimulus, explains the success of parasitic brown-headed cowbird chicks in yellow warbler nests. *Proceedings of the Royal Society of London, Series B* **265**, 249-254.

Long, J.S. 1997. *Regression Models for Categorical and Limited Dependent Variables.* London: Sage Publications Limited.

Lotem, A. 1998a. Brood reduction and begging behaviour in the swift *Apus apus*: no evidence that large nestlings restrict parental choice. *Ibis* **140**, 507-511.

Lotem, A. 1998b. Higher levels of begging behavior by small nestlings: a case of a negatively correlated handicap. *Israel Journal of Zoology* **44**, 29-45.

Lotem, A. 1998c. Differences in begging behaviour between barn swallow, *Hirundo rustica*, nestlings. *Animal Behaviour* **55**, 809-818.

Lovie, A.D. 1997. Commentary on Michell, quantitative science and the definition of measurement in psychology. *British Journal of Psychology* **88**, 393-394.

Manly, B.J.F. 1991. *Randomisation and Monte Carlo Methods in Biology.* London: Chapman & Hall.

Marcus-Roberts, H.M. & Roberts, F.S. 1988. Meaningless statistics. *Journal of Educational Statistics* **12**, 383-394.

Michell, J. 1987. Measurement scale and statistics: a clash of paradigms. *Psychological Bulletin* **100**, 398-407.

Mock, D.W. & Ploger, B.J. 1987. Parental manipulation of optimal hatch asynchrony in cattle egrets: an experimental study. *Animal Behaviour* **35**, 150-160.

Mondloch, C.J. 1995. Chick hunger and begging effort affect parental allocation of feedings in pigeons. *Animal Behaviour* **49**, 601-613.

Noreen, E.W. 1989. *Computer-Intensive Methods for Testing Hypotheses*. New York: John Wiley & Sons.

Ottosson, U., Bäckman, J. & Smith, H.G. 1997. Begging affects parental effort in the pied flycatcher, *Ficedula hypoleuca*. *Behavioral Ecology and Sociobiology* **41**, 381-384.

Peterman, R.M. 1990. Statistical power analysis can improve fisheries research and management. *Canadian Journal of Fisheries and Aquatic Sciences* **47**, 2-15.

Ploger, B.J. & Mock, D.W. 1986. Role of sibling aggression in food distribution to nestling cattle egrets (*Bubulcus ibis*). *The Auk* **103**, 768-776.

Price, K. 1998. Benefits of begging for yellow-headed blackbird nestlings. *Animal Behaviour* **56**, 571-577.

Price, K. & Ydenberg, R. 1995. Begging and provisioning in broods of asynchronously-hatched yellow-headed blackbird nestlings. *Behavioral Ecology and Sociobiology* **37**, 201-208.

Price, K., Harvey, H. & Ydenberg, R. 1996. Begging tactics of nestling yellow-headed blackbirds, *Xanthocephalus xanthocephalus*, in relation to need. *Animal Behaviour* **51**, 421-435.

Redondo, T. & Castro, F. 1992. The increase in risk of predation with begging activity in broods of magpies *Pica pica*. *Ibis* **134**, 180-187.

Roulin, A., Kölliker, M. & Richner, H. 2000. Barn owl (*Tyto alba*) siblings vocally negotiate resources. *Proceedings of the Royal Society of London, Series B* **267**, 459-463.

Rydén, O. & Bengtsson, H. 1980. Differential begging and locomotory behaviour by early and late hatched nestlings affecting the distribution of food in asynchronously hatched broods of altricial birds. *Zeitschrift für Tierpsychologie* **53**, 209-224.

Saino, N., Ninni, P., Calza, S., Martinelli, R., De Bernardi, F. & Møller, A.P. 2000. Better red than dead: carotenoid-based mouth coloration reveals infection in barn swallow nestlings. *Proceedings of the Royal Society of London, Series B* **267**, 57-61.

Schwarz, G. 1978. Estimating the dimension of a model. *American Statistician* **6**, 461-464.

Smiseth, P.T., Amundsen, T. & Hansen, L.T.T. 1998. Do males and females differ in the feeding of large and small siblings? An experiment with the bluethroat. *Behavioral Ecology and Sociobiology* **42**, 321-328.

Smith, H.G. & Montgomerie, R. 1991. Nestling American robins compete with siblings by begging. *Behavioral Ecology and Sociobiology* **29**, 307-312.

Stamps, J., Clark, A., Kus, B. & Arrowood, P. 1987. The effects of parent and offspring gender on food allocation in budgerigars. *Behaviour* **110**, 177-199.

Stamps, J., Clark, A., Arrowood, P. & Kus, B. 1989. Begging behavior in budgerigars. *Ethology* **81**, 177-192.

Stevens, S.S. 1946. On the theory of scales of measurement. *Science* **103**, 677-680.

Stine, W.W. 1989. Meaningful inference: the role of measurement in statistics. *Psychological Bulletin* **105**, 147-155.

Stoehr, A.M. 1999. Are significance thresholds appropriate for the study of animal behaviour? *Animal Behaviour* **57**, F22-F25.

Teather, K.L. 1992. An experimental study of competition for food between male and female nestlings of the red-winged blackbird. *Behavioral Ecology and Sociobiology* **31**, 81-87.

Thomas, L. & Juanes, F. 1996. The importance of statistical power analysis: an example from Animal Behaviour. *Animal Behaviour* **52**, 856-859.

Thomas, L. & Krebs, C.J. 1997. A review of statistical power analysis software. *Bulletin of the Ecological Society of America* **78**, 126-139.

Whittingham, L.A. & Robertson, R.J. 1993. Nestling hunger and parental care in red-winged blackbirds. *The Auk* **110**, 240-246.

Zar, J.H. 1996. *Biostatistical Analysis, 3rd Edtn*. London: Prentice-Hall.

INDICES

SPECIES INDEX

SUBJECT INDEX

GPSR Compliance

The European Union's (EU) General Product Safety Regulation (GPSR) is a set of rules that requires consumer products to be safe and our obligations to ensure this.

If you have any concerns about our products, you can contact us on ProductSafety@springernature.com

In case Publisher is established outside the EU, the EU authorized representative is:

Springer Nature Customer Service Center GmbH
Europaplatz 3
69115 Heidelberg, Germany

Batch number: 10809436

Printed by Printforce, the Netherlands